Lecture Notes in Artificia  4

Edited by J. G. Carbonell and J. Sie

Subseries of Lecture Notes in Computer Science

Lecture Notes in Artificial Intelligence 4604

Edited by J. G. Carbonell and J. Siekmann

Subseries of Lecture Notes in Computer Science

Uta Priss   Simon Polovina
Richard Hill (Eds.)

Conceptual Structures: Knowledge Architectures for Smart Applications

15th International Conference
on Conceptual Structures, ICCS 2007
Sheffield, UK, July 22-27, 2007
Proceedings

 Springer

Series Editors

Jaime G. Carbonell, Carnegie Mellon University, Pittsburgh, PA, USA
Jörg Siekmann, University of Saarland, Saarbrücken, Germany

Volume Editors

Uta Priss
Napier University, School of Computing
10 Colinton Road, Edinburgh, EH10 5DT, UK
E-mail: u.priss@napier.ac.uk

Simon Polovina
Sheffield Hallam University, City Campus, Harmer Building
Howard Street, Sheffield, S1 1WB, UK
E-mail: s.polovina@shu.ac.uk

Richard Hill
Sheffield Hallam University, City Campus, Stoddart Building
Howard Street, Sheffield, S1 1WB, UK
E-mail: r.hill@shu.ac.uk

Library of Congress Control Number: 2007930463

CR Subject Classification (1998): I.2, G.2.2, F.4.1, F.2.1, H.4

LNCS Sublibrary: SL 7 – Artificial Intelligence

ISSN	0302-9743
ISBN-10	3-540-73680-8 Springer Berlin Heidelberg New York
ISBN-13	978-3-540-73680-6 Springer Berlin Heidelberg New York

Springer is a part of Springer Science+Business Media

springer.com

© Springer-Verlag Berlin Heidelberg 2007

Typesetting: Camera-ready by author, data conversion by Scientific Publishing Services, Chennai, India
Printed on acid-free paper   SPIN: 12092222   06/3180   5 4 3 2 1 0

Preface

This volume contains the proceedings of the 15th International Conference on Conceptual Structures (ICCS 2007), which is an annual event that, for the first time, was hosted in the UK. Conceptual structures focus on the representation and analysis of concepts, events, actions and objects with applications in research, software engineering, manufacturing and business. The conference brings together researchers in computer science, information technology, artificial intelligence, philosophy and a variety of applied disciplines to explore novel ways that information technologies can be leveraged to assist human reasoning and interaction for tangible business or social benefits. Conceptual structures can be used to augment human intelligence by facilitating knowledge integration, decision making, the creation of intelligent software systems and the exploration of implicit structures.

The theme for this year's conference was "Conceptual Structures: Knowledge Architectures for Smart Applications." Knowledge architectures give rise to smart applications that allow enterprises to share meaning across their interconnected computing resources and to realize transactions that would otherwise remain as lost business opportunities. Conceptual structures and smart applications integrate the creativity of individuals and organizations with the productivity of computers for a meaningful digital future. A focus of ICCS 2007 was on papers that apply conceptual structures in business and technological settings. Other submitted papers covered research in conceptual structures, which is supported by mathematical and computational theory, including formal concept analysis, algorithm design and graph theory, and a variety of software tools.

The conference had a rigorous refereeing process. All papers were reviewed by one Editorial Board member and two Program Committee members. About 50% of the submitted papers were accepted as full papers to be presented at the conference. A few additional papers were accepted as short or position papers. In addition, four invited papers and one introductory paper by Simon Polovina are included in this proceedings volume.

We wish to express our thanks to all the authors of submitted papers, to the members of the Editorial Board and Program Committee, to the Workshop and Tutorial Chair, B. Akhgar, the Industry Chair, J. Schiffel, and the Sponsorship Chair, D. Corbett, and to our sponsors!

July 2007

Uta Priss
Simon Polovina
Richard Hill

Organization

The International Conference on Conceptual Structures (ICCS) is the annual conference and principal research forum in the theory and practice of conceptual structures. The conference has been held since 1993 in various locations: Universitè Laval (Quebec City, 1993), University of Maryland (1994), University of California (Santa Cruz, 1995), Sydney (1996), University of Washington (Seattle, 1997), Montpellier (1998), Virginia Tech (Blacksburg, 1999), Technische Universität Darmstadt (2000), Stanford University (2001), Borovets (Bulgaria, 2002), Technische Universität Dresden (2003), University of Alabama (Huntsville, 2004), Universität Kassel (2005) and Aalborg University (Denmark, 2006).

General Chair

Simon Polovina   Sheffield Hallam University, UK

Program Chairs

Uta Priss   Napier University, Edinburgh, UK
Richard Hill   Sheffield Hallam University, UK

Other ICCS Chairs

Jeffrey Schiffel (Industry Chair)
Babak Akhgar (Workshop Chair)
Daniel Corbett (Sponsorship Chair)

Editorial Board

Galia Angelova (Bulgaria)   Michel Chein (France)
Frithjof Dau (Germany)   Aldo de Moor (The Netherlands)
Harry Delugach (USA)   Bernhard Ganter (Germany)
Pascal Hitzler (Germany)   Mary Keeler (USA)
Sergei Kuznetsov (Russia)   Guy Mineau (Canada)
Bernard Moulin (Canada)   Marie-Laure Mugnier (France)
Peter Øhrstrøm (Denmark)   Heather Pfeiffer (USA)
Henrik Schärfe (Denmark)   John Sowa (USA)
Gerd Stumme (Germany)   Rudolf Wille (Germany)
Karl Erich Wolff (Germany)

Program Committee

Radim Bělohlávek (Czech Republic)
Tru Cao (Vietnam)
Pavlin Dobrev (Bulgaria)
David Genest (France)
Udo Hebisch (Germany)
Wolfgang Hesse (Germany)
Christian Jacquelinet (France)
Pavel Kocura (UK)
Robert Kremer (Canada)
Leonhard Kwuida (Switzerland)
Robert Levinson (USA)
Carsten Lutz (Germany)
Claudio Masolo (Italy)
Jørgen Fischer Nilsson (Denmark)
Ulrik Petersen (Denmark)
Gary Richmond (USA)
Sebastian Rudolph (Germany)
Janos Sarbo (The Netherlands)
GQ Zhang (USA)

Anne Berry (France)
Dan Corbett (Australia)
Peter Eklund (Australia)
Ollivier Haemmerle (France)
Joachim Hereth Correia (Germany)
Andreas Hotho (Germany)
Adil Kabbaj (Morocco)
Yannis Kalfoglou (UK)
Markus Krötzsch (Germany)
Michel Leclère (France)
Michel Liquière (France)
Philippe Martin (Australia)
Engelbert Mephu Nguifo (France)
Sergei Obiedkov (South Africa)
Anne-Marie Rassinoux (Switzerland)
Olivier Ridoux (France)
Éric Salvat (France)
William Tepfenhart (USA)

Further Reviewers

Simone Braun (Germany)
Rainer Osswald (Germany)
Yimin Wang (Germany)

L. John Old (UK)
Quan Thanh Tho (Vietnam)

Sponsoring Institutions

Cultural, Communication and Computing Research Institute (C3RI), Sheffield Hallam University, UK
EU IST MOSIACA Project
Institute of Engineering and Technology, South Yorkshire Network, UK
Natural Language Processing Research Group, Department of Computer Science, University of Sheffield, UK
British Computer Society, South Yorkshire Branch

Table of Contents

Invited Papers

An Introduction to Conceptual Graphs 1
  Simon Polovina

Trikonic Inter-Enterprise Architectonic 15
  Gary Richmond

Hypermedia Discourse: Contesting Networks of Ideas and Arguments ... 29
  Simon Buckingham Shum

Dynamic Epistemic Logic and Knowledge Puzzles 45
  H.P. van Ditmarsch, W. van der Hoek, and B.P. Kooi

Peirce on Icons and Cognition 59
  Christopher Hookway

Conceptual Graphs

Using Cognitive Archetypes and Conceptual Graphs to Model Dynamic
Phenomena in Spatial Environments 69
  Hedi Haddad and Bernard Moulin

A Datatype Extension for Simple Conceptual Graphs and Conceptual
Graphs Rules... 83
  Jean-François Baget

A Knowledge Management Optimization Problem Using Marginal
Utility in a Metric Space with Conceptual Graphs.................. 97
  Jeffrey A. Schiffel

Conceptual Graphs as Cooperative Formalism to Build and Validate a
Domain Expertise... 112
  Rallou Thomopoulos, Jean-François Baget, and Ollivier Haemmerlé

An Inferential Approach to the Generation of Referring Expressions 126
  Madalina Croitoru and Kees van Deemter

A Conceptual Graph Description of Medical Data for Brain Tumour
Classification 140
  *Madalina Croitoru, Bo Hu, Srinandan Dashmapatra, Paul Lewis,
  David Dupplaw, and Liang Xiao*

A Conceptual Graph Based Approach to Ontology Similarity
Measure 154
  *Madalina Croitoru, Bo Hu, Srinandan Dashmapatra, Paul Lewis,
  David Dupplaw, and Liang Xiao*

A Comparison of Different Conceptual Structures Projection
Algorithms   165
  Heather D. Pfeiffer and Roger T. Hartley

A Conceptual Graph Approach to Feature Modeling   179
  Randall C. Bachmeyer and Harry S. Delugach

From Conceptual Structures to Semantic Interoperability of Content . . .   192
  Pavlin Dobrev, Ognian Kalaydjiev, and Galia Angelova

Formal Concept Analysis

Faster Concept Analysis   206
  Adam D. Troy, Guo-Qiang Zhang, and Ye Tian

The Design Space of Information Presentation: Formal Design Space
Analysis with FCA and Semiotics   220
  Michael May and Johannes Petersen

Reducing the Representation Complexity of Lattice-Based
Taxonomies   241
  Sergei Kuznetsov, Sergei Obiedkov, and Camille Roth

An FCA Perspective on n-Distributivity   255
  Heiko Reppe

Towards a Semantology of Music   269
  Rudolf Wille and Renate Wille-Henning

Analysis of the Publication Sharing Behaviour in BibSonomy   283
  Robert Jäschke, Andreas Hotho, Christoph Schmitz, and
  Gerd Stumme

The MILL – Method for Informal Learning Logistics   296
  Andreas Faatz, Manuel Goertz, Eicke Godehardt, and
  Robert Lokaiczyk

Bilingual Word Association Networks   310
  Uta Priss and L. John Old

Using FCA for Encoding Closure Operators into Neural Networks   321
  Sebastian Rudolph

Conceptual Structures

Arc Consistency Projection: A New Generalization Relation for
Graphs   333
  Michel Liquiere

Mining Frequent Closed Unordered Trees Through Natural
Representations   347
  José L. Balcázar, Albert Bifet, and Antoni Lozano

Devolved Ontology for Smart Applications 360
    Iain Duncan Stalker, Nikolay Mehandjiev, and Martin Carpenter

Historical and Conceptual Foundation of Diagrammatical Ontology 374
    Peter Øhrstrøm, Sara L. Uckelman, and Henrik Schärfe

Learning Common Outcomes of Communicative Actions Represented
by Labeled Graphs 387
    Boris A. Galitsky, Boris Kovalerchuk, and Sergei O. Kuznetsov

Belief Flow in Assertion Networks 401
    Sujata Ghosh, Benedikt Löwe, and Erik Scorelle

Conceptual Fingerprints: Lexical Decomposition by Means of
Frames – A Neuro-cognitive Model 415
    Wiebke Petersen and Markus Werning

Constants and Functions in Peirce's Existential Graphs 429
    Frithjof Dau

Revelator Game of Inquiry: A Peircean Challenge for Conceptual
Structures in Application and Evolution 443
    Mary Keeler

Short Papers

Helping System Users to Be Smarter by Representing Logic in
Transaction Frame Diagrams 460
    David Cox and Simon Polovina

Quo Vadis, CS? On the (non)-Impact of Conceptual Structures on the
Semantic Web 464
    Sebastian Rudolph, Markus Krötzsch, and Pascal Hitzler

A Framework for Analyzing and Testing Overlapping Requirements
with Actors in Conceptual Graphs 468
    Bryan J. Smith

Implementation of SPARQL Query Language Based on Graph
Homomorphism 472
    Olivier Corby and Catherine Faron-Zucker

Cooperative CG-Wrappers for Web Content Extraction 476
    Fotis Kokkoras, Nick Bassiliades, and Ioannis Vlahavas

Conceptual Graphs and Ontologies for Information Retrieval 480
    Catherine Comparot, Ollivier Haemmerlé, and Nathalie Hernandez

Representation Levels Within Knowledge Representation 484
    Heather D. Pfeiffer and Joseph J. Pfeiffer Jr.

Supporting Lexical Ontology Learning by Relational Exploration 488
    Sebastian Rudolph, Johanna Völker, and Pascal Hitzler

Characterizing Implications of Injective Partial Orders   492
    José L. Balcázar and Gemma C. Garriga

DVDSleuth: A Case Study in Applied Formal Concept Analysis for
Navigating Web Catalogs   496
    Jon Ducrou

Navigation in Knowledge-Based System for Helpdesk Based on FCA . . .   501
    Vladimír Sklenář, Martin Radvanský, and Michal Dobeš

Functorial Properties of Formal Concept Analysis   505
    Hideo Mori

Towards an Ontology to Conceptualize Solution Analysis Tasks in
CSCL Environments   509
    Rafael Duque, Crescencio Bravo, and Manuel Ortega

Author Index   513

An Introduction to Conceptual Graphs

Simon Polovina

Culture, Communication and Computing Research Institute (C3RI)
Faculty of Arts, Computing, Engineering & Sciences
Sheffield Hallam University, UK S1 1WB
s.polovina@shu.ac.uk

Abstract. This paper provides a lucid introduction to Conceptual Graphs (CG), a powerful knowledge representation and inference environment that exhibits the familiar object-oriented features of contemporary enterprise and web applications. An illustrative business case study is used to convey how CG adds value to data, including inference for new knowledge. It enables newcomers to conceptual structures to engage with this exciting field and to realise "Conceptual Structures: Knowledge Architectures for Smart Applications", the theme of the 15th Annual International Conference on Conceptual Structures (ICCS 2007, www.iccs2007.info).

1 Introduction

Conceptual Graphs (CG, www.conceptualgraphs.org) provide a powerful knowledge representation and inference environment, whilst exhibiting the familiar object-oriented and database features of contemporary enterprise and web applications. CG capture nuances in natural language whilst being able to be implemented in computer software. CG were devised by Sowa from philosophical, psychological, linguistic, and artificial intelligence foundations in a principled way [8, 9]. Hence CG are particularly attractive as they are built upon such a strong theoretical and wide-ranging base.

There is an active CG community, evidenced by the annual International Conferences on Conceptual Structures (ICCS, www.iccs.info), now in its 15th year (ICCS 2007, www.iccs2007.info), not to mention the annual CG workshops beforehand (www.conceptualstructures.org/confs.htm). There is also the CG discussion list (cg@conceptualgraphs.org). Its participants happily support newcomers to CG e.g. in answering queries; www.conceptualgraphs.org provides information on how to join this list, as well as a comprehensive catalogue of software CG tools.

CG is core to the ISO Common Logic standard (http://cl.tamu.edu/)[1]. The CG community has furthermore grown to embody a wider notion of Conceptual Structures (CS, www.conceptualstructures.org, and the title of Sowa's seminal 1984 text [9]). This is typified by the large scale and valued contributions that Formal Concepts Analysis (FCA, www.upriss.org.uk/fca) brings to ICCS each year [4].

---

[1] Assigned to WG2 (Metadata) under SC32 (Data Interchange) of ISO/IEC JTC1.

U. Priss, S. Polovina, and R. Hill (Eds.): ICCS 2007, LNAI 4604, pp. 1–14, 2007.

A strong case therefore exists for bringing an awareness of CG to an even wider community. Through an illustrative business case study, the following explication of CG[2] aims to achieve this objective so that many more researchers and industry professionals can realise the benefits of CG and, as the theme of ICCS 2004 highlighted, put them to work [10].

2   Concepts and Relations

CG are based upon the following general form:

This may be read as: "*The* relation *of a* Concept-1 *is a* Concept-2". The direction of the arrows assists the direction of the reading. If the arrows were pointing the other way, then the reading would be the same except that Concept-1 and Concept-2 would exchange places (i.e. "*The* relation *of* Concept-2 *is a* Concept-1").

As an alternative to the above 'display' form[3], the graphs may be written in the following 'linear' text-based form:

    [Concept_1] -> (relation) -> [Concept_2].

The full stop '.' signals the end of a particular graph. Consider the following example:

    [Funding_Request] -> (initiator) -> [Employee].

This example will form a part of an illustrative case study about requests for funding new business projects in the fictitious enterprise 'P-H Co.'. The example graph reads as "The initiator of a funding request is an employee". This may create readings that may sound long-winded or ungrammatical, but is a useful mnemonic aid in constructing and interpreting any graph. It is easier to state "An employee initiates a funding request".

Concepts can have *referents*, which refers to a particular instance, or individual, of that concept[4]. For example consider the concept:

    [Employee: Simon].

This reads as "The employee known as Simon". The referent is a *conformity* to the *type label* in a concept. This example shows that Simon conforms to the type label 'Employee'.

---

[2] That also draws on an earlier introduction to CG in 1992 [7].

[3] The display form CG throughout this paper were produced using the *CharGer* software (http://charger.sourceforge.net/), one the many useful CG software tools that are catalogued at www.conceptualgraphs.org.

[4] There are other kinds of referents, such as plural ('sets' of) referents (which are rather like collections in object-orientated classes, and scalars ('measures') [8, 9]. In passing, as well as Concepts and Relations in CG there are Actors (which incidentally are not to be confused with UML Actors! [6]) Delugach is a proponent for the use of CG Actors [3].

A concept that appears without an individual referent has a *generic* referent. Such 'generic concepts' should be denoted as [*Type_Label*: *]. Writing [*Type_Label*] is merely a convenient shorthand.

Generic concepts may take up an individual referent. A unique identifier can be used to make a concept distinct. Thereby the generic concept [Funding_Request] might become [Funding_Request: #1234] with respect to [Employee: Simon]. This would yield:

    [Funding_Request: #1234] -> (initiator) -> [Employee: Simon].

If there are two or more employees with the name Simon we would need to make them distinct from one another e.g. [Employee: Simon#122014].

3 The Type Hierarchy

In CG, type labels belong to a type hierarchy. Thereby:

    Manager < Employee. ("A manager is an employee")

This means Manager is a more specialised type of the type Employee i.e. Manager is a *subtype* of Employee. Alternatively, this can be stated as Employee is a *supertype* of Manager. (Subtypes and supertypes are analogous to subclasses and superclasses in object-orientation, thus a subtype inherits the characteristics of its supertypes.) Similarly, the remainder of the hierarchy may be:

    Employee < Person.
    Person < Animal.
    Animal < Entity.
    Entity < T.
    Funding-Request < Request.
    Request < Act.
    Act < Event.
    Act < T.

Sowa provides a conceptual catalog that includes a representative set of hierarchical concepts, as well as relations [9]. It also shows the context in which a type label should be used. For example an Act is an Event with an Animate agent:

    [Act] -> (agent) -> [Animate].

The type denoted as 'T' means the universal supertype. It has no supertypes and is therefore the most general type.

A subtype can have more than one immediate supertype. For example, consider the concept [Animal], which has a more detailed set of supertypes than indicated above. This concept has a type label Animal which may be defined as [9]:

    Animal < Animate, Mobile_Entity, Physical_Object, ¬Machine.
    Animate < Entity.
    Mobile_Entity < Entity.
    Physical_Object < Entity.

An animal therefore inherits the characteristics of being animate, a mobile-entity, and a physical object, but it is not a machine (¬ means 'not').

4 Projection

Projection in CG extends the notion of projection in SQL in database systems [1, 8, 9]. It is used in conjunction with specialised graphs. A CG becomes more specialised when one or more of the following happens:

a) more concepts, types and relations are added to it to narrow the scope further, e.g. an employee initiates a funding request that is the outcome of a company policy (as opposed to the earlier where it is only shown that an employee initiates a funding request) i.e.:

```
[Funding_Request] -
  (initiator) -> [Employee]
  (outcome) <- [Company_Policy].
```

This larger graph illustrates the use of hyphen '-', which allows the relations of a concept (being Funding-Request in this case) to be listed on subsequent lines. In the visual display form of CG this is not needed as the following equivalent CG shows:

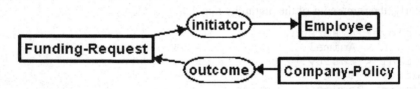

b) it acquires non-generic referents, e.g. [Employee] becomes [Employee: Simon], or
c) subtypes are substituted for (super)types e.g. replacing [Employee] with [Manager]

The following example illustrates a combination of these:

```
[Funding-request] -
  (initiator) -> [Manager: Susan]
  (outcome) <- [Company-Policy: #CP76321].
```

Therefore each specialised graph may have a more general graph from which it was derived. Likewise a general graph can have a number of specialised variants. Thus *projection* is where we take a graph and try to see if another graph is a generalisation of it. It there is such a fit we have determined that the graph is indeed a specialised variant, and that the other graph is indeed a generalisation of it. We can then state that the general graph 'projects' into the specialised one.

Projection plays an important role in the inference for new knowledge with conceptual graphs. If a graph projects into another, then a particular pattern may have been identified. Projection may result in a new graph being asserted. *Inference*, the operation that describes these assertions, is discussed later. Projection also features in the *combining* of graphs to form larger graphs. This subject is discussed first, and will serve to illustrate projection and help us to understand inference better.

5   Combining Graphs

The joining of graphs facilitates inference because more projections can be made into bigger graphs. *Maximal join*, which extends the notion of join in SQL in database systems [1, 8, 9], defines the optimal method by which graphs are joined. Maximal join occurs when graphs are joined on the most common, or *maximally extended*, projection. This is illustrated by the P-H Co. case study in **Fig. 1** as follows:

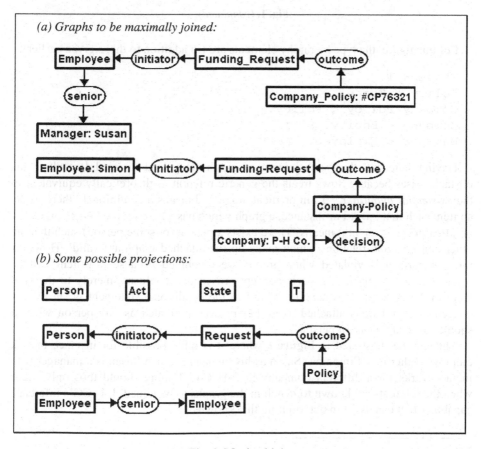

Fig. 1. Maximal join

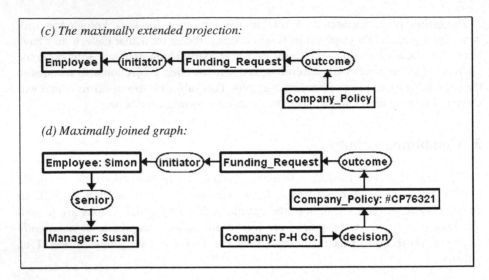

(c) The maximally extended projection:

(d) Maximally joined graph:

Fig. 1. (*continued*)

For this figure, the type hierarchy elements are (in addition to those given earlier):

```
State < T.
Policy < State.
Company-Policy < Policy.
Company < Entity.
Manager < Employee.
```

Having stated all this, such joining of graphs may lead to invalid results. This obstacle arises because Sowa treats the generic referent as theoretically equivalent to the *existential quantifier*, '∃', in predicate logic[5]. ∃ means a conditional "there exists an item such that ..... ". For instance, a graph which has [Person], [Person: *], or [Person: *x] in it means 'There exists a person (or some person) such that the other concepts and relations which make up the attached graph are valid'. However this comparison is violated when graphs are combined because *any* item will be suitable as a referent in a generic concept provided it conforms merely to the type label in the concept. This pays no regard to the conditional statement given by any concepts and relations attached to it. [Person] is treated as *any* person when it should mean an *unknown* person.

This can be illustrated in **Figure 1** where it is quite possible for instance that an employee other than Simon has Susan as his manager, or that Susan is a manager who in fact works for a different company to P-H Co.! Joining should thus only occur when the referents are known to match in the graphs to be joined. It is CG's inference capability that can assist in determining this knowledge.

---

[5] Conceptual graphs are, in fact, an existential notation, as there is a direct mapping between conceptual graphs and first order predicate logic.

6 Inference

Inference in conceptual graphs theory is based upon the existential graphs logic of Charles Sanders Peirce[6]. This 'Peirce logic' is developed by Sowa in the same principled way to provide a comprehensive inference capability in conceptual graphs. Peirce logic, cited by its founder as 'the logic of the future', is described by Sowa as an enhancement of the traditional propositional and predicate logic of Peano, Russell, and Whitehead [9].

Consider the following example (where we have simply referred to graphs as 'Graph 1' and 'Graph 2' as this allows us to focus on how the inference operates; a fuller example involving actual CG will be described later):

if Graph 1 **then** Graph 2.

This may be read as "If Graph 1 can project into any outer graphs, then Graph 2 can be asserted".

In Peirce logic, 'if-then' can be rewritten as: **not** (Graph 1 **and not** Graph 2).

This can be written graphically in Peirce logic as:

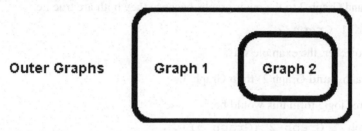

This visual form illustrates the contextual domination of graphs over other graphs. A graph is dominated by another graph if the dominated graph is ringed by what is known as a *negative context*[7], whereas the dominating graph is outside of that ring. Here, the outer graphs dominate Graph 1 which dominates Graph 2.

Any graph which projects into a graph which dominates it may be 'rubbed out' or *deiterated*. To assert Graph 2, Graph 1 must project into the outer graphs. Should this occur, Graph 1 can be deiterated leaving two rings around Graph 2 thus:

The term '**not**(**not** Graph 2)' equates to the term 'Graph 2' so the empty outside ring and the inside ring cancel out, or *double negate*. This frees Graph 2 out of its

---

[6] Pronounced as 'purse'.
[7] Peirce referred to these as 'cuts' [9].

contexts and thereby means it has been asserted as a new graph in the outer graphs. Graph 2 is thus true. This example demonstrates that a true antecedent in an 'if-then' rule means its consequent is always true. This is the general inference rule of *modus ponens* and has been demonstrated here using Peirce logic.

The linear equivalent of the above is:

```
¬[Graph 1 ¬[Graph 2]].
```

As we know, the symbol '¬' means 'not'. A '¬[....]' forms a negative context ring. Thus '¬[*Graph*]' means that graph is not true. The term 'not true' in this sense equates to that graph being *false*. Therefore it is also possible by the appropriate use of negative contexts and nested negative contexts to build a knowledge base consisting of both true graphs, false graphs, and various inferences of those graphs. For the sake of clarity, this discourse will substitute '(....)' in place of '¬[....]' to denote negative contexts written in the linear form.

Using this preference, the linear equivalent of the above example is:

```
(Graph 1 (Graph 2)).
```

Consider the next example: Graph 1 **and** Graph 2. This is merely a case of adding Graph 1 and Graph 2 to the outer graphs because they both are true i.e.:

```
Graph 1 Graph 2.
```

Say, however, the example was:

if (Graph 1 **and** Graph 2) **then** Graph 3.

In Peirce logic form this would be:

```
(Graph 1 Graph 2 (Graph 3)).
```

Assuming that Graph 1 and Graph 2 existed in the outer graphs, then they can be deiterated and Graph 3 double negated thereby asserting it as a new graph. Now say that the outer graphs happened to include the graph (Graph 3) instead – i.e. Graph 3 is false. Regarding the above rule, (Graph 3) can be deiterated from it leaving:

```
(Graph 1 Graph 2).
```

This shows that because Graph 3 is false then *both* Graph 1 and Graph 2 are false. It is still possible for *either* Graph 1 *or* Graph 2 to be in the knowledge base *but not both*. If they were or some derivative that would state they were, this would show there is an inconsistency in that knowledge base. Return to the first 'if Graph 1 then Graph 2' example:

```
(Graph 1 (Graph 2)).
```

Should (Graph 2) be in the outer graphs, then (Graph 1) would be asserted. This demonstrates another general inference rule of *modus tollens*, or that if the consequent ('then' part) of an 'if-then' rule is false then so is its antecedent ('if' part). The illustration also shows that if the antecedent is false, then the consequent cannot be determined from it. If (Graph 1) was in the knowledge base to begin with, there is no possible way to assert either Graph 2 or (Graph 2) from (Graph 1) alone.

It should be noted that if a graph is false then so will be any of its specialisations. If "the employee Simon" is false, then "the employee Simon who works for P-H Co." (or indeed any employer) must also be false.

As a final example, consider: Graph 1 **or** Graph 2. Logically, 'or' can be rewritten as: **not** (**not** (Graph 1) **and not** (Graph 2)). This maps to the Peirce logic form:

```
((Graph 1) (Graph 2)).
```

By the above discussed Peirce logic operations, if either Graph 1 or Graph 2 was false then Graph 2 or Graph 1 would be true respectively. It also shows that 'or' is the same as 'if not ... then'. Thus '**if not** Graph 1 **then** Graph 2' is the same as 'Graph 1 **or** Graph 2', as is '**if not** Graph 2 **then** Graph 1'.

An example using actual CG is given by **Fig. 2**:

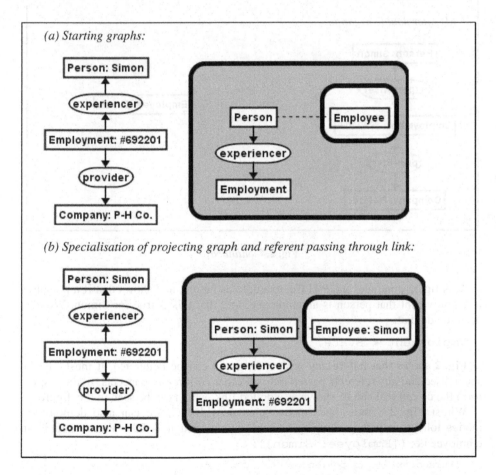

Fig. 2. An actual conceptual graphs inference

10   S. Polovina

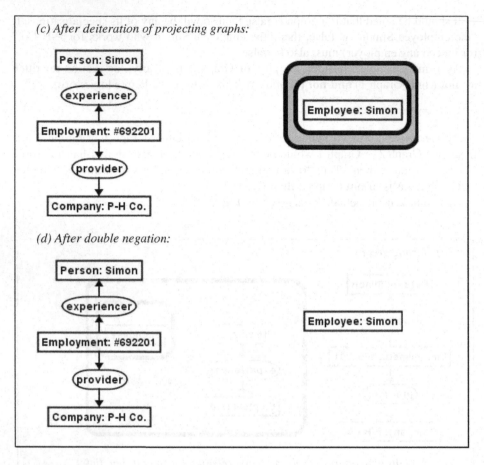

Fig. 2. (*continued*)

This figure continues the P-H Co. example as before, and shows that if a person is in employment that person is an employee, and that this is true for Simon. We also need to add that:

```
Employment < State.
```

Fig. 2 shows that before any projecting graph can be deiterated, it must first be specialised and any referents passed onto all other *co-referent* concepts. In the display form the co-referent can be shown as a dotted link, which can be seen in this figure.

Whilst **Fig. 2** demonstrates modus ponens with CG, we can also demonstrate modus tollens with the following CG that shows that it is false that Simon is an employee i.e. ([Employee: Simon]):

```
([Person:*x] <- (experiencer) <- [Employment] ([Employee:*x])).
```

Here the co-referent is shown by the alternative of matching *x values. After specialisation:

```
([Person: Simon] <- (experiencer) <- [Employment] ([Employee: Simon]])).
```

After deiteration:

```
([Person: Simon] <- (experiencer) <- [Employment]).
```

Hence as it is false that Simon is an employee, he cannot be experiencing employment!

In summary, Peirce logic shows contexts of knowledge elements visually dominating others and inference is performed by attempting to reduce those contexts. As we can also see, it also sets the conditions by which referents can be correctly instantiated thus addressing the issue identified by the earlier **Fig. 1** discussions.

7 A More Comprehensive Illustration

Whilst the foregoing examples demonstrate the elegance of CG, real-world information systems would need to handle much more complex problems. The following is a more realistic illustration of a typical problem, again referring to the P-H Co. funding request scenario as the ongoing case study. Note that even this example has to be necessarily simple so as to get the ideas across, but that it more fruitfully suggests how CG can be put to work in the real world.

Let us therefore refer to the following statement as our starting point:

P-H Co. Company Policy # PHCP69692. *There have been no guidelines to help P-H Co. allocate its budget for funding requests by employees for new business projects. There is a need for a guideline so that the decisions of the company are consistent, to encourage less senior members of staff to make such requests, and to evidence this rationale. P-H Co. therefore has decided on this company policy, which is that funds will be allocated in the following order (highest priority first): Junior Staff, Senior Staff, Manager, and lastly Director.*

The actual employee seniority relationships in P-H Co. are given by **Fig. 3**:

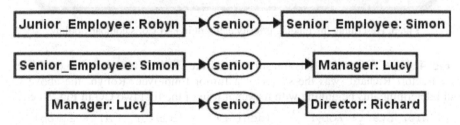

Fig. 3. The seniority relationships in P-H Co

Added to the type hierarchy are:

```
Junior_Employee < Employee.
Senior_Employee < Employee.
Director < Employee.
```

We can also describe in CG the general applicable concept of seniority, **Fig. 4**.

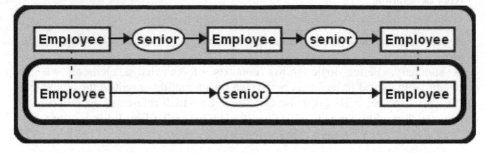

Fig. 4. The generally applicable concept of seniority

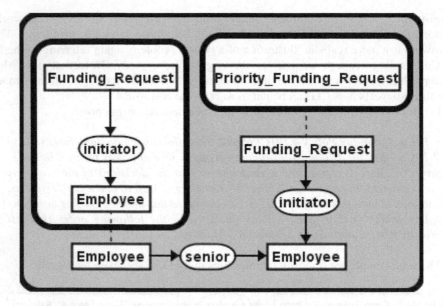

Fig. 5. Describing the funding priority element in P-H Co.'s company policy

Fig. 4 shows that seniority is transitive in nature. Thus if we wished to establish if the Director 'Richard' was the senior of a Junior Employee 'Robyn' the above CG could determine this beginning with maximal join of the first two CG in **Fig. 3** i.e.:

```
[Junior_Staff: Robyn] -> (senior) -> [Senior_Staff: Simon]
-> (senior) -> [Manager: Lucy].
```

This is a legitimate maximal join as we had denoted that the referents match through being co-referent. The antecedent component of **Fig. 4** can project onto this CG and specialise to match it:

```
[Junior_Staff: Robyn] -> (senior) -> [Senior_Staff: Simon]
-> (senior) -> [Manager: Lucy].
```

Continuing, the lines of identity (co-referent links) would specialise the consequent part of **Fig. 4** to:

```
[Junior_Staff: Robyn] -> (senior) -> [Manager: Lucy].
```

The antecedent can then be deiterated and, after double negation, the consequent would be asserted. It can, in turn, be maximally joined with the third CG in **Fig. 3** to give:

```
[Junior_Staff: Robyn] -> (senior) -> [Manager: Lucy]
 -> (senior) -> [Director: Richard].
```

This forms a new antecedent CG in **Fig. 4** (after specialisation, deiteration then double negation) to give the asserted CG:

```
[Junior_Staff: Robyn] -> (senior) -> [Director: Richard].
```

Thus Richard is senior to Robyn. Using this background knowledge we can draw a CG rather like **Fig. 5** to describe the company policy. This more comprehensive CG models the priority of funding that favours less senior employees. Essentially, if a funding request is initiated by a more senior employee then that will become a priority funding request provided a less senior employee has not initiated a funding request. To appreciate the lucidity of this CG (and the previous ones), try them out with various scenarios (including modus ponens and modus tollens), following the steps of projection, maximal join, specialisation, deiteration and double negation as appropriate.

To assist, Aalborg University provide a comprehensive online course on CG (www.huminf.aau.dk/cg). For a practical implementation of Peirce logic in CG, refer to Heaton [5]; for further theoretical discussion see Sowa and Dau [2, 9].

Also **Fig. 5** might be adapted in the light of new circumstances: P-H Co's funding policy would need to be adapted if there was more than one employee of the same grade; the new CG may then need to reflect that, for example, priority is then given according to the proposal being ranked by a panel of judges. We can therefore see that this simple funding example is beginning to take on real-world dimensions.

8 Concluding Remarks

We can see that CG are variable-sized, hierarchical structures. Projection, maximal join and inference show how value is being added to data as CG capture the underlying concepts behind data, relate data to other data, and find patterns from them. By capturing the meaning behind data, CG therefore capture knowledge in a way that is more useful to people whilst bring able to be directly represented in software.

Whilst many in the CG and CS community will find little that's new in this introduction (albeit it might provide a useful refresher), it is hoped that it will encourage newcomers to engage with this exciting field and to help realise "Conceptual Structures: Knowledge Architectures for Smart Applications", the theme of ICCS 2007.

References

1. Connolly, T.M.: Database Systems: A Practical Approach to Design, Implementation, and Management, 4th edn. International Computer Science Series. Addison-Wesley, Harlow (2005)
2. Dau, F.: The Logic System of Concept Graphs with Negation. LNCS (LNAI), vol. 2892. Springer, Heidelberg (2003)
3. Delugach, H.: Active Knowledge Systems for the Pragmatic Web. GI P-89, 67–80 (2006)
4. Ganter, B., Stumme, G., Wille, R. (eds.): Formal Concept Analysis. LNCS (LNAI), vol. 3626. Springer, Heidelberg (2005)
5. Heaton, J.E.: Goal Driven Theorem Proving using Conceptual Graphs and Peirce Logic. Doctoral Thesis, Loughborough University (1994)
6. Mattingly, L., Rao, H.: Writing Effective use Cases and Introducing Collaboration Cases. The Journal of Object-Oriented Programming 11, 77–87 (1998)
7. Polovina, S., Heaton, J.: An Introduction to Conceptual Graphs. AI Expert 7, 36–43 (1992)
8. Sowa, J.F.: Knowledge Representation: Logical, Philosophical and Computational Foundations. Brooks-Cole (2000)
9. Sowa, J.F.: Conceptual Structures: Information Processing in Mind and Machine. Addison-Wesley, London, UK (1984)
10. Wolff, K.E., Pfeiffer, H.D., Delugach, H.S. (eds.): ICCS 2004. LNCS (LNAI), vol. 3127. Springer, Heidelberg (2004)

Trikonic Inter-Enterprise Architectonic

Gary Richmond

City University of New York
New York, USA
garyrichmond@rcn.com

Abstract. There is a need for information, application, and other enterprise architectures which are robust and flexible enough to meet the challenges of today's heterogeneous, rapidly changing, digitally networked environment. Developing advanced architectures may prove essential for achieving emerging research, business, and social goals. Indeed, the profoundly changed landscape suggests that a new paradigm may be needed, an *inter-enterprise architectonic* (I-EA) informing architectures capable of integrating all key components and processes in an increasingly interconnected environment. To meet this challenge, a systems architectonic is outlined that is based on the trichotomic category theory of Charles S. Peirce. *Trikonic Inter-Enterprise Architectonic* involves a pragmatic approach to the observation and manipulation of diagrams as models of enterprise and inter-enterprise processes.

1 Introduction

Peter Skagestad in "'The Mind's Machines: The Turing Machine, the Memex, and the Personal Computer" [18] considers the history of Artificial Intelligence (AI) in relation to Intelligence Augmentation (IA) and concludes that the American scientist, logician and philosopher, Charles S. Peirce, provided a theoretical basis for IA analogous to Turing's for AI. Besides being keenly interested in the possibility of the evolution of human consciousness as such, Peirce seems even to have anticipated Doug Engelbart's notion of the co-evolution of man and machine. In another paper on 'virtuality' as a central concept in Peirce's pragmatism Skagestad goes so far as to suggest that "in Peirce's thought . . . we find the most promising philosophical framework available for the understanding and advancement of the project of augmenting human intellect through the development and use of virtual technologies"[1] [19].

Whatever the exact intellectual genealogy of the AI-IA connection may prove to be, there can be little question that in our digital networked era there appears to be a marked interpenetration of "man and machine" at least in the sense that it has become something of a truism that information technology is having a significant impact on our personal and professional lives, especially by profoundly influencing the structure

---

[1] Skagestad notes, however, that for Peirce "reasoning in the fullest sense of the word could not be represented by an algorithm, but involved observation and experimentation as essential ingredients" [19].

U. Priss, S. Polovina, and R. Hill (Eds.): ICCS 2007, LNAI 4604, pp. 15–28, 2007.
© Springer-Verlag Berlin Heidelberg 2007

and functioning of many organizations and institutions. For example, most large enterprises are to some extent already infra-structurally distributed computing systems. Meanwhile a new ecosystem of "pervasive networks, reusable services, and distributed data" [28] is changing the way nearly all enterprises operate in a deeply networked environment. In addition, the ubiquitousness and power of the internet has brought about a substantial increase in the participation of consumers of information through web-based services. Looking creatively to the future, evolving networks seem even to have the potential for catalyzing the growth of new forms of cross-disciplinary research and new models of inter-enterprise collaboration such as are implied by the idea of a Pragmatic Web [3, 4, 17]. New architectures may be needed in order to help guide the creation of the conditions which would allow for enterprise and, in particular, inter-enterprise endeavors to respond quickly and creatively to difficult challenges and fresh opportunities in a highly volatile environment. It is likely that in the future IT researchers and technologists will need to work closely with business leaders and other decision makers to more fully integrate the technical and semantic aspects of nets with the purposes of the users of these technologies.

Many businesses and other enterprises are finding that a good deal of what they are providing today is 'services' dependent on information technologies [7]. It has been suggested that because of this service orientation we will need more than ever to "apply technology, engineering and disciplined thinking and design to the people aspect of businesses" [27]. For the business sector *service oriented architecture* (SOA) has been a creative response to the new context, while even those at the forefront of SOA development have had to admit that much remains to be done. For example, SAP acknowledges, in consideration of its own Enterprise Services Architecture (ESA) which is meant to be a "blueprint for how enterprise software should be constructed to provide the maximum business value," that "the current state of the art is a long way from ESA" [28]. To move things forward a new architectural paradigm may be needed, one affording overarching design principles for creating and assembling all components in a landscape involving myriad diverse distributed users in a wide variety of inter-connected activities. Such a model would be in effect a veritable *inter-enterprise architectonic* (I-EA) capable of guiding the development of powerful new architectures for bringing about the coherence of all key components, processes, and user functions in, especially, large-scale projects involving several enterprises and perhaps hundreds of thousands of users when we include—as we now must—digitally connected customers and clients.

The architectonic to be discussed here is based on Peirce's *trichotomic category theory* and in the present case involves the creation and observation of diagrams of the pertinent *patterns of processes* factoring into the functioning and growth of especially inter-connected enterprises. In [14] we outlined a diagrammatic approach to the category theory of Peirce, *Trikonic,* as a more iconic representation of his science of *Trichotomic*, an applied science with considerable untapped potential to contribute to the development of new models and architectures needed in all fields. *Trikonic* is developed here in the direction of a type of tricategorial *vector cycle analysis-synthesis* employing diagrams of key structures and processes important to enterprise and inter-enterprise development. At the heart of this approach is the principle that the growth of ideas in complex systems is facilitated by individual and group diagram observation and manipulation, potentially eliciting novel approaches and strategies for stimulating in particular inter-enterprise projects and partnerships. Here we will expand the argument made in [15] that diagram observation supports domain and cross-domain analysis and, going beyond analysis, tends towards the synthesis of the patterns and structures needed for project and enterprise development.

Architectures that are fully responsive to tomorrow's landscape will allow for flexible and rapid system modifications addressing changing enterprise and inter-enterprise goals and requirements. It will be argued that *trikonic architectonic* could contribute to the development of architectures powerful and flexible enough to meet this challenge, moving beyond building collections of infrastructural functionality towards conceiving entire inter-enterprise ecosystems architectonically. Section 2 examines Peirce's *systems architectonic* built on his category theory and explicated in his semeiotic. The system of his *classification of the sciences* is considered as a preliminary but significant step in a tricategorial analysis with implications for cross-disciplinarity in today's networked landscape. Section 3 shows how trikonic offers a "more iconic" approach to Peirce's trichotomic analysis. Section 4 considers how trikonic might assist in the development of the kinds of inter-enterprise architectures which will be needed in the future. Building on this foundation *Trikonic Inter-Enterprise Architectonic* (I>*k I-EA) is outlined. Section 5 introduces a variety of *vector cycle analysis-synthesis* involving the creation, observation, and manipulation of design templates for analyzing possible structures and strategies, patterns and processes involved in distributed settings such as inter-enterprise partnerships. Section 6 concludes with prospects for the future.

2 Architectonic Developed Tricategorially

Few thinkers have emphasized what might be termed systems architectonic more than C.S. Peirce [10, 11]. His essentially trichotomic *classification of the sciences* (to be discussed below) represents one important facet of his architectonic thinking. The classification has been acknowledged as not only a significant contribution to the philosophy of science, but as anticipating contemporary cross-disciplinarity, especially regarding the sharing of methods[2] in inter-disciplinary inquiry [10]. Peirce holds that "systems ought to be constructed architectonically" [CP 6.9] and, indeed, his widely influential *philosophical pragmatism* is itself both a product of and a moment in his vast trichotomic architectonic. The very first trichotomy of his classification schema is a structural division into three grand sciences: *science of discovery* (pure, theoretical science), *science of review* (including philosophy of science and the classification itself), and practical science (that is, applied arts and sciences).

But turning now to the main focus of this section, within discovery science the third and final normative science, logic as semeiotic, itself has three divisions culminating in *methodology* (which Peirce also refers to as *methodeutic* or *pure*

---

[2] Peirce comments that that "which constitutes science . . . is not so much correct conclusions, as it is a correct method. But the method of science is itself a scientific result" [CP 6.428].

rhetoric). At the heart of his methodology is the marriage of the *pragmatic maxim*[3] with the tripartite structure of inquiry, namely, *abduction* of a hypothesis, *deduction* of the implications of the hypothesis for testing, and *induction* in the sense of an actual experimental testing. Thus we see the architectonic genesis of Peirce's pragmatism: *"Pragmatism was ... designed and constructed ... architectonically ... [so that] in constructing [it] . . . the properties of all indecomposable concepts were examined [respecting] the ways in which they could be compounded"* [CP 5.5]. The grounds of these "indecomposable concepts" are *universal categories* of possible objects of thought: *"Peirce found three which he came to call Firstness, Secondness, and Thirdness ... [T]he definition of such concepts is the first step in erecting an architectonic philosophy"* [11].

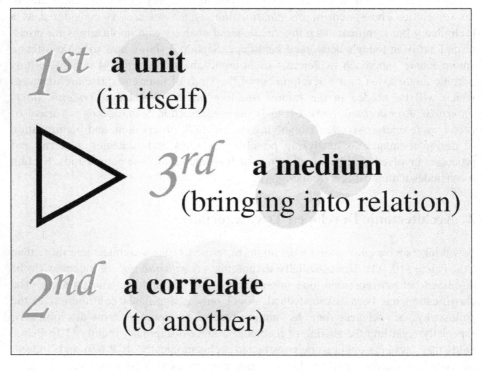

Fig. 1.

Firstness (1^{ns}) may be characterized as *qualitative possibility* (something "may be"), secondness (2^{ns}) as *actuality*, that is, existential action-reaction ("something exists"), and thirdness (3^{ns}) as *mediation* bringing the other two into 'lawful' relationship; and it is by its 'lawfulness'—that is, by the tendency to take regular habits which can express themselves intelligibly *in futuro*—that 3^{ns} is able to mediate between 1^{ns} and 2^{ns} (see Fig. 1).

---

[3] "C. S. Peirce's Pragmatic Maxim is that one best clarifies a conception by representing it in terms of conceivable experience on which the conception's truth would have some conceivable practical bearing" [26].

Models built on such simple and essentially mathematical ideas could have significant implications for intellectual/cultural evolution as providing templates which might persist, 'reproduce', and then be combined and recombined, modified and 'manipulated' in ways potentially contributing to the generation of emergent phenomena such as creative solutions to significant institutional and organizational problems. This is so because such models allow us to "look for the same phenomena in different contexts [in order to] separate features that are always present from features that are tied to context" [8]. Peirce constructs his entire systems architectonic (including his vast semeiotic) upon his three categories, admittedly "conceptions so very broad and consequently indefinite that they are hard to seize and may be easily overlooked" [CP 6.32]. In his view science is essentially trichotomic: 'First science' in science of discovery, *mathematics,* has three divisions (finite collections, infinite collections, true continua); 'second science', *cenoscopic philosophy*[4], involves three sciences (trichotomic *phenomenology,* the three *normative sciences* of theoretical esthetics, practics, and logic as semeiotic, and lastly a scientific *metaphysics*); 'third science' includes all the physical and psychical *special sciences,* themselves arranged trichotomically (as descriptive, classificatory or nomonological)   All the above trichotomies represent *tricategorial relations* and not mere triadic groupings. Retrospectively, a trichotomic structure can be seen at the very beginning of science in the mathematics of logic as a kind of *mathematical valency theory* in consideration of "the simplest mathematics" viewed in light of Peirce *reduction thesis*[5]. However, the three categories are first *observed* in phenomenology where the character of each is experienced *as such,* that is, in its firstness. Significantly, Peirce's vast trichotomic classification is arrived at through a kind of diagram observation, a topic we turn to next.

3 Trikonic Is "More Iconic" Than Trichotomic

Stjernfelt [23] distinguishes two complementary notions of *iconicity* in Peirce's analysis, the *operational* and the *optimal.* These ideas are tied to Peirce's movement towards an *extreme realism* which includes 'real possibilities', what he calls 'would-bes' in the sense that they would be realized in the future if certain conditions were brought about favoring their emergence. The *operational criterion* involves not only the idea that an icon resembles its object in any given diagram, but also the somewhat surprising notion that the construction of a kind of diagram is involved in virtually *all* reasoning. Most important for the thesis of this paper is the principle that through a certain kind of diagram observation and manipulation we may obtain *new* information.

As valuable as this operational conception is, Peirce concludes that it results in too broad a definition of iconicity for certain purposes. For example, in logic the alpha and beta parts of Peirce's *existential graphs* (EGs) are strictly equivalent to

---

[4] Cenoscopy is "philosophy, which deals with positive truth . . . yet contents itself with observations such as come within the range of every man's normal experience" [CP 1.241n1].
[5] The reduction thesis holds "that all higher order polyads can be reduced to triads; conversely, all higher order polyads can be constructed from triads" [11]. It has been given a strict mathematical proof in [2].

propositional logic and first order predicate logic respectively. Yet Peirce, who in fact earlier invented the linear version of these logics[6], found his graphical form, EGs, to be "more iconic" than the linear one. The concept of *optimal iconicity* emphasizes the observation of *graphical diagrams* optimally suited to visually displaying pertinent relationships. Exercising "careful probing, moving back and forth between conditions and phenomena," through diagram observation we can see existent patterns, and through diagram manipulation may even begin to provoke emergent patterns of relationship [8].

According to Peirce one of the essential things one needs to observe in considering the construction of a scientific architectonic is the relations of the various disciplines to each other. For Peirce this is developed as a classification of the sciences, something which he worked on over several decades as a *natural classification* in which the various sciences are observed as "the actual living occupation" of groups of people following some particular research goal and using unique methods, procedures, manners and devices of observation [10]. Beyond domain specific problems, Peirce held that one field can stimulate another toward solving its own seemingly intractable problems. Indeed entire branches of science can participate in this mutual stimulation as when, for example, *pure research science* lends its principles to the *special* and *applied sciences* which, in turn, "incessantly egg on researches into theory" [CP 7.72].

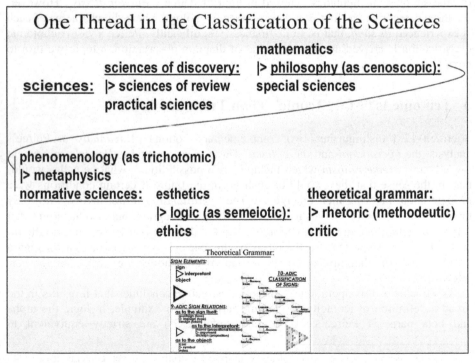

Fig. 2.

---

[6] Peirce was the first to invent a 'symbolic logic" although he is rarely credited with it [20].

Following Comte, Peirce organizes the sciences so that those earlier in the classification offer *principles* for those which follow, while those occurring later in the schema provide *examples and cases* for the former [CP 1.180]. The so-called "perennial classification" represents Peirce's final view as to the structure of the scientific enterprise taken as a whole [9]. While in some ways linear outlines of the classification[7] articulate the most general features of Peirce's systematic architectonic, they are of somewhat limited value in offering but an abstract and, as it were, static view of the structure of science. Trikonic diagrams can reveal significant tricategorial relations such as those obtaining in this diagram string of one thread of science of discovery culminating in Peirce's 10-adic classification of signs (see Fig. 2).

4 Trikonic Inter-Enterprise Architecture (|>*k I-EA)

"More iconic" approaches to knowledge representation such as Existential Graphs (EGs), Conceptual Graphs (CGs), and *Trikonic* (|>*k) may prove especially helpful in offering 'relational perspicuity' in that one *directly observes* the relationship. Contemporary technology has the capability of building tools for distributed diagram creation, observation and manipulation (in conjunction with consensus seeking and report authoring tools) which could lead to rich and, as it were, 'fractal-like' analysis of the categorial relationships important within an enterprise or research project. Observing and manipulating genuine tricategorial relations important to the structure and function of collaborative projects could influence the very evolution of the systems involved.

Since they articulate the ways in which the various components of a system are organized and integrated, all large scale enterprises have implicit and typically explicit architectures representing structures and processes important to successful functioning and growth. What appears to be increasingly needed is architecture capable of catalyzing rapid modification of the system for addressing emergent goals and requirements. Our networked era requires subtle and complex (but also 'user-friendly') architectures modeling overarching design for cohesion and coherence of all systems as these relate to users. New architectures might help guide the conception, design, and assemblage of all components of an enterprise's superstructure. Some have even suggested that it may be that the success of complex inter-enterprise operations will increasingly depend on robust architectures being efficiently and effectively envisioned, designed, and strategically employed [24]. Ultimately architecture should be able to fully analyze the fundamental structures, components, roles and significant relationships involved in enterprise systems. This structural knowledge can then assist in redesigning and reintegrating systems to meet—and "on the fly" as it were—new goals and requirements, becoming a kind of 'evolutionary architectonic' catalyzing the growth of emerging technologies, processes, business strategies, and so forth.

To create, develop, test, and implement such an architectonic will undoubtedly require leadership willing to communicate the vision of a common framework encapsulating the processes of all domains yet focusing on the goals of the enterprise

---

[7] See, for example [1].

as a whole [24]. It would seem especially important to achieve a balance between the vision of those leading such a paradigm shift and the design specialists working creatively from their individual and domain expertise. This is but one of the many challenges to developing and deploying an I-EA capable of integrating complex systems in an evolving inter-enterprise context. Yet whether we consider an individual enterprise or an inter-enterprise system-of-systems, the evolution of any overarching system will require understanding the principles governing the architecture of *all* its systems and those to which it stands in relation. While even the analysis of this is clearly no small task, the creation and deployment of the requisite design and development tools presents an even greater challenge.

Trikonic I-EA represents a conceptual structure with the potential for developing methodologies and integrated artifacts for modeling critical enterprise and inter-enterprise activities analyzed in terms of their significant tricategorial relationships. Relative to the needs outlined above, representations of conceptual knowledge tending to foster inter-enterprise development will connect conceptual modeling, knowledge management, information and web technologies and much else. The task is to develop elegant and effective approaches to integrating the power and efficiency of computers with the creativity and resourcefulness of people, what Douglas Engelbart calls *intelligence augmentation* (IA) [6]. *Trikonic architectonic* is designed so that these two aspects—the human and the computational—may interpenetrate in mutually productive ways. It is thus closely aligned with the conception of an emergent Pragmatic Web [3, 13].

Trikonic diagrammatically explicates and vectorially expands *Trichotomic*, Peirce's applied science of tricategorial analysis. While it was originally conceived in the interest of facilitating scientific inquiry and philosophical discourse, it is here directed towards the creation, observation, and manipulation of diagrams of significant relational structures and patterns in complex organizational systems. Yet, however it is employed such diagram observation ought to occasion a moment of applied *critical commonsense*, an idea at the heart of Peirce's theory of inquiry and by which is meant that kind of thinking which finds critical analysis and the development of a thorough going 'reasonableness' essential for real learning—including organizational learning—to occur [5]. Pragmatism strongly suggests that we are more likely to reach agreement when we employ a group observational method, when we "look together" at the same data, related patterns, etc., creating and manipulating diagrams of the relationships of the component elements. Naturally diagram observation needs to be accompanied by critical discussions of what participants say can be *objectively* seen there.

A more iconic and thoroughly architectonic approach would also tend to encourage the introduction of new ideas and hypotheses by individuals and teams. It has been suggested [8] that we need models which use if → then rules to assist in creating and designing possible scenarios for emergent phenomena. We need to be able to better "see and manipulate the mechanisms and interactions *underlying* … models, using [our] intuition to move the models into plausible regimes," what Peirce called abduction (or, retroduction to a plausible hypothesis). Diagram manipulation allows participants to explore new, even risky territory and, like a flight simulator, lets them 'push the envelope' without committing themselves to dangerous overt actions. It is certain that introducing such a novel architectural style will require new rules clearly and unambiguously stated as standards for effecting enterprise/inter-enterprise collaboration. Although it is impossible to fully define these standards in advance, SOA-centric companies are tending towards open-standards, portable components, and increased interoperability [13].

5 The Telic Vector Cycle for Systems Architecture

In [15] six *trikonic vectors* were introduced representing movement through possible trichotomic relations[8], especially as groups and threads of linked tricategorial structure/process relationships. Diagramming patterns involved in processes of potential importance to researchers and organizations is potentially one of the most promising applications of trikonic. This paper introduces the *telic cycle* for modeling enterprise and, in particular, I-E processes. The leading idea here is to bring about "a framework that uses a simple set of architectural artifacts to address the needs of enterprise architecture" [24]. Developing the architecture needed in this complex landscape is non-trivial when one looks at all the aspects and artifacts of analysis, synthesis, design and implementation which need to be considered "all together one after another," such architectures becoming decisive in the sense that the "models become the requirements" [27].

Fingar [7] outlines the inter-enterprise development cycle in a richly imagined scenario from which the following diagrams abstract the key concepts and relationships. There is no way to here represent any of the details which would need to be considered in an actual inter-enterprise development cycle, so that even were they highly abstracted and abbreviated, the elements/activities addressed in each of the six vectorial moments are too multitudinous and too complex to include in a short paper. Therefore the ensuing discussion merely introduces the telic cycle *as such* (the interested reader is referred to the elaborated scenario just mentioned.)

The *telic cycle* involves two complementary 'wings' organized in relation to the categorial position at which each of the six trichotomic vectors in the cycle arrives: in a word, the vectors are structured *teleologically*, that is, as to ends (Fig. 3). The first three vectors represent the *problem* side of the cycle, while the remaining three represent the *solution* side. Further, the whole cycle (or parts of it) may and typically would iterate over the life of an inter-enterprise endeavor. Individual trikonic analyses, trikonic group and string analyses, as well as the employment of other vector cycles (such as the *chiral cycle[9]*) could be employed at appropriate moments in an actual I-E development cycle. Certain activities (such as quality control) should be seen as occurring at many or even every stage of the cycle. The goal of the I-E teleological cycle model is to encapsulate each of the six phases of a development

---

[8] Trikonic makes much of vectorial permutations of the three categorial relations; there are, of course, six possible paths of movement [12].

[9] The *chiral vector cycle* is introduced in [15] and was employed in the analysis of a software engineering problem in [21].

cycle in architectural diagrams observed and manipulated by members of the development team. Only a bare bones framework can be presented here[10].

---

[10] Elements of the kinds of content to be expected in perhaps most inter-enterprise component-based development cycles following the telic cycle are briefly outlined below [see 7, chapter 7]. It is necessary here to abstract and simplify the important considerations at each phase. In addition, the actual tricategorial relations occurring in each of the six phases must be completely passed over because of limitations of space and the complexity of the topic. Yet, when one considers that, say, **Phase 5**, for example, represents the equivalent of the three categorially distinct stages of a complete inquiry (hypothesis formation, deduction of implications for testing, inductive testing) one may begin to imagine just how much has here been omitted.

Beginning at the **Problem** side of the *telic cycle* with 1) a determination of the requirements which leads logically to 2) a functional analysis of these requirements in relation to the project needs, this wing culminates in 3) a translation of these needs into the design of the I-E system. Then in the **Solution** side 4) the ordering of the phases of project development is followed by 5) testing and piloting culminating in the actual 6) launching of the project. Here, as at other points in the process, vectors, vector pairs, and other vector cycles may also be employed and iterated.

Phase 1—Requirements gathering: In the requirements gathering stage, some important considerations are: What are the roles of and who are the intended users of the proposed system? What access privileges are needed? What are the points of integration between I-E systems and how are these to be integrated? For example, which I-E business processes need to be mapped and for whom in real time? Also, what is to be placed in a repository of use cases binding system development? Finally, the development life cycle steps for quality assurance and testing purposes need to be considered at this phase.

Phase 2—Analysis: The most important question of the analysis phase is: What are the functional requirements? In addition there are considerations of the ways in which context level use cases may be elaborated as well as how to best detail specific systems requirements. Another key question is how the logical applications are to be developed.

Phase 3—Design: The design phase represents the core of the I-E design process. Its central problem is how to best move from a *problem space* to a *solution space* for both business objects and user interface design. Specific questions include: What functional modules will be most effective? What is the projected flow of operations between functional modules? How do we map analysis models to target platforms? How should deployment models be packaged as reusable components in an I-E environment? What are the possible effects of user task requirements on the applications flow? Graphics and usability groups need to create prototypes relating to user experience. When can the object model and design be finalized and the component repository yield reuse objects? Finally, it is important to consider how the system design document will be updated and how and when the specifications are to be distributed to the development team.

Phase 4—Development: While in one sense the design phase melds into the solution side of the cycle, the particular challenge for developers is how best to order the component assembly. The crucial consideration is how the glue code built to assemble components is to be tested (both unit and integrated testing) in the interest of interoperability. It is only at this stage that the application begins, as it were, "to come to life".

Phase 5—Testing: While various forms of quality assurance will necessarily have been involved from the very beginning, the question of how to ensure that functionality in the application meets the requirements needs becomes paramount at the testing level. Here

6 Summary and Prospects

An architectonic capable of relating all systems at all stages of inter-enterprise activity could have a significant impact on the ways enterprises would tend to operate in the future. Providing a coherent framework for managing inter-organizational complexity, it has been argued that *Trikonic* could catalyze the creation, development and deployment of new architectures which will almost certainly be needed for I-E analysis, creative synthesis and collaboration, as well as providing a basis for negotiations and decision making at all stages of inter-enterprise development. For example, distributed 'inter-team' diagram observation and manipulation could facilitate negotiations in difficult but crucial decision making processes such as selecting and integrating tools and procedures for increasing interoperability and security in partnered operations [13]. It is anticipated that both systems development and maintenance could be enhanced using appropriate vector cycle diagram observation and manipulation. The creation of reusable, 'evolving' templates of significant vectorial patterns could catalyze the development process.

At the heart of this approach is the esthetic of a *shared reasonableness* being seen as of intrinsic value by all parties involved in a given inter-enterprise activity. This in turn implies an ethics of fairness (involving the idea of *critical commonsense*) to complement the logics needed to help structure the required architectures. Critical commonsense, pragmatic semeiotic, and tripartite inquiry are applied to organizational/inter-organizational development through a methodology which respects both individuals and the organizations involved. As challenging as the development of such an architectonic framework may in fact be, yet the potential increase in social/business value would seem to make it worth taking up the challenge. It is through the ability to better model patterns and processes that we can have a realistic hope of gaining a modicum of control in the evolution of the new environment since it is "by inferring lawlike connections between salient, repeating features [that] we can bring past observations to bear on current conditions [and so] anticipate and control future occurrences" [8].

The view that an inter-enterprise architectonic could possibly be developed to optimize the way enterprises develop and operate internally and in relation to each other can be made attractive to leaders and decision makers to the extent that they become convinced that it has the potential for significantly benefiting their organizations should they choose to embrace it. A promising sign is that e-commerce has already begun to address some of the issues discussed here, and such organizations as Oracle and SAP seem dedicated to furthering the development of the requisite architectures. In any event, the expansion of business technology has resulted in a distributed, inter-enterprise, user and consumer-driven landscape which is both novel and ubiquitous, vast both in size and complexity, offering challenges

---

quality assurance is central to the development process. Inevitably this includes consideration of how bugs and system change requests are to be tracked.

Phase 6—Piloting and Launching: In the concluding piloting and launching phase we are concerned with what form and when the integration templates will be shipped out. In piloting, the most important questions concern what I-E pilots ought to be initiated. Finally, towards launching, critical questions include when and in what form the new I-E system will be extended to partners.

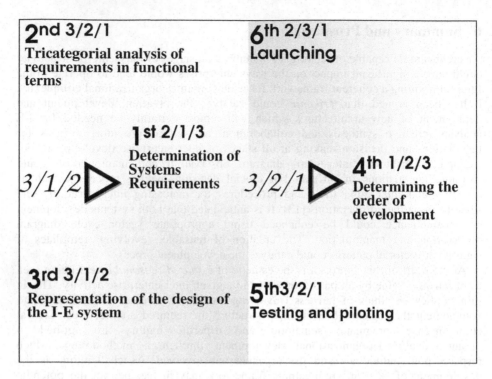

Fig. 3.

and opportunities to those who would act boldly and creatively. In a global environment as unpredictable as is ours, good models can provide "a way of compensating for the perpetual novelty of the world" [8].

It has been argued here that the emerging landscape requires a new paradigm, a veritable inter-enterprise architectonic which is itself capable of evolving. This may prove to be decisive as "ultimately, significant innovation depends on the 'long line': the ability to go beyond cut-and-try recombinations . . . to the more distant combinatorial horizon" [8]. Peirce's category theory and architectonic, especially as diagrammatically represented in *trikonic vector cycle diagram analysis-synthesis*, may prove to be of some considerable heuristic value in evolving a new collaborative paradigm.

Acknowledgements

I would like to acknowledge and thank the IET South Yorkshire Local Network http://www.iee.org/OnComms/Branches/UK/England/NorthE/SYo rks/ for their generous support. I am indebted to Simon Polovina for recommending many pertinent books and papers and to Benjamin Udell for his help in preparing the manuscript for publication.

References

[CP] *Collected Papers of Charles Sanders Peirce*, 8 vols. Edited by Charles Hartshorne, Paul Weiss, and Arthur Burks (Harvard University Press, Cambridge, Massachusetts, 1931-1958).

1. Arkins, R.K.: Restructuring the Sciences: Peirce's Categories and His Classification of the Sciences. Transactions of the Charles S. Peirce Society. Fall 2006 42(4) (2005)
2. Correa, J.H., Poschel, R.: The Teridentity and Peircean Algebraic Logic. In: Schärfe, H., Hitzler, P., Øhrstrøm, P. (eds.) ICCS 2006. LNCS (LNAI), vol. 4068, Springer, Heidelberg (2006)
3. de Moor, A.: Patterns for the Pragmatic Web. In: Dau, F., Mugnier, M.-L., Stumme, G. (eds.) ICCS 2005. LNCS (LNAI), vol. 3596, Springer, Heidelberg (2005)
4. de Moor, A., Keeler, M., Richmond, G.: Towards a Pragmatic Web. In: Priss, U., Corbett, D.R., Angelova, G. (eds.) ICCS 2002. LNCS (LNAI), vol. 2393, Springer, Heidelberg (2002)
5. de Tienne, A.: Learning qua semiosis. In: S.E.E.D, vol. 3(3) (2003)
6. Engelbart, D.C.: Augmenting Human Intellect: A Conceptual Framework (1962), http://www.invisiblerevolution.net/engelbart/full_62_paper_augm_hum_int.html
7. Fingar, P., Kumar, H., Sharma, T.: Enterprise E-Commerce. Meghan-Kiffer, Tampa (2000)
8. Holland, J.A.: Emergence: From Chaos to Order. Perseus, Reading (1998)
9. Kent, B., Peirce, C.S.: Logic and the Classification of the Sciences. McGill-Queen's University Press, Kingston (1987)
10. Nubiola, J.: The Classification of the Sciences and Cross-Disciplinarity. Transactions of the Charles S. Peirce Society, XLI(2) (Spring 2005)
11. Parker, K.A.: The Continuity of Peirce's Thought. Vanderbilt University Press, Nashville & London (1998)
12. Parmentier, J.: Signs Place in Medias Res: Peirce's Concept of Semiotic Mediation. In: Elizabeth, M., Parmentier. (ed.) Semiotic Mediation: Sociocultural and Psychological Perspectives (1985)
13. Richmond, G.: Interoperability as Desideratum, Problem, and Process. In: Conceptual Structures Tool Interoperability Workshop of 14th International Conference on Conceptual Structures, Aalborg University Press (2006)
14. Richmond, G.: Outline of trikonic I > *k: Diagrammatic Trichotomic. In: Dau, F., Mugnier, M.-L., Stumme, G. (eds.) ICCS 2005. LNCS (LNAI), vol. 3596, Springer, Heidelberg (2005)
15. Richmond, G.: Trikonic Analysis-Synthesis and Critical Common Sense on the Web. In: Schärfe, H., Hitzler, P., Øhrstrøm, P. (eds.) ICCS 2006. LNCS (LNAI), vol. 4068, Springer, Heidelberg (2006)
16. Richmond, G. (with B. Udell): Trikonic, slide show in ppt format of PORT-ICCS Huntsville, Alabama. http://members.door.net/arisbe/menu/library/aboutcsp/richmond/trikonicb.ppt
17. Schoop, M., de Moor, A., Dietz, J.L.G.: The Pragmatic Web: A Manifesto. In: Communications of the ACM, vol. 49(5), ACM Press, New York (2006)
18. Skagestad, P.: The Mind's Machine: The Turing Machine, the Memex, and the Personal Computer. Semiotica, 111(3/4) (1996)
19. Skagestad, P.: Peirce, Virtuality, and Semiotic in Paideia, http://www.bu.edu/wcp/Papers/Cogn/CognSkag.htm

20. Sowa, J.: Signs, Processes, and Language Games: Foundations for Ontology, http://www.jfsowa.com/pubs/signproc.htm
21. Spence-Hill, C., Polovina, S.: Enhancing Software Engineering with Trikonic for the Knowledge Systems Architecture of CG Tools. In: Conceptual Structures Tool Interoperability Workshop of 14th International Conference on Conceptual Structures, Aalborg University Press (2006)
22. Spyns, P., Meersman, R.: From knowledge to Interaction: from the Semantic to the Pragmatic Web. Technical Report 05, STAR Lab, Brussels. (2003), http://www.starlab.vub.ac.be/publications/STAR-2003-05.pdf
23. Stjernfelt, F.: Two Iconicity Notions in Peirce's Diagrammatology. In: Schärfe, H., Hitzler, P., Øhrstrøm, P. (eds.) ICCS 2006. LNCS (LNAI), vol. 4068, Springer, Heidelberg (2006)
24. Theuerkorn, F.: Lightweight Enterprise Architectures. Auerbach, London & New York (2005)
25. Trikonic article in Wikipedia: http://en.wikipedia.org/wiki/Trikonic
26. Udell, B.: (weblog: The Tetrast in the August 24, post accessed March 21, 2007. http://tetrast.blogspot.com/2006_08_01_tetrast_archive.html#pragmax
27. Wladawsky-Berger, I.: And Now a Syllabus for the Service Economy in The New York Times (December 3, 2006)
28. Woods, D., Mattern, T.: Enterprise SOA: Designing IT for Business Innovation. O'Reilly, Beijing and Cambridge (2006)

Hypermedia Discourse:
Contesting Networks of Ideas and Arguments

Simon Buckingham Shum

Knowledge Media Institute, The Open University, Milton Keynes, UK
www.kmi.open.ac.uk/sbs

Abstract. This invited contribution motivates the *Hypermedia Discourse* research programme, investigating the reading, writing and contesting of ideas as hypermedia networks grounded in discourse schemes. We are striving for *cognitively and computationally tractable conceptual structures*: fluid enough to serve as augmentations to group working memory, yet structured enough to support long term memory. I will describe how such networks can be (i) mapped by multiple analysts to visualize and interrogate the claims and arguments in a literature, and (ii) mapped in real time to manage a team's information sources, competing interpretations, arguments and decisions, particularly in time- pressured scenarios where harnessing collective intelligence is a priority. Given the current geo-political and environmental context, the growth in distributed teamwork, and the need for multidisciplinary approaches to wicked problems, there has never been a greater need for sensemaking tools to help diverse stakeholders build common ground.

1 Introduction

I want to talk about the challenge of our generation. [...] Our challenge, our generation's unique challenge, is learning to live peacefully and sustainably in an extraordinarily crowded world. [...] The way of solving problems requires one fundamental change, a big one, and that is learning that the challenges of our generation are not us versus them, they are not us versus Islam, us versus the terrorists, us versus Iran, they are us, all of us together on this planet against a set of shared and increasingly urgent problems. [...] But we are living in a cloud of confusion, where we have been told that the greatest challenge on the planet is us versus them, a throwback to a tribalism that we must escape for our own survival.

Jeffrey Sachs: 2007 Reith Lectures: http://www.bbc.co.uk/radio4/reith2007

With these "minds", a person will be well equipped to deal with what is expected, as well as with what cannot be anticipated; without these minds, a person will be at the mercy of forces that he or she can't understand, let alone control. [...] The disciplined mind... the synthesizing mind... the creating mind... the respectful mind... the ethical mind.

Howard Gardner: *Five Minds for the Future.* Harvard Univ. Press, 2006: p.2

U. Priss, S. Polovina, and R. Hill (Eds.): ICCS 2007, LNAI 4604, pp. 29–44, 2007.
© Springer-Verlag Berlin Heidelberg 2007

The context in which we find ourselves presents problems on a global scale which will require negotiation and collaboration across national, cultural and intellectual boundaries. At the same time we are in a climate which questions claims to knowledge, and in which the quality of discourse is often poor. This, I suggest, presents both major challenges and unique opportunities for us as a community dedicated to understanding how to provide computational support for negotiating the construction of coherent, conceptual structures. We have choices about the kinds of problems we work on, the way in which we do our modelling, and the functionalities of the systems we offer. What do we have to offer?

My thesis is that part of the solution could be discourse-oriented tools to help capture, comprehend, and manage competing interpretations and arguments for action. There is a particular need to provide languages for communities to *agree and disagree* in principled ways. This paper considers the challenge of evolving interactive tools that are flexible enough to mediate and capture discourse between stakeholders with different perspectives, yet introduce sufficient structure to provide computational services. The *Hypermedia Discourse* research programme[1] is focused on co-evolving the semantics, user interfaces, technical infrastructure, and human work practices to embed such tools in highly pressured, real time sensemaking scenarios, face-to-face and over the internet, as well as to support extended, asynchronous discourse lasting from a few days to many years.

Discourse means different things in different fields. It is used here in a broad sense to cover the diversity of verbal and written workplace communication that we want to support, which would include the framing of problems, review of solutions, and argumentation. *Discourse communities* refers to communities of practice [15] and other networks of people who "make and take perspectives" [2].

The paper is organised as follows. I start by motivating the need for tools to assist with sensemaking in socially complex scenarios, in particular, to manage discourse when tackling wicked problems [22]. The attributes required of tools to support the expression, exploration and contesting of perspectives in shifting, contentious domains defines a new class of tool for Hypermedia Discourse. The Compendium methodology and tool is then introduced as a relatively mature exemplar, before concluding with directions for future research.

2 Sensemaking

The world, indeed our lives, make sense to the extent that we can sustain a coherent narrative about who we are and why we matter. If the story fragments, our identity crumbles if we cannot re-integrate it into our narrative [3]. When we are confronted by breaches in normality, Karl Weick draws our attention to *sensemaking* as literally "the making of sense": sharing interpretations using different representations of the situation. He proposes that: *Sensemaking is about such things as placement of items into frameworks, comprehending, redressing surprise, constructing meaning, interacting in pursuit of mutual understanding, and patterning. [30], p.6*

---

[1] Hypermedia Discourse project: http://kmi.open.ac.uk/projects/~hyperdiscourse

Weick's concern is to characterise what people do in socially complex situations, when confronted by incomplete evidence and competing interpretations : *The point we want to make here is that sensemaking is about plausibility, coherence, and reasonableness. Sensemaking is about accounts that are socially acceptable and credible. [...] It would be nice if these accounts were also accurate. But in an equivocal, postmodern world, infused with the politics of interpretation and conflicting interests and inhabited by people with multiple shifting identities, an obsession with accuracy seems fruitless, and not of much practical help, either.* [30], p.61

In other words, when there is uncertainty, *what else is there to do* but through discourse, construct a narrative to fill in the gaps?

3 Argumentative Discourse

Sensemaking wrestles with conflicting interpretations, tracks technical facts with emerging issues and ideas as the problem is reframed, and tries to reconcile socio-political arguments. This is a formidable functional requirements specification for a software tool to satisfy. Elsewhere [4, 5] we trace the work of design and policy planning theorist Horst Rittel, whose characterisation in the 1970's of "wicked problems" has continued to resonate since: *Wicked and incorrigible [problems]...defy efforts to delineate their boundaries and to identify their causes, and thus to expose their problematic nature.* [22]

Rittel concluded that many problems confronting policy planners and designers were qualitatively different to those that could be solved by formal models or methodologies, classed as the 'first-generation' design methodologies. Instead, an *argumentative* approach to such problems was required: *First generation methods seem to start once all the truly difficult questions have been dealt with. ...[Argumentative design] means that the statements are systematically challenged in order to expose them to the viewpoints of the different sides, and the structure of the process becomes one of alternating steps on the micro-level; that means the generation of solution specifications towards end statements, and subjecting them to discussion of their pros and cons.* [22]

This intersects with Doug Engelbart's 40+ year mission to develop software tools to augment human intellect, our "collective capability for coping with complex, urgent problems" [14]. Our work in a variety of domains has led to the definition of a class of 'augmentation system' to assist argumentative design in Rittel's terms, and other modes of workplace discourse more broadly.

4 Hypermedia Discourse

Discourse modelling is at once both useful and limited. It is limited in the sense that, like any model, it captures only key features of the world's richness, in our case, the richness of textual prose and verbal discourse.[2] However – if done appropriately – stripping out detail to focus on underlying structure can yield cognitive, computational and theoretical benefits:

- **Cognitive:** a well designed external representation exploits the human perceptual and cognitive system to direct attention to relevant information;
- **Computational:** a formal model also provides machines with structure to reason with;
- **Theoretical:** the removal of detail may assist in identifying generalisable patterns across diverse contexts (see discussion of Cognitive Coherence Relations later).

The function of a *medium* is to make it possible for people to express, and work with, structure. Sensemaking calls for a particular kind of discourse, expressed through one or more media. *Hypermedia* can be thought of as the craft, art, science and engineering of managing structure, specifically, relationships, making it the primary discourse modelling medium for several reasons:

- **Modelling discourse relations:** an utterance only has meaning in a context, that is, when juxtaposed with others before and after it, and in relation to other possible utterances that make its selection significant.
- **Expressing different perspectives on a conceptual space:** diverse stakeholders are usually needed to define and resolve wicked problems, so support tools need to provide support for modelling flexibly, to show agreements and differences between viewpoints.
- **Supporting the incremental formalization of ideas:** as understanding develops, so that patterns can be captured using representations that are intuitive, fast in real time usage scenarios, and expressive enough to enable computational support.
- **Rendering structural visualizations:** to assist users in grasping complex interconnections between ideas and information.
- **Connecting heterogeneous content:** the *content* that stakeholders refer to during sensemaking can range from media fragments which offer little or no obvious structure, to material sufficiently structured to support forms of machine reasoning; similarly, *relationships* may range from associations expressed spatially or as untyped links, to being formally grounded in a known semantic schema.

4.1 Key Characteristics

Bringing these concepts together, we can define a class of tools designed to model discourse as hypermedia networks, with the objective of making the process and product of discourse tangible and manipulable through the combination of:

- **A discourse ontology:** A set of explicit constructs that express a subset of the richness of human verbal or written communication. An example (discussed

---

[2] As described later, there are ways to compensate for the terseness of modelling by integrating source texts, audio and video as richer resources for humans (and possibly machines) to supplement the discourse model.

below) is IBIS; another that we have been developing is the ScholOnto discourse schema [7].

- **One or more notations:** Symbol system(s) for rendering the ontology. For instance, IBIS can be rendered as a textual outline, and as a directed graph flowing from left to right, or from top to bottom. Each has different affordances which can complement each other as coupled visualizations.
- **An intuitive user interface:** These tools are intended for knowledge workers in diverse sectors of society, not only for discourse modellers, knowledge engineers or information scientists. The notations are therefore just part of designing the overall cognitive and aesthetic experience of working with the tool.
- **Computational services:** The above come together as augmentation of human capability through software implementation. For instance, "services" would include more efficient *capture, interpretation, sharing, retrieval, discovery and integration* of discourse modelled in the 'knowledge repository'. Interoperability not only with other relevant tools, but also compatibility with existing work practices will contribute to the overall service augmentation.
- **Literacy and fluency:** The tool's functionality is only part of the story, however. We must also examine the capabilities assumed on the part of the user, which we will do under the heading of *literacy*, the ability to read and write ideas in the new medium in a manner appropriate to the context, ideally moving towards *fluency*.

5   Compendium

Having defined the key characteristics of a Hypermedia Discourse system, we focus now on the most mature approach we have developed, in terms of its dissemination and breadth of use. This has provided a longitudinal case study to reflect on issues of knowledge technology adoption and practice [9].

Compendium is a dialogical medium for modelling the discourse around problems. We are aiming for a tool which in the hands of skilled users, can facilitate the capture and structuring ideas, not only to model discourse, but also to model problem domains in a manner that invites and structures contributions, whether this is in a synchronous or asynchronous discussion. It is optimised for use in what is arguably the most demanding context of deployment for a knowledge representation tool: real time collaborative modelling. The software is a free Java application for all platforms, including the source code. Downloads and other community resources are coordinated via the not-for-profit Compendium Institute: www.CompendiumInstitute.org

5.1   Ontology

Compendium is a direct descendent of Conklin's gIBIS prototype [13] and the 1990's QuestMap product. Its ontology expresses Rittel's IBIS and similar Design Rationale schemes such as MacLean *et al's* Questions-Options-Criteria (QOC) [16]. The focus is on capturing key issues, possible responses to these, and relevant arguments. Users can define their own ontology if they wish, or map concepts in a completely unconstrained manner. Entities are described in free text, while labels may be free text or grounded in a predefined scheme. Additional semantics can be expressed textually by defining one or more *Tag* groups, which operate as flat keyword spaces, analogous to web-based tagging, whereby tag combinations can be used to define different searchable views of the database. Semantics can, additionally, be expressed visually, either by predefining a palette of icons, or by selecting images to reflect ideas as they emerge in discussion (eg. from a library, or by searching the Web).

5.2 Notation

Some people use Compendium to support their preferred style of concept mapping [20]. However, following the gIBIS system, Compendium is designed specifically to render IBIS as a directed graph, normally with a root issue on the left, with the structure of the developing conversation about this issue growing to the right of the screen. User customizable icons distinguish different entities, and link colours with optional labels indicate relational semantics. Links typically point from right to left, to reflect the conversational dynamic that new contributions (added to the right) *respond-to* existing ones.

The discourse-orientation of the approach, and the demands of real time participatory modelling to capture the progress of meetings, have led to a number of notational strategies. A root *Issue* (signalled with a **?** question mark icon) provides the orientation to a map, establishing the problematic context for the discussion: *Why are we here? To tackle this issue.* Two discourse modelling methodologies have developed around the capabilities of Compendium. *Dialogue Mapping* is a set of skills developed by Conklin [12] for mapping IBIS structures in real time during a meeting in order to support the analysis of wicked problems, as defined by Rittel. In Dialogue Mapping, Issues are usually unconstrained freetext expressions summarising an agenda item or a participant's contribution, with Ideas responding to them, and any associated arguments (Fig. 1).

Conversational Modelling [23] incorporates and extends Dialogue Mapping by deriving Issues from a modelling methodology (or for instance, an organizational procedure/best practice). Issue nodes can be saved as reusable *issue-template* structures to seed different kinds of discussions. Fig. 2 shows a fragment of one template, with *Idea* icons serving as placeholders for responses. These lead to consequent *Issues* to be considered (on the right).

In addition, the modelling methodology specifies that the placeholder Ideas appear in three different views, indicated by the numeral 3 on each Idea icon. Rolling the mouse over this numeral displays a menu of hyperlinks to these other views. When views are labelled informatively, this facility provides rich context at a glance to the different 'conversations' in which a node is being discussed. Node label auto-completion assists the reuse of these granular chunks, offering users a menu of existing nodes which they can select from as they type.

With the addition of *catalogues* of reusable nodes, metadata *tagging* and multiple linked *issue-templates*, Compendium provides generic building blocks to construct a discourse-oriented modelling environment for team deliberation (Tate et al [28] document the customisation of Compendium in an hour from receipt of a planning

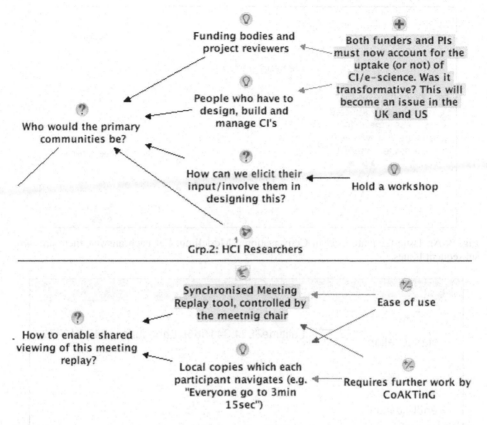

Fig. 1. Fragments from two Dialogue Maps using IBIS. In the top example exploring requirements for a website, a Pro argument of a political nature is highlighted, backing two Idea nodes. In the lower example, a QOC-style design discussion examines Option tradeoffs against more formally expressed design Criteria.

methodology). Conversational Modelling enables the real time capture of both expected, well-structured information through the use of issue templates, with the flexibility to capture unexpected, ad hoc information and discussions as they arise.

From a more formal knowledge representation perspective, we represent semantics using a variety of conventions. In a NASA field trial (Fig. 3), science metadata was represented using templates which look like visual forms, with each Issue inviting the team to answer (or if necessary debate) the values of the 'slots'.

An issue-template such as this provides a user-friendly way to engage in participatory modelling which permits argumentation if necessary, and results in a set of semantic assertions amenable to automated analysis (data entry into a simulation engine in this case). Each *Issue* in fact embodies the relational semantic connecting its answer to the entity represented by the containing map. However, rather than ask the team to complete sets of semantic triples, they are offered a set of question mark icons to which they need to link lightbulb icons. Thus, Fig. 2 provides an interface to elicit

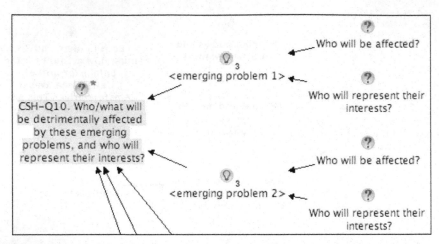

Fig. 2. An Issue-template used in Conversational Modelling. For each answer, there are two subsequent Issues.

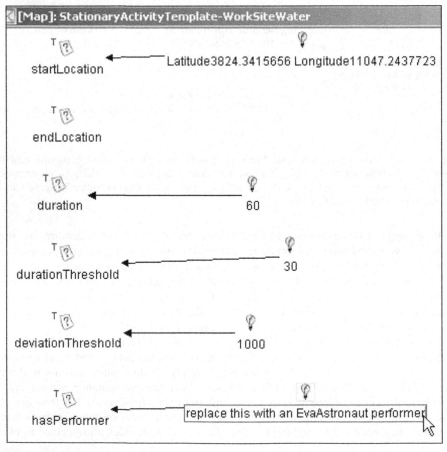

Fig. 3. The science team completes a template which will be later read by a software agent

the structured assertion <*user's answer*> *will_be_affected_by* <*emerging problem 1*>, while FIG. 3 will elicit <*WorkSiteWater*> *hasPerformer* <*user's answer*>.

Relational semantics are also expressed in the link types, but for speed – a key requirement in real time mapping under pressure – link types are set to be unlabelled by default, with the semantics loaded on the nodes' iconic language. Every link can be classified and labelled if desired using the default IBIS linkset, or a user defined linkset.

5.3 Intuitive User Interface

There are many improvements that could be made to Compendium, but as the preceding figures show, it looks familiar to users of concept mapping or graph-editing applications. It comes with IBIS preloaded, and hypermedia functionality which makes it simple to (i) create navigational links to a given database view, and (ii) reuse a hypertext node simultaneously in different views by copying and pasting. A keyword tagging scheme combined with search assists with filtering nodes across many maps.

Complete beginners can learn to map simple but well-formed IBIS structures after working through a tutorial on the Compendium Institute website. End users can express quite sophistcated data and relationships without needing to perform complicated technical actions or remember arcane commands. The user feedback on the website reflects the personal sense of satisfaction that users have with the tool.

5.4 Computational Services

We earlier defined "services" as the set of affordances at the intersection of ontology, notation, user interface, and the human and machine reasoning these enable. Compendium's display has a number of visual affordances which enable one to read off information about the state of an analysis that is not immediately obvious, either in conventional text documents or other concept mapping approaches. This includes unresolved issues, competing ideas, the extent to which explicit evidence is used to back ideas, and the 'depth' of node reuse and tagging (an indicator of the degree of modelling utilised).

When Compendium is interfaced to other tools, its database can be automatically populated or reasoned about. Examples include the use of software agents to autonomously read data and pass this to a simulation and planning engine, and also to populate the database with multimedia data for subsequent analysis by scientists [10]; the exchange of issues with a planning tool which could analyse the option space exhaustively or raise new issues [28]; the export of populated issue templates to different notational formats for other stakeholders to work on [26].

Most recently, we have automated the exchange of Compendium data with an RDF triplestore, in order to deliver a video conferencing capture and semantic replay tool [8]. Fig. 4 illustrates the complementary use of video from meetings to 'fill in the gaps' that a terse conceptual graph cannot possibly express; conversely, Compendium provides semantic indexing within and across meetings, enabling users to jump to the point in a meeting when, for instance, an argument was made.

5.5   Literacy and Fluency

Advanced tools are more effective when used expertly. The concept of services must, therefore, be qualified by the degree of literacy and fluency that the user brings. Our research agenda is directed towards understanding the whole learning curve associated with reading and writing in this new medium. We have analysed the cognitive tasks that a beginner must learn [6] and there are training programmes to help with initial adoption of the tool, but equally, we need to characterise expert, 'fluent' use of the tool in the most demanding contexts we work in, namely, supporting real time sensemaking in time pressured teams (e.g. [10, 28]). Constructing a language for fluency should help to expand the boundaries of expertise, improve the apprenticing of new practitioners, foreground new functionalities that the tool should provide, and illuminate an emerging literacy in this new medium.

Selvin [24, 25] has begun to explore the nature of fluency in what he terms *Participatory Hypermedia Construction.* Detailed analysis of screen recordings from teleconferences and face-to-face meetings is providing an account of the representational moves that Compendium mappers make, and the different roles they can play in meetings.

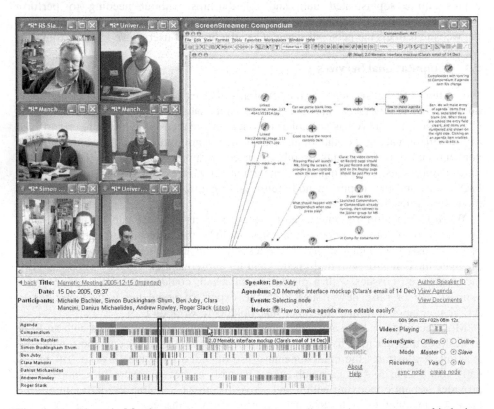

Fig. 4. The Memetic Meeting Replay tool, using Compendium nodes as a means of indexing and navigating meeting videos

6 Semantic Scholarly Publishing and Annotation

A second instantiation of the Hypermedia Discourse concept is the suite of tools developed in the Scholarly Ontologies project.[3] Unlike Compendium, which simply offers Web exports and supports the embedding of websites in IBIS conversational models, these tools were conceived from the start as distributed Web applications. The design rationale is the need for representational infrastructure to evolve the current prose document and associated practices for publishing and contesting research results and – equally significant – authors' *interpretations of their significance*. Within current research into 'e-Science' (UK) and 'Grid/cyberinfrastructure' (USA), this is a neglected part of the scholarly lifecycle, which is ironic: we engage in research in order to substantiate *knowledge level claims*. Perhaps, however, the absence of activity in this latter stage of research should not surprise us, because we are of course dealing with the difficult issue of computational support for an intrinsically *pragmatic* process, by which a discourse community (in this case, research peers) negotiates what some reported facts should be taken to *mean*. The emerging Pragmatic Web community has as a primary focus the interplay between formal representation and context, conversations and commitments to action, and it will be interesting to see how this takes shape.

We detail elsewhere [27, 29] the design and evaluation of ClaiMaker and the associated suite of tools for authoring (ClaiMapper), querying (ClaimFinder) and the collaborative, semantic annotation (ClaimSpotter) of research claims and argumentation. These are less mature than Compendium, proof of concept research tools which are not publicly available. Space precludes as detailed a treatment as Compendium, but ClaiMaker's 'hypermedia discourse profile' below conveys the essence of the approach:

- **Discourse ontology:** A two-layer relational taxonomy which provides base relational classes in which 'dialects' from different discourse communities are grounded (Fig. 5).
- **Notation:** A conceptual graph of claims that can be visualized using different schemes to show discourse connections between concepts annotated onto the literature.
- **User interface:** We have investigated a variety of interaction paradigms for annotation tools, in order to help untrained users create semantic annotations.
- **Computational services:** The use of a richer discourse scheme than IBIS enables us to offer more powerful services. For instance, the semantic citation maps can be filtered in response to queries such as, *What documents report data that challenges this author's hypothesis?What is the lineage of this concept: the key ideas on which this work builds?* (Fig. 6)
- **Literacy and fluency:** Being less mature than Compendium, we do not yet have a large enough user community to provide a good description of what it means to read and write such argumentative networks, particularly beyond initial learning. Our empirical studies provide insight into how untrained and more expert users construct and query claim networks [27, 29].

---

[3] Scholarly Ontologies project: http://kmi.open.ac.uk/projects/scholonto

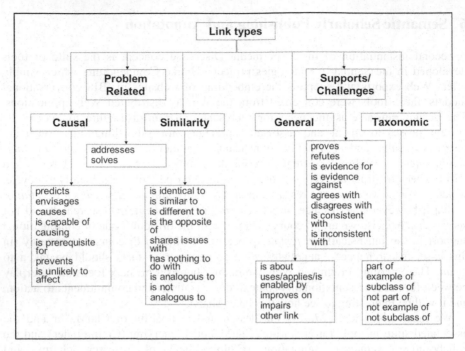

Fig. 5. ClaiMaker's discourse scheme, which groups the 'dialect' of a discourse community under more primitive relational classes

Fig. 6. ClaimFinder's *Lineage* query traces the 'intellectual roots' of a concept. displayed at the top. The conceptual graph is analysed and filtered to show potentially significant relational types such as *uses/applies/is enabled by, improves on*, and *solves*.

7   Conclusions and Future Work

The complexity of the dilemmas we face at an organizational, societal and global scale forces us into sensemaking activity. The requirements on tools to support such work have motivated basic and applied action research into a new class of Hypermedia Discourse tool to mediate, structure and augment the expressing and contesting of perspectives that may agree and disagree in principled ways. Such tools are hybrids borrowing from concept mapping, information visualization, discourse relations and decision-support. We need tools flexible enough for real time use in meetings, structured enough to help manage longer term memory, and powerful enough to filter the complexity of extended deliberation and debate on an organizational or global scale.

I suggest that this focus on the intersection of discourse and hypermedia provides insights into a number of pressing problems:

- **We have to talk.** The only way that anything is accomplished in this world is by people talking, building trust and sufficient common ground that they can frame problems in mutually meaningful ways, and commit to action in mutually acceptable ways. The challenge for a community such as ours is understand how to weave software support into the social fabric without ripping it, but possibly in the process, enriching that fabric to exploit the new threads we have to offer. The work summarised here points to possible ways to evolve network-native infrastructures for synchronous and asynchronous discourse, that step out of the shadow of the printing press and conventional meetings (building on their strengths, but transcending their limitations).

- **Modelling in the absence of consensus.** Knowledge-based systems (including for our purposes the data models and ontologies underpinning the Semantic Web) encapsulate *consensus models* of the problem domain, and how to reason about it. How can we provide computational services *in the absence of consensus*, when one group's assumption is another group's problem? This is the domain of discourse, especially argumentation, in which we provide a language for stakeholders to agree and disagree in principled ways. Compendium uses a semiformal network representation optimised for real time use. ClaiMaker uses finer grained semantics for modelling asynchronously in a more detailed manner.

- **Negotiating the knowledge capture bottleneck.** In knowledge engineering, but also in less formal approaches to Knowledge Management (KM), Organizational Memory and Design Rationale (DR), the cost/benefit tradeoff must be negotiated to acquire useful abstractions of naturally occurring activity, and experts' descriptions thereof. The Compendium approach emphasises the collaborative modelling of information, ideas and argument in order to add immediate value to the users (useful *working memory*), as well as seeding the *long term memory* required for KM. This has, for instance, provided a way of tackling the DR capture bottleneck [9].

Future work will continue to co-evolve tools and practices, study the skills associated with high performance discourse modelling, and develop conceptual frameworks that recognise the complexity of modelling, mediating and mapping real discourse about wicked problems. Specific challenges we are working on include:

- **Distributed, online apprenticeship in hypermedia discourse.** The Compendium community now has members who are recognised 'expert mappers', but they are a scarce resource. A very applied concern is how to use the internet to spread this literacy through the creation of e-learning resources and 'e-apprenticeship'.
- **Social networks and folksonomic tagging.** Behind a conceptual structure are people. We are integrating our social networking tools with our conceptual networking tools to support *Open Sensemaking Communities*, learners and educators who must self-organise around open source learning resources, but by extension, any epistemic community on the internet. Based on the ScholOnto project, we have prototyped and formatively evaluated a next generation social bookmarking tool for linking tags via discourse connectives, moving from the annotation of isolated keywords on web reosources, to a mode knowledge construction and negotiation: *from tag clouds to tag webs* [27].
- **Hypermedia discourse engines as computational theory.** We are investigating the potential of modelling and reasoning over an upper level relational ontology, derived from linguistics coherence relations research [18]. If it is the case that we perceive 'coherence' in a medium because it structures elements according to a small, bounded set of relational primitives, then it should be possible to model and reason over such structures in a manner which is 'coherent' across different domains of discourse, languages and even cultures. Such an engine would be a formal expression, and test, of the hypotheses generated by this theory.

To return to our opening quote from Gardner's *Five Minds for the Future*, perhaps Hypermedia Discourse tools provide a way to move fluidly between the different minds: a way to provide representational scaffolding for disciplined modelling, but permitting the creative breaking of patterns when needed and the forging of new syntheses; a way to show respect for diverse stakeholders' concerns by explicitly integrating them into the conversation; a way to bring into an analysis 'messy' requirements such as ethical principles, as well as hard data and constraints. We have some evidence from our case studies that we're on the right track, but there remains much to do.

Acknowledgements. I am grateful to Al Selvin, Clara Mancini, Jack Park and David Kolb whose comments improved earlier versions of this paper. The evolution of Compendium has been a long term action research programme with Al Selvin, Maarten Sierhuis and Jeff Conklin, with programming by Michelle Bachler. Its development has been funded by the UK's research councils EPSRC, ESRC and JISC. The Meeting Replay tool is joint work with the Universities of Manchester and Southampton as part of the JISC Memetic project. The Scholarly Ontologies project was funded by the EPSRC, and is the product of work with Victoria Uren, Gangmin Li, Clara Mancini, Bertrand Sereno, Enrico Motta and John Domingue. The William and Flora Hewlett Foundation is now supporting the work through the Open University's *OpenLearn* initiative.

References

1. Boden, D.: The Business of Talk. Polity, Cambridge (1994)
2. Boland, R.J.J., Tenkasi, R.V.: Perspective Making and Perspective Taking in Communities of Knowing. Organization Science 6(4), 350–372 (1995)
3. Bruner, J.S.: Acts of Meaning. Harvard University Press, Cambridge, MA (1990)
4. Buckingham Shum, S.: The Roots of Computer Supported Argument Visualization. In: Kirschner, P.A., Buckingham, S., Carr, C. (eds.) Visualizing Argumentation, pp. 3–24. Springer, London (2003)
5. Buckingham Shum, S., Hammond, N.: Argumentation-Based Design Rationale: What Use at What Cost? Int. J. Human-Computer Studies 40(4), 603–652 (1994)
6. Buckingham Shum, S., et al.: Graphical Argumentation and Design Cognition. Human Computer Interaction 12(3), 267–300 (1997)
7. Buckingham Shum, S., et al.: Modelling Naturalistic Argumentation in Research Literatures: Representation and Interaction Design Issues. Int. J. Intelligent Systems (Special Issue on Computational Modelling of Natural Argument) (in Press)
8. Buckingham Shum, S., et al.: Memetic: An Infrastructure for Meeting Memory. In: Proc. 7th Int. Conf. on the Design of Cooperative Systems. 2006 of Conf. Carry-le-Rouet (2006)
9. Buckingham Shum, S.J., et al.: Hypermedia Support for Argumentation-Based Rationale: 15 Years on from gIBIS and QOC. In: Dutoit, A., et al. (eds.) Rationale Management in Software Engineering, pp. 111–132. Springer, Heidelberg (2006)
10. Clancey, W.J., et al.: Automating CapCom Using Mobile Agents and Robotic Assistants. In: 1st Space Exploration Conf. 2005 of Conf. Orlando, FL (2005), http:// eprints.aktors.org/375
11. Conklin, J.: Dialogue Mapping: Reflections on an Industrial Strength Case Study. In: Kirschner, P.A., Buckingham, S., Carr, C. (eds.) Visualizing Argumentation, Springer, London (2003)
12. Conklin, J.: Dialogue Mapping: Building Shared Understanding of Wicked Problems. Wiley, Chichester (2005)
13. Conklin, J., Begeman, M.L.: gIBIS: A Hypertext Tool for Exploratory Policy Discussion. ACM Transactions on Office Information Systems 4(6), 303–331 (1988)
14. Engelbart, D.C.: A Conceptual Framework for the Augmentation of Man's Intellect. In: Howerton, P., Weeks. (eds.) Vistas in Information Handling, pp. 1–29. Spartan Books, Washington, DC, London (1963)
15. Lave, J., Wenger, E.: Situated Learning: Legitimate Peripheral Participation. Cambridge University Press, Cambridge (1991)
16. MacLean, A., et al.: Questions, Options, and Criteria: Elements of Design Space Analysis. Human-Computer Interaction 6(3 & 4), 201–250 (1991)
17. Mancini, C.: Cinematic Hypertext. Investigating a New Paradigm. In: Frontiers in Artificial Intelligence and Applications. IOS Press, Amsterdam (2005)
18. Mancini, C., Buckingham Shum, S.: Modelling Discourse in Contested Domains: A Semiotic and Cognitive Framework. Int. J. Human-Computer Studies 64(11), 1154–1171
19. McCall, R.: Fundamentals - Rationale Representation, Capture and Use. In: Dutoit, A., et al. (eds.) Rationale Management in Software Engineering, pp. 49–52. Springer, Heidelberg (2006)
20. Novak, J.D.: Learning, Creating, and Using Knowledge: Concept Maps as Facilitative Tools in Schools and Corporations. LEA, Mawah, NJ (1998)
21. Park, J.: Topic Mapping: A View of the Road Ahead. In: Maicher, L., Park, J. (eds.) TMRA 2005. LNCS (LNAI), vol. 3873, Springer, Heidelberg (2006)

22. Rittel, H.W.J.: Second Generation Design Methods. Interview in: Design Methods Group 5th Anniversary Report: DMG Occasional Paper, vol. 1, pp. 5–10 (1984). In: Cross, N. (ed.) Developments in Design Methodology (reprinted), pp. 317–327, J. Wiley & Sons, Chichester (1972)
23. Selvin, A.: Supporting Collaborative Analysis and Design with Hypertext Functionality. Journal of Digital Information 1(4) (1999)
24. Selvin, A.: Fostering Collective Intelligence: Helping Groups Use Visualized Argumentation. In: Kirschner, P.A., Buckingham, S., Carr, C. (eds.) Visualizing Argumentation, Springer, London (2003)
25. Selvin, A.: Aesthetic and Ethical Implications of Participatory Hypermedia Practice, Technical Report KMI-05-17, Knowledge Media Institute, Open University (2006)
26. Selvin, A.M., Buckingham Shum, S.J.: Rapid Knowledge Construction: A Case Study in Corporate Contingency Planning Using Collaborative Hypermedia. Knowledge and Process Management 9(2), 119–128 (2002)
27. Sereno, B., Buckingham Shum, S., Motta, E.: Formalization, User Strategy and Interaction Design: Users' Behaviour with Discourse Tagging Semantics. In: Workshop on Social and Collaborative Construction of Structured Knowledge. 16th Int. World Wide Web Conference, Banff, Canada, May 8-12, 2007 (2007), [http://www2007.org/workshops/paper_30.pdf]
28. Tate, A., et al.: Co-OPR: Design and Evaluation of Collaborative Sensemaking and Planning Tools for Personnel Recovery, Technical Report KMI-06-07, Knowledge Media Institute, Open University (2006)
29. Uren, V., et al.: Sensemaking Tools for Understanding Research Literatures: Design, Implementation and User Evaluation. Int. J. Human Computer Studies 64(5), 420–445 (2006)
30. Weick, K.E.: Sensemaking in Organizations. Sage Publications, Thousand Oaks, CA (1995)

Dynamic Epistemic Logic and Knowledge Puzzles

H.P. van Ditmarsch[1,*], W. van der Hoek[2], and B.P. Kooi[3]

[1] University of Otago, New Zealand
hans@cs.otago.ac.nz
[2] University of Liverpool, United Kingdom
wiebe@csc.liv.ac.uk
[3] University of Groningen, Netherlands
b.p.kooi@rug.nl

Abstract. We briefly give an overview of Dynamic Epistemic Logic (DEL), mainly in semantic terms. We focus on the simplest of epistemic actions in DEL, called public announcements. We also sketch the effect of more complex epistemic actions, and briefly show how als factual change can be modelled in the same framework. We then apply the logic of public announcements in DEL to the analysis of a knowledge puzzle, called 'What Sum'.

Keywords: multiagent systems, epistemic logic, dynamic epistemic logic, belief revision.

1 Dynamic Epistemic Logic

- A: As you know, ICCS 2007 takes place in Sheffield.
- B: Is that so? When I google ICCS 2007, I end up in Beijing: http://www.iccs-meeting.org/, where the International Conference on Computational Science 2007 is organized.
- C: Now is ICCS in Sheffield, or not?
- A: Ah, but I am talking about the 15th international conference on conceptual structures, http://www.iccs.info/, something entirely different from the Beijing thing! And it certainly takes place in Sheffield.
- C: OK. We now all know that the conceptual structures conference is indeed in Sheffield.

1.1 Single-Agent Knowledge

One way of resolving an uncertainty is being explicit about acronyms. In general, given uncertainty about the truth of a propositional statement 'ICCS is in Sheffield', abbreviated as Sheff, and from now referring to the conceptual structures conference, one can associate two different 'worlds', or 'states' with it: one

---

* Material from section 2 is similarly found in [21], to which we acknowledge Ji Ruan's contributions.

U. Priss, S. Polovina, and R. Hill (Eds.): ICCS 2007, LNAI 4604, pp. 45–58, 2007.

wherein it is true, and one wherein it is false. The reader of course knows it is true. But an agent a who cannot distinguish the actual state where it is true from the state where it is false, is said not to *know* that Sheff. Knowing a statement to be true means that in all states that one considers possible, that statement is true. As, according to a's information, there is a possible state where Sheff is true and one where it is false, Sheff is therefore not known. We write $\neg K_a$Sheff for that; K_a stands for 'agent a knows'. A simple and abstract representation of that situation, called *epistemic model* or *information state*, (formal definitions of language and semantics will be given in Section 2.1) is

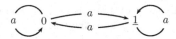

The two states are named 0 and 1. In state 0 the proposition Sheff is false, and in state 1 it is true. The arrows, labelled with a, stand for what agent a considers possible. First, we reason from the actual state of affairs: ICCS is in Sheffield. This is true in state 1. To indicate that 1 is the actual state, we have underlined it in the figure. In state 1, agent a, *Anne*, considers it possible that Sheff is true. Therefore there is a reflexive arrow from 1 to itself. But she also considers it possible that Sheff is false. Therefore there is an arrow from 1 to 0. There are more arrows in the picture! The reason for this extra structure is that we assume that Anne is aware of her ignorance of Sheff. In terms of our simplifying assumptions such awareness can be equated with knowledge. We then get: Anne knows that she does not know that ICCS is in Sheffield. Formally, this is: $K_a\neg K_a$Sheff. Such reflection is called a higher-order aspect of knowledge. In practice this means stacking K-operators to greater depth than just one. To represent this higher-order knowledge, Anne should be able to reason from the perspective that the actual state *were* 0, where Sheff is false. From that perspective, it is also both conceivable that Sheff is true and that Sheff is false. Therefore, there are also a-arrows from 0 to 1 and from 0 to itself.

1.2 Multi-Agent Knowledge

Let us introduce a second agent b, say *Bill*. We can model that Bill knows that Anne is ignorant of Sheff but he himself is aware of the truth, and that even Anne knows that. For a more realistic setting than the initial one, imagine the truth about Sheff to be written on a sheet of paper in a closed envelope that is handed to Bill, in the presense of Anne, by some third party stating 'this envelope contains the truth about Sheff,' after which Bill opens the envelope and reads its contents still in the presence of Anne. (Epistemic logic is full of such scenarios.) The resulting situation is now perfectly modelled by the above assumptions. And in fact by something stronger: Bill knows that Anne knows that Bill knows the truth about Sheff, and Anne knows that, etc. We say that Anne and Bill have *common knowledge* about the situation where Anne is ignorant and Bill knowledgeable about Sheff: $C_{ab}[\neg(K_a$Sheff $\lor K_a\neg$Sheff$) \land (($Sheff $\land K_b$Sheff$) \lor (\neg$Sheff $\land K_a\neg$Sheff$))]$, depicted as:

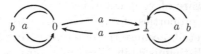

Common knowledge formalizes what agents in a group know about each other, and about each other's knowledge, and so on ad infinitum. We can also think of common knowledge as *background knowledge* describing the structure of information states. In that sense, the model above is precisely described by the common knowledge formula we gave. This aspect will reappear when we present knowledge puzzles.

The *epistemic logic* we have now used to formalize knowledge, and how to interpret this in relational structures as above, is generally said to have started with [8]. The aspect of knowledge iterations, and how the concept of common knowledge formalizes arbitrary finite iterations, is from somewhat later date. The standard reference is [10] and another early source is within economics, [3]. Recent introductions into epistemic logic are [6,13].

1.3   Public Announcements

What happens if the truth about Sheff becomes known to Anne, for example by way of a public announcement of Sheff, or by Bill saying "I know that ICCS is in Sheffield"? We consider the first, although in this example the informational consequences of both are the same. After the announcement of Sheff, both Anne and Bill now know that Sheff is true, and that the other knows, ad infinitum. Structurally, the result of a public announcement is the restriction of the information state to those states where the announcement is true, and such that all arrows are kept between these remaining states. We get

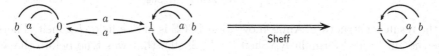

On the right it is common knowledge that Sheff: C_{ab}Sheff. From this also follows that Anne knows Sheff: K_aSheff. In the information state before the announcement, on the left, Sheff is true, and $\neg K_a$Sheff is true. We can also express the dynamic effect of the announcement in the logical language, by stating that initially (on the left) $\neg(K_a$Sheff $\vee K_a\neg$Sheff$) \wedge [$Sheff$]K_a$Sheff is true, for 'Anne does not know whether ICCS is in Sheffield and ('but') after the announcement that ICCS is in Sheffield, she knows that ICCS is in Sheffield." The operator K_a is called an epistemic operator, and the operator [Sheff] is called a dynamic operator. It is interpreted by means of a transition from one information state to another one, as above: to interpret [Sheff]K_aSheff on the left, first do the transition to the right, and interpret K_aSheff there. As this is true, so is, on the left, [Sheff]K_aSheff.

Standardly accepted properties of knowledge are that: known information is true ($K\varphi \rightarrow \varphi$, for all φ), you are aware of ('know') your knowledge ($K\varphi \rightarrow$

$KK\varphi$), and you are aware of your ignorance ($\neg K\varphi \rightarrow K\neg K\varphi$). These correspond to the structural properties of reflexivity, transitivity, and euclidicity. We do not explain this in detail, because the outcome is that together these properties ensure that the indistinguishability relation above is always an *equivalence relation*. We can therefore think of the domain of states as being partitioned into *epistemic classes*, such that every class consists of epistemically indistinguishable states for each agent. For equivalence relations, a simpler visualization is sufficient, where we link states in the same class, and where singleton classes, consisting of one state, are implicit. We then get:

$$0 \overline{\quad\quad} a \overline{\quad\quad} \underline{1} \qquad\qquad \xRightarrow[\text{Sheff}]{\quad\quad\quad\quad} \qquad \underline{1}$$

We will extensively use this visualization in the next section. The dynamics so far comes under the name of 'public announcement logic'. Standard references are [16,4,7,23].

1.4   Sentences That Become False Because They Are Announced

In public announcement logic, not all formulas remain true after their announcement, in other words, $[\varphi]\varphi$ is *not* a principle of the logic. Some formulas involving epistemic operators become **false** after being announced! Given the information state again wherein Anne is ignorant about Sheff but Bill not, consider Bill saying to Anne: "ICCS is in Sheffield but you don't know that." This is formalized as $K_b(\text{Sheff} \wedge \neg K_a\text{Sheff})$ (the initial K_b can be equated with 'Bill says (truthfully)'). Following the same recipe of restricting the information state to those of its elements where the new information is true, we again get

$$0 \overline{\quad\quad} a \overline{\quad\quad} \underline{1} \qquad\qquad \xRightarrow[K_b(\text{Sheff} \wedge \neg K_a\text{Sheff})]{\quad\quad\quad\quad} \qquad \underline{1}$$

In the resulting structure, Anne knows that ICCS is in Sheffield: $K_a\text{Sheff}$. Therefore the announced formula $K_b(\text{Sheff} \wedge \neg K_a\text{Sheff})$, that was true before the announcement, has become false after the announcement. One can also say that the statement about Anne's ignorance led her to factual knowledge (namely of Sheff). In the somewhat different setting that formulas of the form $p \wedge \neg K_n p$ cannot be consistently known, this phenomenon is called the Moore-paradox [14,8]. In the underlying dynamic setting it has been described as an *unsuccessful update* [7,23,17]. Similarly, statements about ignorance in the knowledge puzzles to be discussed next, may lead to factual knowledge about the numbers these puzzles are dealing with.

1.5   More Complex Dynamics

More complex informative scenarios are also conceivable. A variation of the scenario where a closed envelope containing Sheff or $\neg$Sheff is handed to Bill, is that, after the delivery, Bill is out of Anne's sight for a moment such that, when she returns, Bill *may have* quickly opened the envelope and read its contents. But Anne is not sure whether Bill has done that, or not (and again, we assume some common awareness of this scenario between Anne and Bill). Now, even given that Sheff is true, there are two outcomes, that are indistinguishable for Anne: either Bill opened the envelope, and now knows whether Sheff, or he didn't, in which case he remains ignorant. This informational transition can be depicted as

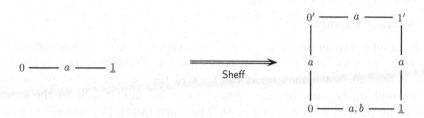

wherein in both 0 and $0'$, Sheff is false and in both 1 and $1'$ Sheff is true. In words: state 0 is the result when Sheff being false and b not reading the contents of the enveloppe, $0'$ is the result when Sheff is false but b *does* read this. We assume (but do not draw) transitivity of indistinguishability relations, such that 0 and $1'$, and $0'$ and $\underline{1}$, are also indistinguishable for Anne. In fact none of the four states can now be distinguished by her. Note that the above represents that Bill actually did *not* look at the contents for the envelope, as in the underlined state he still considers the alternative where Sheff is false! Such more complex informative actions have been investigated by [4,7,19] and are also a major topic in [23].

1.6   Belief and Plausible Reasoning

The difference between knowledge and belief is that beliefs may be false, whereas knowledge is supposed to be true. Imagine that we do not have two agents Anne and Bill, but just Anne. One way to model tentative belief is to associate that with preferences in a structure. For example, given the states 0 and 1 where Sheff is false and true, Anne considers it more likely that it is false: she prefers 0 over 1. But given just state 1 wherein Sheff it true, she prefers 1 (over nothing, given the absence of other states). Using labels K_a to denote epistemic indistinguishabity and B_a doxastic preference, we can depict the transition from Anne believing that Sheff is false to Anne believing that Sheff is true, as

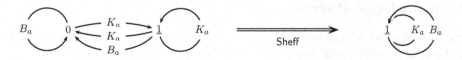

The B_a-arrow on the right, the was not there on the left, reflects that Anne now prefers 1. The belief operator is written as B. On the left, Anne does not know that Sheff, and even believes that Sheff is false: $\neg K_a \text{Sheff} \wedge B_a \neg \text{Sheff}$.

Whereas on the right, she knows (and therefore also believes) that Sheff is true: K_aSheff $\land$ B_aSheff. Such a transition is remarkable because her factual beliefs are now the opposite from what they were before. As this is a typical operation 'revision' in the area indeed known as (AGM) belief revision [1], these settings allow for what is known as dynamic (higher-order) belief revision, on which fairly recently progress has been made [2,20,5].

1.7   Factual Change

Let us consider a different scenario involving Anne (a). Let Crack stand for 'the Ming vase over there has a crack.' An does not know whether Crack is true: she does not see one, but it may be at the back of the vase, from her perspective. Now Barteld walks past, brushes against the vase, and it falls to the ground where it shatters in a thousand pieces. The uncertainty is removed: it is now most certainly cracked. Not by someone looking at the back of the vase, leaving the state of the world unchanged, but by someone changing the world. Given ignorant Anne and knowledgeable Bill again, who observe Barteld's antics, this action can again be described as an information state transition; if we let 1 stand for 'Crack is true, as:

$$0 \text{------} a \text{------} \underline{1} \qquad \xRightarrow{\quad\text{Crack}\quad} \qquad 1 \text{------} a \text{------} \underline{1}$$

Note that Bill is omniscient: all his equivalence classes are singletons. More significantly, observer that the state 0 on the left hand side transforms in a state 1: the truth of Crack changes! Again, it is now common knowledge that Crack is true. In this case, we can write this as $\neg K_a$Crack$\land$[Crack := $\top$]K_aCrack: initially, Anne did not know whether the vase had a crack but after the vase shattered (an *assignment*), she knows: it is now true in *both* states she considers possible. Informative and factual changes can also be combined into more complex actions. Such matters are being investigated in [22,18,9].

  This ends our sweeping overview of dynamic epistemic logic. We now proceed to the analysis of a knowledge puzzle. We can do this without the formalities of the logical language and semantics, and focus on structural transitions for information states as the above. But let it be known that this formal level certainly hovers at the back of anything we playfully introduce! Indeed, such formalisations are behind various model checking operations that can be performed to verify our less formal statements below [21].

2   Knowledge Puzzles

The following riddle (transcribed in our terminology) appeared in Math Horizons in 2004, as 'Problem 182' in a regular problem section of the journal [11].

> *Each of agents Anne, Bill, and Cath has a positive integer on its fore-head. They can only see the foreheads of others. One of the numbers is the sum of the other two. All the previous is common knowledge. The agents now successively make the truthful announcements:*

i. Anne: "I do not know my number."
ii. Bill: "I do not know my number."
iii. Cath: "I do not know my number."
iv. Anne: "I know my number. It is 50."
What are the other numbers?

You know your own number if and only if you know which of the three numbers is the sum. We therefore call the riddle: 'What Sum'. It combines features from wisemen or Muddy Children puzzles [15] with features from the Sum and Product riddle [12]. A common feature in such riddles is that we are given a multi-agent system, and that successive announcements of ignorance finally result in its opposite, typically factual knowledge. In a global state of such a system [6] each agent or processor has a local state, and there is common knowledge that each agent only knows its local state, and what the extent is of the domain. If the domain consists of the full cartesian product of the sets of local state values, it is common knowledge that agents are ignorant about others' local states. In that case an ignorance announcement has no informative value. For ignorance statements to be informative, the domain should be more restrictive than the full cartesian product (here: we only have triples (x, y, z) in which one number is the sum of the other two). As in Muddy Children, we do not take the 'real' state of the agent (the number on its forehead) as its local state, but instead the information seen on the foreheads of others (the other numbers). 'Sum and Product' is also about numbers, and even about sums of numbers, and the announcements are similar. (However, the structure of the background knowledge is very different: in 'Sum and Product', there are only three distinct abc-classes [23].)

Other epistemic riddles involve cryptography and the verification of information security protocols ('Russian Cards', see [23]), or involve communication protocols with private signals involving diffusion of information in a distributed environment ('100 prisoners and a lightbulb', see [25]). The understanding of such riddles is facilitated by the availability of suitable specification languages. For 'What Sum' we propose the logic of public announcements. But '100 prisoners and a lightbulb' requires the complex dynamics and factual change, as discussed in Subsections 1.5 and 1.7. Verification tools, such as DEMO, an epistemic model checker [24], can be used when analyzing such puzzles.

2.1 Details on Public Announcement Logic

We now formally reintroduce the language of public announcement logic, the epistemic structures in which it can be interpreted, and sufficient details of the semantics to understand epistemic statements and transitions caused by announcements. Given a finite set of agents N and a finite or countably infinite set of atoms P, the language of public announcement logic is inductively defined as

$$\varphi ::= p \mid \neg\varphi \mid (\varphi \wedge \psi) \mid K_n\varphi \mid C_B\varphi \mid [\varphi]\psi$$

where $p \in P$, $n \in N$, and $B \subseteq N$ are arbitrary. Other propositional and epistemic operators are introduced by abbreviation. For $K_n\varphi$, read 'agent n knows formula

φ'. For example, if Anne knows that her number is 50, we can write $K_a 50_a$, where a stands for Anne and some set of atomic propositions is assumed that contains 50_a to represent 'Anne has the number 50.' For $C_B \varphi$, read 'group of agents B commonly know formula φ'. For example, we have that $C_{abc}(20_b \rightarrow K_a 20_b)$: it is common knowledge to Anne, Bill, and Cath, that if Bill's number is 20, Anne knows that (because she can see Bill's number on his forehead)—instead of $\{a, b, c\}$ we often write abc. For $[\varphi]\psi$, read 'after public announcement of φ, formula ψ (is true)'. "After Anne announces (I know my number. It is 50.) it is common knowledge that Bill's number is 20" is formalised as $[K_a 50_a]C_{abc}20_b$.

The basic structure is an epistemic model, which was already informally used in the previous section. It is a Kripke model, wherein all accessibility relations are equivalence relations. An *epistemic model* $M = \langle S, \sim, V \rangle$ consists of a *domain* S of (factual) *states* (or 'worlds'), *accessibility* $\sim : N \rightarrow \mathcal{P}(S \times S)$, where each $\sim_n$ is an equivalence relation, and a *valuation* $V : P \rightarrow \mathcal{P}(S)$. For $s \in S$, (M, s) is an *epistemic state*. Given two states s, s' in the domain, $s \sim_n s'$ means that s is indistinguishable from s' for agent n on the basis of its information. For example, at the beginning of the riddle, triples $(2, 14, 16)$ and $(30, 14, 16)$ are indistinguishable for Anne but not for Bill nor for Cath. Therefore, assuming a domain of natural number triples, we have that $(2, 14, 16) \sim_a (30, 14, 16)$. Given a state s, Anne knows a statement if it is true in all states she renders indistinguishable from s: given $(2, 14, 16)$, Anne knows that Bill has 14, $K_a 14_b$, because Bill has 14 in both $(2, 14, 16)$ and $(30, 14, 16)$, the two states that she cannot distinguish. The group accessibility relation $\sim_B$ is the transitive and reflexive closure of the union of all accessibility relations for the individuals in B: $\sim_B \equiv (\bigcup_{n \in B} \sim_n)^*$. This relation is used to interpret common knowledge for group B. Instead of '$\sim_B$ equivalence class' ($\sim_n$ equivalence class) we write B-class (n-class). For example, if Anne sees two even numbers she knows that it is common knowledge that all numbers are even (a full description would be cumbersomely long in this case).

The dynamic modal operator $[\varphi]$ is interpreted as an epistemic state transformer. Announcements are assumed to be truthful, and this is commonly known by all agents. It results in the restriction of the domain to all states where the announcement is true, retaining all uncertainty on that restricted domain. We will only present this informally in the subsequent analysis, by way of such resulting restrictions.

2.2   Formalisation of 'What Sum'

The set of agents $\{a, b, c\}$ represent Anne, Bill and Cath, respectively. Atomic propositions i_n represent that agent n has natural number i on its forehead. Therefore the set of atoms is $\{i_n \mid i \in \mathbb{N}^+ \text{ and } n \in \{a, b, c\}\}$. If Anne sees (knows) that Bill has 20 on his forehead and Cath 30, we describe this as $K_a(20_b \wedge 30_c)$. If an upper bound max for all numbers *were* specified in the riddle, the number of states would be finite and "knowing the others' numbers" would be described as a disjunction $\bigvee_{y, z \leq \max} K_a(y_b \wedge z_c)$. In 'What Sum' no upper bound is given, so strictly we now have an *infinitary* disjunction $\bigvee_{y, z \in \mathbb{N}^+} K_a(y_b \wedge z_c)$, such that

Anne saying: "I don't know my number" is similarly described as $\neg \bigvee_{x \in \mathbb{N}^+} K_a x_a$ (or $\bigwedge_{x \in \mathbb{N}^+} \neg K_a x_a$). Infinitary descriptions are, unlike infinitely large models, not permitted in this (propositional) logic. For the moment, therefore, we restrict our analysis of the updates mainly to a semantic one.

The epistemic model $\mathcal{T} = \langle S, \sim, V \rangle$ is defined as follows $(x, y, z \in \mathbb{N}^+)$:

$$S \equiv \{(x, y, z) \mid x = y + z \text{ or } y = x + z \text{ or } z = x + y\}$$

$$\forall (x, y, z), (x', y', z') \in S \; (x, y, z) \sim_a (x', y', z') \text{ iff } y = y' \text{ and } z = z'$$
$$\forall (x, y, z), (x', y', z') \in S \; (x, y, z) \sim_b (x', y', z') \text{ iff } x = x' \text{ and } z = z'$$
$$\forall (x, y, z), (x', y', z') \in S \; (x, y, z) \sim_c (x', y', z') \text{ iff } x = x' \text{ and } y = y'$$
$$\forall k \in \mathbb{N}^+ \; V(k_a) = \{(x, y, z) \in S \mid x = k\}$$
$$\forall k \in \mathbb{N}^+ \; V(k_b) = \{(x, y, z) \in S \mid y = k\}$$
$$\forall k \in \mathbb{N}^+ \; V(k_c) = \{(x, y, z) \in S \mid z = k\}$$

For the sequel, it is important to realise that, for each agent n, the set of states in which (s)he does *not* know his or her own numbers, is

$$\{(x, y, z) \in S \mid \exists (x', y', z') \neq (x, y, z) \text{ and } (x, y, z) \sim_n (x', y', z')\}$$

A relevant question is what the background knowledge is that is available to the agents, i.e., what the abc-classes in the model are (an abc-class, or $\{a, b, c\}$ equivalence class, of a state s in the model consists of all states t such that $s \sim_{\{a,b,c\}} t$, where $\sim_{\{a,b,c\}} = (\sim_a \cup \sim_b \cup \sim_c)^*$, as above).

An abc-class in $\mathcal{T}$ can be visualised as an infinite binary tree. The depth of the tree reflects the following order on number triples in the domain of $\mathcal{T}$: $(x, y, z) > (u, v, w)$ iff $(x > u \text{ and } y = v \text{ and } z = w)$ or $(x = u \text{ and } y > v \text{ and } z = w)$ or $(x = u \text{ and } y = v \text{ and } z > w)$. If $(x, y, z) > (u, v, w)$ according to this definition, (x, y, z) is a child of (u, v, w) in the tree. Every node except the root has one predecessor and two successors, as in Figure 1.

The root of each tree has label $(2x, x, x)$ or $(x, 2x, x)$ or $(x, x, 2x)$. In each such a root, at least one of the agents knows his or her own number. The idea now is that after an announcement by agent n of the form "I don't know my number", such a root where n knows his number can be eliminated, and the tree gets split in two subtrees. Starting with an arbitrary state (x, y, z) in the tree, such that one is the *sum* of the other two, replace that sum by the *difference* of the other two; one of those other two has now become the sum; if you repeat the procedure, you *always* end up with two equal numbers and their sum. An agent who sees two equal numbers, immediately infers that its own number must be their sum (twice the number that is seen), because otherwise it would have to be their difference 0 which is not a positive natural number. It will be obvious that: the structure truly is a forest (a set of trees), because each node only has a single parent; all nodes except roots are triples of three *different* numbers; and all trees are infinite. All abc-trees are isomorphic modulo (i) a multiplication factor for the numbers occurring in the arguments of the node labels, and modulo (ii) a permutation of arguments and a corresponding swap of agents, i.e., swap of arc labels. For example, the numbers occurring in the tree with root $(6, 3, 3)$ are thrice the

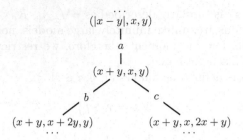

As an illustration, consider (30,14,16) again. It has the following path to the root: (30,14,16) $\sim_a$ (2,14,16) $\sim_c$ (2,14,12) $\sim_b$ (2,10,12) $\sim_c$ (2,10,8) $\sim_b$ (2,6,8) $\sim_c$ (2,6,4) $\sim_b$ (2,2,4). The only sibbling of (30,14,16) is (2,18,16): it is $\sim_b$-connected to the shared parent (2,14,16).

Fig. 1. (Left) Modulo agent symmetry, all parts of the model $\mathcal{T}$ branch as here. Arcs connecting nodes are labelled with the agent who cannot distinguish those nodes. (Right) A description of part of the tree with (30,14,16).

corresponding numbers in the tree with root $(2,1,1)$; the tree with root $(2,1,1)$ is like the tree for root $(1,2,1)$ by applying permutation (213) to arguments and (alphabetically ordered) agent labels alike. The left side of Figure 3 shows the trees with roots $(2,1,1)$, $(1,2,1)$, and $(1,1,2)$. For simplicity, we write 211 instead of $(2,1,1)$, etc. In the left tree, for Bill $(2,1,1)$ is indistinguishable from $(2,3,1)$ wherein his number is the sum of the other two instead of their difference; for Anne triple $(2,3,1)$ is indistinguishable from $(4,3,1)$, etc.

Processing Announcements. The result of an announcement is the restriction of the model to all states where the announcement is true. We can also apply this to the ignorance announcements of agents in 'What Sum'. Consider an *abc*-tree T in $\mathcal{T}$. Let n be an arbitrary agent. Either the root of T is a singleton n-class, or all its n-classes consist of two elements: a two-element class represents the agent's uncertainty about its own number. An ignorance announcement by agent n in this riddle corresponds to removal of all singleton n-classes from the model $\mathcal{T}$. This means that *some* of the model's trees are split into two subtrees (with both children of the original root now roots of infinite trees).

An ignorance announcement may have very different effects on *abc*-classes that are the same modulo agent permutations. For example, given *abc*-classes in $\mathcal{T}$ with roots 121, 112, and 211, the effect of Anne saying that she does not know her number *only* results in elimination of 211, as only the first *abc*-class contains an *a*-singleton. Given 211, Anne knows that she has number 2 (as 0 is excluded). But triple 112 she cannot distinguish from 312, and 121 not from 321. Thus one proceeds with all three announcements. See also Figure 2.

Solving the riddle. We have now sufficient background to solve the riddle. We apply the successive ignorance announcements to the three classes with roots $(2,1,1)$, $(1,2,1)$, and $(1,1,2)$, determine the triples wherein Anne knows the numbers, and from those, wherein Anne's number divides 50. See Figure 3— note that in triple $(8,3,5)$ Anne also knows her number: the alternative $(2,3,5)$ wherein her number is 2 has been eliminated by Cath's, last, ignorance announcement. The *unique* triple wherein Anne's number divides 50 is $(5,2,3)$. In other words, the unique *abc*-tree in the *entire* model $\mathcal{T}$ where Anne knows that she

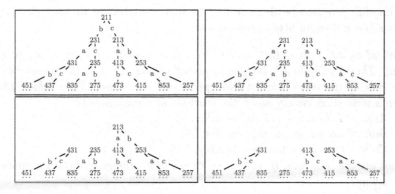

Fig. 2. The results of three ignorance announcements (by a, b, and c, respectively) on the abc-class with root $(2, 1, 1)$

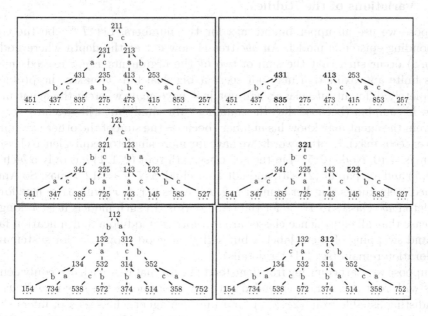

Fig. 3. On the left, abc-classes of the model $\mathcal{T}$ with root 211, 121, and 112. Any other abc-class is isomorphic to one of these, modulo a multiplication factor. The results of the (combined) three ignorance announcements on those abc-classes are on the right. The triples in bold are those where Anne knows her number.

has 50 after the three ignorance announcements, is the one with root $(10, 20, 10)$. The solution to the riddle is therefore that Bill has 20 and Cath has 30. After the three announcements in the abc-class with root $(10, 20, 10)$, the triple $(50, 20, 30)$ remains wherein Anne knows that Bill has 20 and Cath 30.

The original riddle could have been more restrictive: in the quoted version [11] it is *not* required to determine who holds which other number, but as we have seen this can also be determined. It also occurred to us that the original riddle has the following attractive variant:

> Each of agents Anne, Bill, and Cath has a positive integer on its fore-head. They can only see the foreheads of others. One of the numbers is the sum of the other two. All the previous is common knowledge. The agents now successively make the truthful announcements:
> i. Anne: "I do not know my number."
> ii. Bill: "I do not know my number."
> iii. Cath: "I do not know my number."
> What are the numbers, if Anne now knows her number and if all numbers are prime?

Consulting Figure 3, it will be obvious that the answer should be: '5, 2, and 3'.

2.3 Variations of the Riddle

Suppose we use an upper bound max for the numbers. Let $\mathcal{T}^{max}$ be the corresponding epistemic model. An abc-tree is now cut at the depth where nodes (x, y, z) occur such that the sum of two of the arguments x, y, z exceeds max. This finite approximation may not seem a big deal but it makes the problem completely different: abc-classes will not just have *roots* wherein the agent may know his number (because the other numbers are equal) but will also have *leaves* wherein the agent may know his number (because the sum of the other two numbers exceeds max). In other words, we have far more singleton equivalence classes. Let max $= 10$. Node $(2, 5, 7)$ in the abc-class with root $(2, 1, 1)$ has only a b-child $(2, 9, 7)$ and a c-parent $(2, 5, 3)$, and not an a-child, as $5 + 7 = 12 > $ max. So Anne immediately knows that her number is 2. All roots $(2x, x, x)$ with $3x > $ max form singleton abc-classes in $\mathcal{T}^{max}$, for the same reason. In such models it is no longer the case that all equivalence classes are isomorphic modulo a multiplication factor and swapping of agent labels—but sufficient good properties for systematic exploration remain (see [21] for details).

Suppose we start counting from 0 instead of 1. In that case each abc-equivalence class with root $(2x, x, x)$ is extended with one more node: the new root $(0, x, x)$ is indistinguishable from $(2x, x, x)$ for Anne. An agent who sees a 0, infers that his number must be the other number that (s)he sees. If there is a 0, two of the three agents see that. Therefore, the root has just one child $(2x, x, x)$; if the triple is $(0, x, x)$ Bill and Cath know that their number is x.

Now consider the following version of the riddle: If the three numbers have an upper bound, and if 0 is allowed, for which range of the upper bound does Anne now *always* know the numbers after the three announcements? Let max be the upper bound. The requested range includes max $= 10$. Figure 4 shows that from abc-class with root 011 the triples 211 and 213 remain. There is one other abc-class in the epistemic model $\mathcal{T}_0^{10}$ (for $0 \leq x, y, z \leq 10$) that remains non-empty after the three announcements, namely the one with root 022: the triples 242 and 246 then remain. Therefore, whatever the numbers, Anne now knows her's.

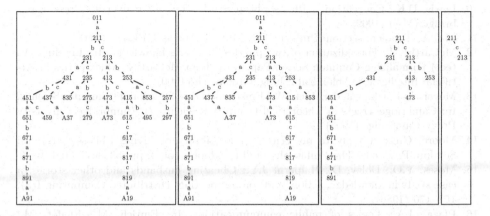

Fig. 4. Successive announcements in the *abc*-class with root 011 in model $\mathcal{T}_0^{10}$. The horizontal order of branches has no meaning. Symbol A represents 10.

The solution range is: $8 \leq \mathsf{max} \leq 13$. This means that if $\mathsf{max} = 7$, the three announcements cannot be made (without lying). And if $\mathsf{max} = 14$, it is not always the case that Anne knows her number: for example, if Bill has 1 and Cath has 3, Anne cannot determine whether her number is 2 or 4; 213 and 413 are in that case the *only* two triples where Anne is still uncertain. For this sort of finetuning a model checker was helpful and even essential [21]. The interplay between designing riddles, model checking, and dynamic epistemic logical analyses, is a good playing ground between theory and practice of multi-agent system dynamics.

References

1. Alchourrón, C.E., Gärdenfors, P., Makinson, D.: On the logic of theory change. Journal of Symbolic Logic 50, 510–530 (1985)
2. Aucher, G.: A combined system for update logic and belief revision. In: Barley, M.W., Kasabov, N. (eds.) PRIMA 2004. LNCS (LNAI), vol. 3371, pp. 1–17. Springer, Heidelberg (2005)
3. Aumann, R.J.: Agreeing to disagree. Annals of Statistics 4(6), 1236–1239 (1976)
4. Baltag, A., Moss, L.S., Solecki, S.: The logic of public announcements, common knowledge, and private suspicions. In: Gilboa, I. (ed.) TARK 98, pp. 43–56 (1998)
5. Baltag, A., Smets, S.: Dynamic belief revision over multi-agent plausibility models. In: Proceedings of LOFT 2006 (to appear)
6. Fagin, R., Halpern, J.Y., Moses, Y., Vardi, M.Y.: Reasoning about Knowledge. MIT Press, Cambridge, MA (1995)
7. Gerbrandy, J.D.: Bisimulations on Planet Kripke. PhD thesis, University of Amsterdam, ILLC Dissertation Series DS-1999-01 (1999)
8. Hintikka, J.: Knowledge and Belief. Cornell University Press, Ithaca, NY (1962)
9. Kooi, B.P.: Expressivity and completeness for public update logics via reduction axioms. Journal of Applied Non-Classical Logics, 2007 (to appear)

10. Lewis, D.K.: Convention, a Philosophical Study. Harvard University Press, Cambridge (MA) (1969)
11. Liu, A.: Problem section: Problem 182. Math Horizons 11(324) (2004)
12. McCarthy, J.: Formalization of two puzzles involving knowledge. In: Lifschitz, V. (ed.) Formalizing Common Sense: Papers by John McCarthy. Ablex Series in Artificial Intelligence, Ablex Publishing, Greenwich (1990)
13. Meyer, J.-J., Ch., van der Hoek, W.: Epistemic Logic for AI and Computer Science. In: Cambridge Tracts in Theoretical Computer Science 41. Cambridge University Press, Cambridge (1995)
14. Moore, G.E.: A reply to my critics (The Library of Living Philosophers). In: Schilpp, P.A. (ed.) The Philosophy of G.E. Moore, vol. 4, pp. 535–677 (1942)
15. Moses, Y.O., Dolev, D., Halpern, J.Y.: Cheating husbands and other stories: a case study in knowledge, action, and communication. Distributed Computing 1(3), 167–176 (1986)
16. Plaza, J.A.: Logics of public communications. In: Emrich, M., Pfeifer, M., Hadzikadic, M., Ras, Z. (eds.) 4th Int. Symp. on Methodologies for Intelligent Systems, pp. 201–216 (1989)
17. Qian, L.: Sentences true after being announced, www.stanford.edu/group/nasslli/student/qian.ps
18. van Benthem, J.F.A.K., van Eijck, J., Kooi, B.P.: Logics of communication and change. Information and Computation 204(11), 1620–1662 (2006)
19. van Ditmarsch, H.P.: Knowledge games. PhD thesis, University of Groningen, ILLC Dissertation Series DS-2000-06 (2000)
20. van Ditmarsch, H.P.: Prolegomena to dynamic logic for belief revision. Synthese (Knowledge, Rationality & Action) 147, 229–275 (2005)
21. van Ditmarsch, H.P., Ruan, J.: Model checking logic puzzles. In: Quatrièmes Journées Francophones MODÈLES FORMELS de l'INTERACTION, Paris, Cahiers du Lamsade, 2007 (to appear)
22. van Ditmarsch, H.P., van der Hoek, W., Kooi, B.P.: Dynamic epistemic logic with assignment. In: AAMAS 05, pp. 141–148. ACM Inc. New York (2005)
23. van Ditmarsch, H.P., van der Hoek, W., Kooi, B.P.: Dynamic Epistemic Logic. In: Synthese Library, vol. 337, Springer, Heidelberg (2007)
24. van Eijck, J.: Dynamic epistemic modelling. Technical report, Centrum voor Wiskunde en Informatica, Amsterdam, (2004) CWI Report SEN-E0424
25. Wu, W.: 100 prisoners and a lightbulb (2001), http://www.ocf.berkeley.edu/~wwu/papers/100prisonersLightBulb.pdf

Peirce on Icons and Cognition

Christopher Hookway

Department of Philosophy, University of Sheffield
C.J.Hookway@sheffield.ac.uk

Abstract. Charles Sanders Peirce often emphasized the importance of iconic representations in all cognition. After addressing some general issues about the analysis of iconic representation, the paper examines the cognitive role of diagrammatic representations, and explores the implications for his theory of perception of Peirce's claim that ideas are like composite photographs. The discussions also consider the role of imagination in cognition.

Keywords: Peirce, icons, diagrams, metaphor, perception.

1 Introduction

When Peirce discussed his classifications of signs, the first threefold classification he introduced would often be that which distinguishes icons, indices and symbols. As is well known, this classification concerns the different relations between signs and their objects that we exploit when we interpret signs and which thus determine just what the object of a given sign is. Icons can function as signs of objects on the basis of resemblance: there is a similarity or likeness between sign an object, a feature or property which both share. Indices can function as signs because of a real 'existential' relation between sign and object. And a symbol denotes its objects because there is a practice, habit or convention that involves interpreting other tokens of that symbol as signs of that object.

This paper is concerned with the role of *iconic* signs in cognition, and the introductory section identifies some of the questions this raises and makes some preliminary observations. The first of these is that even iconic signs will usually have a conventional or symbolic aspect: conventions are required to identify which sorts of resemblance are *relevant* to interpreting the sign, although they do not determine the object of the iconic sign unaided.

The second observation concerns two rather different ways in which iconic signs can be understood. When looking for examples of iconic signs, we tend to think of pictures, photographs, maps and related phenomena. And ordinary uses of 'icon' also suggest that the word picks out distinctive kinds of pictures. Maps, photographs and diagrams typically look like the things they represent: a map records the shape and relations of various features of the terrain as they might look when viewed from some particular position. In interpreting such signs, we can exploit familiar recognitional capacities, and we can use the signs to adapt or refine recognitional capacities that we possess. If we begin by reflecting upon such examples, it is natural to suppose that iconic representations are to be distinguished from other kinds of representation, from

U. Priss, S. Polovina, and R. Hill (Eds.): ICCS 2007, LNAI 4604, pp. 59–68, 2007.

linguistic representations and from mathematical and other kinds of notations. If propositional representation is distinguished from *pictorial* representation, then it can also be distinguished from iconic representation. If we adopt this way of thinking about things, then it is natural to conclude that, when Peirce introduced his existential graphs, as a tool for logic and for the representation of thought, he was making a very radical break with earlier work in logic.   Earlier notations employed non-iconic representations of thoughts, while the existential graphs make an innovative attempt to provide what Peirce once called 'moving pictures of thought'.

There is plenty of evidence that this is *not* how Peirce saw things, although, as we shall see, some of his emphases and formulations do seem to encourage emphasise the sorts of features of the iconic described above. All logical notations are, in varying degrees, iconic, and the existential graphs are to be preferred, not because it is better to have iconic than non-iconic ways of representing thoughts but because they provide *better* iconic representations than their predecessors. All reasoning involves working with mental diagrams, experimenting upon them, transforming them in legitimate ways, and so on. Both mathematical and every day reasoning employs an 'icon or schematic *image* of a general predicate; and from the observation of this *icon* we are supposed to construct a new general predicate' (Peirce 1998: 303). During the 1880s, Peirce wrote that all propositional representations contain elements that are iconic, indexical *and* symbolic, and that predicate expressions always function as iconic signs. In that case when we distinguish propositional signs from pictorial ones, we are concerned with two classes of representations that are similar in being iconic, indexical and symbolic. Those familiar with the philosophy of the early twentieth century will be familiar with this in the form of the 'picture theory of the proposition' that is defended in Ludwig Wittgenstein's *Tractatus Logico-Philosophicus*. Pictorial representations may be able to convey more information than propositional ones; and they may do so in ways that lend themselves to different methods of interpretation. Many of us may have to struggle to identify the features that a mathematical formalism shares with its object, while being able to recognize what is depicted in a photograph easily through exercise of familiar recognitional capacities. We often see *immediately* what a map or photograph depicts while we would not think of ourselves as able to *see* what is represented in a sophisticated mathematical representation.

The apparently contrasting ways of thinking about icons are reflected in some of Peirce's explanations of what is involved in being an icon. In lectures delivered in 1903, Peirce wrote 'An *icon* is a representamen which fulfils the function of a representamen by virtue of a character which it possesses in itself, and would possess just the same though its object did not exist.' (*CP* 5.73)   Such passages, which emphasise only that there is a feature or character which is possessed both by the sign and its object leave open the possibility that this shared feature is very abstract, being something that can only be grasped by hard reflective thought. And for many cases, it is important that this is compatible with being an iconic sign. Other explanations appear to ignore this possibility. In a letter written in the same year, Peirce wrote that an iconic sign is represented 'by virtue of its being an immediate image, that is to say by virtue of characters which belong to it in itself as a sensible object.' (*CP* 8.335).

These differences are ones of degree, of course, and they are reflected in some of Peirce's classifications of icons: an *image* shares some simple quality with its object and a *diagram* is such that that there is are relations between the parts of he sign which are analogous to the relations between the parts of its object. A *metaphor* is more abstract still: metaphors represent ' the representative character of [the sign] by representing a parallel in something else'. The interpretation of images tends to be immediate; the interpretation of diagrams will often involve extensive inquiry; and the interpretation of metaphor calls for the use of the imagination. (*CP* 2.277, and also see Tiles 1988).

But it will be useful to keep them in mind in what follows. I am interested in identifying some different areas in which Peirce has identified a role for schematic iconic representations in cognition. In the following section (section 2), I shall discuss some of his claims about the role of *diagrams* in cognition, relating them to the claim that predicate expressions function as icons. Then (section 3), I shall introduce his claim that our ideas (and presumably our concepts) can be thought of as 'composite photographs', distinguishing some different elements in this metaphor and relating it to questions about the role of icons in cognition. Following this (in section 4), I shall discuss the role of 'composite photographs', and thus of schematic iconic representations, in perception. These issues raise questions about *immediacy* and *imagination* that were noted earlier in this section.

2 Diagrams and Predicates

Although a map often provides a kind of image of a geographical region, so that the shapes used in representing a country, for example, will match the 'outline shape' that the country would have if viewed from an appropriate position, it will be useful to use maps as examples to illustrate what we do with diagrams. A map is, for present purposes, an arrangement of marks on paper. In order to use a map, we require information of at least three kinds. First, we must know about the conventions governing the use of symbols on the map, the scale of the map, the kind of projection used, and so on. The use of these conventions reflects some ways in the map is a conventional sign. The map must also be anchored, so that we know which towns are indicated by particular patches or dots, for example. This anchoring involved recognizing the dots and patches as indexical signs of those towns. Then we can use the map because the relations between the dots and patches are 'analogous to' the relations between the corresponding towns. The map is an iconic sign because these analogies ensure that the map and the terrain share the feature of having elements that stand in these analogous relations. Reflection on the example of 'map' helps us to see just how abstract these resemblances can be: there is no requirement that the shapes on the map 'look like' the shape of the terrain viewed from any position at all.

Interpreting something as a map will often involve using it in particular ways, and, from two perspectives, two features are relevant. First, we can obtain information about the terrain (the object) by making observations of the sign (the map): we exploit the known similarity in order to infer from properties of the sign to properties of the object. We can also infer from properties of the terrain to the properties of what we describe as the map, although, in that case, we are, presumably, using the terrain as a 'map' of the marks on the paper. Second we can make experiments upon the map to extend our knowledge of the terrain: if we want to evaluate plans for adding a house to the terrain, we can amend the map to include a dot corresponding to the house and then investigate how this will affect the other properties of the terrain (Hookway 2000: 197-203).

I have noted that Peirce often tells us that predicates, in natural and artificial languages, function as diagrams. This is reflected in the fact that we can learn more about the world by inspecting sentences, advance our knowledge further by 'experimenting' on sentences in accord with logical laws. This exploits the fact that the inferential relations between sentences can be analogous to the real relations between elements of reality, norms of inference matching laws of nature. We might put this by saying that the inferential role of a sentence or proposition can be 'analogous to' the nomological role of a corresponding state of affairs. Indexical expressions contained in the sentence will pick out elements of the state of affairs, and our grasp of the predicate expressions (simple or complex) enable us track the relations between the objects of those indexical expressions.

There are some further complexities here. Some must be left until later sections of the paper, but one should be noted here. First, a diagram such as a map provides a schematic representation of the terrain: this means that it leaves out lots of details, selecting what should be included in several different ways. One consideration is: what information will we be looking for when we use the map. No more details need be included than will be required by users. Secondly, there are considerations of ease of use: the more detailed the map is, the harder it is to read or use. So the user of a map needs information about its limits, about when it can be trusted and when not. For example when we use the underground map in London, we may need to know that it cannot be trusted as a source of information about the relative distances between different stations. Something similar occurs when we consider the meanings of predicates: our 'schematic diagrams' of their objects are likely to omit lots of details. This can be both for ease of use and because of gaps in our own understanding.

In the remainder of this section, I want to introduce a point that has been made by J E Tiles (1988). Having explained the third kind of icon that Peirce identifies, namely *metaphor*, he argues that 'when an icon, of the sort that Peirce calls a diagram, is used in physical (or natural), as opposed to mathematical science, it shares [a significant] feature with paradigmatic examples of metaphors. His example of a metaphor is someone who claims that his memory is a green pond, and observes that the speaker 'is using a parallelism between the way things disappear beneath the surface of an algae covered pond until dredged up by something acting on the pond and the way the contents of (his) memory are not manifested until something 'stirs' it (Tiles 1988 173). According to Peirce's account, the parallelism 'represents the representative character' of a sign. Once we appreciate the parallelism, we can arrive at a more developed sign of the object. Tiles suggests that we interpret the speaker's memory as 'a container with an astonishing capacity, only a tiny proportion of the contents of which are manifest at any one time, and requiring external stimulus to make individual items manifest'. The metaphor of memory as a green pond encourages us to develop the image of a green pond in this way. One notable feature of interpretation in *this* sort of case is that it involves imaginative work, exploring what the parallel can offer us.

It is Tiles's conjecture that when we use and understand scientific concepts or predicates, similar imaginative work is required. The idea that our grasp of concepts for theoretical concepts in the sciences depends upon the use of models and metaphors has been employed by a number of philosophers of science. And there is evidence that Peirce favoured this view in his own writings. The example that Tiles offers is familiar: he considers the formulation of the Boyle-Charles law. 'Statistical mechanics, using various diagrammatic devices to represent the behaviour of aggregates of gas particles, has provided us with an interpretation of that law.' The explanatory framework provided by the model helps to tell us what the law means.

Peirce's example is in a draft paper called 'The logic of drawing history from ancient documents', written in the first years of the twentieth century. The kinetical theory of gases 'began with a number of spheres almost infinitesimally small occasionally colliding. It was afterwards so far modified that the forces between the spheres, instead of merely separating them, were mainly attractive, that the molecules were not spheres, but systems, and that the part of space within which their motions are free is appreciably less that the entire volume of the gas. There was no new hypothetical element in these motivations.' (*CP* 7.127) Successive version of the theory are understood, Peirce seems to suggest, by exploiting a parallelism between the original formulation when gases were seen as systems of infinitesimal indivisible spheres reacting against each other, and the case where the elements are systems which attract as well as reacting. The simpler story offers a way of thinking about the more complex one; and employing the interpretative strategies that are characteristically used in developing our understanding of metaphor has a fundamental role in the development of our grasp of the objects of scientific concepts.

The conclusions to be taken from this section are as follows. Diagrams form a distinctive kind of iconic sign that has a fundamental role in cognition. This is because we can learn more about the object of an icon by examining (and experimenting upon) the diagram and, indeed, reasoning typically takes this form. Diagrams have a schematic character, providing templates or ideal types to be used in understanding the meanings of predicates. Moreover, in many interesting cases, interpreting diagrams involves the imaginative development of the meanings of the concepts in question, employing a process similar to what goes on when we use, and understand, metaphors.

3   Composite Photographs

In 1893, in a discussion of the grammar of judgment for an unpublished *Short Logic,* Peirce wrote that, when we make a judgment 'we cause an image, or *icon*, to be associated, in a particularly strenuous way, with an object presented to us by an *index*. (*CP* 2.435). Amplifying this claim, he continued:

> Suppose … I detect a person with whom I have to deal in an act of dishonesty. I have in my mind something like a "composite photograph" of all the persons that I have known and read of that have *had* that character, and at the instant I make the discovery concerning that person, who is distinguished from others for me by certain indications, upon that index down goes the stamp of RASCAL, to remain indefinitely. (*CP* 2.435).

This is the first passage I know in which Peirce used the term 'composite photograph' to describe the 'image or icon' that is expressed by a predicate. He is explicit shortly after. When I say 'It rains' 'the icon is the mental composite photograph of all the rainy days the thinker has experienced' (*CP* 2.438)

The metaphor of a composite photograph is a powerful one. Peirce concurred: he continued to use this phrase on many occasions until at least 1908 (Hookway: 2002). Generally speaking, he assumed that his readers understood what it meant and could see how to apply to the idea to the examples he was interested in. Talk of composite photographs is, presumably, intended to have three merits. First, since it is a photograph, the composite is an iconic representation: it resembles its objects. Second, through being composite, the idea has a general character: it captures what is common to all rascals rather than just being a sign of one particular rascal. And third, again because it is a photograph, it possesses a kind of secondness: its character depends upon the causal impacts that various rascals have had upon us. Its content is somehow fixed by our experience of various rascals in the past.

The 1893 manuscript offers more helpful suggestions. The 'composite photograph' picture has wide application: 'any image is a "composite photograph" of innumerable particulars'. (*CP2*.441). Even instantaneous photographs are really 'composite of the effects of intervals of exposure more numerous by far than the sands of the sea': it is, perhaps, better to say that no photograph or image is really instantaneous.

Two applications are particularly noteworthy. Considering the proposition "A sells B to C for the price D", he treats the symbol "- sells – to – for price – " as a 'predicate', and as something that 'refers to a mental icon, or idea of the act of sale', and thus to an image which, like all images, can be described as a composite photograph (*CP* 2.439). Presumably what is involved in something being a sale does not require the existence of this mental icon; and many theorists would treat that as the 'referent' of the sign. Rather, the point must be that the most primitive way of thinking about such things must involve the use of mental images or icons, so understood; composite photographs are the vehicles of our thoughts or judgments about acts of sale.

Secondly, Peirce tells us that icons can be complex (*CP* 2.441). Consider the universally quantified proposition *All men are mortal*. Peirce views this as including a disjunction:

Take anything you please, and it will either not be a man or will be mortal.

This involves the complex predicate '- is not a man or is mortal'. And, just as in the other cases, this involves the 'combination' of the two alternatives, *not being a man* and *being mortal.* According to Peirce, they are combined in just the way that experiences of different rascals give rise to the "composite photograph" of a rascal. This seems very implausible. In one case we form what appear to be simple ideas through experience of many of their instances; in the other case, we use logical operators to form logically complex icons out of simpler ones. I mention the latter cases because I shall not discuss them further: my concern is with what the 'composite photograph' idea contributes to our understanding of concepts that are not explicitly logically complex.

Peirce would have been very familiar with the composite photograph idea, not least from the work of Francis Galton and his followers. According to the *Century Dictionary*, a composite photograph is:

> A single photograph produced from more than one subject. The negatives from the individuals who are to enter faces are made s to show the faces as nearly as possible of the same size and lighting and in the same position, These negatives are then printed together upon the same piece of paper, each being exposed to the light for the same fraction of the full time required for printing.

Galton hoped that such photographs would be illuminating. If we compose a number of photographs of rascals in the manner described, we will obtain a picture of 'a single type which will represent the whole group'. If the pictures depict the type *Rascal*, we obtain 'an essentially new face, - a type representing *par excellence* the peculiar characteristic for which the originals were grouped together.'

In Galton's work, then, we hope to obtain a single picture of a face whose character is determined by the faces of all the particular rascals (or criminals or academicians) whose portraits contributed to the construction of the composite. And, strikingly we obtain something that can function as an iconic sign: it resembles, in a crucial respect, each member of the kind; and it represents a general type of which each of the original pictures is a token or replica. If all of our ideas were, indeed, composite photographs. We should have a satisfying explanation of how the ideas expressed by predicates were images or icons; and we can see how we can use our idea in determining whether someone we have met is a rascal – we consider whether their appearance is a token of the rascal-type. The problem is, surely, that if we interpret the composite photograph idea in this way, the account it offers of our understanding of general terms is, simply, hopeless. The idea that by combining all our experiences of rascals as the individual photographs are, we come up with a sort of visual image which resembles every rascal more closely than it resembles any non-rascal, is wholly implausible. It is very hard to believe that Peirce was unaware of this problem: it is closely related to the difficulties that Berkeley raised for Locke's theory of abstract ideas, and it was acknowledged by Galton's followers who concluded that such photographs could be given for, at most, a very small number of ideas. Since Peirce never wavered in his liking for the composite photograph idea, we should work on the assumption that it is to be understood in a more sophisticated way than has just been considered. The challenge is to arrive at a plausible understanding of his metaphor (as we must now take it to be) that does not require us to give up the view that ideas are iconic representations.

If we take it that the talk of composite photographs is, indeed, metaphorical, then we should consider how it should be interpreted. If Peirce's claims about metaphor are taken seriously, this should involve identifying parallels between ideas and composite photographs. Which parallels should we pay most attention to? First, the appeal to photographs suggests that our ideas have been determined by our experience of a variety of the objects to which the idea applies. My idea of a rascal has been shaped by my interactions with those that have been recognized as rascals. Photographs are traces left when objects causally interact with cameras, and these elements can be preserved. As in a composite photograph, the idea is a single thing that is produced by these encounters. However we should not expect the composite photograph to carry those traces in a single combined 'look': the idea must somehow 'fit' all of those experiences, without itself being something straightforwardly experience like. But second, the idea must share with photographs the fact that the idea provides the basis of a sort of recognitional capacity. Just as a photograph of an escaped criminal can guide us in recognizing him when we see him, so the idea provides guidance in recognizing rascals when we encounter them. Composite photographs and ideas are generated in broadly analogous ways – although, no doubt, there will also be a lot of differences. And they are used in similar ways too: possession of the idea puts us into a position to recognize the objects to which it occurs. Indeed, in many cases, our ideas enable us to recognize things on the basis of how they look. When I see a book, for example, I will often recognize it as a book immediately, without having to collect information about the object or arrive at my identification as the conclusion of conscious inference and deliberation.

4   Icons and Perception

In this section I shall draw attention to some important uses of what Peirce calls 'ideas' in perception. As I have noted before (2002), Peirce was influenced by a section of Kant's *Critique of Pure Reason* commonly referred to as the 'schematism' in which he emphasized a role for schemata in guiding the imagination in applying concepts to objects of experience. Composite photographs fitted the role assigned to these schemata, and this will help us to understand why the ideas themselves can be seen as iconic signs.

First, we should register some of the claims Peirce made about perception in the early 1900s, most notably in his Harvard lectures on Pragmatism but also in a manuscript on Telepathy (For further discussion see Hookway 1985: 155-166). Sometimes he distinguished the 'percept' (the immediate object of perception) from the perceptual judgment (the first judgment we make about the content of the percept). On other occasions, when being more careful, he treated these as inseparable components of a unified whole that he called the 'percipuum'. One conclusion to be drawn from this is that the content of perception is shaped by the exercise of concepts; and it is here, I suggest, that composite photographs, the schemata of the imagination, have a role.

The impact of concepts in perception is manifested in the fact that the content of experience itself involves elements of anticipation. He describes perception as 'the limiting case of an abductive inference': what and how we see reflects our best guess about what we are seeing. And, like Wittgenstein nearly fifty years later, Peirce draws upon cases of aspect shift to provide an intermediate case between ordinary perception and cases where it is clear that we are drawing inferences on the basis of what is seen. Peirce himself uses the Schroeder stair as an example (*CP* 7.647), but using Wittgenstein's example of the duck-rabbit is a nice way of registering some historical dependencies: the duck-rabbit figure is usually attributed to Joseph Jastrow who had been Peirce's research collaborator and student at Johns Hopkins in the late 1870s. Reflecting upon such figures shows that *how something looks* (and not just what we believe it is) depends upon our expectations. Another example used by

Peirce concerned the experience of looking out of a train window and mistakenly taking it that this train is moving, when, in fact, it is the rain on the next platform that is starting to move. The experience can continue to have this deceptive character even when he knows that it is not his own train that is moving (*CP* 5.181). What we experience is not just a clash between our beliefs and our experience; we often experience incoherence within the experience itself, which simultaneously involves anticipations and thwarts those very anticipations. The fact that, in these cases, 'the perceptual judgment, and the percept itself, seems to keep shifting from one general aspect to another and back again' (*CP* 5.183) shows that the percept is not 'entirely free from ... characters that are proper to *interpretations*' (*CP* 5.184).

In cases of aspect shift, we may be able to control what we see, but in perception, this is not normally the case: 'The perceiver is aware of being compelled to perceive what he perceives' (*CP* 4.541). The percept 'neither offers any reasons for (its) acknowledgement or makes any pretence to reasonableness'. 'It acts upon us, it forces itself upon us; but it does not address the reason, nor *appeal* to anything for support'. (*CP* 7.622). The 'abductive suggestion' here comes to us 'in a flash' and (at the time it occurs) is 'absolutely beyond criticism'. But in spite of this, 'it is an act of *insight*, although of extremely fallible insight'. (*CP* 5.181) So we face a question about how this process of automatic, educable and insightful abductive interpretation of perceptual experience is possible. Success in cognition depends upon some of our concepts and ideas taking a form that enables them to serve this role.

This is one more area where the composite photograph metaphor comes into its own. Our ideas provide templates that can provide a sort of recipe for imaginative, automatic anticipation of the future run of experience. We might say that, unconsciously, they guide us in determining how things *should* look and in constructing experience in the light of that; and they do this without the intervention of careful reflective deliberation. And it is easy to see that this is the sort of guidance that can be provided by an iconic representation – for example by a photograph – but not by a paragraph or two of careful description. Our ideas provide diagrams that can structure our experience and give it a form that enables it to inform our further inquiries and deliberations.

5 Conclusion

There is an important difference between icons involved in perception and icons that function as described in section 2: the former are, in a sense, invisible. When I use a map in order to learn more about some geographical area, I am under no illusion that I can see the area of land represented in the map. I obtain knowledge of the map and, relying upon the fact that the map and the terrain possess similar structures, I am able to draw inferences about the properties of the terrain. The geographical knowledge that I obtain is indirect. The same occurs when I exploit a metaphor. This can be the case even if the shapes on the map look like the shapes of the countries and maps portrayed. And this is all the more likely to be the case when the resemblances between the diagram and its object are extremely abstract, in the cases where the sign doesn't look or sound like an iconic representation.

When 'composite photographs' direct the kinds of abductions that are 'beyond criticism', this is not the case. We take ourselves to perceive external objects and their properties directly. We are not aware of the image or percept as something that acts as an intermediary between us and the things that we see. Describing a case where he sees a chair, Peirce wrote that 'The chair I appear to see makes no professions of any kind ..., does not stand for anything else, nor 'as' anything' (*CP* 7.619). When I reflect upon my experience. I cannot distinguish the green expanse that fills my experience from the green expanse that is the surface of the chair. Joseph Ransdell has made a lot of this, arguing that the iconic character of experience is essential to perception having this kind of immediacy or directness. In this case there is a character of being a distinctive shade and a distinctive outline shape that are the same in both perceptual experience and chair experienced. The experience and the chair have the same phenomenal colour property, and this is why the experience can be (directly) of that chair and its colour (Ransdell 1979). Peirce also gives the impression that the object seen – in this case the chair – is itself also a component of the percept. The percept is thus both the immediate object of perception and also includes what Peirce would call the dynamic object of perception. And this means that philosophers do not have the task of explaining how we get *from* our experience *to* its external object, although, of course, it has to be allowed that our perceptual experience is fallible.

Bibliography

Hookway, C.: Peirce. Routledge and Kegan Paul, London (1985)

Hookway, C.: Truth, Rationality, and Pragmatism. Oxford University Press, Oxford (2000)

Hookway, C.: ...a Sort of Composite Photograph: Pragmatism, Ideas and Schematism. Transactions of the Charles S Peirce Society XXXVIII, 29–48 (2002)

Peirce, C.S.: The Essential Peirce: Selected Philosophical Writings. In: Peirce, C.S.(ed.) Edition Project. vol. 2, Indiana University Press, Indiana (1998)

Peirce, C.S.: Collected Papers, eight volumes, Hartshorne, C., Weiss, P., Burks, A. (eds). Harvard University Press, Cambridge, MA, (1931-1956). Following standard scholarly practice, references in the text are given by volume and numbered paragraph to CP; thus CP2.439 refers to paragraph 439 of vol. 2

Ransdell, J.: The Epistemic Function of Iconicity in Perception. In: Peirce Studies no. 1, Institute for Studies in Pragmaticism, Lubbock Texas, pp. 51–68 (1979)

Tiles, J.E.: Iconic thought and the Scientific Imagination. Transactions of the Charles S Peirce Society XXIV, 161–178 (1988)

Using Cognitive Archetypes and Conceptual Graphs to Model Dynamic Phenomena in Spatial Environments

Hedi Haddad and Bernard Moulin

Computer Science Department, 3904 Pav. Pouliot, Laval University,
Québec, G1K 7P4, Canada
hedi.haddad.1@ulaval.ca, bernard.moulin@ift.ulaval.ca

Abstract. In this paper we propose a qualitative model to represent and reason about dynamic phenomena in a geographic space. Our model is based on linguistic cognitive archetypes, ontological definitions of geographic space and Conceptual Graphs (CGs). In a first part, we present the main concepts of the model and how we define them using CGs. In a second part, we present an overview of how this model is applied to the multiagent geosimulation domain in the context of the MAGS-COA project. Our model is original for two main reasons. First we use a linguistic approach to qualitatively model dynamic situations in a geographic environment. Second, we use CGs to represent the knowledge associated with such situations. Using CGs makes our model computationally feasible and useable to carry out spatial qualitative reasoning.

1 Introduction

Spatial modeling aims to propose solutions to describe space and to reason about it. Research works in spatial modeling may be divided into two broad areas: Qualitative Spatial Reasoning (QSR) [3] and Geographic Information Systems (GIS) [18]. In both areas, space is essentially a static view of the geographical realm [7, 26]. Considering the fact that reality is essentially dynamic, there is a need for an approach that addresses the computational aspects of space and time change modeling [22]. In fact, there is a growing body of work showing that handling dynamic aspects of geographic phenomena is a necessity: for example, transportation and urban analysis [17], ecological systems analysis [23], etc. Recently, a new domain emerged, Multi-Agent Geosimulation [19], which consists of simulating agents' behaviors in geographic environments. Geosimulation is characterized by using GIS to represent the geospatial features of the simulation environment. When using Multi-Agent Geosimulation approaches to simulate and analyze dynamic geographic phenomena, we need new models that can represent complex real geographic spaces as well as different kinds of dynamic phenomena that may occur in these spaces and that may involve different kinds of entities (people, physical objects, etc.). These models need also to be computationally feasible in order to qualitatively simulate and/or to reason about these phenomena in order to make predictions and inferences.

In this paper we propose a qualitative model to represent dynamic phenomena in a geographic space. Although several researchers introduced models to describe dynamic phenomena, especially in the geographic ontologies and GIS communities,

U. Priss, S. Polovina, and R. Hill (Eds.): ICCS 2007, LNAI 4604, pp. 69–82, 2007.

to our knowledge no one has modeled dynamic phenomena using linguistic situations and Conceptual Graphs (CGs) as we do.

In Section 2 we present the linguistic approach that we use as a theoretical foundation for our solution. In Section 3 we present the main elements of our model, and in Section 4 we apply it to the domain of Geosimulation. Finally, we conclude by presenting some related works, stressing the original aspects of our model and presenting future works.

2 Cognitive Archetypes' Model for Dynamic Situations

The study of dynamic phenomena is widely tied to the concepts of *states*, *events* and *processes*, collectively named *dynamic situations*. Dynamic situations have been mainly studied by three research communities, respectively temporal logic, geographic ontologies and linguistic communities. Although temporal logics, such as situation calculus [16], event calculus [14] and interval temporal logics [2] refer to the same concepts of state, event and process, they propose different definitions for them. Another important shortcoming of these logics is that they use simplifying constraints that make them unable to model and to reason about complex real dynamic situations. For example, relationships between actions, such as concurrency, cannot be expressed in situation calculus, processes with duration cannot be represented in event calculus, etc. In addition, the semantic relationships between agents and situations are not usually represented, and moreover, the space is not made explicit. In the geographic ontologies community [10], the aim is to "produce an account of the entities existing in the world, of the types or categories under which these entities fall, and of the different sorts of relations which hold between them" [10]. Geographic ontologies put the emphasis on the classification of different types of geographic phenomena rather than on models to reason about them. Thus, these works remain at an abstract conceptual level and cannot be directly used in spatial decision support tools or in a GIS, because there is a mismatch between the reality as perceived by geographers and GIS data models [22].

For many years linguists studied dynamic phenomena from a perspective of the aspectual analysis of enunciations and proposed interesting semantic classifications of verbs. Among the different linguistic theories the *temporal constitution theory* and the *participative constitution theory* [8] provide two complementary perspectives of dynamic situations which are of interest to our project. In fact, the temporal constitution theory structures the semantics of verbs using aspectual polarities such as dynamic and static property. In contrast, the theory of participative constitution studies different roles (agent, patient, etc.) that an object involved in a situation may play. Finally, several computational models have been proposed to represent and to reason about these dynamic situations. An interesting example of these models was proposed by Vilnat [24] using the CG formalism and elaborating a new classification of French verbs of action based on the concepts of *state*, *process* and *action*. Of interest in this model is the simplicity of the classification. In fact, a *state* describes a relationship between entities, a *process* describes a change and is considered as "*states in becoming* or *in finishing*", and an *action* is a process which makes explicit the entity that causes the change. For example, the action corresponding to "changing place" is represented as follows:

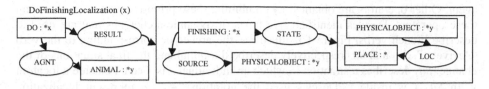

Of interest to our work is Desclès' representation of dynamic situation based on cognitive archetypes [4] for two main reasons.

First, it is based on the *localist hypothesis* [15] according to which a state change is equivalent to a change in space. Hence, states are considered as abstract localizations. Based on this hypothesis, Desclès distinguished *static*, *cinematic* and *dynamic* situations. Static situations express stability, i.e. the absence of change. Desclès distinguished two types of static situations: *situations of localization* (spatial and temporal) and *situations of attribution* (affect a property to an object). In contrast, cinematic and dynamic situations "introduce change on a static background". The difference between cinematic and dynamic situations is determined by the degree of control that exercises an object or an actor on the situation.

Second, Desclès introduced the concept of *cognitive archetype* to describe these situations. Cognitive archetypes are based on a topologic visualization of space and time [4]. They are cognitive representations constructed on a visual perception of space (position of objects relatively to localities, interiorities / exteriorities, etc.) and perception of stabilizations (*states*) or changes (*processes / events*) in time [4]. Practically, an archetype describes abstract relationships between objects and situations. These relationships are expressed in function of the degree of control that an object or an actor has on a situation. Static archetypes describe static situations and are defined using a set of *static operators* (*localization, possession, identification*, etc.). Dynamic archetypes describe dynamic situations. The most general dynamic archetype of a cinematic situation has the following form:

$$\text{Initial situation} \xrightarrow{\ transition\ } \text{Final situation}$$

A *transition* introduces modifications in the *Universe* (composed of objects and situations). The transition is a process that makes the Universe evolve from an initial situation Sit_1 to another posterior situation Sit_2. The transition comprises three temporal zones: *before transition (Sit_1), during transition* from Sit_1 to Sit_2, and *after transition (Sit_2)*. Dynamic archetypes are described using *cinematic* and *dynamic operators* considered as primitives of the cognitive system. These operators are organized in a hierarchy (Fig.1.A). At the top of the hierarchy, the CINEM operator expresses transitions between two static situations. In transitions of type MODIF, initial and final situations are different. In CONSV transitions, the two situations are identical. The operator MODIF may describe an oriented spatio-temporal movement toward a target (MOUVOR), a non-oriented spatio-temporal movement (MOUVNOR), or a state change of an entity (CHANGT) [5]. Transitions are introduced by forces that control changes. These forces are described by another type of primitives, called *agentive operators,* which express the degree of control that an agent exercises on the change. Desclès proposed some primitive agentive operators. For example, the primitive CONTR means that a modification is controlled by an agent. Other operators are obtained by combining these primitive operators. Based on these operators, Desclès identified five types of dynamic archetypes that are used to describe French verbs. Figure 1.B illustrates the archetype corresponding to the verb "Arrive at". Sit_1 and Sit_2 respectively describe the initial and the final situations of the change. Sit_1 describes a situation in which an individual x is located outside a location y. In Sit_2 x is inside y. Desclès used the notation " $\in_0$ " to represent localization relationships between objects and places. $Ex(location)$ and $in(location)$ are two topologic operators that respectively describe the outside and the inside of a location. The MOUVT operator indicates that the transition between Sit_1 and Sit_2 is a movement in space. The operator CONTR indicates that the individual x takes the control of the transition.

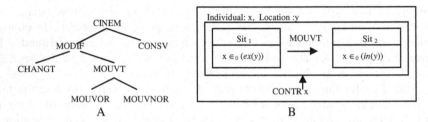

Fig. 1. A) Hierarchy of dynamic operators B) Archetype of the verb "Arrive at"

Desclès' cognitive archetypes provide an elegant and efficient way to represent dynamic situations in a spatial environment, and are particularly suitable to represent movements perceived in such an environment. Because initial and final situations of a change are explicitly represented in the model, it is easy to represent a situation corresponding to a movement from a source A to a destination B.

However, from a knowledge representation perspective, Desclès' model presents some shortcomings. First, time is not explicitly represented, although Desclès indicated that situations are indexed by time. Second, the model does not explicit constraints that may apply to objects involved in situations, such as for example "an instrument must be a non-animated entity". Third, agentive operators are not an easy and intuitive way to define different roles such as instrument, agent and patient. Finally, it is not possible to directly use these archetypes for reasoning purposes: they must be first transformed into functions or predicates. Desclès mentioned that this transformation is a difficult task [4]. We think that all these shortcomings can be remedied by using CGs as a knowledge representation formalism.

3   A Model for Dynamic Situations in Geographic Space

Our goal is to model dynamic phenomena in a virtual geographic environment in which there are different kinds of complex spatial entities (rivers, buildings, etc.). There are also different kinds of objects (people, cars, etc.) which may move in this geographic environment and modify its state (block a road, etc.). In addition, different "happenings" may occur in the environment (explosions, floods, etc.) and may influence it (destroying a bridge may block a road and disrupt a river, etc.). The model that we propose uses cognitive archetypes to capture a qualitative view of dynamic phenomena. We also take advantage of ontological works on geographic space and geographic objects [6] to define the structure of space in our model. Finally, we use CGs to formalize our model for several reasons. First, they allow for a great deal of expressive power, especially with the use of nested contexts and their logical formalism. Second, they offer an easy way to define new concepts and relations. Third, they are well suited to natural language processing.

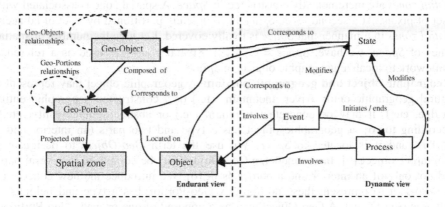

Fig. 2. Concepts and relationships used to model dynamic situations in a spatial environment

According to Grenon and Smith [10], we may distinguish two modes of existence for entities populating the world. The first mode corresponds to an *'endurant' view* according to which there are entities "that have continuous existence and a capacity to endure through time even while undergoing different sorts of changes" [10]. The second mode corresponds to an *occurrent view* that describes occurrent entities that "occur in time and unfold themselves through a period of time" [10]. Similarly to this classification, we define two views in our model: the *endurant* view and the *dynamic* view (Fig.2). The endurant view is composed of the geographic space and the physical objects embedded in it. In the highest level of granularity, a geographic space is composed of geographic objects (*Geo-Object*). Each Geo-Object is decomposed into one or several parts, called *Geo-Portions*. In addition, for spatial referencing purposes, each Geo-Portion is projected onto a spatial zone. Physical objects of the world are represented using the *Object* concept. At a given moment, an object is located in only one Geo-Portion. There are different relationships (as for example spatial relationships) that may hold between Geo-Portions and Geo-Objects. We propose a dynamic view composed of concepts similar to states, events and processes as defined by Desclès. A state may correspond to the description of an object, a Geo-Portion, a Geo-Object or a relationship between these concepts. Both, event and process may have an effect that modifies a state and may involve one or several objects playing different roles (agent, patient, etc.). In addition, a process is characterized by an event that marks its beginning and an event corresponding to its end. These concepts are detailed in the following paragraphs.

3.1 Endurant View

This view describes the structure of space and the physical objects that may be located in it. We define and use the following concepts:

-Space and Spatial zone: We adopt the definition of *Space* and *spatial zone* proposed in [10]. *Space* is the entire spatial universe (the maximal spatial region) and all spatial zones are parts of it. However, we use a different partition of *Space*. At a first elementary level, the *Space* is partitioned by a set of regular cells called pixels. Then, *spatial zones* are incrementally constructed in *Space*. A spatial zone is associated with a set of pixels. At a second level, *Space* is completely partitioned by a set of adjacent *spatial zones* in a manner that *Space* is totally covered. Let n be the number of spatial zones of *Space*, we have: $Space = \cup_i^n (z_i)$. Spatial zones are used as a reference framework to localize geographic objects in *Space*.

-Geographic object and geographic portion: A geographic object may represent a natural geographic entity (river, mountain, etc.) or a constructed geographic entity (bridge, etc.). It may be concrete (mountain, etc.) or abstract (municipality, etc.). According to [6], a geographic object has a type and two parts (an interior and a border), and it is located in *Space*. We use the term *Geo-Object* to designate a geographic object [6]. In our model, a Geo-Object may be composed of several parts and not only of an interior and a border as in [6]. We introduce the new concept of *Geo-Portion* to represent these portions. A Geo-Portion has a type and belongs to only one Geo-Object. A Geo-Object may be composed of one or many Geo-Portions. The decomposition of a Geo-Object into one or several Geo-Portions depends on the spatial scale at which we reason. For example, a lake may be represented as a Geo-Object composed of only one Geo-Portion or composed of several Geo-Portions. A Geo-Portion is projected on only one spatial zone in *Space*. Spatial zones are used to locate Geo-Portions in *Space*. The form and the size of a spatial zone depend on the form and the size of its equivalent Geo-Portion which, in turn, depends on the used spatial model (for example, vector or raster model in a GIS). We use two primitive relations proposed by Vilnat [24] to represent mereology (whole/part) and localization relationships:

type Ingredience (x,y) is {[ENTITY : *x]->(PART)->[ENTITY : *y]}
type Localization (x,y) is {[ENTITY : *x]->(LOC)->[PLACE : *y]}

The relationships between Geo-Object, Geo-Portion and Spatial-Zone are represented as follows:

A Geo-Object may be associated with constraints applied to the type of Geo-Portions that compose it, or to their relationships. For example, a Geo-Object with type River must be composed of Geo-Portions of the same type:

type RiverGeoObject(x) is [GEO-OBJECT: x]->(PART)->[RIVER-GEOPORTION: {*}@n]

In addition, each of these Geo-Portions must be adjacent to two other ones. This constraint cannot be formalized using CGs only. We formalize it using an algorithm with the following pseudocode: Let g be a Geo-Object composed of n Geo-Portions. Let p_i (i increasing from 1 to n) be the identifier of a Geo-Portion. For $i = 1$ to $i = n\text{-}1$ check

[GEO-OBJECT: g]->(PART)->[GEO-PORTION: p_i]
[GEO-OBJECT: g]->(PART)->[GEO-PORTION: p_{i+1}]
[GEO-PORTION: p_i]->(ADJ)->[GEO-PORTION: p_{i+1}]

As we will see later, we implement this algorithm using the Amine platform [12] that allows combining CG rules and Java code.

-Geo-Object / Geo-Object and Geo-Portion / Geo-Portion relationships: We distinguish two main categories of relationships: spatial and interaction relationships.

a) Spatial relationships describe the relative spatial positions of geo-entities.

-Topological relationships: Although several models integrate topological relationships between two simple surfaces [3], representing topological relationships between Geo-Objects composed of several Geo-Portions is not an easy task. Since Price's model [21] applies to objects composed of parts which match our definition of Geo-Object and Geo-Portion, we use it to define our set of primitives and to create topologic relationships (*Separate, Equal, Contain, Inside*, etc. [21]).

-Superposition relationship: Superposition is an important relationship when reasoning about geographic space. Providing a formal definition of the superposition relationship is not an easy task [4, 1]. We adopt the solution proposed in [10] which consists of adding another dimension in the projection function to specify that a Geo-Portion is located *over, under* or *on* a spatial zone. We define three primitive relationships of superposition: *Over, Under* and *On*.

-Proximity relationships: We use another model [13] based on object's *influence areas* to define a set of proximity relationships between Geo-Objects and Geo-Portions. This model allows us to compute proximity relationships such as *Close to* (*near*) and *Distant* (*far from*) as well as orientation relationships such as *In-Front-of*.

b) Interaction Relationships are used to qualitatively specify the effects of "happenings" that may occur in the geographic space. In fact, the effect of a happening that occurs on a specific Geo-Portion may not be limited to this Geo-Portion, but it usually propagates to other *neighboring* Geo-Portions. For example, if it rains on a mountain, the river that springs from this mountain may be flooded after a certain delay. The rain is a happening that occurs on the mountain and modifies its state (for example, 'the mountain is flooded'). However, the effect of the rain propagates and affects the river. We model this issue using a specific neighboring relationship *Spring from* between the river and the mountain. We use attributes of Geo-Portions and their spatial relationships to specify the propagation effect of such happenings. First, a happening modifies the state of the Geo-Portion in which it occurs. Second, the new state of this Geo-Portion modifies the states of neighbor Geo-Portions which have certain spatial relationships with this Geo-Portion. For example, suppose that there is a road that crosses a river over a bridge. So, the bridge is OVER the portion of the river that is UNDER it. There is a portion of the road which is ON the bridge. If the bridge is destroyed (State change), the corresponding road portion will be destroyed and the river portion UNDER the bridge will be congested. Using conceptual graphs, this situation is represented by the following rule:

If [[BRIDGE-GEOPORTION:*b]->(OVER)->[RIVER-GEOPORTION: *r]] And [[ROAD-
   GEOPORTION:*x]->(OVER)->[BRIDGE-GEOPORTION:*b]]   And   [[BRIDGE-
   GEOPORTION:*b]->(ATT)->[STATE : Destroyed]]
Then   [[ROAD-GEOPORTION:*x]->(ATT)->[STATE: Destroyed]]   And   [[RIVER-
   GEOPORTION: *r]->(ATT)->[STATE : Congested]]

-Object: Objects are used to specify endurant entities other than Geo-Objects and
Geo-Portions such as people and cars. An object has a type and can be stationary or
mobile. In our model, objects are located in Geo-Portions.

-Trafficability: Trafficability is an important concept to consider when reasoning
about movements in a spatial environment. Trafficability depends on Geo-Portions
and objects types. For example, a person may walk to the top of a rocky hill, but not a
car. In our model, an object can cross a Geo-Portion if 1) it has the ability to cross the
Geo-Portion and 2) there is no obstacle in this Geo-Portion. The ability of an object
depends on its physical characteristics and its resources. For example, a person can
cross a river if she knows how to swim or if she can use a boat. The concept of
obstacle depends widely on the context. For example, a truck may be an obstacle if it
has broken down on the road. We define trafficability as a function that uses
knowledge to check if a Geo-Portion is crossable or not for a given object.

Let E be the set of animated living beings (humans, animals, etc.), C be the set of
movement capabilities' types (swimmer, climber, etc.), R be the set of resource
transportation's types (Plane, car, etc.), G be the set of Geo-Portions' types (forest,
etc.) and N be the set of values that may takes the trafficability function (R for
restricted, U for Unrestricted and SR for Severely Restricted [7]). Formally,
trafficability is defined using the following three *Crossable* functions: *Crossable1*:
(G, R) --> N; *Crossable2*: (G, C) --> N; *Crossable3*: (G, E) --> N. *Crossable1* and
Crossable2 are predefined functions (*Crossable1(Forest, Car)* = R, *Crossable2(Lake,
Swimming)* = U, etc.). *Crossable3* is a computed function. An animated living being
can cross a Geo-Portion if this Geo-Portion is not blocked by an obstacle (*Blocked*
predicate) and if the living being possesses the physical capabilities or the transport
resources to do so (*Possess* predicate). Formally, *Crossable3* is defined as follows:
$\forall e \in E$, $\forall g \in G$, Crossable3(g, e) = U $\Leftrightarrow \exists c \in C$ (Crossable2(g, c) = U) $\wedge$ Possess$(e,$
$c) \wedge \neg$ Blocked$(g)) \vee (\exists r \in R$, Crossable1$(g, r) \wedge$ Possess$(e, r) \wedge \neg$ Blocked$(g)))$.

3.2 Dynamic View

We adopt Desclés' definitions of static, cinematic and dynamic situations, and we use
the model of *temporal situation* proposed by Moulin [20]. A temporal situation is
associated with a time interval which characterizes its temporal location on a time
axis. An elementary time interval is specified by a list of parameters, essentially the
begin-time BT, the end-time ET, the time scale TS and the time interval duration DU.
Formally, a temporal situation is a triple <SD, SPC, STI> where:

-The situation description SD is a pair [situation-type, situation-descriptor] used to
identify the temporal situation. The situation type is used to semantically distinguish
different kinds of temporal situations: states, events and processes. The situation
descriptor identifies an instance of a situation and is used for referential purposes.
-The situation propositional content SPC is a non-temporal knowledge structure
described by a conceptual graph. It makes explicit situation's semantic characteristics.

-The situation time interval STI is a structure which aggregates the temporal information associated with the temporal situation.

Temporal situations are related by *temporal relations*. Based on Allen's temporal relations [2] we consider three primitive relations called "BEFORE", "DURING" and "AFTER". Given two time intervals X and Y, the relation BEFORE(X, Y, Lap) holds if we have the following constraints between the begin- and end-times of X and Y compared on a time scale with the operators {>, <, =}: BT(X) < ET(X); BT(Y) < ET(Y); BT(X) < BT(Y); ET(X) < ET(Y); BT(Y) − ET(X) = Lap. The Lap parameter is a real number that measures the distance between the beginning of interval Y and the end of interval X on their time scale. DURING and AFTER relations are defined in the same way.

Using this notation we formalize three kinds of situations: state, event, and process.
-State: We adopt François' definition [8] which distinguishes between the state of the world and the state of an entity. A state of the world is the description of the world at a time instance *t* resulting from the assignment of a *value* to each *entity* and to each relationship between entities of the world. In our model, the entities are objects, Geo-Objects, Geo-Portions and Situations. Different relationships may be defined between these entities. In Section 3.1 we defined relationships between Geo-Objects and Geo-Portions. We define the different kinds of relationships between objects using Vilnat's static primitives (*possession*, etc.) [24]. We define relationships between situations and objects using Desclès' agentive operators which describe the different roles that an object may have in a situation (agent, patient, etc.). Because of space limitations, we do not detail these relationships in this paper. Figure 3 illustrates a simple example of a state. Since a static situation "describes the absence of change where neither begin- nor end-time are observed" [5], temporal information is not represented.

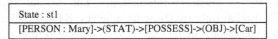

State : st1
[PERSON : Mary]->(STAT)->[POSSESS]->(OBJ)->[Car]

Fig. 3. An example of state specifying that Mary possesses a car

-Event: We adopt Desclès's definition [5]. An event expresses a temporal occurrence that appears in a static background, and that may change or not the state of the world. It marks a break between the "before-event" and the "after-event". Its duration is negligible.
-Process: We adopt Desclès's definition of a process [5]. A process expresses a change initiated by an event that marks the beginning of the process, and may have an end state and a resulting state. A process makes the universe transit from an initial situation corresponding to the "before-process" to a final situation describing the "after-process". In contrast to an event, a process has a non negligible duration, and we can talk about a situation holding during the process. Based on the temporal relations BEFORE, AFTER and DURING, we define three relationships BEFORE-SITUATION, DURING-SITUATION and AFTER-SITUATION corresponding respectively to the initial situation before the beginning of a process, the intermediate situation during the process, and the final situation after the end of the process.

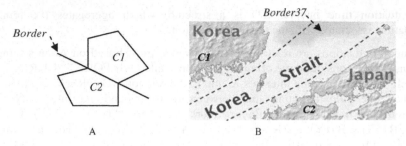

Fig. 4. A) The frontier between A and B may be a line B) or a surface (Photo taken from en.wikipedia.org)

We apply our model to describe movements in space. The interpretation of a dynamic situation of movement as an event or a process depends on the structure of the space. Let us consider the following situation: "Plane P3724 crosses the frontier to go from country *C1* to country *C2*". If the frontier is represented by a line (Fig.4.A), the situation is interpreted as an event. In Fig.5 we present the situation describing the fact that "The plane P3724 takes only an instant (point in time) to cross the frontier in order to go from a source country *c1* to a destination country *c2*". The situation is an Event identified by *ev1*. Its propositional content makes explicit the agent, the location, the source and the destination of the movement. Its time interval parameters are: BT: 10:00:00; ET: 10:00:01; TS: Time; DU: 1 and DS: Second. The event triggers a change from a "before event situation" to an "after event situation". The first situation is a State identified by *st1*. It has only two time parameters: ET: 09:59:59 and TS: time. Its propositional content describes the fact that the plane P3724 is located in the country *c1* and not in the country *c2*. This state is related to the event *ev1* by the Before-Situation relationship. The second situation is a State identified by *st2*. It also has only two time parameters: BT: 10:00:02 and TS: time. Its propositional content describes the fact that P3724 is located in the country *c2* and not in the country *c1*. This state is related to the event *ev1* by the After-Situation relationship.

However, if the frontier is represented by a Geo-Object with a surface, such as the Korea Strait, a 200 Km wide sea passage between South Korea and Japan (Fig. 4.B),

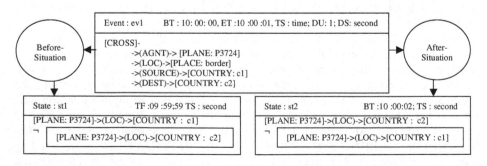

Fig. 5. The movement is interpreted as an event

the situation is interpreted as a process. In Fig.6 we present the situation describing the fact that "The plane P3724 takes 10 minutes to cross the frontier in order to go from a source country *c1* to a destination country *c2*". The situation is a Completed-Process, identified by *cp1*. It has an initial and a final situations specified similarly to those that we presented for the event. In addition, a process has a situation describing the state that holds during its progress. This situation is a State identified by *st3*. It has the same temporal parameters as the process *cp1*. Its propositional content describes the fact that P3724 is located in the place called Border37 and neither in *c1* nor in *c2*. It is related to the process *cp1* by the During-Situation relationship.

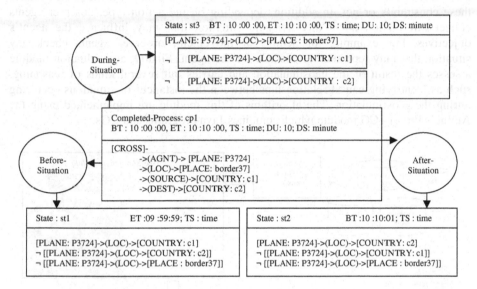

Fig. 6. The movement is interpreted as a process

4   MAGS-COA

We apply our model to represent and reason about dynamic situations in multiagent geosimulations. Our team is developing MAGS-COA, a software that uses a multiagent geosimulation approach to assess courses of actions (COA) (Fig.7). In this system, the user first specifies a scenario of a COA. The scenario describes how and where a set of assets (i.e. planes) must coordinate to achieve a given mission. The scenario is then simulated in a virtual geographic environment [19]. The assets are software agents that try to achieve their objectives, assigned to them by the scenario. We record the unfolding of the geosimulation in a log enhanced by semantic annotations in the form of the dynamic situation model that we presented in Section 3. We use *observer agents* to collect and record information about relevant situations occurring in the geosimulation software environment. The output of the geosimulation is then analyzed in order to produce a qualitative evaluation of the dynamic execution of the scenario, using domain knowledge. More specifically, if an agent fails to reach its goal, we trace back the possible reasons. We use an enhanced version of MAGS

[19] as a multi-agent geosimulation environment and the Amine platform [12] to specify domain knowledge and as an inference engine based on Prolog+CG. Using Amine's CG formalism, we create the ontologies of our different concepts (geospatial entities, dynamic situations, objects, domain actions, etc.) and their constraints (such as, for example, "an action of movement requires that the *agent* must be mobile and the Geo-Portion must be crossable"). We also use the Amine platform to represent and reason about different kinds of constraints specified in the scenario. For example, let us suppose that a plane must go to a destination without coming close to some areas. First, the user identifies relevant danger areas for the plane. Second, for these areas, she affects observer agents to check during the simulation if the plane violates these constraints or not. In addition, depending on the action type, observer agents collect information about any dynamic situation that may influence the agent's objectives. For example, for actions of movement, observer agents check any situation that may represent an obstacle for the agent. Finally, an evaluation module assesses the result of the geosimulation and carries out several kinds of reasoning, such as identifying causal relationships between the instances of situations occurring during the geosimulation. The algorithms of this module are implemented using the Amine's Prolog+CG module which combines Java programs and CGs.

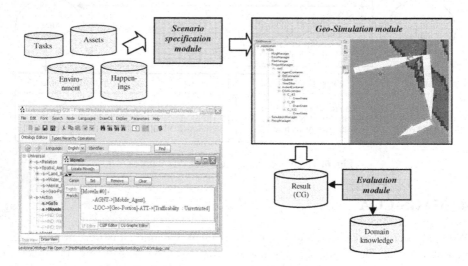

Fig. 7. General workflow of MAGS-COA

5 Conclusion and Future Works

In this paper we presented a model based on linguistic cognitive archetypes, ontological definitions of geographic space and CGs to represent dynamic phenomena in a geographic environment. We presented the linguistic foundations of our model and we argued how appropriate they are in the context of modeling dynamic situations in a spatial environment. We introduced the endurant and dynamic views of our model and their main concepts. In a next step, we presented an overview of how this model is applied to the multiagent geosimulation domain in the MAGS-COA project. In recent years, the solutions that have been proposed to model dynamic phenomena in a geographic space have been proposed by the GIS and geographic ontologies research's communities. Hornsby and Egenhofer [11] used the notion of *lifeline* as a way to track objects through series of changes conceived in terms of transitions between successive states, and a similar approach is used by Frank [9]. In these approaches, "changes are not considered as entities in their own right" [10]. In addition, these works do not use a knowledge representation that allows to reason about geographic phenomena. Worboys studied the construction of a modeling approach for dynamic geospatial domains from an information system perspective [25, 26]. He proposed a model based on objects and events that extends the traditional object-based paradigm. Our model is original for two main reasons. First we use a linguistic approach to qualitatively model dynamic situations in a geographic environment. Second, we use CGs as a knowledge representation to describe such situations. In fact, CGs have been rarely applied to qualitative spatial reasoning, and we do not know of any work that uses them in the context of representing and reasoning about multiagent geosimulations as we do. In contrast to the other proposed models, the use of CGs makes our model computationally feasible and useable to carry out qualitative spatial reasoning.

Using the MAGS-COA software we are currently applying our model to simulate and evaluate a real scenario of search and rescue of planes in the north west of Ontario, Canada.

Acknowledgments

This research is supported by GEOIDE, the Canadian Network of Centers of Excellence in Geomatics (MUSCAMAGS Project), by the Canadian National Sciences and Engineering Research Council, and by Valcartier Defence R&D Canada (DRDC), Quebec, Canada (MAGS-COA Project).

References

1. Adams, E.: Surfaces and Superposition: Field Notes on some Geometrical Excavations. Center for the Study of Language and Information. Lecture Notes (2001)
2. Allen, J.F.: Maintaining Knowledge about temporal intervals. Communications of the ACM 26, 832–843 (1983)
3. Cohn, A., Hazarika, S.: Qualitative Spatial Representation and Reasoning: An Overview. Fundamental Informaticae 46(1-2), 1–29 (2001)
4. Desclés, J.-P.: Langages applicatifs, langues naturelles et cognition. Hermès, Paris (1990)
5. Desclès, J.-P.: Quelques Concepts Relatifs au Temps et à l'Aspect pour l'Analyse des Textes. In: Stanislaw et al. (eds.) Études Cognitives: Sémantique des Catégories d'aspect et de Temps, vol. 1. Warszawa, pp. 57–88 (1994)
6. Fonseca, F., Egenhofer, M., Davis, C., Câmara, G.: Semantic Granularity in Ontology-Driven Geographic Information System. Annals of Mathematics and Artificial Intelligence 36(1-2), 121–151 (2002)
7. Forbus, K., Usher, J., Chapman, V.: Qualitative Spatial Reasoning about Sketch Maps. In: Proceedings of the 15th annual CIAAI. Acapulco, Mexico (2003)

8. François, J.: Prédications d'état, d'événement et d'action: esquisse d'une méta classification du sémantisme verbal. Exposé original présenté devant le groupe interdisciplinaire Cognition et Langage (Paris 8), le 9 novembre (1988)

9. Frank, A.: Ontology for Spatio-Temporal Databases. In: Sellis, T., Koubarakis, M., Frank, A., Grumbach, S., Güting, R.H., Jensen, C., Lorentzos, N.A., Manolopoulos, Y., Nardelli, E., Pernici, B., Theodoulidis, B., Tryfona, N., Schek, H.-J., Scholl, M.O. (eds.) Spatio-Temporal Databases. LNCS, vol. 2520, Springer, Heidelberg (2003)

10. Grenon, P., Smith, B.: SNAP and SPAN: Towards dynamic spatial ontology. Spatial Cognition and Computation 4(1), 69–103 (2004)

11. Hornsby, K., Egenhofer, M.: Modeling Moving Objects over Multiple Granularities. Annals of Mathematics and Artificial Intelligence, vol. 36, pp. 177–194. Kluwer, Dordrecht (2002)

12. Kabbaj, A., Bouzouba, K., El Hachimi, K., Ourdani, N.: Ontologies in Amine Platform: Structures and Processes. In: Schärfe, H., Hitzler, P., Øhrstrøm, P. (eds.) ICCS 2006. LNCS (LNAI), vol. 4068, pp. 300–313. Springer, Heidelberg (2006)

13. Kettani, D., Moulin, B.: A Spatial Model Based on the Notions of Spatial Conceptual Map and of Object's Influence Areas. In: Proc. of the International Conference on Spatial Information Theory: Cognitive and Computational Foundations of Geographic Information Science (1999)

14. Kowalski, R.A., Sergot, M.J.: A logic-based calculus of events. New Generation Computing 4, 67–95 (1986)

15. Lyons, A.: Sémantique linguistique. Paris: Larousse. vol. 2 de Semantics (1980)

16. McDermott, D.: A temporal logic for reasoning about processes and plans. Cognitive Science. 6, 101–155 (1982)

17. Miller, H.: What about people in geographic information science? Computers, Environment and Urban Systems 27(5), 447–453 (2003)

18. Miller, H., Wentz, E.: Representation and spatial analysis in geographic information systems. Annals of the Association of American Geographers. 93(3), 574–594 (2003)

19. Moulin, B., Chaker, W., Perron, J., Pelletier, P., Hogan, J., Gbei, E., MAGS,: Project: Multi-agent geosimulation and crowd simulation. In: Kuhn, W., Worboys, M.F., Timpf, S. (eds.) COSIT 2003. LNCS, vol. 2825, pp. 151–168. Springer, Heidelberg (2003)

20. Moulin, B.: Temporal contexts for discourse representation: An extension of the conceptual graph approach. Applied Intelligence. 7, 225–227 (1997)

21. Price, R., Tryfona, N., Jensen, C.S.: Modeling Topological Constraints in Spatial Part-Whole Relationships. In: Kunii, H.S., Jajodia, S., Sølvberg, A. (eds.) ER 2001. LNCS, vol. 2224, pp. 27–40. Springer, Heidelberg (2001)

22. Pullar, D.: Integrating dynamic spatial models with discrete event simulation. In: Proc. Simulation Technology and Training (SimTechT2000) Conf. Sydney (2000)

23. Salles, P., Bredeweg, B.: Modeling population and community dynamics with qualitative reasoning. Ecological Modeling journal, 114-128 (2006)

24. Vilnat, A.: Dialogue et analyse de phrases. Mémoire préparé en vue de l'obtention de l'Habilitation à diriger des recherches. Groupe Langues, Information, Représentations, Université Paris-Sud, Paris (2005)

25. Worboys, M., Hornsby, K.: From objects to events: GEM, the geospatial event model. In: Egenhofer, M.J., Freksa, C., Miller, H.J. (eds.) GIScience 2004. LNCS, vol. 3234, pp. 327–344. Springer, Heidelberg (2004)

26. Worboys, M.: Event-oriented approaches to geographic phenomena. International Journal of Geographical Information Science. 19(1), 1–28 (2005)

A Datatype Extension for Simple Conceptual Graphs and Conceptual Graphs Rules

Jean-François Baget

INRIA Rhône-Alpes & LIG / LIRMM
jean-francois.baget@inrialpes.fr

Abstract. We propose in this paper an extension of Conceptual Graphs that allows to use datatypes (strings, numbers, ...) for typing concept nodes. Though the model-theoretic semantics of these datatypes is inspired by the work done for RDF/RDFS, keeping sound and complete projection-based algorithms for deduction has led to strong syntactic restrictions (datatyped concept nodes of a target graph cannot be generic). This restriction, however, allows us to smoothly upgrade our extension to rules, and to introduce functional relations (that compute the value of datatyped concept nodes) while keeping sound and complete reasonings.

1 Introduction

Many extensions of simple conceptual graphs [Sow84] have answered the necessity to use datatypes (strings, integers) to type concept nodes. Among these extensions, actors [LM98] also use procedural relations between nodes. We study such an extension in this paper, the originality of our approach resides in our definition of model theoretic semantics for this datatype extension of simple conceptual graphs. As in [Hay04], values of a datatype will be interpreted by themselves and, as in SPARQL [PS06], a procedural relation between values will be evaluated to true when the result of the application of that procedure returns a result consistant with these values. We present two different semantics for these graphs: though the full semantics (where all interpretations contain all possible values) are more interesting from a knowledge representation point of view, partial semantics will lead to decidable (NP-complete) reasonings akin to the ones used in the simple conceptual graphs formalism. This projection-like mechanism allows to extend the proposed formalism with rules of form "if ... then ... ".

2 D-Graphs: Syntax

Datatype simple conceptual graphs extend simple conceptual graphs by allowing access to types and procedures from a programming language library. Instead of defining exactly this library (an important work for normalization), we decided to characterize the properties it must satisfy for our results to remain valid.

2.1 Programming Language Library

We consider a *library* $\mathcal{L}$, written in some programming language (we have adopted a "Scheme-like" terminology, borrowed from [KCE98]), allowing access to:

U. Priss, S. Polovina, and R. Hill (Eds.): ICCS 2007, LNAI 4604, pp. 83–96, 2007.

Strings and identifiers. Two infinite, countable, disjoint sets $\mathcal{I}$ and $\mathcal{S}$, respectively of *identifiers* and *strings*. The precise rules for forming identifiers and strings depend upon the programming language used. We consider here that an identifier is a sequence of letters, digits and extended alphabetic characters (such as !, ?, but not the double quote "); and that a string is a sequence of characters enclosed in double quotes. By example, `Number`, `Integer`, `Boolean`, `String`, `+`, `not`, `Person`, `John` are identifiers, and `"7"`, `"true"` and `"Bob"` are strings.

Types, values and procedures. Three infinite, countable, pairwise disjoint sets $\mathcal{T}$, $\mathcal{V}$ and $\mathcal{P}$, respectively of *types* (or classes), of *values* (or objects), and of *procedures* (or methods). Though classes are objects in most object-oriented programming languages, we impose disjointness to avoid self-containing sets in models of the library $\mathcal{L}$ (see Sect. 3.1). This restriction can be easily enforced in any library by considering only a subset of the classes and objects available.

Binding identifiers to types and procedures. We consider a bijection $bind : \mathcal{I}' \subseteq \mathcal{I} \rightarrow \mathcal{T} \cup \mathcal{P}$. Identifiers bound to types are called *type identifiers* (we note $bind^{-1}(\mathcal{T}) = \mathcal{I}_T$) and identifiers bound to procedures are called *procedure identifiers* (we note $bind^{-1}(\mathcal{P}) = \mathcal{I}_P$). In the previous example, `Number`, `Integer`, `Boolean` and `String` are type identifiers, and `+` and `not` are procedure identifiers.

Functions is-a? and eqv?. We consider an equivalence relation over values (encoding that they should normally be regarded as the same) and a relation over $\mathcal{V} \times \mathcal{T}$ (encoding that a value belongs to, is an instance of, a type). They are respectively computed by the functions `is-a?` $: \mathcal{V} \times \mathcal{T} \rightarrow \{\#t, \#f\}$ and `eqv?` $: \mathcal{V} \times \mathcal{V} \rightarrow \{\#t, \#f\}$. Note that these functions are not procedures of $\mathcal{P}$.

Specification 1 (Booleans). *The library $\mathcal{L}$ must satisfy these properties:*

- `Boolean` *is a type identifier of* $\mathcal{I}$*, and we note* $B = $ `bind(Boolean)`*;*
- $\#t$ *and* $\#f$ *are two values of* $\mathcal{V}$ *(they correspond to "true" and "false");*
- `is-a?`$(\#t, B) = \#t$ *and* `is-a?`$(\#f, B) = \#t$*;*
- $\forall x \in \mathcal{V},$ `is-a?`$(x, B) = \#t \Rightarrow ($`eqv?`$(x, \#t) = \#t$ *or* `eqv?`$(x, \#f)) = \#t$*.*

Specification 2 (About eqv?). *Let x and y be two equivalent values of $\mathcal{V}$ (i.e. such that* `eqv?`$(x, y) = \#t$*). Then* $\forall t \in \mathcal{T},$ `is-a?`$(x, t) = \#t \Leftrightarrow$ `is-a?`$(y, t) = \#t$*.*

Signature and application of a procedure. Two mappings `isig` $: \mathcal{P} \rightarrow \mathcal{T}^+$ and `osig` $: \mathcal{P} \rightarrow \mathcal{T}$ respectively define the *input* and *output signatures* of a procedure p. The *arity* of p is the size of `isig`(p). A procedure can be applied to values in order to obtain another value, Thanks to the *partial* function `apply` $: \mathcal{P} \times \mathcal{V}^+ \rightarrow \mathcal{V}$. The functions `isig`, `osig` and `apply` are not procedures of $\mathcal{P}$.

Specification 3 (Signature). *Let $p \in \mathcal{P}$ be a procedure,* `isig`$(p) = (t_1, \ldots, t_k)$ *and* `osig`$(p) = t$*. Then* $\forall (v_1, \ldots, v_k) \in \mathcal{V}^k$*, with* `is-a?`$(v_i, t_i) = \#t$ *for* $1 \le i \le k$*,* `apply`$(p, v_1, \ldots, v_k)$ *is defined and returns a value v such that* `is-a?`$(v, t) = \#t$*. If* $\exists\, 1 \le i \le k$ *such that* `is-a?`$(v_i, t_i) = \#f$*,* `apply`$(p, v_1, \ldots, v_k)$ *is not defined.*

Note that `apply` is a (partial) function in a mathematical sense: the value it may return is entirely determined by the parameters p and v_i.

Specification 4 (Equivalent parameters). *Let* $p \in \mathcal{P}$ *be a procedure,* $\text{isig}(p)$ $= (t_1, \ldots, t_k)$ *and* $\text{osig}(p) = t$. *Then* $\forall (v_1, \ldots, v_k) \in \mathcal{V}^k$, $\forall (v'_1, \ldots, v'_k) \in \mathcal{V}^k$, *with* $\text{is-a?}(v_i, t_i) = \#t$ *and* $\text{eqv?}(v_i, v'_i) = \#t$ *for* $1 \leq i \leq k$, $\text{apply}(p, v_1, \ldots, v_k) = v$ *and* $\text{apply}(p, v'_1, \ldots, v'_k) = v' \Rightarrow \text{eqv?}(v, v') = \#t$ *(applying a procedure on equivalent parameters must yield an equivalent result).*

External representation, read and write. A partial function $\text{read} : \mathcal{I}_T \times \mathcal{S} \to \mathcal{V}$ builds, when possible, a value v belonging to a type t from the type identifier bound to t and a string s (s is then called an *external representation* of v for the type t). The partial function $\text{write} : \mathcal{I}_T \times \mathcal{V} \to \mathcal{S}$ is used to obtain one of these external representations (called the *canonical representation*) from a value. As before, none of these functions are procedures of $\mathcal{P}$.

Specification 5 (External representations). *The following properties must be satisfied by the functions* `read` *and* `write`:

- *Let* n *be a type identifier of* $\mathcal{I}_T$ *and* s *be a string of* $\mathcal{S}$. *Then* $\text{read}(n, s)$ *is defined and returns a value* $v \Rightarrow \text{is-a?}(v, \text{bind}(n)) = \#t$.
- *Let* t *be a type of* $\mathcal{T}$ *and* v *be a value of* $\mathcal{V}$. *Then* $\text{write}(t, v)$ *is defined and returns a value* s *if and only if* $\text{is-a?}(v, t) = \#t$. *In that case, we have* $\text{equiv?}(v, \text{read}(\text{bind}^{-1}(t), s)) = \#t$.
- *if* v *and* v' *are two values of* $\mathcal{V}$ *such that* $\text{eqv?}(v, v') = \#t$ *and* $\text{is-a?}(v, t) = \#t$, *then* $\text{write}(t, v) = \text{write}(t, v')$.

2.2 Vocabulary

The vocabulary (or canon, support) is the structure traditionally used to encode the ontological knowledge in the conceptual graphs formalism.

Definition 1 (Vocabulary). *A vocabulary is a tuple* $((T_C, \leq_C), (T_R^1, \leq_1), \ldots, (T_R^k, \leq_k))$ *where* $T_C, T_R^1, \ldots, T_R^k$ *are pairwise disjoint sets of identifiers (respectively of concept types, relation types of arity 1 to k, and* $\leq_C, \leq_1, \ldots, \leq_k$ *are partial orders on these sets.*

Unlike concept types, type identifiers of $\mathcal{L}$ are not explicitly ordered by a subtype relation. As shown in Sect. 3.1, type identifiers are interpreted by sets that are uniquely determined by $\mathcal{L}$. Inclusion of these sets thus determines the subtype relation. However, expliciting this relation could cause a difference (in case of a bad programming of $\mathcal{L}$) between the interpretations of type identifiers and the interpretation of the (redundant) subtype relation.

Finally, we want to be able to decide if an identifier is a type of the vocabulary or a type or procedure identifier in the library:

Definition 2 (Compatibility). *A vocabulary V and a library $\mathcal{L}$ are compatible if their sets of identifiers are disjoint (i.e.* $(T_C \cup T_R^1 \cup \ldots T_R^k) \cap (\mathcal{I}_T \cup \mathcal{I}_P) = \emptyset$).

Note that it is always possible to obtain compatibility between the vocabulary and the library, up to a renaming of identifiers.

2.3 D-Graphs

Definition 3 (D-graph). *A datatyped simple conceptual graph (D-graph), defined on a vocabulary $\mathcal{V}$ and a compatible library $\mathcal{L}$, is a tuple $G = (E, R, \gamma, \tau, \mu)$ where E and R are two disjoint sets (respectively of* entities *and* relations*), γ : $R \rightarrow E^+$ maps each relation to a tuple of entities (its* arguments*), if $\gamma(r) = (e_1, \dots, e_k)$ we say that* $\mathrm{degree}(r) = k$*, and τ and μ are mappings s.t.:*

- *$\forall e \in E, \tau(e)$ (the* type *of e)is either a concept type of T_C (as usually done for simple CGs, e is then called a* concept node*) or a type identifier of $\mathcal{I}_T$ (e is then called a* datatype node*);*
- *$\forall r \in R, \tau(r)$ (the* type *of r) is either a relation type of arity $\mathrm{degree}(r)$ (as usually done for simple CGs, r is then called a* standard relation*) or a procedure identifier n of $\mathcal{I}_P$ such that $|\mathrm{isig}(n)| = \mathrm{degree}(r) - 1$ (in that case, r is called a* computed relation*). If r is a computed relation, then all its arguments must be datatype nodes.*
- *$\forall e \in E, \mu(e)$ (the* marker *of e) can be the generic symbol $*$, which is not an identifier (e is characterized as* generic*) or, if e is a datatype node, a string of $\mathcal{S}$, or, if e is a concept node, an identifier of $\mathcal{I}$ that is not bound in $\mathcal{L}$ and is not a type of $\mathcal{V}$ (in both cases, the node e is characterized as* individual*).*

A D-graph $G = (E, R, \gamma, \tau, \mu)$ is graphically represented as follows: as for simple CGs, concept nodes are represented by rectangles and standard relations by ovals; we represent datatype nodes by parallelograms and computed relations by diamonds (as a reminder of actors). If e is an individual entity, then we write $\tau(e) : \mu(e)$ inside the shape representing e, otherwise we only write $\tau(e)$. If r is a relation, we write $\tau(r)$ inside the shape representing r. If $\gamma(r) = (e_1, \dots, e_k)$, we draw a line between the shape representing r and each of the shapes representing the e_j. If r is a standard relation, these lines are numbered from 1 to k, according to the place of the entity in $\gamma(r)$, if r is a computed relation, then the lines are oriented from "input entities" to relation and relation to "output entity", and only the first $k - 1$ arrows are numbered. Fig. 1 represents such a D-graph. It asserts that it exists a person whose age is greater than 27, the age of Bob. It can be simplified by omitting the output entity with marker "true" of a computed relation whose type is a procedure identifier bound to a procedure with output signature bound by Boolean.

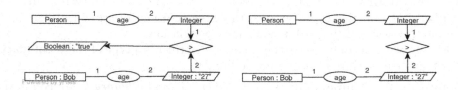

Fig. 1. Two graphical representations of a D-graph G

3   D-Graphs: Semantics

3.1   Interpretations and Models

We extend the usual model-theoretic semantics of simple conceptual graphs (see [ABC06]) to take into account the programming language library. We first define interpretations, then the conditions under which an interpretation is a model of a vocabulary, a library, or a D-graph.

Definition 4 (Interpretation). *Let $S \subseteq \mathcal{S}$ be a set of strings and $\mathcal{I}$ be the set of identifiers. An* interpretation *of $(\mathcal{I}, S)$ is a triple $I = (\Delta, \iota, \sigma)$ where Δ is a non empty set called the* interpretation domain, *and ι and σ are the* interpretation functions; *$\iota : \mathcal{I} \to \Delta \cup 2^{\Delta} \cup 2^{\Delta \times \Delta} \cup \ldots \cup 2^{\Delta^k}$ maps each string to an element of the domain or a set of tuples of elements of the domain; and $\sigma : \mathcal{I} \times S \to \Delta$ is a partial mapping. I is said* full *if $S = \mathcal{S}$.*

Definition 5 (Model of a vocabulary). *Let $S \subseteq \mathcal{S}$ be a set of strings, and $\mathcal{I}$ be the set of identifiers. An interpretation $I = (\Delta, \iota, \sigma)$ of $(\mathcal{I}, S)$ is a model of the vocabulary $V = ((T_C, \leq_C), (T_R^1, \leq_1), \ldots, (T_R^k, \leq_k))$ (we note $(\mathcal{I}, S) \models V$) iff:*

- *if c is a concept type of T_C, then $\iota(c) \in 2^{\Delta}$;*
- *if r is a relation type of T_R^i, then $\iota(r) \in 2^{\Delta^i}$;*
- *if t and t' are two concept or relation types such that $t \leq t'$ (where $\leq$ can be $\leq_C$ or one of the $\leq_i$), then $\iota(t) \subseteq \iota(t')$;*

Note that this definition is a generalization of the usual model-theoretic semantics of the vocabulary of CGs: $I = (\Delta, \iota, \sigma)$ is a model of V if and only if (Δ, ι) is a model of V (in the usual sense of [ABC06]).

Definition 6 (Model of a library). *Let $S \subseteq \mathcal{S}$ be a set of strings, and $\mathcal{I}$ be the set of identifiers. An interpretation $I = (\Delta, \iota, \sigma)$ of $(\mathcal{I}, S)$ is a model of the library $\mathcal{L}$ (and we note $(\mathcal{I}, S) \models \mathcal{L}$) if and only if:*

- *$\forall s \in S$, $\forall i \in \mathcal{I}$, $\sigma(i, s)$ is defined if and only if i is a type identifier of $\mathcal{I}_T$ and* **read**$(i, s) = v$ *is defined. In that case, $\sigma(i, s) =$ **read**$(i, $**write**$($**bind**$(i), v))$ (the value associated with the canonical representation of all equivalent ones).*
- *$\forall t \in \mathcal{I}_T, \iota(t) = \{\delta \in \Delta \mid$ **is-a?**$(\delta,$ **bind**$(t)) = \#t\}$.*
- *$\forall p \in \mathcal{I}_P$, with* **isig**$($**bind**$(p)) = (t_1, \ldots, t_k)$, $\iota(p) = \{(\delta_1, \ldots, \delta_k, \delta) \in \Delta^{k+1}$ *s.t.* **apply**$(p, \delta_1, \ldots, \delta_k) = \delta\}$.

Definition 7 (Models of a D-graph). *Let $S \subseteq \mathcal{S}$ be a set of strings, and $\mathcal{I}$ be the set of identifiers. An interpretation $I = (\Delta, \iota, \sigma)$ of $(\mathcal{I}, S)$ is a model of the D-graph $G = (E, R, \gamma, \tau, \mu)$ (and we note $(\mathcal{I}, S) \models G$) if and only if there exists a mapping $\alpha : E \to \Delta$ (alpha is called a* proof *of G in I) such that:*

- *for each individual concept node $e \in E$, $\alpha(e) = \iota(\mu(e)) \in \iota(\tau(e))$;*
- *for each individual datatype node $e \in E$, $\alpha(e) = \sigma(\tau(e), \mu(e))$;*
- *for each generic node $e \in E$, $\alpha(e) \in \iota(\tau(e))$;*
- *for each relation $r \in R$, $\gamma(r) = (e_1, \ldots, e_k) \Rightarrow (\alpha(e_1), \ldots, \alpha(e_k)) \in \iota(\tau(r))$.*

Note that if a D-graph is a simple conceptual graph (it has only concept nodes and standard relations), these constraints correspond to the usual constraints on models of a simple conceptual graph.

3.2 Full vs. Partial Semantics

Definition 8 (Satisfiability, validity, consequence). *Let G and G' be two D-graphs defined over a vocabulary V and a compatible library $\mathcal{L}$. An interpretation I of $(\mathcal{I}, S)$ covers G if every string labelling nodes of G belong to S. Then:*

- *G is* partially *(resp.* fully*) satisfiable if there exists an interpretation (resp. a full interpretation) I covering G such that $I \models V$, $I \models \mathcal{L}$ and $I \models G$. Otherwise it is said* partially *(resp.* fully*) unsatisfiable.*
- *G is* partially *(resp.* fully*) valid if for any interpretation (resp. full interpretation) I covering G, $I \models V$ and $I \models \mathcal{L} \Rightarrow I \models G$. Otherwise it is said* partially *(resp.* fully*) invalid.*
- *G is a* partial *(resp.* full*) consequence of G' if for any interpretation (resp. full interpretation) I covering G and G', $I \models V$, $I \models \mathcal{L}$ and $I \models G' \Rightarrow I \models G$. We note $G' \vdash G$ (resp $G' \Vdash G$).*

With simple conceptual graphs, there is no distinction between full and partial semantics (since there is no string). Every graph is satisfiable, and the only valid graph is the empty graph. Things are more complicated with D-graphs.

Property 1. Let G and G' be two D-graphs defined over a vocabulary V and a compatible library $\mathcal{L}$. Then (since full interpretations are interpretations):

1. G is fully satisfiable $\Rightarrow$ G is partially satisfiable;
2. G is partially valid $\Rightarrow$ G is fully valid;
3. $G' \vdash G \Rightarrow G' \Vdash G$.

There are more consequences with full semantics, and they are relevant (according to the library). To explain the differences between these semantics, let us point out that full interpretations interpret all strings of S, and since all values of $\mathcal{V}$ have an external representation, full interpretations impose $\mathcal{V}$ (more precisely equivalence classes for `eqv?`) to be a subset of the domain of interpretation Δ. With partial semantics, some values may not belong to Δ.

Let us consider the D-graphs of Fig. 2. The empty D-graph G_3 is both partially valid and fully valid. The D-graph G_2 (there exists an integer) is fully valid because Δ contains, by example, the value whose external representation is "7" and α can map the node of G_2 to that value. However, G_2 is not partially valid, since Δ may not contain any value v such that `is-a?`$(v, \text{bind}(\text{Integer})) = \#t$. The D-graph G_1 ("7 is an integer") is both partially valid and fully valid: even if

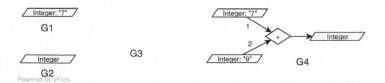

Fig. 2. Examples of D-graphs

an interpretation is not full, it must cover G_1, and thus the value whose external representation is "7" (a string of G_1) belongs to Δ. Finally, the D-graph G_4 ("there exists an integer that is the sum of 7 and 9") is fully valid, but is not partially valid (the value represented by "16" may not belong to Δ).

Property 2. Let G and G' be two D-graphs defined over a vocabulary V and a compatible library $\mathcal{L}$. Then G is fully satisfiable $\Leftrightarrow G$ is satisfiable.

Proof. Thanks to Prop. 1, we only have to prove that G satisfiable $\Rightarrow G$ fully satisfiable. Since $G = (E, R, \gamma, \tau, \mu)$ is satisfiable, there exists a partial interpretation $I = (\Delta, \iota, \sigma)$ and a mapping $\alpha : E \rightarrow \Delta$ that satisfies the constraints of Def. 7. The partial interpretation I can be "completed" to a full interpretation $I' = (\Delta', \iota', \sigma')$ (by adding all values of V to Δ, while respecting the constraints of Def. 6), and $\alpha : E \rightarrow \Delta \subseteq \Delta'$ still satisfies all constraints of Def. 7. $\qquad \Box$

As seen in these examples, full semantics are more interesting than partial semantics. However, they are a tremendous source of computational complexity.

4 Inference Mechanisms for D-Graphs

4.1 Full Validity of D-Graphs

Though we are mainly interested in computating consequence, we discuss the problems linked to computing full validity (a particular case of full consequence, since G is fully valid iff G is the full consequence of the empty D-graph).

Some examples of full validity computation. Let us first consider an "easy example", the D-graph G_4 of Fig. 2. To prove that G_4 is partially valid, we have to prove that there is an integer that is the sum of 7 and 9: we only have to use the library to compute $\texttt{apply}(\texttt{bind}(+), \texttt{read}(\texttt{Integer},"7"), \texttt{read}(\texttt{Integer},"9"))$. If $\texttt{bind}(+)$ is decidable, then this method is decidable.

Let us now consider the D-graph G_1 of Fig. 3 (there exists an integer that, added to 7, has result 9). To prove its full validity, we cannot use $\texttt{apply}$ as in the previous example. And since the library is a "black box", we cannot guess that this integer can be obtained by substracting 7 to 9. So we have to try to find a value v in $V \subseteq \Delta$ such that $\texttt{apply}(\texttt{bind}(+), \texttt{read}(\texttt{Integer},"7"), v) = \texttt{read}(\texttt{Integer},"9")$. Though we do not have a direct access to values of V ("black box", again), we can enumerate this countable, infinite set by enumerating all strings s of $\mathcal{S}$ and computing the value, if it exists $\texttt{read}(\texttt{Integer}, s)$. As soon as we enumerate the string "2", we can assert that G_1 is fully valid.

Finally, let us now try to prove that the D-graph G_2 (there exists an integer that is the result of itself added to 7) of Fig. 3 is fully valid. As in the previous example, we enumerate strings until a satisfying value is found, but, this time, no such value exists: the algorithm will run forever...

Characterization of full validity

Definition 9 (Pure D-graph). *A pure D-graph is a D-graph whose only entities are datatype nodes and whose only relations are computed relations.*

Fig. 3. Examples of D-graphs

Property 3. Only pure D-graphs can be valid or fully valid.

Proof. If a D-graph G is not pure, we can build a full interpretation that is a model of V and $\mathcal{L}$, but not a model of G: build a full interpretation that is a model of $\mathcal{L}$, then for each concept or relation type t of V, define $\iota(t) = \emptyset$. □

Theorem 1 (Characterization of full validity). *Let G be a pure D-graph defined over a vocabulary V and a compatible library $\mathcal{L}$. Then G is fully valid if and only if there exists a (pure) D-graph G' obtained from G by replacing all generic markers in G by strings of S such that:*

- *for each entity e of G', $f(e) = \mathtt{read}(\tau(e), \mu(e))$ is defined;*
- *for each relation r of G', with arguments $\gamma(r) = (e_1, \ldots, e_k, e)$, we have* $\mathtt{eq?}(\mathtt{apply}(\mathtt{bind}(\tau(r)), f(e_1), \ldots, f(e_k)), f(e)) = \#t.$

This is a generalization to any pure D-graph of the algorithm explained for the D-graphs of Fig. 3. However, this algorithm requires the enumeration of all tuples of strings in $\mathcal{S}^k$, where k is the number of generic entities in G. As suggested by the first example (D-graph G_4 of Fig. 2), there is a way to reduce the number of strings to be tested.

Definition 10 (Resolving a D-graph). *Let G be a D-graph, and r be a determined relation of G, i.e. a computed relation of G whose input arguments $e_1, \ldots, e_k$ are individual datatype nodes and whose output argument is a generic datatype node e. Then $\mathrm{resolve}(G, r)$ is the graph obtained from G by replacing the generic marker of e by the string $\mathtt{write}(\mathtt{bind}(\tau(r)), \mathtt{apply}(\mathtt{bind}(\tau(r)), \mathtt{read}(\tau(e_1)),$ $\mu(e_1)), \ldots, \mathtt{read}(\tau(e_k), \mu(e_k)))).$*

Property 4. G and $G' = \mathrm{resolve}(G, r)$ are fully equivalent ($G \Vdash G'$ and $G' \Vdash G$), G is a consequence of G', but G' is not necessarily a consequence of G.

As an immediate consequence, we can compute full validity of a D-graph G by applying, as long as there exists a determined relation, a succession of resolve, then compute full validity of the obtained D-graph G' (where G' is called a resolve of G). Finally, if G can be resolved into a D-graph where all datatype nodes are individual, and if all computed relations are decidable, full validity is decidable.

4.2 Satisfiability of D-Graphs

Fig. 4 shows two D-graphs that are not satisfiable. The D-graph G_1 expresses that "the sum of 2 and 7 is 11" and the graph G_2 that "there is an integer that is

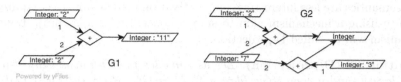

Fig. 4. Examples of D-graphs

both the sum of 2 and 7 and the sum of 3 and 7". This latter example highlights the main difference between our approach and the one used in conceptual graphs actors: though actors consider it as a "convergence problem", our semantics-based approach defines this D-graph as unsatisfiable.

Property 5. A D-graph G is satisfiable iff its pure D-graph restriction G' (obtained by removing all concept nodes and all standard relations) is satisfiable.

Proof. ($\Rightarrow$) is immediate, the restriction of a proof of G in I is a proof of G' in I. For ($\Leftarrow$), let us now consider a proof α of G' in an interpretation (Δ, ι, σ). Then α is a proof of G' in $(\mathcal{V}, \iota, \sigma)$. We build a proof α' of G in $I' = (\Delta', \iota', \sigma')$ where Δ' is the union of $\mathcal{V}$ and of concept nodes of G, and define, for all concept types t, $\iota(t) = \Delta'$, and, for all relation types r of arity k, $\iota(r) = \Delta'^k$.   □

Property 6. A pure D-graph G is either fully valid or unsatisfiable.

Proof. The truth value of a pure D-graph is only determined by the library.   □

We conclude that, as for full validity, satisfiability is decidable when the D-graph G can be resolved into a D-graph G' where all datatype nodes are individual, and computed relations are decidable.

4.3   Partial and Full Consequences of D-Graphs

D-graphs consequence is the problem we are most interested in: is a D-graph Q (a query) consequence of a D-graph, or a set of D-graphs (the knowledge base), or is there an answer to Q in a given knowledge base ?

Problems with full consequence. A D-graph G is fully valid iff it is the full consequence of the empty D-graph $\emptyset$. So a sound and complete algorithm for full consequence must tackle with full validity of a query. However, we have seen (in Sect. 4.1) that a condition to be able to compute full validity was that G could be resolved into a D-graph without any generic datatype node. To compute full consequence, this condition should be translated to a restriction on queries, which we believe to strong: the only generic datatype nodes would be determined by the query itself, a useless feature in the query language.

An extension of projection. As discussed above, full semantics would lead either to undecidable calculus of consequence, or unacceptable syntactic restrictions for queries. This is why we focus on partial semantics of D-graphs. Though

these semantics are less interesting, our motivation is to be able to compute consequence using a mechanism akin to simple conceptual graphs projection; this
mechanism will be called D-projection.

Definition 11 (D-projection). *Let us consider two D-graphs defined over a
vocabulary V and a compatible library $\mathcal{L}$, $G = (E_G, R_G, \gamma_G, \tau_G, \mu_G)$ and $H = (E_H, R_H, \gamma_H, \tau_H, \mu_H)$. A D-projection from H into G is a mapping π from
the entities of H onto the entities of G such that (where we note $\rho(H, c) = \mathtt{read}(\tau_H(c), \mu_H(c))$):*

- *for each concept node c of H, $\tau_G(c) \leq_C \tau_H(c)$;*
- *for each individual concept node c of H, $\mu_H(c) = \mu_G(\pi(c))$;*
- *for each individual datatype node c of H, $\mathtt{eq?}(\rho(H, c), \rho(G, \pi(c))) = \#t$;*
- *for each generic datatype node c of H, $\mathtt{eq?}(\mathtt{read}(\tau_H(c), \mu_G(\pi(c))), \rho(G, \pi(c))) = \#t$;*
- *for each standard relation r of H, s.t. $\gamma_H(r) = (e_1, \ldots, e_k)$, there is a standard relation r' of G, with $\tau_G(r') \leq_k \tau_H(r)$ and $\gamma_G(r') = (\pi(e_1), \ldots, \pi(e_k))$;*
- *for each computed relation r of H, s.t. $\gamma_H(r) = (e_1, \ldots, e_k, e)$, $\mathtt{eq?}(\rho(G, \pi(e)),$
$\mathtt{apply}(\mathtt{bind}(\tau_H(r)), \rho(G, \pi(e_1)), \ldots, \rho(G, \pi(e_k)))) = \#t$.*

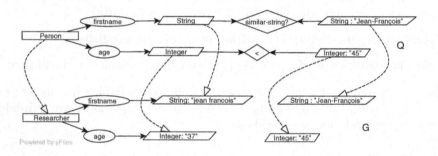

Fig. 5. An example of D-projection

Fig. 5 features an example of D-projection. Dashed arrows represent the mapping
π from the query Q (is there a person whose first name is a string similar
to "Jean-François" and whose age is lesser than 45), where `similar-string?`
is bound to a predicate returning $\#t$ when the distance between two strings
is small, to a knowledge base G. The two individual datatype nodes marked
"`Jean-François`" and "`45`" have to be present in G for a D-projection to exist.

Property 7 (Soundness). If there is a D-projection from Q into G, then Q is a
partial consequence (and thus full consequence) of G.

However, completeness of D-projection is not achieved in the general case. Let
us consider some causes of incompleteness:

1. *Normality:* as for simple conceptual graphs, the knowledge base G must be
put into its equivalent normal form;

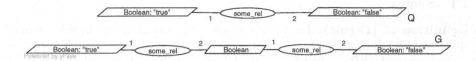

Fig. 6. There is no D-projection from Q into G

2. *Assertion of individual datatype nodes:* as seen in the previous example (Fig. 5), all necessary values must be represented by individual datatype nodes in the knowledge base G;
3. *Satisfiability:* If the knowledge base G is not satisfiable, then anything is consequence of G, and in particular queries that have no projection into G;
4. *Branching:* [LM07] show that adding atomic negation to simple conceptual graphs leads to incompleteness of projection. We adapt their counterexample to D-graphs: there is no D-projection from Q to G in Fig. 6, but there is one whether we consider that the generic node of G represents $\#t$ or $\#f$.

The first cause of incompleteness is handled by updating the normalization operation, the second by adding individual datatype nodes of the query to the knowledge base. However, to answer to the third and fourth causes of incompleteness, we need to adopt a severe restriction.

Definition 12 (Normalization). *A D-graph G is put into its normal form nf(G) by fusioning all concept nodes having the same individual marker, and all datatype nodes e e' such that eq?($\mathtt{read}(\tau(e), \mu(e)), \mathtt{read}(\tau(e'), \mu(e'))) = \#t$.*

Theorem 2 (Soundness and completeness). *Let G and Q be two D-graphs over a vocabulary V and a compatible library $\mathcal{L}$. Let us consider that G can be resolved into a D-graph G_r without generic datatype node. Then Q is a partial consequence of G if and only if G is unsatisfiable or there is a D-projection into the D-graph G' that is the normal form of the disjoint union of G_r and of all individual datatype nodes of Q.*

Because of lack of space, we do not include the proof of this theorem in this paper. The proof framework used is the one in [Bag05].

We have characterized partial consequence of D-graphs as a kind of graph homomorphism, akin to projection. To do so, we gave up the more interesting full semantics and restricted allowed target D-graphs (representing the knowledge base) to the ones that can be resolved into D-graphs without generic datatype nodes (a slight improvement with respect to databases that forbid generic nodes). As an added benefit, this characterization as a graph homomorphism allows us to extend the proposed formalism to datatyped conceptual graphs rules.

5   Datatyped Conceptual Graphs Rules (D-Rules)

We present in this section an extension of D-graphs to rules, as done for simple conceptual graphs [Sal98]. We present their syntax, their partial semantics, and briefly show how the derivation mechanism of rules can be updated for D-rules.

5.1   Syntax

Definition 13 (D-rule). *Let V be a vocabulary, and $\mathcal{L}$ be compatible library. A datatyped conceptual graph rule over V and $\mathcal{L}$ is a D-graph R where one partial subgraph of R is identified as the* hypothesis. *The other part of R (its not necessarily a D-graph) is called its* conclusion.

In the rest of this paper we will represent D-rules in the same way as D-graphs, but the shapes representing entities and relations of the hypothesis will be shaded in gray. The graph R of Fig. 7 represents such a D-rule, asserting that "for every even integer x, there exists an integer obtained by dividing x by two".

5.2   Semantics

To update the partial semantics of D-graphs to D-rules, we have to define under which conditions an interpretation is a model of a D-rule, as well as the (partial) consequence problem with D-rules:

Definition 14 (Models of a D-rule). *Let $S \subseteq \mathcal{S}$ be a set of strings, and $\mathcal{I}$ be the set of identifiers. An interpretation $I = (\Delta, \iota, \sigma)$ of $(\mathcal{I}, S)$ is a model of a D-rule R if every proof α that I is a model of the hypothesis of R can be extended to a proof α' that I is a model of R.*

Definition 15 (D-rules (partial) consequence). *Let G and G' be two D-graphs defined over a vocabulary V and a compatible library $\mathcal{L}$, and $\mathcal{R}$ be a set of D-rules defined over V and $\mathcal{L}$. We say that G' is a partial consequence of G and $\mathcal{R}$ if all (partial) interpretations that are models of $V, \mathcal{L}, G$ and of all D-rules in $\mathcal{R}$ is also a model of G'. We note $G, \mathcal{R} \vdash G'$.*

5.3   D-Derivation

With simple conceptual graphs rules, consequence can be computed with forward chaining: finding a sequence of transformations (rule applications) of the target graph G such that G' can be projected into G. With D-graphs, we must first point out that, to be able to find a D-projection corresponding to the semantics into any of these derived graphs, these D-graphs must be resolved into D-graphs without generic datatype nodes. We will then restrict ourselves to D-rules that ensure that property to be true at each step of the derivation.

Definition 16 (Resolvable D-rule). *A D-rule is* resolvable *iff all generic datatype nodes in its conclusion are the last argument of a computed relation.*

Definition 17 (Rule application). *Let G be a D-graph and R be a D-rule. R is said* applicable *to G if there exists a D-projection π from the hypothesis of R into G. In that case, the application of R on G following π is the D-graph obtained by making the disjoint union of G and of the conclusion of R then, for each relation r of the conclusion such that $\gamma_i(r) = e$ is an entity of the hypothesis, replace $\gamma_i(r)$ by $\pi(e)$. Finally, we resolve the obtained graph (see Fig. 7).*

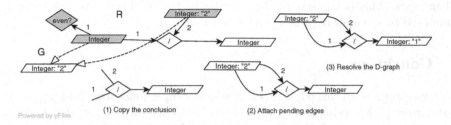

Fig. 7. Applying a D-rule on a D-graph

Property 8. The application of a resolvable D-rule on a D-graph that can be resolved into a D-graph without generic datatype nodes is also a D-graph that can be resolved into a D-graph without generic datatype nodes.

A second problem, that must be handled during derivation, is that the derivation process may generate D-graphs that are not satisfiable.

Theorem 3 (Soundness and completeness). *Let V be a vocabulary, $\mathcal{L}$ be a library, G and Q be two D-graphs and $\mathcal{R}$ be a set of rules over V and $\mathcal{L}$. Then $G, \mathcal{R} \vdash Q$ if and only if there exists a D-graph G' obtained from the disjoint union of G and of all individual datatype nodes in $\mathcal{R}$ and Q by a sequence of rule applications or normalizations such that either G' is unsatisfiable or there exists a D-projection from Q into G'.*

The proof in [BS06] is easily adapted to D-rules. As deduction with simple conceptual graphs rules, partial consequence with D-rules is semi-decidable.

5.4 Computing Factorial with D-Rules

The rule R_1 in Fig. 8 expresses that "if $y = x + 1$ and the *fact* of x is z, then the *fact* of y is $y * z$" (*fact* is a standard relation). With full semantics, this rule is sufficient to compute factorials: if the knowledge base contains the D-graph "the *fact* of 1 is 1", then all queries of form "what is the *fact* of an individual integer n?" can be correctly answered.

However, with partial semantics, nothing ensures that the values required for the recursive computation are in the interpretation domain. One way to overcome this limitation is to force these values to be interpreted, by generating

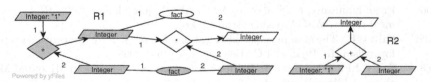

Fig. 8. D-rules that compute the factorial procedure

all integers. This is the role of the rule R_2 of Fig. 8. These two rules can be compared to the ones in [Min98] to compute factorial.

6   Conclusion

We have proposed in this paper an extension of simple conceptual graphs [Sow84] and of rules [Sal98] using datatypes (as in [Hay04]) and procedural relations (similar to the actors of [LM98])). We have proposed two model-theoretic semantics, based upon a library written in some programming language.

While full semantics are more interesting from a KR point of view, we can compute partial consequence with D-graphs and D-rules (provided that the knowledge base does not contain generic datatype nodes). The last example provides a hint on ways to overcome the limitations of partial semantics.

Further work will consist in an implementation of D-graphs and D-rules (on top of CoGiTaNT), and optimizations of reasonings (inspired by [BS06]).

References

[ABC06]   Aubert, J.-P., Baget, J.-F., Chein, M.: Simple conceptual graphs and simple concept graphs. In: Schärfe, H., Hitzler, P., Øhrstrøm, P. (eds.) ICCS 2006. LNCS (LNAI), vol. 4068, pp. 87–101. Springer, Heidelberg (2006)

[Bag05]   Baget, J.-F.: Rdf entailment as a graph homomorphism. In: Gil, Y., Motta, E., Benjamins, V.R., Musen, M.A. (eds.) ISWC 2005. LNCS, vol. 3729, pp. 82–96. Springer, Heidelberg (2005)

[BS06]   Baget, J.-F., Salvat, E.: Rules dependencies in backward chaining of conceptual graphs rules. In: Schärfe, H., Hitzler, P., Øhrstrøm, P. (eds.) ICCS 2006. LNCS (LNAI), vol. 4068, pp. 102–116. Springer, Heidelberg (2006)

[Hay04]   Hayes, P.: RDF Semantics. Technical report, W3C Recommendation (2004)

[KCE98]   Kelsey, R., Clinger, W., Rees, J. (eds.): Revised[5] Report on the Algorithmic Language Scheme. ACM SIGPLAN Notices 33(9), 26–76 (1998)

[LM98]   Lukose, D., Mineau, G.W.: A comparative study of dynamic conceptual graphs. In: Proc. of KAW'98 (1998)

[LM07]   Leclere, M., Mugnier, M.-L.: Some algorithmic improvments for the containment problem of conjunctive queries with negation. In: Schwentick, T., Suciu, D. (eds.) ICDT 2007. LNCS, vol. 4353, pp. 401–418. Springer, Heidelberg (2006)

[Min98]   Mineau, G.W.: From actors to processes: The representation of dynamic knowledge using conceptual graphs. In: Mugnier, M.-L., Chein, M. (eds.) ICCS 1998. LNCS (LNAI), vol. 1453, pp. 65–79. Springer, Heidelberg (1998)

[PS06]   Prudhommeaux, E., Seaborne, A.: SPARQL query language for RDF. Technical report, W3C Working Draft (2006)

[Sal98]   Salvat, E.: Theorem proving using graph operations in the conceptual graphs formalism. In: Proc. of ECAI'98, pp. 356–360 (1998)

[Sow84]   Sowa, J.F.: Conceptual Structures: Information Processing in Mind and Machine. Addison-Wesley, London, UK (1984)

A Knowledge Management Optimization Problem Using Marginal Utility in a Metric Space with Conceptual Graphs

Jeffrey A. Schiffel

The Boeing Company, Wichita, KS 67277 USA
jeffrey.a.schiffel@boeing.com

Abstract. Knowledge management has emerged as a field blending a systems approach with methods drawn from organizational management and learning. In contrast, knowledge representation, a branch of artificial intelligence, is grounded in formal methods. Research in the separate behavioral and the structural disciplines - knowledge management and knowledge engineering - have not traditionally cross-pollinated, preventing the development of many practical uses. Organization managers lack guidance in where to direct improvement efforts targeted at specific groups of knowledge workers. Demonstrated here is Knowledge Improvement Measurement System, an optimization solution that employs marginal utility theory in a metric space, and formal reasoning via software agents realized in conceptual graphs. This allows for repeated evaluation of knowledge improvement measurements. The KIMS method can measure activities that organize and encourage knowledge sharing to achieve competitive advantage. The solution takes into account the body of knowledge related to human understanding and learning, and formal methods of knowledge organization.

1 Introduction

A practical knowledge management (KM) problem exists in that managers lack specific organizational metrics to make effective selections from among possible alternatives for KM initiatives that leverage the collective knowledge of collaborating groups of knowledge workers. Collaborating groups share a culture [20]. Values permeating organizational culture guide decision-making and provide a basis for measurement [26]. Knowledge is key to decision-making: a higher quality of knowledge and more shared knowledge lead to better decision-making [24]. Knowledge management is thus intimately tied to corporate culture and values, to strategy, and to competitive advantage. It should be measured since managers wish their organizations to achieve competitive advantage [43]. Like other asset management, usefulness is tied to asset utility, but managers find it difficult to determine when progress is no longer being made compared to the resources expended.

This paper develops Knowledge Improvement Measurement System, a practical knowledge improvement metric derived from two separate disciplines, knowledge

U. Priss, S. Polovina, and R. Hill (Eds.): ICCS 2007, LNAI 4604, pp. 97–111, 2007.

management and knowledge engineering. Knowledge management is associated with management theory and information technology for competitiveness via problem solving efficiency [19][27][28][32][33][34][38][63]. Knowledge engineering is systems analysis aimed toward formalism in the management of well-defined, explicit knowledge and is associated with artificial intelligence to build models of logic and ontology for some useful purpose [5][17][51][55]. After 40 years of research, it is likely that knowledge representation has matured to the point where the systems aspects of knowledge engineering can be reconciled with the mental and behavioral aspects of applying knowledge [14][56].

The Knowledge Improvement Measurement System metrics described later in this paper will use knowledge engineering technology to organize, encourage and measure progress in knowledge sharing for competitive advantage and value creation. The goal is to develop a simple measurement system derived from principles of management and knowledge representation to measure progress in organizing and encouraging knowledge sharing for competitive advantage. Based on knowledge management principles from behavioral science and economics, drawing from economic transaction theory, and using active agents in conceptual graphs, this optimization metric employs marginal utility theory in a metric space to allow formal reasoning via software agents. This yields distance calculations using the city-block technique over the metric space of measurement vectors. This metric may be applied to any set of desired measurements. The measurements to be quantified into metrics may be similar or dissimilar, based on benchmarks or financially oriented intellectual property measurements. This will enable quantifiable evaluation of knowledge improvement measurements.

The organization of this paper is as follows: Section 2 summarizes knowledge management, section 3 reviews the strategic requirements that justify Knowledge Improvement Measurement System tool development, sections 4 and 5 describe the measurement basis and marginal utility calculations needed for the software agents, and section 6 develops the conceptual graphs and the actors that implement the software agents. Section 7 then develops the Knowledge Improvement Measurement Systems (KIMS), which integrates these constituents into a set of conceptual graphs, enabling measurements of the results of knowledge management improvement programs.

2   Knowledge

Knowledge is information assembled, assimilated, and put to use to solve problems [3][31][48][49][51][55]. Knowledge cannot be assessed directly; only the results of applying the knowledge can be assessed. Knowledge assessment must necessarily be indirect, since knowledge is usually applied while in its tacit form. Indirect measurements include measuring the rate of problem solving, customer satisfaction, internal morale, and cost vs. productivity measurements. The level of knowledge may also be inferred by comparison to external criteria such as quality baselines available through agencies such as the Software Engineering Institute or the Project Management Institute, or from benchmarks of similar industries.

Some definitions of knowledge management focus on the operational use and maintenance of organizational knowledge to solve problems [3][16][22][31][55]. Others emphasize the human mental activity [2][6][40]. The Knowledge Improvement Measurement System tool should be applicable to both, since the management of knowledge can be seen primarily as a way to enhance problem-solving [19], and knowledge as a corporate asset [27][29][38][59][63]. Knowledge is "information in motion" - information used by people to solve some problem. It follows that knowledge is only in the mind of the problem-solver. This is the usual definition of tacit knowledge. Nonaka and Taguchi [29] adapted Polanyi [40] in terms of two knowledge types. Polanyi, however, did not use the term "explicit," but his epistemology focused on interior knowledge.

This formal proposal synthesizes the many definitions knowledge management as (a) The identification and analysis of available and required knowledge assets needed to solve problems, (b) The identification and analysis of the processes related to knowledge acquisition and use, and (c) The planning and control of actions to develop both the assets and the processes so as to fulfill organizational objectives.

3 Strategy

To be useful over the widest variety of strategic concepts, a measurement system should be applicable to both imposed strategies that create structure from the top down in the organization, and synthesized, bottom-up strategies that generate structure from individual business units [7][11]. It should avoid excessive complications from elaborate theories, and be direct and practical for planning and analysis [3][41][61]. It should be oriented toward customer satisfaction in the allocation of knowledge resources and value [4][12][15][30][33][36]. A measurement system should be adaptable. Assessments using metrics based on past data should aid the selection of alternative future courses of action [47][57][58]. If future choices result from evaluation of propositions - sets of possible futures - the metrics should be quantifiable results of propositions. Such a metric is to allow comparisons of improvements over time to groups of knowledge workers and communities of common practice. Regardless of the strategic orientation of the organization - top-down or bottom-up strategy formulation - this knowledge improvement metric tool will present graphical and numerical sets of measurements over time. Comparisons can be made to previous iterations, or against a pre-determined set of goals.

Knowledge-based views of organizational structure require that the internal resources be described in terms of knowledge content. Similarly, knowledge-based views suggest that external opportunities in competition also be supported by knowledge [4][9][30][34][45][46]. This further supports the need for an adaptable measurement system.

The prevailing organizational culture defines the structure of decision-making. It is key to diagnosing and improving the state of knowledge in an organization. This is a consequence of the emotional, social, economic, and political context of its possessors [39][54]. The culture is the context for policy; policy is the frame for decision-making. Just as the prevailing culture is conceptual, so too knowledge is conceptual (as compared to plant, property, and equipment). It can be managed only indirectly.

Knowledge also differs from other resources because it is renewable. Like other intellectual assets, knowledge does not suffer the scarcity problems of tangible assets, but is self-renewing [8][36][44][52]. A measurement system should accommodate the prevailing culture where it is employed, and must be sensitive to the nature of knowledge as a resource.

4 Measurements and Metrics

A measurement is a quantification of changes in some system. The act of measurement consists of the set of operations for the purpose of determining the value of a quantity [21]. The quantifiers of measurements are metrics, derived attributes of measurements [42]. A measurement can be direct or indirect. Direct measurements use a graduated reference standard and are already in metical form. Indirect measurements transform the observed data to some useful form [23].

To support a chosen strategy, its tactics, and the operations of the tactics, a measurement system should be adaptable, allowing for assessments not only of the past, but provide guidance into selection of alternative future courses under uncertainty [47][57][58]. Predictive metrics are better than descriptive metrics, especially when directly attempting to optimize workers' knowledge capital. Predictive metrics show what may happen, rather than describing the recent past.

The assessments should be quantifiable results of propositions. Propositions, in this sense, may be construed as sets of possible world scenarios. There are three aspects of knowledge management: (a) the identification and analysis of available and required knowledge assets needed to solve problems, (b) the identification and analysis of the processes related to knowledge acquisition and use, and (c) the planning and control of actions to develop both the assets and the processes so as to fulfill organizational objectives.

Widdows defines a geometric space adaptable for knowledge management purposes as follows [62]. For a set A, and the function d defined on the cross product $(d : A \, X \, A \rightarrow \Re)$, d is a metric under four conditions.

$$d(a,b) \geq 0 \; for \; all \; a, \, b \in A \quad (positive \; or \; 0). \tag{1}$$

$$d(a, \, b) = 0 \;\; if \; and \; only \; if \;\; a = b \quad (identity). \tag{2}$$

$$d(a, \, b) = d(b, \, a) \quad (symmetry). \tag{3}$$

$$d(a, \, c) \leq d(a, \, b) + d(b, \, c) \quad (triangle \; inequality.) \tag{4}$$

Different sets, each containing some aspect of knowledge management, can be formed into a metric space where a notion of distance between elements of a given set is defined to be the difference between the knowledge management goal for that aspect and the currently observed state. This allows the cross product of several dissimilar sets, each with different measurements, to be treated together in a single metric space. This solution uses the "city-block" metric to measure distances between desired and current values of KM metrics, so called due to the analogy of measuring distance by driving about city streets with right-angled intersections. Reaching a desired minimum space, judged by a calculating the diminishing marginal return of the individual metrics in the total knowledge management metric space, is equivalent to reaching the knowledge improvement goals.

Also called the "Manhattan," or "taxi-cab" metric for the typical movement in a city, the city-block is a simple case of the Minkowski metric, a distance geodesic tensor given by [18][60]. (Minkowski, a mathematician, realized around 1907 that a four-dimensional space-time tensor could formulate the special theory of relativity developed by Einstein in 1905.) The advantage of this metric is that it allows several distinct and independent distances to be combined orthogonally into a metric space.

$$d_m = \sqrt[m]{\Sigma_i |\, x_{i1} - x_{i2}\,|^k} \tag{5}$$

For the case k = 2, the Minkowski metric simplifies to the usual Euclidean distance:

$$d_e(x_1, x_2) = \sqrt{\Sigma_i (x_{i1} - x_{i2})^2} \tag{6}$$

For the case k = 1, the Minkowski metric simplifies to the city-block metric:

$$d_c(x_1, x_2) = \Sigma_i |\, x_{i1} - x_{i2}\,| \tag{7}$$

The city-block metric may be also weighted:

$$d_c(x_1, x_2) = \Sigma_i w_i (|\, x_{i1} - x_{i2}\,|) \tag{8}$$

In relation to other metrics, the weighting can emphasize the metric ($w_i > 1$) or de-emphasize the metric ($0 < w_i < 1$). In a knowledge metric space of two or more vectors, a given metric can be temporarily removed from the model by setting $w_i = 0$.

A metric for knowledge management compares a currently observed value to a desired goal value. Together, these two values form the end-points of a vector. For example, if the current cost for an engineering trade study is determined to be $1,800, and the goal is the same quality of trade study content for $1,200, the vector is < 1200, . . ., 1800 >, and the unweighted city block metric is |1800 - 1200| = 600.

For example, two different metrics form a two-dimensional space, for example, using the same trade study costs as before, and adding a desire to model a CADCAM solid part in four days compared to the current seven days, the two vectors < 1200, . . ., 1800 > and < 4, . . ., 7 > form the 2-dimensional space in the first quadrant as shown in Figure 1.

A set of n different metrics therefore generates an n-space, wherein the goal is to minimize the overall volume to a desired minimum. Since each knowledge management metric in the space is of a different kind, the values and measurements for the respective vectors can vary without affecting the calculation and optimization of the complete space. Further, since some knowledge management measurements may be more important than others, the metric for that measurement may be weighted as shown by equation (8). Reaching a desired minimum can be judged by calculating the diminishing marginal return of the individual metrics in the total knowledge management metric space, as described in the next section.

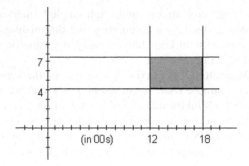

Fig. 1. Knowledge management 2-dimensional metric space

5 Marginal Utility

The general definition of the utility function may be defined for knowledge management as follows. Let X be an outcomes set, a set of all alternatives some particular KM improvement could conceivably reach. The KM improvement's utility function assigns a progress score to each alternative in the outcomes set. If $u(x) > u(y)$, then x is strictly a better outcome than y.

For example, suppose an improvement outcomes set is X = *(nothing, gain, gain, gain and gain, large gain, very large gain)*, and its utility function is $u(nothing) = 0$, $u(gain) = 1$, $u(gain) = 2$, $u(gain and gain) = 4$, $u(large gain) = 2$, and $u(very large gain) = 3$. Then this KM improvement prefers gain to gain, and also prefers one gain each rather than a single very large gain. Dollars expended on knowledge improvement programs may provide the cash equivalent commodity for transaction costs. Like other funds expended in expectation of greater value return, these dollars are subject to marginal utility evaluation. The derivation of marginal utility is shown in the following equations.

(1) Let d_n = the city block distance metric for some deviation of observed knowledge measurement k_n compared to a desired goal.

(2) Let c_n = the cost of the KM program related to that knowledge measurement d_n.

Therefore, c_n represents the cost of executing a portion of the KM program that is allocated to some knowledge area and is measured by the metric d_n. The pair (d_n, c_n) is the deviation from desired knowledge for the expenditure c_n during the nth iteration of the decision cycle.

Now assume the next decision cycle iteration is again run. Comparing the new d_{n+1} to the previously observed d_n yields a new pair as shown by $(d_n + \Delta d, c_{n+1})$, where $\Delta d = (d_{n+1} - d_n)$.

This is the marginal improvement resulting from the just-executed cycle. Initially, utility was $U(d_n, c_n)$ and after the change utility is $U(d_n + \Delta d, c_{n+1})$.

Change in total utility is

$$\Delta U = U(d_n + \Delta d, c_{n+1}) - U(d_n, c_n) \tag{9}$$

and, therefore, the rate of change in utility is

$$\Delta U/\Delta d_n = (U(d_n + \Delta d_n, c_{n+1}) - U(d_n, c_n))/\Delta d_n \tag{10}$$

for a cost of

$$|c_n - c_{n+1}| \tag{11}$$

6 Conceptual Graphs and Actors

Cognitive maps in the form of conceptual graphs can be used to categorize and formalize external representations of the state of knowledge. Sowa's conceptual graphs, derived from Peirce's existential graphs, are formalized cognitive maps for first order logic [25][50][51]. Conceptual graphs formalize natural language and can easily map concepts of an underlying ontology. They support extensions to first order logic such as lambda typing, and sets and set membership. (Note that first order logic as expressed in the usual predicate calculus does not encompass set membership. Variables are imprecise in sentences such as "Isaac is my ancestor." Isaac is a member of an unstated set of ancestors. Symbolically, in $x \in X$ the element x must be quantified as a variable, but x is not quantifiable, since the set X is not quantifiable. This is the problem of ontological commitment.)

The use of logical joins to unify conceptual graphs from one or more already existing allows discovery or extraction of further domain knowledge [10][37]. By embedding conceptual graphs in others as referents, complex propositions may be expressed. Exploiting this will allow the generation of the metric space necessary for the knowledge management measurements. Dataflow graphs - conceptual graphs with actor nodes - formalize behavioral and state relationships for KM metrics. Logically joined dataflow graphs contain the marginal utility calculations.

Assume a situation where a group of knowledge workers produce a product according to some process. One wishes to generate metrics regarding some aspect of that process, and also metrics about the products. This generic situation may be expressed as a set of conceptual graphs. The graphs use a somewhat arcane form of English, which is intended to constrain natural English into a somewhat controlled form. The form chosen here follows the ontology lattice developed by Sowa [51]. The statements are, "There is a situation where a group of knowledge workers use a process to produce a product. There is a proposition that measures of the process are metrics. There is a proposition that measures of the product are metrics." Figure 2 is shows the dataflow graph of this idea. *Agnt* (agent) and *Use* are relations; *Measure* is an actor, which calculates metric sets *mprod** and *mproc**. Concepts such as *KnowledgeWorker* have as referents variables such as the set **kw*. The actor *GenSpace* calculates a metric space from the individual metrics. Finally, the concepts *Situation* and *Proposition* have conceptual graphs as referents. The conceptual graphs were developed using CharGer [13] before outputting through Microsoft Office software [35]. This, and similar dataflow graphs, form the basis for instances of conceptual organization and computation, as shown in the next section.

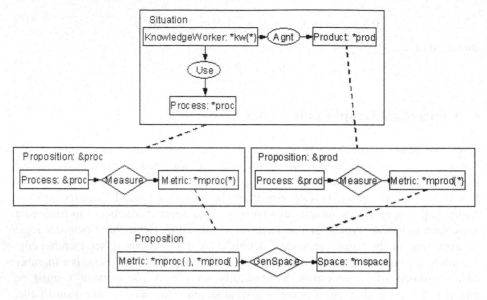

Fig. 2. Conceptual graphs, display form

6.1 Conceptual Structure of the Knowledge Management Program

At this point in the development of the Knowledge Improvement Measurement System Model, knowledge, knowledge management, metrics, and the graphical method to reason about the metrics have been defined. (A form of FOL, conceptual graphs process concepts and relations.) The concepts must be provided to the conceptual graphs to define what will be measured. The last element for the model is a surface ontology, to express the relevant concepts and their attributes for use in the conceptual graphs reasoner.

The ontology is derived from two sources, the goals of the strategy as they relate to the promotion of knowledge exchange for competitive advantage, and the programs, plans, and methods used in knowledge management. The ontology concepts must also be delineable by metrics.

Three levels of KM exist within the business organization: (a) the strategic, (b) the tactical, and (c) the operational [1]. The strategic level defines why knowledge is connected to strategy and the business model. The tactical level is concerned with connecting people to each other for tacit to tacit knowledge exchange, and connecting people to information to extract tacit knowledge from explicit knowledge. The operational level contains the methods to link knowledge and KM activities to business objectives. The operational level contains the means to codify the methods to share knowledge, defines which people connect to other people, and provides resources necessary to access knowledge, including repositories of information, training in "best practices" and making e-tools available.

Figure 3 shows the relationship of strategy and knowledge management as an enabler of strategic goals. The relationship between goals and levels is a subset of their cross-product, as is the relationship between levels and measurements. The relationship between measurements and metrics, and between methods and metrics is functional. Figure 3 is a surface ontology. Its concepts and relationships may be instantiated as concrete, attributed concepts and named relationships when applied to a given problem domain. Instantiated concepts and named relationships provide the arguments for the conceptual graphs.

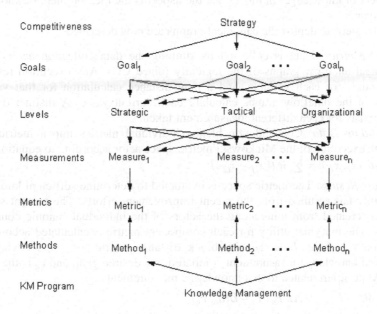

Fig. 3. Knowledge management as an enabler of strategic goals

7   The Knowledge Improvement Measurement System Model

The Knowledge Improvement Measurement System model for knowledge improvement metrics is presented in this section, which integrates the previous discussions.

A metric space using city block measurements of marginal utilities may be generated using the dataflow form of conceptual graphs. The linear form of the display graph of Figure 2 is

> [Situation : [KnowledgeWorker : *kw{ }] –> (Agnt) –> [Product : *prod]
> –> (Use) –> [Process : *proc]].
> [Proposition *proc : [Product : &prod] –> <Measure> –> [Metric : *mprod]].
> [Proposition *prod : [Product : &proc] –> <Measure> –> [Metric : *mproc]].
> [Proposition : [Metric : &mproc, &mprod] –> <GenSpace> –> [Mspace :
> *mspace{}]].

If the conceptual graphs are applied to a service organization, the concept *[Product : *prod]* is merely replaced by a service concept, *[Service : *serv]*.

The graphs form a canon, a framework for knowledge organization. This set of graphs are unified by joining on the co referents, that is, the concept referents that appear in more than one concept, or in more than one conceptual graph. The joins yield one composite canonic graph. It is canonic because it is abstract, and forms the basis for any number of instantiations. Each instantiation uses the canon to represent some aspect of knowledge. In this usage, the aspect is the measurement of knowledge improvement.

The four steps to deploy the canonical graphs are now described.

1. *Collect Metrics.* Data is collected by running the data collection actors in the individual dataflow graphs of the logically joined CGs. Any execution results in new values for each variable, and a new distance calculation for that variable. Actors in the dataflow graphs calculate each current value. A distinct dataflow graph exists for each different measurement taken.
2. *Consolidate Data.* Consolidation blends individual metrics into a metric space through execution of the Minkowski metric space actor according to equation (7):

$$d_c(x_1, x_2) = \sum_i w_i(|x_{i1} - x_{i2}|)$$

3. *Evaluate Results.* The metric space is evaluated to determine sufficient knowledge capability has resulted from the recent improvement efforts. The current solution space is created from joins along the actors of the individual canonic conceptual graphs. The marginal utility for each component metric is calculated according to equation 8, where d_n is the city block distance metric for some deviation of observed knowledge measurement compared to a desired goal, and c_n is the cost of the KM program related to that knowledge measurement d_n:

$$\Delta U = U(d_n + \Delta d, c_n+1) - U(d_n, c_n)$$

The rate of change in utility is found from equation (10):

$$\Delta U/\Delta d_n = (U(d_n + \Delta d_n, c_n+1) - U(d_n, c_n))/\Delta d_n$$

This leads to a cost calculation compared to the last loop iteration in equation 11:

$$|cn - cn+1|$$

In this stage the tactical steps to be taken for knowledge improvement are determined. Resources allocated to knowledge improvement may be re-evaluated or re-allocated.

4. *Decide.* Here the decisions of the previous stage are acted upon by evaluating the goals and allowing strategic adjustment. Prior to again executing the model, the weights of individual metric weight vectors may be adjusted to emphasize or de-emphasize the relative importance of the vectors in the metric space. A metric calculation may be zeroed out if the goal for that metric has been met. External events, such as a changing business environment, the values of external benchmarking, business forecasts, and other such external stimuli may affect the overall strategy, leading to continuation of the knowledge improvement program.

7.1 Validating the KIMS Model

KIMS can be validated through a demonstration based on historical data from a software engineering improvement program. Validation is a process to evaluate the

KIMS model to insure it complies with its requirements; that the model design meets its intended use. This definition aligns with most standard definitions, such as the Institute for Electrical and Electronic Engineers and the Software Engineering Institute.

Senior managers of the information technology group directed the IT managers and staff to begin work toward compliance with the Capability Maturity Model (CMM). CMM is a framework developed and maintained by the Software Engineering Institute (SEI) for project management, software engineering, and engineering support intended to improve software product quality and development process capability. The goals were to increase efficiency and effectivity in software development of new software systems, and also in modification and sustainment of existing software applications. Increasing efficiency meant decreasing project cost, decreasing project cycle time (schedule), and increasing the quality of products and work products. Increasing effectivity meant improving customer satisfaction, primarily by improved quality of delivered software applications and increased responsiveness to customer desires by the IT staff.

Senior management set as the goals those advertised by the SEI as characteristics of organizations that reach level 3 of the 5-level CMM model. A return on investment in CMM activities was expected. The principle characteristics include 20 percent less project cost, 30 percent improved schedule time, and up to 60 percent fewer software defects. The improvement program, a Capability Maturity Model exercise, was planned as a means to increase software engineering knowledge. Increased knowledge was expected regarding how the approximately 150 software engineers approached customer support, software design, change management, quality control, and project management. Such increased knowledge among the software engineers - clearly knowledge workers - was to be reflected in shared common practices, improvements to software quality, reduce development or maintenance cost, and decreases in the cycle time of development or maintenance schedules. Management set performance goals to be met, developed a strategy to attain the goals, modified the organization for a knowledge breakthrough, and monitored progress in applying CMM. The tactical plan, measurements, and methods were assigned to a software engineering process group. This corresponds to knowledge management as an enabler of strategy shown in figure 3.

The data thrown off by the software engineers' CMM efforts was collected monthly for five years, beginning in 1999. The data was charted and used by management to adjust the knowledge improvement program. This was also shared with all software engineering teams. For illustration in this paper, and for brevity, only year-end data is used. This is shown in Table 1, and is the input to successive runs of KIMS. Each run will indicate the knowledge improvement for the year as measured by the stated management goal areas. Each run following the initial year run will be compared to the previous year to find the marginal improvement for that year, if any. This demonstration will confirm the efficacy of KIMS by instantiating the model with real-world data from a subset of the collected measurements shown in Table 1.

108   J.A. Schiffel

Table 1. Software Engineering Performance Improvement – January 1999 to January 2001

Month	Jan-1999	Jan-2000	Jan-2001
Defect Density			
• Post-Release (0 lowest))	3.02	1.38	0.87
• Pre-Release (0 is lowest)	1.46	0.67	0.41
Customer Satisfaction (1=low to 5=high)	3.86	3.59	4.29

In the dataflow graphs, *KnowledgeWorker* refers to the individual project teams, i.e., the set **kw{*}*. *Product* and *Process* are each instantiated with project team data with Table 1 data from a given year, in **prod* and **proc* respectively. The actor *Measure* then calculates metric sets *mprod{*}* and *mproc{*}* from which *GenSpace* calculates the KIMS space. Specifically for Jan-1999, **prod* is loaded with the array |3.86 - 1| representing product data; **proc* is loaded with 3.02 and 1.46 for process data. Repeating this for each of the following two years produces the three KIMS spaces shown in figure 4.

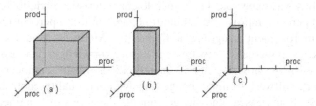

Fig. 4. Sample KIMS spaces for three years

8 Conclusion

This paper developed the Knowledge Improvement Measurement System, a model to calculate the marginal utilities generated by successive iteration of the processes of an organization's knowledge management program. The underlying principles and mechanics of KIMS need not be taught to users, providing them a straight-forward method to assess the results of knowledge management programs. KIMS, a tool that blends the principles of knowledge management and knowledge representation, is easily incorporated into an organization's knowledge management program to achieve competitive advantage by increasing shared knowledge.

References

[1] Allee, V,: Three levels of knowledge management. Walnut Creek, CA: Integral Performance Group. Updated November 17, 2001(2001) Retrieved January 3, 2002 from http://www.vernaallee.com/home/Library.htm
[2] Bateson, G.: Mind and nature: A necessary unity (1st edn.) New York: Dutton (1979)

[3] Bhatt, G.: Knowledge management in organizations: Examining the interaction between technologies, techniques, and people. Journal of Knowledge Management 5(1), 68–75 (2001)

[4] Bovel, D., Martha, J.: From supply chain to value net. Journal of Business Strategy 21(4), 24–28 (2000)

[5] Brachman, R., Levesque, H.: Knowledge representation and reasoning. Elsevier, Inc. San Francisco, CA (2004)

[6] Brown, J.S., Duguid, P.: The Social life of information. Harvard Business School Press, Boston (2000)

[7] Burgelman, R.: A model of the interaction of strategic behavior, corporate context, and the concept of strategy. The Academy of Management Review 8(1), 61–70 (1983)

[8] Chase, R.: Knowledge management benchmarks. Journal of Knowledge Management 1(1), 83–92 (1997)

[9] Conner, K., Prahalad, C.: A resource-based theory of the firm: Knowledge versus opportunism. Organization Science 7(5), 477–501 (1996)

[10] Corbett, D.: Reasoning and unification over conceptual graphs. Kluwer Academic Publishers, New York (2003)

[11] Cyert, R.M., March, J.G.: A behavioral theory of the firm, 2nd edn. Blackwell Business (Prentice-Hall), Englewood Cliffs, NJ (1992)

[12] De Long, D., Seeman, P.: Confronting conceptual confusion and conflict in knowledge management. Organizational Dynamics 29(1), 16–23 (2000)

[13] Delugach, H.S.: CharGer (Version 3.5b1) [conceptual graph software tool]. Huntsville, AL: University of Alabama - Huntsville (2005)

[14] Doyle, J.: (June 2-5, 2006) On mechanisation of thought processes. In: 10th International Conference on Principles of Knowledge Representation and Reasoning, Lake District, UK (2006)

[15] Drucker, P.: The age of social transformation. The Atlantic Monthly 274, 53–86 (1994)

[16] Esposito, F., Fanizzi, S., Basile, T., Di Mauro, N.: Multistrategy operators for relational learning and thier cooperation. Fundamenta Informaticae 69(4), 389–409 (2006)

[17] Fagin, R.: Combining fuzzy information from multiple systems. Journal of Computer and System Sciences 58(1), 83–99 (1999)

[18] Gardenfors, P.: Conceptual spaces: The geometry of thought. MIT Press, Cambridge, MA (2000)

[19] Guns, W., Valikangas, L.: Rethinking knowledge work: Creating value through idiosyncratic knowledge (Business Intelligence Program, D98-2138). Menlo Park, CA: SRI Consulting (1998)

[20] Hofstede, G.: Cultures and organizations: Software of the mind. McGraw-Hill, New York (1997)

[21] Howarth, P., Redgrave, F.: Metrology in short, 2nd edn. EUROMET publication (2003), http://www.euromet.org/docs/pubs/docs/Metrology_in_short_2nd_edition_may_2004.pdf

[22] Jonassen, D.: Computers as mindtools for schools, 2nd edn. Prentice-Hall, Upper Saddle River, NJ (2000)

[23] Juran, J.M., Gryna, F.: Quality planning and analysis, 2nd edn. McGraw-Hill, New York (1980)

[24] Kaner, M., Karni, R.: A capability maturity model for knowledge-based decisionmaking. Information Knowledge Systems Management 4(4), 225–252 (2004)

[25] Kayed, A., Colomb, R.: (January 28 - February 2, 2002) Using ontologies to index conceptual structures for tendering automation. In: Proceedings of the Thirteenth Australasian Conference on DatabaseTechnologies, Melbourne, Australia (2002)

[26] Keeney, R.: Using values in operations research. Operations Research 42(5), 793–814 (1994)

[27] Kelley, J.: Prospecting for knowledge. Cross Talk: The Journal of Defense Software Engineering 16(4), 24–27 (2003)

[28] Kreiner, K.: Tacit knowledge management: the role of artifacts. Journal of Knowledge Management 6(2), 112–123 (2002)

[29] Krogh, G., Ichijo, K., Nonaka, I.: Enabling knowledge creation: How to unlock the mystery of tacit knowledge and release the power of innovation. Oxford University Press, Oxford (2000)

[30] Krogh, G., Nonaka, I., Aben, M.: Making the most of your company's knowledge: A strategic framework. Long Range Planning 34(4), 421–439 (2001)

[31] Marakas, G.: Decision support systems in the twenty-first century. Prentice Hall, Upper Saddle River, NJ (1999)

[32] McElroy, M.: Double-loop knowledge management. Systems Thinker 10(8), 1–5 (1999a)

[33] McElroy, M.: The second generation of knowledge management. Knowledge Management Magazine, 86–88 (1999b)

[34] McElroy, M.: The new knowledge management: Complexity, learning, and sustainable innovation. Amsterdam: Butterworth-Heinemann, USA (2003)

[35] Microsoft Corp. Microsoft Word 2000 (Version 9.0.4402 SR-1) [Word processing software]. Bellevue, WA: Microsoft Corp. (1999)

[36] Morris, M.: Intangible assets and their role in corporate value (CFA Report). Irving, TX: Value, Inc. (2001)

[37] Nguyen, P., Corbett, D.: A basic mathematical framework for conceptual graphs. IEEE Transactions on Knowledge and Data Engineering 18(2), 261–271 (2006)

[38] O'Leary, D.: Enterprise knowledge management. Computer 31(3), 54–61 (1998)

[39] Parker, K.: The esthetic grounding of ordered thought. In: 12th International Conference on Conceptual Structures, Huntsville, AL (July 19-23) (2004)

[40] Polanyi, M.: The logic of tacit inference. In: Grene, M. (ed.) Knowing and being, pp. 138–158. University of Chicago Press, Chicago (1969)

[41] Porter, M.: Competitive advantage: Creating and sustaining superior performance. The Free Press, New York (1998)

[42] Price, G.: Terms in transition: Software testing terminology. Crosstalk: The Journal of Defense Software Engineering, Ogden Air Logistics Center, Hill Air Force Base, Utah (July 1994) Retrieved November 22, 2005 from http://www.stsc.hill.af.mil/crosstalk/1994/07/xt94d07l.asp

[43] Qureshi, B.R., Hlupic, V.: Value creation from intellectual capital: Convergence of knowledge management and collaboration in the intellectual bandwidth model. Group Decision and Negotiation 15(3), 197–220 (2006)

[44] Robinson, G., Kleiner, B.: How to measure an organization's intellectual capital. Managerial Auditing Journal 11(8), 36–39 (1996)

[45] Saloner, G., Shepard, A., Podolny, J.: Strategic management. John Wiley & Sons, Inc. New York (2001)

[46] Schrage, M.: Serious play: How the world's best companies simulate to innovate. Harvard Business School Press, Boston (2000)

[47] Schwartz, P.: The art of the long view: Paths to strategic insight for yourself and your company. Currency Doubleday, New York (1996)

[48] Smith, G.: Corporate valuation: A business and professional guide. John Wiley & Sons, New York (1998)

[49] Smith, P.: An introduction to knowledge engineering. International Thomson Computer Press, London (1996)
[50] Sowa, J.: Conceptual structures: Information processing in mind and machine. Addison-Wesley Publishing Company, Reading, MA (1984)
[51] Sowa, J.: Knowledge representation: Logical, philosophical, and computational foundations. Pacific Grove, CA: Brooks/Cole (2000)
[52] Stahle, P., Hong, J.: Dynamic intellectual capital in global rapidly changing industries. Journal of Knowledge Management, 177–189 (2002)
[53] Strassmann, P.: The politics of information management. The Information Economics Press, New Canaan, CT (1995)
[54] Swanson, R.: Analysis for improving performance. Berrett-Koehler Publishers, San Francisco (1994)
[55] Turban, E., Aronson, J.: Decision support systems and intelligent systems. Prentice-Hall, Inc, Upper Saddle River, NJ (2001)
[56] Uschold, M.: Keynote address: Ontologies, ontologies everywhere – But who knows what to think? In: 9th International Protégé Conference, Stanford, CA. (July 23-26, 2006) Retrieved from the author as pdf from http://protege.stanford.edu/conference/2006/index.html
[57] van der Heijden, K.: Scenarios: The art of strategic conversation. John Wiley & Sons, New York (1996)
[58] van der Heijden, K.: Scenarios, strategies and the strategy process. Breukelen, The Netherlands: Centre for Organisational Learning and Change, Nijenrode University (2007)
[59] Villegas, R.: Knowledge management white paper. Indianapolis, IN: New Century Marketing Concepts (2000)
[60] Weisstein, E.: Minkowski metric. From mathWorld–A Wolfram Web Resource. Updated May 9, 2005. Retrieved May 9, 2005 from http://mathworld.wolfram.com/MinkowskiMetric.html
[61] Whittington, R.: What is strategy - And does it matter. London: Routledge (1993)
[62] Widdows, D.: Geometry and meaning. CSLI Publications, Stanford, CA (2004)
[63] Wiig, K.: Knowledge management: An introduction and perspective. Journal of Knowledge Management 1(1), 6–14 (1997)

Conceptual Graphs as Cooperative Formalism to Build and Validate a Domain Expertise

Rallou Thomopoulos[1,2], Jean-François Baget[3,2], and Ollivier Haemmerlé[4]

[1] INRA, UMR1208, F-34060 Montpellier cedex 1, France
`rallou.thomopoulos@supagro.inra.fr`
[2] LIRMM (CNRS & Université Montpellier II), F-34392 Montpellier cedex 5, France
[3] LIG/INRIA Rhône-Alpes, F-38334 St-Ismier cedex, France
`jean-francois.baget@inrialpes.fr`
[4] IRIT, Université Toulouse le Mirail, F-31058 Toulouse cedex, France
`ollivier.haemmerle@univ-tlse2.fr`

Abstract. This work takes place in the general context of the construction and validation of a domain expertise. It aims at the cooperation of two kinds of knowledge, heterogeneous by their granularity levels and their formalisms: expert statements represented in the conceptual graph model and experimental data represented in the relational model. We propose to automate two stages: firstly, the generation of an ontology (terminological part of the conceptual graph model) guided both by the relational schema and by the data it contains; secondly, the evaluation of the validity of the expert statements within the experimental data, using annotated conceptual graph patterns.

1 Introduction

Cooperation of heterogeneous knowledge is considered here in the case of different kinds of knowledge that do not have the same statute: one of the sources contains synthetic knowledge, at a general granularity level, it provides generic rules and is considered as intuitive to understand by humans; the other sources, on the contrary, are at a very detailed granularity level, precise and reliable, but too detailed to be directly exploitable by humans. In this study, the representation formalisms used for the different sources are adapted to the kind of knowledge to be represented: (i) expert statements that express generic knowledge rising from the experience of domain specialists and describing commonly admitted mechanisms. This knowledge is represented in the conceptual graph model [1], chosen for its graphical representation of both knowledge and reasoning, relatively intuitive for the experts (see [2,3] for more details). The formalization of simple conceptual graphs and of their extension to rules adopted in this paper is that of [4]; (ii) experimental data from the international literature of the domain, represented in the relational model. These numerous data describe in detail, in a quantified way, experiments carried out to deepen the knowledge of the domain. They may confirm the knowledge provided by the expert statements – or not.

U. Priss, S. Polovina, and R. Hill (Eds.): ICCS 2007, LNAI 4604, pp. 112–125, 2007.

The cooperation of both kinds of knowledge aims at testing the validity of the expert statements within the experimental data, with the longer-term objective to refine them and to consolidate the domain expertise.

Two major differences between the two formalisms are the following. Firstly, the conceptual graphs represent knowledge at a more generic scale than the relational data. Secondly, the conceptual graph model includes an ontological part (hierarchized vocabulary that constitutes the support of the model), contrary to the relational model. We propose as a first stage the generation of an ontology, guided by the structure and the data of the relational model, which in the considered case preexist to the knowledge represented in the conceptual graph model. Some of the questions to answer are the following: how can one identify, within the relational schema and the data it contains, the concepts which can be considered as relevant at a more general granularity level, for the expression of expert statements? How can one organize the identified concepts into a hierarchy, although the relational model does not explicitly take into account the "kind of" relation? Can one go further in the suggestion of relevant complementary concepts? The proposed method is semi-automatic, expert validation is required.

As a second stage, we introduce a process that allows one to test the validity of expert statements within the experimental data, that is, to achieve the querying of a relational database by a system expressed in the conceptual graph formalism. This stage is automatic. Besides the definition of the evaluation of expert statements validity, the problem to solve concerns the automation of the generation of SQL queries on the basis of conceptual graphs whose form and content can vary. The process is based on the use of conceptual graph patterns.

Section 2 describes the generation of an ontology, guided by information about the structure and the data of the relational model. Section 3 presents a method to evaluate the validity of expert statements within the experimental data. Section 4 illustrates the results within a concrete case concerning food quality control, studied at the INRA French institute for agronomical research. In this application the objective is to highlight major trends concerning the impact of food process operations (e.g. milling, storage, extrusion, hydration, etc.) on end-product quality markers (e.g. vitamins, minerals, lipids, etc.).

2   Generation of an Ontology

Our work takes place in the case where a collection of detailed experimental data represented in the relational model preexists to the expression of expert knowledge of a higher granularity level. Our goal is to automate as far as possible the generation of a simple ontology. That ontology is constituted of the set of concept types belonging to the terminological part of the knowledge in the conceptual graph model, by using the existing relational schema and data.

In this section, after a presentation of related work, we describe three steps of the generation of the ontology: the identification of high-level concept types, the organization of these concept types into a hierarchy, and the proposition of complementary concept types.

2.1   Related Work

As the distinction between relevant and non-relevant high-level concept types entails to a large extent human expertise, a completely automatic method to generate the ontology cannot be considered [5]. Our goal differs from concept learning as proposed in the FCA approach – Formal Concept Analysis [6] – which relies on the existence of properties shared by subsets of data in order to group them into new concepts. Here, our main goal is to identify and hierarchize relevant concepts for the expression of expert knowledge, among those which preexist in the data in a non-explicit form or with an inadequate structure. The search for a new structure for specific goals in already structured data, which is our objective here, is not that frequent. Close works are those which concern the cohabitation of heterogeneous vocabularies, like model transformation [7] and ontology alignment [8]. In ontology alignment, mappings are established between pre-existing vocabularies while in our study, the ontology results from the data.

From conceptual graphs to databases. The mapping between simple conceptual graphs and conjunctive queries in databases is well-known [9,4]. Let $\mathcal{V}$ be a vocabulary, and G and Q two simple graphs on $\mathcal{V}$. G and Q are transformed (into G' and Q') in the following way: the concept types are transformed into unary relations, and each concept of type t becomes a non-typed concept incident to a unary relation typed t. For each relation r of type t, for each supertype t' of t, we add a new relation r' of type t' such that $\gamma(r) = \gamma(r')$.[1] Consequently, $G \models_{\mathcal{V}} Q$ iff $\Phi(G') \models \Phi(Q')$ (we do not need the formulas which translate the support anymore since their consequences have been translated in the graphs). Since $\Phi(G')$ and $\Phi(Q')$ are positive conjunctive formulas, we can define $\mathcal{B}$ as the tables having $\Phi(G')$ as associated logical formula and A as the query having $\Phi(Q')$ as associated logical formula. So we have $G \models_{\mathcal{V}} Q$ iff there exists an answer to A in $\mathcal{B}$. Nevertheless, that mapping relies on an identification between the vocabulary of the conceptual graphs and the database schema, which is a too strong hypothesis as we shall see in the following.

Sym'Previus. The Sym'Previus system has been developed in a French research project on the assessment of the microbiological risk in food products [10]. The tool relies on three distinct databases which have been added successively during the evolution of the project: a classic relational database, a conceptual graph database and an XML database. The three bases are queried simultaneously by means of a unique interface based on a single ontologyand on a query language close to the relational formalism (in Section 3 we will deal, on the contrary, with the querying of a relational database by conceptual graph queries). Contrary to the approach proposed in this paper, that ontology was built manually, when the conceptual graph database was added to the system. A relational database schema and its data were pre-existent. To build the ontology, the set of the attributes corresponding to meaningful entities of the application were divided into two parts: the attributes for which the values could be hierarchized according to the "kind of" relation (substrate, pathogenic germ ...)

---

[1] We denote γ the function that associates, with each relation, a tuple of concepts (its arguments). The size of the tuple is the degree of the relation.

and the attributes for which the values were "flat" sets (the names of the authors of publications for example). All the meaningful attributes were added to the Sym'Previus ontology as concept types. The hierarchized attribute values were inserted as concept subtypes in the ontology. Their precise position in the hierarchy was determined manually by experts.

2.2   Identification of High-Level Concept Types

In this stage, our goal is to identify high-level concept types (concept types located at a general granularity level). We identify two kinds of entities which we consider as potentially relevant high-level concept types: (i) those whose occurrences have a name, that is to say which have an attribute "name" (or "label", or which contain the string "name", etc.). We assume that these entities have a most general nature, by opposition to secondary entities whose occurrences are not named but only identified by a numeric label. Only the first ones are useful to express expert assertions: the experts use notions designated by a name (e.g. process operation names, food names, etc.), they do not use information that are only relevant in specific circumstances (e.g. parameter values that are specific to each experiment); (ii) those which can be divided into subcategories. In order to identify them, we search for entities with an attribute "category" (or "family", "type", etc.). We assume that these entities, due to the classification induced by their subcategories, provide relevant concept types for the ontology.

The border between these two cases is not strict and depends on the kind of modelization used. They will be considered in a homogeneous way in the following. In order to simplify, we do not indicate the exhaustive list of the considered attributes ("name", "category", "family", etc.) but we refer to them under the term of *flag attributes*. Those attributes are of type string.

Definition 1. *We call* **flag attribute** *each attribute whose name belongs to a predefined list composed of terms expressing denomination or classification. Such an attribute is considered to belong to an entity of general granularity level.*

Use of the relational schema. In a first step, we use the schema of the relational database. From a database engineering point of view, after a modelization for example in the entity-association model, we know that a relation (or table) of the relational database schema corresponds: (i) either to an entity of the considered domain – then it contains its attributes. It can also contain the identifiers of other entities (with which it was linked by an association), more rarely association attributes; (ii) or to an association (of type many to many) between entities – it has their identifiers as attributes, more rarely association attributes. The resulting table is generally labelled by the name of the corresponding entity or association. In order to identify high-level concept types, we make the following simplifying assumptions: (i) the entities – rather than the associations – carry the main concepts of the considered domain. Then the high-level concept types must be searched in the names of entities, that is to say among the names of the tables of the relational schema; (ii) the case of an association having a flag attribute is considered as exceptional.

Definition 2. *We consider as* **high-level concept types extracted from the relational schema** *the names of the tables which contain a flag attribute. These identified high-level concept types are added to the ontology.*

Example 1. In our application, examples of high-level concept types extracted from the database schema are the following: *Food product, Change, Component, Method, Operation, Property, Variable,* ... On the other hand, *Experiment, Default value, Experimental value,* for example, have not been considered as high-level concept types. The case of *Experiment* for instance was validated by the experts as not being a relevant high-level concept type because it refers to an experimental scale – judged too specific – and not to the scale of general mechanisms governing the domain.

Use of the relational data. In a second step, we consider the values taken by the flag attributes. We have made the hypothesis that the flag attributes can take as values subcategories of the entity they belong to. Thus taking into account the relational data allows us to propose the values of the flag attributes as high-level concept types. Their hierarchical organization is specified in Section 2.3.

Definition 3. *The values taken by the flag attributes of the database are considered as* **high-level concept types extracted from the data**. *These identified high-level concept types are added to the ontology.*

Example 2. In our application, the following high-level concept types have been extracted from the data: *Increase, Decrease, Protein, Lipid, Vitamin, Vitamin B, Quality, Content,* ...

2.3   Organization of the Concept Types into a Hierarchy

Two organization levels are proposed: (i) between high level concept types extracted from the data and high-level concept types extracted from the schema: the value taken by a flag attribute of a table (high-level concept type extracted from the data) is considered as a specialization of the concept type that has the name of this table (high-level concept type extracted from the schema). For example, *Vitamin* is a specialization of *Component*; (ii) among high-level concept types extracted from the data: this level is based on the inclusion of the labels of the concept types. For example, *Vitamin B* is a specialization of *Vitamin*.

Definition 4 summarizes the stages 2.2 and 2.3, validated by experts.

Definition 4. *The* **generation of a simple ontology** O *from the relational database is processed in the following way. For each table, whose name is denoted* T, *of the database, if table* T *has at least one flag attribute, then:*
– *the high-level concept type extracted from the schema* T *is added to* O;
– *for each flag attribute of* T, *that takes a set of values* $v_1, \ldots, v_n$:
  • *the high-level concept type extracted from the data* v_i, *subtype of* T, *is added;*
  • *if* v_i *is included in* v_j *$(i, j \in [1, n])$, then* v_j *is a subtype of* v_i.

Example 3. For example the table *Component* has the flag attribute *component_name*, whose values are *Protein, Lipid, Vitamin*, etc. The high-level concept type (extracted from the schema) *Component* is added to *O* and the high-level concept types (extracted from the data) *Protein, Lipid, Vitamin, Vitamin B* are added to *O* as subtypes of *Component*. As "Vitamin" is included in "Vitamin B", the concept type *Vitamin B* is a subtype of *Vitamin* (see Figure 1).

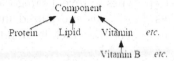

Fig. 1. Example of hierarchical organization between concept types

2.4   Proposition of Complementary Concept Types

The method proposed in this Section so as to complete the ontology with the suggestion of additional relevant concept types, is specific to the form of the expert knowledge considered in the application. We are in the following case. Expert knowledge is expressed by rules of the form "if (hypothesis) then (conclusion)". More precisely, these are causality rules. They express a relation of cause and effect between (i) a set of conditions, described by the hypothesis, interacting to produce and (ii) a resulting effect, described by the conclusion.

For example, a simple expert rule is the following: "if a food product, characterized by a vitamin content, undergoes a cooking in water, then that content decreases". It is represented by the conceptual graph rule of Figure 2.

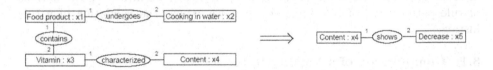

Fig. 2. Example of expert knowledge represented as a conceptual graph rule

The nature of interactions between the concepts that appear in the hypothesis is not always well-known by the experts: these interactions can be due to the interference of other concepts which are not necessarily identified. The objective of this part is to highlight some of these concepts. The method is based on the comparison of textual descriptions of the concepts that appear in the hypothesis. Indeed, the tables of the relational database from which were extracted the concept types that appear in the hypothesis (see Def. 4) sometimes provide textual descriptions, in the values of attributes named for example "description", "comments", etc. For each pair of concept types appearing in the same expert rule hypothesis, and for which such descriptions are available, the proposed method consists in searching for the existence of shared terms in these descriptions.

Example 4. The textual decriptions of some operations (Cooking in water, Steaming, Hydration, Drying) and of some components (Wheat bran, Fiber, Lipid, Vitamin, Polyphenol) share the term "water". Indeed, these unit operations all have an effect on water content (water addition or withdrawal) and these components all have subcategories that have a particular affinity with water (solubility, absorption). Highlighting the shared term "water" led the experts to complete the ontology, firstly, by adding the concept type *Water*, secondly, by specializing existing concept types to reveal categories that have a particular interaction with water: thus *Vitamin* is specialized into *Hydrosoluble vitamin* (super-type, for instance, of *Vitamin B*, which is soluble in water) and *Liposoluble vitamin.*

The obtained results are numerous and must be sorted manually by the experts. The search for shared terms uses techniques from natural language processing, such as the suppression of stopwords and of syntactic variations.

3   Evaluation of the Validity of Expert Statements

Contrary to the previous stage (Section 2) which requires an expert intervention, the method presented in this Section is automatic. The objective is to test whether the expert knowledge expressed as conceptual graph rules (created beforehand manually) is valid within the experimental data of the relational database. This must be achieved without having to define manually, for each rule, the queries to be executed in the database to obtain this information. A validity rate is computed for the tested rule and the data that constitute exceptions to the rule are identified and can be visualized by the user. In this Section, after defining the evaluation of the validity of a rule, we introduce the notions of rule pattern and of rule instance, then we expose the validation of a rule instance.

3.1   Computation of a Validity Rate

Evaluating the validity of an expert rule within the experimental data consists in calculating the proportion of data that satisfy both the hypothesis and the conclusion of the rule, among the data which satisfy the hypothesis of the rule. Let n_H be the number of data that satisfy the hypothesis and $n_{H \wedge C}$ the number of data that satisfy both the hypothesis and the conclusion. The validity rate V of a rule is $V = \frac{n_{H \wedge C}}{n_H} \times 100$, where n_H and $n_{H \wedge C}$ are the results of SQL queries counting the data that respectively satisfy the hypothesis, and both the hypothesis and the conclusion. The problem to solve is the automation of the construction of these queries.

3.2   Rule Patterns, Rule Instances and Associated Properties

Although the expert rules can take various forms, it is possible to group them into sets of rules which follow the same general form.

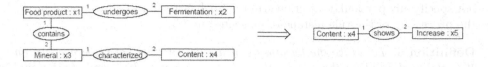

Fig. 3. Example of an expert rule that has the same form than the rule of Figure 2

Example 5. The expert rules given in Figures 2 and 3 have the same form.

The "general form" of a set of expert rules can itself be represented by a rule, called rule pattern. Its structure is identical to that of the expert rules that compose the set, but its concept vertices are more general. In other words, each expert rule of the set has a hypothesis and a conclusion which are specializations (by restriction of the labels) of those of the rule pattern. These expert rules are called rule instances. The hypothesis and conclusion of the rule pattern can thus be projected into those of each of its instances.

Example 6. The rules represented in Figures 2 and 3 are instances of the rule pattern of Figure 4.

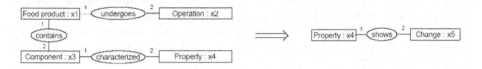

Fig. 4. Example of a rule pattern

The concept types used in a rule pattern have the following generality level: they are high-level concept types extracted from the relational schema. On the contrary, the concept types used in a rule instance can be high-level concept types extracted from the data (the markers can moreover be individual). This characteristic is essential for the validation of a rule instance.

Definition 5. *A* **rule pattern** *is a rule, in the conceptual graph formalism, whose concept vertices have high-level concept types extracted from the relational schema and whose markers are generic. A* **rule instance** *is a rule, in the conceptual graph formalism, obtained by restriction of the labels of the concept vertices of a given rule pattern. The rule instance is said to* **conform** *to this pattern.*

Consequently, the concept types that appear in a rule pattern provide a list of table names of the database (high-level concept types extracted from the schema). The hypothesis (respectively, the conclusion) of a rule pattern can be interpreted, within the database, as the formula of a query that selects the data satisfying the hypothesis (respectively, the conclusion). This formula uses the tables that appear as concept types in the hypothesis (respectively, the conclusion) of the rule pattern. This formula simply specifies a query schema. It does not specify any particular selection criteria. Such criteria will only appear during the processing of the rule instances, presented in 3.3.

Definition 6. *Let H be the hypothesis of a rule pattern. Let Q be a query on the relational database that selects the data satisfying H. In terms of relational calculus, Q can be written as: $\{t|F(T)\}$, where F is a formula, T a tuple variable of F and $F(T)$ an evaluation of F. The answer to the query Q will be a set of tuples $\{t|F(t)\ true\}$. F is built as the conjunction of the following formulas.*
– *Atomic formulas associated with the concepts of H: Let $s_{c_1}, \ldots, s_{c_n}$ be the concepts of H, of types $c_1, \ldots, c_n$ (these are high level concept types extracted from the relational schema and therefore tables of the relational database). As the concepts of H are generic, each concept s_{c_i} provides the atomic formula: $\exists x_i, c_i(x_i)$.*
– *Formulas associated with the relations of H: Let s_r be a relation vertex of H with $\gamma(s_r) = (s_{c_k}, \ldots, s_{c_l})$. Two cases are possible:*
  • *no other tables than those present in H are necessary in the schema of Q to join the tables $c_k, \ldots, c_l$. Each concept $s_{c_k}, \ldots, s_{c_l}$ of $\gamma(s_r)$ provides at least one atomic formula[2] of the form: $x_i.a_i = X_i$, where a_i denotes an attribute of table c_i and X_i a constant or an expression $x_j.a_j$ ($j \in [k, l]$, a_j attribute of c_j).*
  • *additional tables to those present in H are necessary in the schema of Q to join the tables $c_k, \ldots, c_l$. Let $t_m, \ldots, t_p$ be these additional tables. Each of them provides an atomic formula $\exists x_i, t_i(x_i)$ and at least one atomic formula $x_i.a_i = X_i$. The relation vertex s_r thus provides a (non-atomic) formula of the form: $\exists x_m, \ldots, x_p, t_m(x_m) \wedge \ldots \wedge t_p(x_p) \wedge x_k.a_k = X_k \wedge \ldots \wedge x_l.a_l = X_l \wedge x_m.a_m = X_m \wedge \ldots x_p.a_p = X_p$.*
– *Requested attributes: Let $attr_1, \ldots, attr_q$ be the requested attributes, respectively from tables tbl_1, ..., tbl_q ($attr_i$ not necessarily distinct from a_j, $j \in [k, l] \cup [m, p]$ and tbl_i in $\{c_1, \ldots, c_n, t_m, \ldots, t_p\}$). $F(t)$ is constrained by: $t.attr_i = tbl_i.attr_i$ ($i \in [1, q]$).*
  In the general case, $F(t)$ is thus of the form: $\exists x_1, \ldots, x_n, x_m, \ldots, x_p, c_1(x_1) \wedge \ldots \wedge c_n(x_n) \wedge t_m(x_m) \wedge \ldots \wedge t_p(x_p) \wedge x_k.a_k = X_k \wedge \ldots \wedge x_l.a_l = X_l \wedge x_m.a_m = X_m \wedge \ldots x_p.a_p = X_p \wedge t.attr_1 = tbl_1.attr_1 \ldots t.attr_z = tbl_q.attr_q$.

This formula can only partly be generated in an automatic way. Indeed, tables $t_m, \ldots, t_p$, the attributes a_i and the terms X_i cannot always be calculated. The limits of automation are due to the ambiguity of joins between tables and the multiple possibilities that can be encountered in case of intermediate joins between tables. Thus the formula F must be defined by the designer, for the hypothesis of each rule pattern. The formula of the query that selects the data satisfying the conclusion of a rule pattern is built in the same way. Finally, the formula of the query that selects the data satisfying both the hypothesis and the conclusion of a rule pattern is obtained as the conjunction of the formulas associated with the hypothesis and the conclusion.

---

[2] These atomic formulas are not necessarily distinct from those provided by the other neighbours of s_r, for example a neighbour may provide $x_i.a_i = x_j.a_j$ and another one $x_j.a_j = x_i.a_i$.

To allow the evaluation of an expert rule (see Section 3.1), the two required queries are the one that counts the data satisfying the hypothesis and the one that counts the data satisfying both the hypothesis and the conclusion of a rule pattern. Each rule pattern is associated with those two queries by the designer.

Example 7. The formula of the hypothesis of the rule pattern of Fig. 4 is:
$\exists x_1, x_2, x_3, x_4, x_5, x_6,$ *Food product*$(x_1) \wedge$ *Operation*$(x_2) \wedge$ *Component*$(x_3) \wedge$
Property$(x_4) \wedge$ *Result*$(x_5) \wedge$ *Study*$(x_6) \wedge$
$x_1.id\_foodProduct = x_5.id\_foodProduct \wedge x_2.id\_operation = x_6.id\_ operation$
$\wedge\ x_3.id\_component = x_5.id\_subComponent \wedge x_4.id\_property = x_5.id\_ property$
$\wedge\ x6.id\_study = x5.id\_study \wedge t.x_4.id\_result = x_5.id\_result.$
The SQL query associated with the hypothesis of the rule pattern of Fig. 4 is:

```
SELECT CUUNT(result.id_result) FROM result, food_product, component, study, operation
WHERE result.id_foodProduct = food_product.id_foodProduct AND
study.id_operation = operation.id_operation AND result.id_subComponent = component.id_component
AND result.id_property = property.id_property AND result.id_study = study.id_study.
```

Information is associated with each concept of a rule pattern, intended to inform about the specialization of this concept vertex within the rule instances that conform to this pattern: (i) if the concept type of this vertex has subtypes (high-level concept types extracted from the data), these subtypes are values of an attribute (of the corresponding table): which attribute? It is supposed to be the same for all the instances of a given rule pattern; (ii) if the marker of this vertex can be individual within the rule instances, this marker is then a value of an attribute (of the corresponding table): which attribute? Such an attribute is supposed to exist and to be the same for all the instances of a given rule pattern.

Example 8. In the rule pattern of Figure 4, the type *Component* can be specialized by subtypes which are also values of the attribute *name_component* of table *Component*. Hence in Figures 2 and 3, *Vitamin* and *Mineral*, which are specializations of the concept type *Component*, are also values of the attribute *Component.name_component.*

Definition 7. *An* **annotated** *pattern is a rule pattern P associated with:*
– a hypothesis query, that counts the tuples of the database satisfying the hypothesis of P;
– a hypothesis and conclusion query, that counts the tuples of the database satisfying both the hypothesis and the conclusion of P;
– for each of its concept vertices s_c (of type c), two attributes : (i) a type attribute, which indicates the attribute of table c that contains the specializations (denoted c_i') of the concept type c expected in the rule instances conforming to P, for an image of s_c (through the projection operation); (ii) a marker attribute, which indicates the attribute of table c that contains the markers of concept types c or c_i' expected in the rule instances conforming to P, for an image of s_c (through the projection operation).

Remark 1. As the formulas of the queries associated with a rule pattern only specify a query schema, the results of both queries should be equal to the number of data in the database. Thus the validity of a rule pattern must be 100 %.

3.3   Validation of a Rule Instance

In order to test the validity of an expert rule, i.e. of a rule instance, which is the researched objective, two new queries will be automatically built: a query that counts the data satisfying the hypothesis of the rule instance (called hypothesis query) and a quary that counts the data satisfying both the hypothesis and the conclusion of the rule instance (called hypothesis and conclusion query).

These queries are composed of two parts: (i) the first part describes the schema of the query to be executed: this first part corresponds to the query associated with the rule pattern that the rule instance to be evaluated conforms to. This part is thus provided by the annotations of the rule pattern; (ii) the second part allows one to select exclusively the tuples which take the attribute values corresponding to the specializations that appear in the rule instance. This part thus specifies selection criteria, which will be automatically built by using, as selection attributes, the annotations of the rule pattern (type attributes and marker attributes) and as selection values, the concept types and the markers of the rule instance to be evaluated.

Definition 8. *Let P be a rule pattern and I an rule instance to be evaluated, that conforms to P. The hypothesis query (respectively the hypothesis and conclusion query) of I, denoted Q_H (resp. $Q_{H \wedge C}$), is the conjunction of:*
– the hypothesis query (resp. the hypothesis and conclusion query) associated with P;
– the set of of selection criteria of the form attribute = value obtained as follows. Let π be a projection of P into I. Let $sc = [c, m]$ be a concept vertex of the hypothesis of P (resp. of whole P) and $sc' = [c', m']$ its image in I through π:
  • if $c' < c$ (with the meaning of the specialization relation) then a selection criterion is created, whose attribute is the type attribute associated with sc and whose value is c' (high level concept type extracted from the data, that corresponds to a value taken by the type attribute associated with sc). If moreover c' has subtypes, in the set of concept types, then for each of these subtypes c'' a selection criterion is created, whose attribute is the type attribute associated with sc and whose value is c'';
  • if $m' < m$ (with the meaning of the specialization relation) then a selection criterion is created, whose attribute is the marker attribute associated with sc and whose value is m'.

Remark 2. If there are several projections from P into I, a hypothesis query (resp. a hypothesis and conclusion query) of I is obtained for each of these projections. Only the hypothesis query (resp. the hypothesis and conclusion query) that provides the greatest result (greatest number of data) is retained: it is estimated to correspond to the expected specialization of the rule pattern.

Example 9. SQL query of the hypothesis of the rule instance of Figure 2:

```
SELECT COUNT(result.id_result) FROM result, food_product, component, study, operation
WHERE result.id_foodProduct = food_product.id_foodProduct AND
study.id_operation = operation.id_operation AND result.id_subComponent = component.id_component
```

```
AND result.id_property = property.id_property AND result.id_study = study.id_study
// Part of the query which is added to that of the pattern (Fig. 7)
AND operation.name_operation = 'Cooking in water' AND property.name_property = 'Content'
AND (component.name_component = 'Vitamin'
// Part which corresponds to the subtypes of the concept type Vitamin
OR component.name_component='Liposoluble vitamin' OR component.name_component='Vitamin E' ...)
```

The results of the queries Q_H and $Q_{H \wedge C}$ are respectively n_H and $n_{H \wedge C}$, which allows one to calculate the validity rate of the rule instance. The rules whose validity rate is strictly lower than 100 % have exceptions within the database. These exceptions can be visualized by the user.

Example 10. The validity rate V of the rule of Figure 2 is equal to 97.5 %.

4 Application

The presented methods have been applied within a project concerning food quality. The objective is to better control the parameters that impact the nutritional quality of food products. After a presentation of the work environment, we describe the experimental data and the expert knowledge of the project, then we illustrate the validation of expert knowledge.

4.1 Work Environment

The experimental data are stored in a MySQL database. The data can be entered and consulted by domain specialists through a web browser, using PHP forms. The database is composed of about thirty tables and currently contains the detailed results of approximately 600 experiments.

The expert rules are represented using the interface CoGUI (http://www.lirmm.fr/~gutierre/cogui/). About 150 expert rules are available, about twenty are used to test the proposed methodology, starting with the simplest cases.

The communication between the two systems is based on a JDBC connection.

4.2 Experimental Data Description

Designed for scientists and industrials in the domain of food processing, the experimental data acquisition and consultation tool (in English language) gathers scientific data, as exhaustive as possible, from international publications dealing with nutritional qualities of durum wheat based food products, and describing the impact of food processing on these qualities. Such scientific publications ususally include information about experimental measures concerning nutritional composition analysis, results about the impact of unit operations on nutritional qualities, data concerning the influence of parameters of the unit operation and of other unit operations, bibliographical references, etc.

4.3   Expert Knowledge Description

The concept types of the vocabulary used to express expert knowledge were obtained as presented in Section 2. In the vocabulary, most of semantics is expressed through the concept types. The relation types constitute general connectors, as stable as possible. Expert knowledge, represented as conceptual graph rules, expresses qualitatively, for each unit operation that is part of the process of a durum wheat based food product, and for each nutritional component of interest, the known impact of the considered operation on the considered component. The impact can concern a variation in the component content (increase, decrease, stagnation) but also a modification of qualitative properties of the component, such as digestibility, allergenicity, etc.

4.4   Expert Knowledge Validation

The evaluation of expert knowledge can be visualized in two ways by the user: individually, rule after rule, which allows the user to access the experimental data that constitute exceptions to the considered rule; or as a synthetic table containing all the rules available in the application and their validity rates. Figure 5 illustrates the first case. Validity rates spread out from 73 to 100 %.

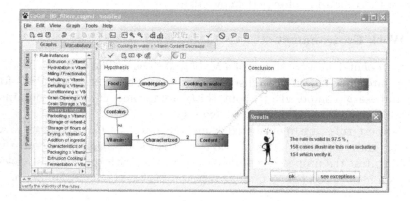

Fig. 5. Evaluation of the validity of an expert rule

5   Conclusion and Perspectives

Given two heterogeneous kinds of information available on a domain (generic expert knowledge expressed as causality rules on the one hand, detailed experimental results on the other hand) represented in two distinct formalisms (respectively the conceptual graph model and the relational model), in this article we proposed two stages for the construction of an expertise on the domain: (i) the generation of an ontology, by identifying high level concept types within the relational diagram and the relational data, and by organizing these concept types into a hierarchy. This stage is automatic but is subject to expert validation;

(ii) the evaluation of the validity of expert knowledge within the experimental data. This stage is based on the notion of rule pattern in the conceptual graph formalism, associated with a corresponding SQL query schema in the relational formalism. The evaluation of a rule instance that conforms to a given rule pattern is then processed by completing the query schema associated with the pattern by selection criteria specific to the considered rule instance. This stage is automatic, which is allowed by annotations of the rule patterns. The proposed methodology thus relies on the cooperation of the two kinds of information and the two heterogeneous formalisms. It is illustrated by a concrete application case.

The longer-term objective of the causality rules is a use for decision-making: given a user's query that expresses a required goal, the issue is to determine which conditions allow one to achieve this goal, by identifying rules whose conclusions would satisfy the required goal, and whose hypotheses would provide sufficient conditions to obtain it.

References

1. Sowa, J.F.: Conceptual Structures: Information Processing in Mind and Machine. Addison-Wesley, London, UK (1984)
2. Bos, C., Botella, B., Vanheeghe, P.: Modeling and Simulating Human Behaviors with Conceptual Graphs. In: Delugach, H.S., Keeler, M.A., Searle, L., Lukose, D., Sowa, J.F. (eds.) ICCS 1997. LNCS, vol. 1257, pp. 275–289. Springer, Heidelberg (1997)
3. Genest, D.: Extension du modèle des graphes conceptuels pour la recherche d'informations. PhD thesis, Université Montpellier II (December 2000)
4. Mugnier, M.-L.: Knowledge Representation and Reasoning based on Graph Homomorphism. In: Ganter, B., Mineau, G.W. (eds.) ICCS 2000. LNCS, vol. 1867, pp. 172–192. Springer, Heidelberg (2000)
5. Pernelle, N., Rousset, M.C., Ventos, V.: Automatic construction and refinement of a class hierarchy over multi-valued data. In: Siebes, A., De Raedt, L. (eds.) PKDD 2001. LNCS (LNAI), vol. 2168, pp. 386–398. Springer, Heidelberg (2001)
6. Tilley, T.A., Cole, R.J., Becker, P., Eklund, P.W.: A survey of formal concept analysis support for software engineering activities. In: Ganter, B., Stumme, G., Wille, R. (eds.) Formal Concept Analysis. LNCS (LNAI), vol. 3626, pp. 250–271. Springer, Heidelberg (2005)
7. Sendall, S., Kozaczynski, W.: Model transformation: The heart and soul of model-driven software development. IEEE Software 20(5), 42–45 (2003)
8. Euzenat, J., Le Bach, T., Barrasa, J., Bouquet, P., De Bo, J., Dieng-Kuntz, R., Ehrig, M., Hauswirth, M., Jarrar, M., Lara, R., Maynard, D., Napoli, A., Stamou, G., Stuckenschmidt, H., Shvaiko, P., Tessaris, S., Van Acker, S., Zaihrayeu, I.: State of the art on ontology alignment. deliverable 2.2.3, Knowledge web NoE (2004)
9. Kolaitis, P.G., Vardi, M.Y.: Conjunctive-Query Containment and Constraint Satisfaction. In: Proceedings of PODS'98 (1998)
10. Haemmerlé, O., Buche, P., Thomopoulos, R.: The MIEL system: uniform interrogation of structured and weakly structured imprecise data. Journal of Intelligent Information Systems (2006)

An Inferential Approach to the Generation of Referring Expressions

Madalina Croitoru and Kees van Deemter

University of Southampton University of Aberdeen

Abstract. This paper presents a Conceptual Graph (CG) framework to the Generation of Referring Expressions (GRE). Employing Conceptual Graphs as the underlying formalism allows a new rigorous, semantically rich, approach to GRE: the intended referent is indentified by a combination of facts that can be deduced in its presence but not if it would be absent. Since CGs allow a substantial generalisation of the GRE problem, we show how the resulting formalism can be used by a GRE algorithm that *refers* uniquely to objects in the scene.

1 Introduction

Generation of Referring Expressions (GRE) is a key task in Natural Language Generation (Reiter and Dale 2000). Essentially, GRE models the human ability to verbally identify objects from amongst a set of distractors: given an entity that we want to refer to, how do we determine the content of a referring expression that uniquely identifies that intended referent?

In the classical approach, a GRE generator takes as input (1) a knowledge base (KB) of (usually atomic) facts concerning a set of domain objects, and (2) a designated domain object, called the *target*. The task is to find some combination of facts that singles out the target from amongst all the distractors in the domain. These facts should be true of the target and, if possible, false of all distractors (in which case we speak of a *distinguishing* description). Once expressed into words, the description should ideally be 'natural' (i.e., similar to human-generated descriptions), and effective (i.e., the target should be easy to identify by a hearer). Many of the main problems in GRE are summarized in Dale and Reiter (1995). (See also Dale and Haddock 1991 for GRE involving relations; Van Deemter 2002 and Horacek 2004 for reference to sets and for the use of negation and disjunction). Here, we focus on logical and computational aspects of the problem, leaving empirical questions about naturalness and effectiveness, as well as questions about the choice of words, aside.

Recently, a graph-based framework was proposed (Krahmer et al. 2003), in which GRE was formalised using labelled di-graphs. A two-place relation R between domain objects x and y was represented by an arc labelled R between nodes x and y; a one-place predicate P true of x was represented by an looping arc (labelled P) from x to x itself. By encoding both the description and the KB in this same format (calling the first of these the *description graph* and

U. Priss, S. Polovina, and R. Hill (Eds.): ICCS 2007, LNAI 4604, pp. 126–139, 2007.

the second the *scene graph*), these authors described the GRE problem in graph-based terms using subgraph isomorphisms. This provides the ability to make use of different search strategies and weighting mechanisms when adding properties to a description. Their approach is elegant and has the advantage of a visual formalism for which efficient algorithms are available, but it has a number of drawbacks. Most of them stem from the fact that their graphs are not part of an expressively rich overarching semantic framework that allows the KB to tap into existing ontologies, and to perform automatic inference.

It is these shortcomings that we addressed in Croitoru and van Deemter (2006), while maintaining all the other advantages of the approach of Krahmer et al. (2003). The core of our proposal is to address GRE using a Conceptual Graph (GRE) framework. CGs provide a simple approach that adds discriminatory power. This emphasizes the important role the underlying representation plays in the generation of referring expressions: if we want to emulate what people do, then we not only need to design algorithms which mirror their behaviour, but these algorithms have to operate over the same kind of data. Another interesting quality of our approach is that the algorithm devised explicitly tracks the focus of attention. Objects which are "closely related" (in the combinatorial structure provided by the CG) to the most recent target object are taken to be more salient than objects which are not in the current focus space. Conceptual Graphs are a visual, logic-based knowledge representation (KR) formalism. They encode ontological ('T Box') knowledge in a structure called *support*. The support consists of a number of taxonomies of the main concepts and relations used to describe the world. The world is described using a bipartite graph in which the two classes of the partition are the objects, and the relations respectively. The CGs semantics translate information from the support in universally quantified formulae (e.g., 'all cups are vessels'); information from the bipartite graph is translated into the existential closure of the conjunction of formulae associated to the nodes (see section 3.2). A key element of CGs is the logical notion of subsumption (as modelled by the notion of a projection), which will replace the graph-theoretical notion of a subgraph isomorphism used by Krahmer et al. (2003).

The main contribution of the present paper is to highlight that the CG framework allows us to replace the GRE-classical content determination approach by an *inferential* approach: *the target is now individualized by a logical formulae which can be deduced from the information associated to the CG-scene, but which can not be deduced from the information associated to the CG-scene without the target*. We believe it is important to draw attention to the deep role played by inference in addressing GRE in a CG framework, which provides a simple and effective mechanism for handling a more realistic setting than those used by the existing work in the field.

The aim of this paper is therefore to present a new and effective application of CGs in the area of Natural Languages Processing (NLP). This reveals also, some new interesting questions related to the combinatorial and algorithmic properties of CGs. For example, we found in a natural way, that the "eccentricity" of a concept node can be considered as its salience in the description provided by the CG. This can be used by a CG layout tool in order to enhance the visual quality of the picture, by placing "central concept nodes" in the middle of the picture. Also, we arrived at the notion of "non-ambiguous description" provided by a CG, that is a description in which no two concept nodes could be confused. Recognizing such a property of a CG is obviously important for the CG models of real world applications. This can be viewed as a certain discipline of modelling in an area which is sometimes dominated by rhetorical metaphors.

2   Conceptual Graphs (CGs)

2.1   Syntax

Here we discuss the (simple) conceptual graph (CG) model and explain how it can be used to formalise the information in a domain (or 'scene') such as Figure 1. In section 3 we show how the resulting CG-based representations can be used by a GRE algorithm that *refers* uniquely to objects in the scene.

The CG model (Sowa (1984)) is a logic-based KR formalism. Conceptual Graphs make a distinction between ontological (background) knowledge and factual knowledge. The ontological knowledge is represented in the support, which is encoded in hierarchies. The factual knowledge is represented by a labelled bipartite graph whose nodes are taken from the support. The two classes of partitions consist of concept nodes and relation nodes. Essentially, a CG is composed of a support (the concept / relation hierarchies), an ordered bipartite graph and a labelling on this graph which allows connecting the graph nodes with the support.

We consider here a simplified version of a support $S = (T_C, T_R, \mathcal{I})$, where: $(T_C, \leq)$ is a finite partially ordered set of *concept types*; $(T_R, \leq)$ is a partially ordered set of *relation types*, with a specified *arity*; $\mathcal{I}$ is a set of *individual markers*.

Formally (Chein and Mugnier (1992)), a (simple) CG is a triple CG= $[S, G, \lambda]$, where:

- S is a support;
- $G = (V_C, V_R, E)$ is an ordered bipartite graph ; $V = V_C \cup V_R$ is the node set of G, V_C is a finite nonempty set of *concept nodes*, V_R is a finite set of *relation nodes*; E is the set of edges $\{v_r, v_c\}$ where
  the edges incident to each relation node are ordered and this ordering is represented by a positive integer label attached to the edge; if the edge $\{v_r, v_c\}$ is labelled i in this ordering then v_c is the i-neighbor of v_r and is denoted by $N_G^i(v_r)$;
- $\lambda : V \to S$ is a labelling function; if $v \in V_C$ then $\lambda(v) = (type_v, ref_v)$ where $type_v \in T_C$ and $ref_v \in \mathcal{I} \cup \{*\}$; if $r \in V_R$ then $\lambda(r) \in T_R$.

For simplicity we denote a conceptual graph CG= $[S, G, \lambda]$ by G, keeping support and labelling implicit. The order on $\lambda(v)$ preserves the (pair-wise extended) order on T_C (T_R), considers $\mathcal{I}$ elements mutually incomparable, and $* \geq i$ for each $i \in \mathcal{I}$. The fact that two concept labels with distinct individual markers is in concordance with the unique name assumption, that is, there is an unique name of naming a specific entity.

Consider the following KB described in Figure 1. The Krahmer et al. (2003) associated scene digraph is illustrated in Figure 2 and the CG scene graph description is given in Figure 3.

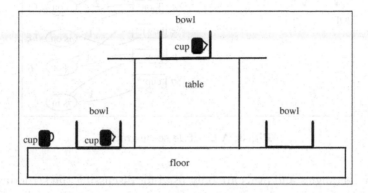

Fig. 1. A scene

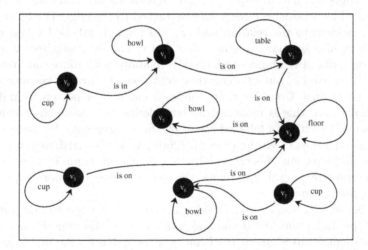

Fig. 2. Krahmer et al. scene digraph

In Figure 3 the concept type hierarchy T_C of the support is depicted on the left. The factual information provided by Figure 1 is given by the labelled bipartite graph on the right. There are two kinds of nodes: rectangle nodes representing concepts (objects) and oval nodes representing relations between concepts. The former are called concept nodes and the second relation nodes. The labels r_i and

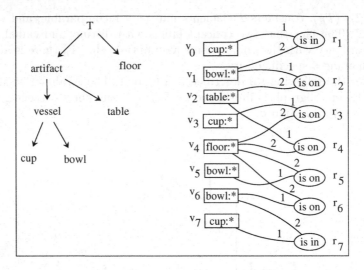

Fig. 3. A CG-style scene graph

v_i outside rectangles and ovals are only used for discussing the structure of the graph, they have no meaning. $\{v_0, \ldots, v_7\}$ are the concept nodes and $\{r_1, \ldots, r_7\}$ are the relation nodes. Each edge of the graph links a relation node to a concept node. The edges incident to a specific relation node are ordered and this ordering is represented by a positive integer label attached to the edge. For example, the two edges incident to the relation node r_1 are $\{r_1, v_0\}$, labelled 1 and $\{r_1, v_1\}$, labelled 2; we also say that v_0 is neighbor 1 of r_1 and v_1 is neighbor 2 of r_1.

In the digraphs of Krahmer et al. (2003), relations with more than two places are difficult to handle, but CGs can represent these naturally, because relation instances are *reified*. Consider that x gives a car y to a person z, and a ring u to v. Using CGs, this is modelled by considering two instances r_1 and r_2 of giving, each of which has a labelled arc to its three arguments. We note also that in the scene digraphs of Krahmer et al. (2003), object's attributes are encoded using labelled loops, and this can conduct to unpleasant complications of for the graphical representation. Using relation nodes of degree 1 is more expressive in the CG representation.

The label of a concept node (inside the rectangle) has two components: a concept type and either an individual marker or $*$, the *generic marker*. The concept node designates an entity of the type indicated by the first component. If the second component is $*$, this entity is arbitrary ; if it is an individual marker then the entity is specific. Intuitively, by using labels, CGs have associated, by definition, a "local" referential mechanism: each concept node refers to an entity belonging to the subset of the universe established by the interpretation of its type. In Figure 3 all concepts have generic markers and the nodes v_0, v_3 and v_7 designate three arbitrary objects of type *cup*, v_4 designates an arbitrary object of type *floor*, etc. For a relation node, the label inside the oval is a relation type from T_R. The arity of this relation type is equal to the number of vertices incident to the relation node r (denoted by $deg(r)$. Intuitively, this means that the objects designated by its concept node neighbours are in the relation designated by the label. In Figure 3 the relation node r_2 asserts that the bowl designated by v_1 is on the table designated by v_2.

Overall the conceptual graph in Figure 3 states that there is a floor on which there are a table, a cup and two bowls; on the table there is a a bowl and in this bowl there is a cup.

2.2   Formal Semantics of CGs

Usually, CGs are provided with a logical semantics via the function Φ, which associates to each CG a FOL formula (Sowa (1984)). If S is a support, a constant is associated to each individual marker, a unary predicate to each concept type and a n-ary predicate to each n-ary relation type. We assume that the name for each constant or predicate is the same as the corresponding element of the support. The partial orders specified in S are translated in a set of formulae $\Phi(S)$ by the following rules: if $t_1, t_2 \in T_C$ such that $t_1 \leq t_2$, then $\forall x(t_2(x) \rightarrow t_1(x))$ is added to $\Phi(S)$; if $t_1, t_2 \in T_R$, have arity k and $t_1 \leq t_2$, then $\forall x_1 \forall x_2 \ldots \forall x_k(t_2(x_1, x_2, \ldots, x_k) \rightarrow t_1(x_1, x_2, \ldots, x_k))$ is added to $\Phi(S)$.

If CG$= [S, G, \lambda]$ is a conceptual graph then a formula $\Phi(\text{CG})$ is constructed as follows. To each concept vertex $v \in V_C$ a term a_v and a formula $\phi(v)$ are associated: if $\lambda(v) = (type_v, *)$ then $a_v = x_v$ (a logical variable) and if $\lambda(v) = (type_v, i_v)$, then $a_v = i_v$ (a logical constant); in both cases, $\phi(v) = type_v(a_v)$. To each relation vertex $r \in V_R$, with $\lambda(r) = type_r$ and $deg_G(r) = k$,

the formula associated is $\phi(r) = type_r(a_{N_G^1(r)}, \ldots, a_{N_G^k(r)})$.

$\Phi(\text{CG})$ is the existential closure of the conjunction of all formulas associated with the vertices of the graph. That is, if $V_C(*) = \{v_{i_1}, \ldots, v_{i_p}\}$ is the set of all concept vertices having generic markers, then $\Phi(\text{CG}) = \exists v_1 \ldots \exists v_p(\wedge_{v \in V_C \cup V_R} \phi(v))$.

If G is the graph in Figure 3, then

$\Phi(G) = \exists x_{v_0} \exists x_{v_1} \exists x_{v_2} \exists x_{v_3} \exists x_{v_4} \exists x_{v_5} \exists x_{v_6} \exists x_{v_7} [cup(x_{v_0}) \wedge bowl(x_{v_1}) \wedge table(x_{v_2}) \wedge cup(x_{v_3}) \wedge floor(x_{v_4}) \wedge bowl(x_{v_5}) \wedge bowl(x_{v_6}) \wedge cup(x_{v_7}) \wedge isin(x_{v_0}, x_{v_1}) \wedge ison(x_{v_1}, x_{v_2}) \wedge ison(x_{v_1}, x_{v_2}) \wedge ison(x_{v_3}, x_{v_4}) \wedge ison(x_{v_2}, x_{v_4}) \wedge ison(x_{v_5}, x_{v_4}) \wedge ison(x_{v_6}, x_{v_4}) \wedge isin(x_{v_7}, x_{v_6})]$.

If (G, λ_G) and (H, λ_H) are two CGs (defined on the same support S) then $G \geq H$ (G subsumes H) if there is a projection from G to H. A projection is a mapping π from the vertices set of G to the vertices set of H, which maps concept vertices of G into concept vertices of H, relation vertices of G into relation vertices of H, preserves adjacency (if the concept vertex v in V_C^G is the ith neighbour of relation vertex $r \in V_R^G$ then $\pi(v)$ is the ith neighbour of $\pi(r)$) and furthermore $\lambda_G(x) \geq \lambda_H(\pi(x))$

for each vertex x of G. A projection is a morphism between the corresponding bipartite graphs with the property that labels of images are decreased. $\Pi(G, H)$ denotes the set of all projections from G to H.

Informally $G \geq H$ means that if H holds then G holds too. This is motivated by the fact that the subsumption relation corresponds to deduction for the fragment of first order logic (FOL) associated to CGs. More precisely, if $G \geq H$ then $\Phi(S), \Phi(H) \models \Phi(G)$ (*soundness*) (Sowa (1984)). *Completeness* (if $\Phi(S), \Phi(H) \models \Phi(G)$ then $G \geq H$) only holds if the graph H is in *normal form*, i.e. if each individual marker appears at most once in concept node labels (Chein and Mugnier (1992)). Using only CGs in normal form is a natural condition for our GRE purposes and this will be assumed implicitly in the following.

For the GRE problem the following definitions are needed to rigorously identify a certain type of a subgraph. If $G = (V_C^G, V_R^G, E)$ is an ordered bipartite graph and $A \subseteq V_R^G$, then the *subgraph spanned by* A *in* G is the graph $[A]_G = (N_G(A), A, E')$ where $N_G(A)$ is the neighbour set of A in G, that is the set of all concept vertices with at least one neighbour in A, and E' is the set of edges of G connecting vertices from A to vertices from $N_G(A)$. It is easy to see that if G is a CG then the subgraph $[A]_G$ and the restriction of λ_G to its vertices is a CG too, the spanned conceptual subgraph of G. Clearly $[A]_G \geq G$ since the identity is a trivial projection from $[A]_G$ to G.

3   CGs for Generation of Referring Expressions

3.1   Stating the Problem

Let us see how the GRE problem can be stated in terms of CGs.

Definition 1. *Let G be a CG and v_0 be a concept node in G. We define that a CG H (on the same support S as G)* **uniquely refers** *to v_0 in G if :*
$$H \geq G \text{ and } H \not\geq G - v_0.$$

Since projection is sound and complete with respect to Sowa's semantics Φ for (normal) CGs, it follows that H *uniquely refers* to v_0 in G if and only if $\Phi(S), \Phi(G) \models \Phi(H)$ and $\Phi(S), \Phi(G - v_0) \not\models \Phi(H)$. This intuitively means that H *uniquely refers* to v_0 in G if and only if the facts stated by H can be logically deduced from the facts stated by scene G, but this is no longer the case if the target v_0 is removed from the scene.

It is easy to see that if H uniquely refers to v_0 in G and H' is any subgraph of H such that $H' \not\geq G - v_0$, then H' also uniquely refers to v_0 in G. Clearly, in the GRE problem we will be interested in obtaining only minimal CGs H that uniquely refers to v_0 in G.

On the other hand, let us note that if H uniquely refers to v_0 in G, then there is π a projection from H to G (since $H \geq G$) and a concept node w in H such that $\pi(w) = v_0$ (otherwise, π is a projection from H to $G - v_0$). Hence, if $\pi(H)$ is the image of H, then $\pi(H)$ is a spanned subgraph of G namely, $[\pi(V_R^H)]_G$, containing v_0. Clearly, $\pi(H) \geq G$ (identity is an obvious projection) and, furthermore, $\pi(H) \not\geq G - v_0$ (if there is a projection π_1 from $\pi(H)$ to $G - v_0$ then $\pi_1 \circ \pi$ is a projection from H to $G - v_0$). Therefore, we have obtained that $\pi(H)$ uniquely refers to v_0 in G.

It follows that (analogous to Krahmer et al. 2003) in the GRE problem we can restrict only to referring graphs 'part of' the scene graph. It is possible to formulate GRE using only the combinatorial structure CG G and the vertex v_0.

Definition 2. *Let G be a* CG *and v_0 be a concept node in G.*
*A v_0-**referring** subgraph of G is the subgraph $G' = (\{v_0\}, \emptyset, \emptyset)$ or any spanned subgraph $G' = [A]_G$ containing v_0 (that is, $A \neq \emptyset$ and $v_0 \in N_G(A)$).*
*A v_0-referring subgraph $[A]_G$ is called v_0-**distinguishing** if $[A]_G \not\leq G - v_0$.*

It is not difficult to verify that a v_0-referring subgraph $[A]_G$ is v_0-distinguishing if and only if v_0 is a fixed point of each projection π from $[A]_G$ to G, that is $\pi(v_0) = v_0 \ \forall \pi \in \Pi([A]_G, G)$.

The GRE problem is now:

> **Instance:** CG$= [S, G, \lambda]$ a conceptual graph representation of the scene; v_0 a concept vertex of G.
> **Output:** $A \subseteq V_R$ such that $[A]_G$ is a v_0-distinguishing subgraph in CG, or the answer that *there is no v_0-distinguishing subgraph in cg.*

Example. Consider the scene described in Figure 3. $A = \emptyset$ is not a solution for the GRE instance (CG, $\{v_0\}$) since $G_1 = (\{v_0\}, \emptyset, \emptyset)$ can be projected to $(\{v_7\}, \emptyset, \emptyset)$ or $(\{v_3\}, \emptyset, \emptyset)$. However, $A = \{r_1, r_2\}$ is a valid output since $G_1 = [\{r_1, r_2\}]_G$ is a v_0-distinguishing subgraph.

Note that the description of the entity represented by v_0 in G_1 has the intuitive meaning *the cup in the bowl on the table*, which does individuates this cup. In our inferential approach this holds since $\Phi(G_1) = \exists x_{v_0} \exists x_{v_1} \exists x_{v_2} (cup(x_{v_0}) \wedge bowl(x_{v_1}) \wedge table(x_{v_2}) \wedge isin(x_{v_0}, x_{v_1}) \wedge ison(x_{v_1}, x_{v_2}))$ can be deduced from $\Phi(G)$ but not from $\Phi(G - v_0)$.

If $G_1 = [A]_G$ is a v_0-distinguishing subgraph in CG, and if we denote by A' the relation nodes set of the connected component of G_1 containing v_0, then $[A']_G$ is a v_0-distinguishing subgraph in CG too. Hence, by the minimality assumption, we consider only connected v_0-distinguishing subgraphs.

On the other hand, intuitively the existence of a v_0-distinguishing subgraph is assured only if the CG description of the scene has no ambiguities.

Theorem 1. *Let (*CG$, \{v_0\}$*) be a* GRE *instance. If $[A]_G$ is v_0-distinguishing then $[A']_G$ is v_0-distinguishing for each $A' \subseteq V_R^G$ such that $A \subseteq A'$.*

Proof. Indeed, since $A \subseteq A'$ and $v_0 \in N_G(A)$ it follows that $v_0 \in N_G(A')$, therefore $[A']_G$ is v_0-referring. If $[A']_G$ is not v_0-distinguishing then there is π a projection from $[A']_G$ to G such that $\pi(v_0) \neq v_0$. But then, π_A, the restriction of π to the subgraph $[A]_G$, has the same property, $\pi_A(v_0) \neq v_0$, contradicting the hypothesis that $[A]_G$ is v_0-distinguishing.
In particular, taking $A' = V_R$, we obtain:

Corollary 1. *There is a v_0-distinguishing subgraph in G iff $G \not\leq G - v_0$.*

Proof. If there is $[A]_G$ a v_0-distinguishing subgraph in G, then (since $A \subseteq V_R$ and $[V_R]_G = G$), by the above theorem, G is v_0-distinguishing and therefore $G \not\leq G - v_0$.

Conversely, if $G \not\leq G - v_0$ then it follows that G is a v_0-distinguishing subgraph.

A concept vertex v_0 which does not have a v_0-distinguishing subgraph is called an *undistinguishable concept vertex* in G. We say that a CG provides an *well-defined scene representation* if it contains no undistinguishable vertices. Testing if a given GRE instance defines such an ambiguous description is, by the above corollary, decidable.

Let $v_0 \in V_C$ be an arbitrary concept vertex. The set of concept vertices of G different from v_0, in which v_0 could be projected, is (by projection definition) contained in the set

$$Distractors^0(v_0) = \{w | w \in V_C - \{v_0\}, \lambda(v_0) \geq \lambda(w)\}.$$

Clearly, if $Distractors^0(v_0) = \emptyset$ then v_0 is implicitly distinguished by its label (type + referent), that is ($\{v_0\}, \emptyset, \emptyset$) is a v_0-distinguishing subgraph.

Therefore we are interested in the existence of a v_0-distinguishing subgraph for concept vertices v_0 with $Distractors^0(v_0) \neq \emptyset$. In this case, if $N_G(v_0) = \emptyset$, clearly there is no v_0-distinguishing subgraph (the connected component containing the vertex v_0 of any spanned subgraph of G is the isolated vertex v_0). Hence we assume $N_G(v_0) \neq \emptyset$.

3.2   Complexity

Some of the main complexity results in GRE are presented in Dale and Reiter (1995). Among other things, these authors argue that the problem of finding a uniquely referring description that contains the minimum number of properties (henceforth, a Shortest Description) is NP-complete, although other versions of GRE can be solved in polynomial or even linear time. As we have argued, CG allows a substantial generalisation of the GRE problem. We proved in Croitoru and van Deemter (2006) that this generalisation does not affect the theoretical complexity of finding Shortest Descriptions. More precisely, we proved that the decision problem associated with minimum cover (Garey and Johnson (1979)) can be polynomially reduced to the problem of finding a concise distinguishing subgraph. If this later problem is

Shortest Description

Instance: G a CG such that $d_G(r) = 1$, for each relation node $r \in V_R$;
a vertex $v_0 \in V_C$; s a positive integer.

Question: Is there a v_0-distinguishing subgraph $[A]_G$ such that $|A| \leq s$?

then we proved (Croitoru and van Deemter (2006)):

Theorem 2. Shortest Description *is* **NP***-complete.*

Note that in the above problem we considered the simple case when all relation vertices $r \in V_R^G$ unary. In other words, G is a disjoint union of stars centered in each concept vertex. Intuitively, this means that each object designated by a concept vertex in the scene represented by G is characterized by its label (type and reference) and by some other possible attributes (properties) and each $r \in V_R^G$ designates an unary relation. This is the classical framework of the GRE problem, enhanced with the consideration of basic object properties (the types) and the existence of a hierarchy between attributes.

In this particular case, if $N_G(v_0) = \{r_1, \ldots, r_p\}$ $(p \geq 1)$ (the properties of the concept designated by v_0) then for each $r_i \in N_G(v_0)$ we can consider:

$X_i := \{w|w \in Distractors^0(v_0)$ such that there is no $r \in N_G(w)$ with $\lambda(r_i) \geq \lambda(r)\}$.

In words, X_i is the set of v_0-distractors which will be removed if r_i would be included as a single relation vertex of a v_0-distinguishing subgraph (since there is no $r \in N_G(w)$ such that $\lambda(r_i) \geq \lambda(r)$ it follows that there is no projection π of the subgraph $[r_i]_G$ to G such that $\pi(v_0) = w$).

The proof of the above theorem is based on the following lemma which we have proved in Croitoru and van Deemter (2006):

Lemma 1. *There is a v_0-distinguishing subgraph in G iff:*

$$\cup_{i=1}^{p} X_i = Distractors^0(v_0).$$

To summarize, if all relation vertices have degree 1, *deciding if a vertex v_0 admits a v_0-distinguishing subgraph can be done in polynomial time.*

However, the above lemma shows that $[A]_G$ is a v_0-distinguishing subgraph if and only if $A \subseteq N_G(v_0)$ and $\cup_{r_i \in A} X_i = Distractors^0(v_0)$. Therefore the problem of finding a v_0-distinguishing subgraph with a **minimum number** of vertices (e.g., Dale and Reiter 1995) is reduced to the problem of finding a minimum cover of the set $Distractors^0(v_0)$ with elements from $X_1, \ldots, X_p$, which is an NP-hard problem.

3.3   A Simple GRE Algorithm

In the general case, each object in the scene represented by G is characterized by its label (type and reference), by some other possible attributes (properties) and also by its relations with other objects, expressed via relation nodes of arity ≥ 2. In this case, if v_0 an arbitrary concept node, it is possible to have vertices in $Distractors^0(v_0)$ which cannot be distinguished from v_0 using individual relation neighbors but which could be removed by collective relation neighbors. Let us consider the scene described in Figure 4 :

Note that relation labels are assumed to be incomparable. Clearly, $N_G(v_0) = \{r_1, r_2\}$ and $Distractors^0(v_0) = \{v_2, v_4\}$. The vertex v_4 can be removed by r_1 (v_4 has no relation neighbor with a label at least *know*) and by r_2 (despite of the existence of a relation neighbor r_5 labelled *is near*, v_4 is the second neighbor of r_5; v_0 is the first neighbor of r_2). The vertex v_2 cannot be removed by r_1 ($[r_1]_G \geq [r_3]_G$)) and by $r_2([r_2]_G \geq [r_4]_G)$, but $\{r_1, r_2\}$ destroys v_2 (there is no projection of $[\{r_1, r_2\}]_G$ mapping v_0 to v_2 and in the same time mapping v_1 to a common neighbor of r_3 and r_4).

This example shows a way to obtain an algorithm for constructing a v_0-distinguishing subgraph in general.

For an arbitrary concept vertex v_0, let us denote $N^0(v_0) = \emptyset$, $N^1(v_0) := N_G(v_0)$ and for $i \geq 2$, $N^i(v_0) = N_G(N_G(N^{i-1}(v_0)))$. Clearly, since G is finite, there is $k \geq 1$ such that $N^i(v_0) = N^k(v_0)$ for each $i \geq k$ ($N^k(v_0)$ is the relation nodes set of the connected component of G which contains v_0). This parameter is called the *eccentricity* of v_0 and is denoted $ecc(v_0)$. The Figure 5 illustrates the construction of this sequence of relation nodes.

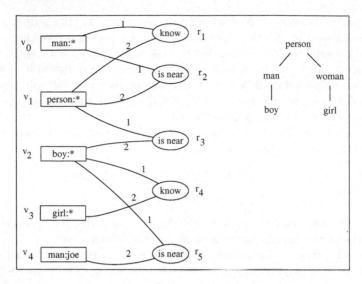

Fig. 4. Scene Illustration

The basic idea is to test successively if the above constructed relation neighbors sets of v_0 destroy the set $Distractors^0(v_0)$.

We can consider, inductively, distractors of higher order for a vertex v_0. We will use the following notation: if G is a CG containing a vertex v and H is a CG containing a vertex w then $G \geq_{v \to w} H$ means that there is a projection π from G to H such that $\pi(v) = w$. Now, $Distractors^i(v_0)$ are defined by:

$Distractors^0(v_0) = \{w | w \in V_C - \{v_0\}, \lambda(v_0) \geq \lambda(w)\}$, and, for each $i = 1, ecc(v_0)$,
$Distractors^i(v_0) = \{w | w \in Distractors^{i-1}(v_0), [N^i(v_0)]_G \geq_{v_0 \to w} [N^i(w)]_G\}$.

Note that $Distractors^0(v_0) \supseteq Distractors^1(v_0) \supseteq \ldots \supseteq Distractors^{ecc(v_0)}(v_0)$. However, only the set $Distractors^0(v_0)$ can be computed in polynomial time. The set $Distractors^i(v_0)$, $i \geq 1$, contains the vertices w from the previous set, $Distractors^{i-1}(v_0)$, which cannot be destroyed by $N^i(v_0)$ and this means that we need to test if $[N^i(v_0)]_G \geq [N^i(w)]_G$. But the last test is, in general, non-polynomial.

Theorem 3. *Let $(G, \{v_0\})$ be a GRE instance, and let i_0 be the first $i \in \{0, \ldots, ecc(v_0)\}$ such that $Distractors^i(v_0) = \emptyset$. If i_0 exists then $[N^i(v_0)]_G$ is a v_0-distinguishing subgraph, otherwise v_0 is an undistinguishing vertex.*

Proof. We can suppose that G is connected. Therefore $[N^{ecc(v_0)}(v_0)]_G = G$.

Also, if $Distractors^0(v_0) = \emptyset$ then the theorem holds trivially. Inductively, we can prove that

(∗)   *If $Distractors^i(v_0) \neq \emptyset$, then $[N^i(v_0)]_G$ is not a v_0-distinguishing subgraph.*

Using the theorem 1 we obtain from (∗) that there is no v_0-distinguishing subgraphs in $[N^i(v_0)]_G$. Therefore, if $Distractors^{ecc(v_0)}(v_0) \neq \emptyset$ then there is no v_0-distinguishing subgraphs in $[N^{ecc(v_0)}(v_0)]_G = G$.

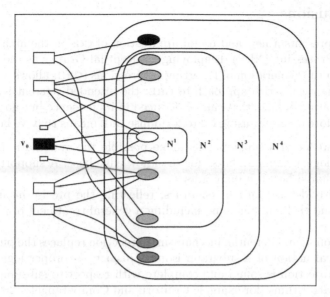

Fig. 5. Successive Relation Neighbors Sets

It follows also from $(*)$ that for each $w \in Distractors^i(v_0)$ there is a projection $\pi \in \Pi_{[N^i(v_0)]_G \to G}$ such that $\pi(v) = w$. If i_0 is the first index i such that $Distractors^i(v_0) = \emptyset$, then, clearly, $[N^{i_0}(v_0)]_G$ is a v_0-distinguishing subgraph.

Therefore, the theorem is completely proved if we show that $(*)$ holds. But this follows easily from the definition of the sets $Distractors^i(v_0)$, using an inductive argument.

The above theorem basically defines a breath first search algorithm for finding a v_0-distinguishing subgraph which can be described as follows.

Input: $CG = [S, G, \lambda]$ a CG representation of the scene; v_0 a concept vertex of G.
Output: $A \subseteq V_R$ such that $[A]_G$ is a v_0-distinguishing subgraph in G,
       or the answer that *there is no v_0-distinguishing subgraph in G.*
{ $D \leftarrow \emptyset$
     for each $w \in V_C^G - \{v_0\}$ **do**
         if $\lambda(v_0) \geq \lambda(w)$ **then** $D \leftarrow D \cup \{w\}$
     $N \leftarrow N_G(v_0)$; *finished* $\leftarrow$ *false*
     while $D \neq \emptyset$ *and not finished* **do**
         {   **for** each $w \in D$ **do**
               if *not* $[N]_G \geq_{v_0 \to w} G$ **then** $D \leftarrow D - \{w\}$
            if $N = N_G(N_G(N))$ **then** *finished* $\leftarrow$ *true*
                  else $N \leftarrow N_G(N_G(N))$
         }
     if $D = \emptyset$ **then return** N
         else return *there is no v_0-distinguishing subgraph in G.*

}

4   Conclusions

This paper presents a new and useful application of CGs in the area of Natural Languages Processing (NLP). Employing Conceptual Graphs as the underlying formalism to the Generation of Referring Expressions (GRE) allows a new, rigorous and semantically rich approach to GRE: the intended referent is indentified by a combination of facts that can be deduced in its presence but not if it would be absent. More precisely, using CG to formalise GRE means that we benefit from:

- The existence of a support. CGs make possible the systematic use of a set of "ontological commitments" for the knowledge base. A support, of course, can be shared between many KBs.
- A properly-defined formal semantics, reflecting the precise meaning of the graphs and their support, and including a general treatment of n-place relations.
- Projection as an inferential mechanism. Projection replaces the purely graph-theoretical notion of a subgraph isomorphism by a proper logical concept (since projection is sound and complete with respect to subsumption). Optimized algorithms (for example Croitoru and Compatangelo (2004)) can be used to improve the new GRE algorithm developed in the present paper.

At the same time, applying conceptual graphs to address the GRE problem raises novel interesting questions related to the combinatorial and algorithmic properties of CGs:

- The "eccentricity" of a concept node which can be used by a CG layout tool in order to enhance the visual quality of the picture.
- "Non-ambiguous descriptions", descriptions in which no two concept nodes could be confused, is obviously important for the CG models of real world applications.

To conclude, the deep role played by inference in addressing GRE in a CG framework provides a simple and effective mechanism to model the way humans refer to objects in a rich inferential setting which has never been used by existing work in the field.

References

[Chein and Mugnier (1992)] Chein, M., Mugnier, M.-L.: Conceptual graphs: Fundamental notions. Revue d'Intelligence Artificielle 6(4), 365 (1992)
[Croitoru and Compatangelo (2004)] Croitoru, M., Compatangelo, E.: A combinatorial approach to conceptual graph projection checking. In: Webb, G.I., Yu, X. (eds.) AI 2004. LNCS (LNAI), vol. 3339, Springer, Heidelberg (2004)
[Croitoru and van Deemter (2006)] Croitoru, M., van Deemter, K.: A Conceptual Graph approach for the generation of referring expressions. In: Proc. of the 20th International Joint Conference on Artificial Intelligence, IJCAI 2007 (to appear)
[Dale and Haddock (1991)] Dale, R., Haddock, N.: Generating referring expressions containing relations. In: EACL-91. Procs. of the 5th Conference of the European Chapter of the ACL (1991)

[Dale and Reiter (1995)] Dale, R., Reiter, E.: Computational Interpretations of the Gricean Maximes in the Generation of Referring Expressions. Cognitive Science 18, 233–263 (1995)

[Garey and Johnson (1979)] Garey, M., Johnson, D.: Computers and Intractability: A Guide to the Theory of NP-Completeness. In: Belz, A., Evans, R., Piwek, P. (eds.) INLG 2004. LNCS (LNAI), vol. 3123, pp. 70–79. Springer, Heidelberg (2004)

[Krahmer et al (2003)] Krahmer, E., van Erk, S., Verleg, A.: Graph-based Generation of Referring Expressions. Computational Linguistics 29(1), 53–72, 2003 (to appear)

[Sowa (1984)] Sowa, J.F.: Conceptual Structures: Information Processing in Mind and Machine. Addison-Wesley, London, UK (1984)

[van Deemter (2002)] van Deemter, K.: Generating Referring Expressions: Boolean Extensions of the Incremental Algorithm. Computational Linguistics 28(1), 37–52 (2002)

[van Deemter and Krahmer] van Deemter, K., Krahmer, E.: Graphs and Booleans: on the generation of referring expressions. In: Computing Meaning, Studies in Linguistics and Philosophy, vol. 3, Kluwer, Dordrecht. (in press)

A Conceptual Graph Description of Medical Data for Brain Tumour Classification

Madalina Croitoru, Bo Hu, Srinandan Dashmapatra, Paul Lewis,
David Dupplaw, and Liang Xiao

University of Southampton

Abstract. HealthAgents proposes an agent-based distributed decision support system for brain tumour diagnosis and prognosis which employs Magnetic Resonance Imaging and Magnetic Resonance Spectroscopy techniques and genomic profiles. From a knowledge representation view point the distributed nature and the heterogeneity of the data to be integrated pose a number of challenging problems. This paper shows how Conceptual Graphs can be employed to describe the data sources in the HealthAgents system. Such knowledge representation based description of data allows for reasoning power when querying and for data modularisation capabilities.

1 Introduction

In this paper we propose a Conceptual Graph [6] based description of the knowledge involved to build a distributed decision system for brain tumour classification (HealthAgents). We present our work formally and demonstrate how a model based semantics description of such highly heterogeneous knowledge, as well as a Conceptual Graph integration of such descriptions can benefit the system by providing modularization and querying power. Our results are theoretical and lay rigorous foundations for future implementation.

HealthAgents [1] is an agent-based, distributed decision-support system (DSS) that employs clinical information, Magnetic Resonance Imaging (MRI) data, Magnetic Resonance Spectroscopy (MRS) data and genomic DNA profile information. The aim of this project is to help improve brain tumour classification by providing alternative, non invasive techniques. A predecessor project, Interpret [7], has shown that MRI and single voxel MRS data can aid in improving brain tumour classification. HealthAgents builds on top of these results and further employs multi voxel MRS data, as well as genomic DNA micro-array information for better classification results. Moreover, HealthAgents is decentralizing the Interpret DSS by building a distributed decision support system (D-DSS). This way, the number of cases to be studied is greatly increased, improving classifier accuracy. Certain differences in patient data, determined by geographic factors, is also easier to identify.

At the moment the data in the HealthAgents system is stored in relational databases at the various participating European clinical centers. A uniform vocabulary needed for interoperability reasons is provided by the means of HADOM

U. Priss, S. Polovina, and R. Hill (Eds.): ICCS 2007, LNAI 4604, pp. 140–153, 2007.

- an ontology containing MRI, MRS and micro-array *domain information* as well as a taxonomy of brain tumours compliant to the WHO(World Health Organisation)[1] classification.

We propose describing the knowledge contained in the sources by the means of Conceptual Graphs. This allows us to build upon the existing ontology while not overcomplicating the ontology with rules to describe data extraction techniques that employ different parameters which greatly influence the outcome data. An immediate advantage of our Conceptual Graphs choice is their graph based reasoning mechanisms which allow versatile querying algorithms [4].

In Section 2 we present the challenges the HealthAgents system poses in terms of knowledge representation and reasoning. This motivates our Conceptual Graph based approach to data description informally introduced in Section 3 and formally presented in Section 4. Section 5 concludes the paper and lays down future work directions.

2   Motivation

From a knowledge representation view point the distributed nature of the HealthAgents system poses a number of problems due to the heterogeneity of the data to be integrated. Once the data acquisition protocols have been agreed upon and the data formats reconciliated, the data has to be managed and queried in an "intelligent" manner. The need – triggered by interoperability issues – for a common vocabulary was already addressed by the HADOM (HealthAgents Domain) ontology which conceptualises the parameters of the employed techniques (MRI, MRS, DNA microarrays etc.), the clinical information needed (age, sex, location etc.) and the known brain tumour classes.[2]

However this is not expressive enough for versatile querying and data integration purposes. This paper shows how Conceptual Graphs (a graph-based, logical knowledge representation formalism) can be employed to describe the data sources in the HealthAgents system. Since Conceptual Graphs are logically equivalent to the existential, positive fragment of First Order Logic, this knowledge based description allows for reasoning power when querying and for data modularisation capabilities which will lead to complete query answering across incomplete data sources.

We claim that a Conceptual Graph (CG) based description of the data within the HealthAgents system adds expressiveness for knowledge representation and versatility for querying. Our choice of knowledge representation (KR) formalism is motivated by the fact that Conceptual Graphs are:

- Expressive enough to be able to represent the data extraction protocols and the rules associated with them.
- Easy to plug in on top of existing ontologies due to the distinction between ontological knowledge (the support) and factual knowledge (bipartite graph).

---

[1] Available from Harvard Medical School at:
   http://neurosurgery.mgh.harvard.edu/newwhobt.htm
[2] According to the WHO classification.

The work we present here is highly technical, addressing the specific Conceptual Graphs problems that occur when describing such data sources. Our work is evaluated theoretically by the soundness and completeness of the proposed definitions.

3   Example

As mentioned in Section 2, the *(i)* heterogeneity of the data to be represented and the *(ii)* distributed nature of the project make knowledge representation a challenging aspect of HealthAgents.

A first step towards addressing the heterogeneity problem was creating an ontology of the main concepts used in the system. In this way a common ontological background was established. This ontology contains a poset (partially ordered set) of the known brain tumour classes according to the WHO classification, a poset of the techniques used for data acquisition characterized by their parameters and the clinical information needed for the patients (age, sex, location, medication, etc.). The HealthAgents data itself is stored anonymously and securely in a distributed network of datamarts in relational databases.

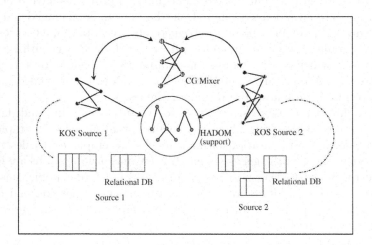

Fig. 1. Conceptual Graph Description of Knowledge

The nature of the acquisition protocols make the output data (the spectra) highly dependent on the parameters employed (for example, multi-voxel MRS techniques require the scanner manufacturer to be known in order for the data to be interpreted in a correct way). Since the data from different clinical centers has to be integrated, a common vocabulary is not enough to represent such knowledge. It is also essential to be able to provide reasoning power between the sources. We propose a "KR annotation" for the relational databases stored at each individual clinical center: Conceptual Graph based descriptions of the data in the sources. These Conceptual Graphs based on the common support of the extension of the HADOM ontology are called a Knowledge Oriented Specification of the source.[3] An example of the Conceptual Graph based approach to data description is shown in Figure 1, while Figure 4 presents a simplified example of two KOSs for MRS and MRI data sources.

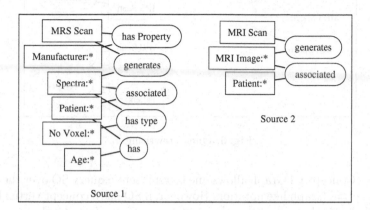

Fig. 2. Knowledge Oriented Specification

In order to query the knowledge Oriented Specification we use query Conceptual Graphs (qCG) [5]. Moreover, once the sources are described with Conceptual Graphs they can be integrated in a CG Mixer. In this "global view" of the system the domain expert specifies exactly what queries can be posed in terms of this integrated schema. Once the query is posed, the relations from the CG Mixer are rewrote to direct the query to the appropriate data sources. Querying a CG Mixer is intuitively depicted in Figure 3.

4 Formalism

In this section we provide the formalization for our approach. The motivation behind such a thorough, step by step rigorous foundation is that, in this way, we benefit from a in depth understanding of the model. This understanding facilitates future implementation.

A couple of definitions are needed to "prepare" linking the proposed Conceptual Graph description to the data sources. We introduce a support model to assign appropriate values from a domain (universe) to each concept type, relation type and marker. An assignment allows to link the concepts of a CG to the domain (universe) of the model defined over its support.

Given a data source, we need to be able to link the information (set of tuples) contained therein with the associated Conceptual Graph and its model. To do this we introduce the notion of a repository.

---

[3] More specifically, we enrich HADOM with a poset of relation types needed for protocol description, patient diagnosis etc.

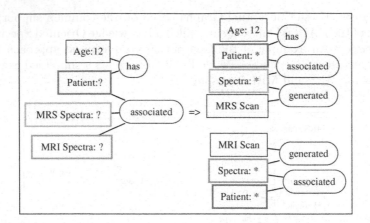

Fig. 3. Query rewriting

A query Conceptual Graph allows one to represent a query SQ over the sources in a Conceptual Graph like notation. However, if SQ has a concept vertex labelled with an individual marker then this vertex can be projected only in a concept vertex of the conceptual graph labelled with the same individual marker, by the definition of a projection. Therefore we introduce a specific querying mechanism, considering legal queries. Lastly, we define an answer to a qCG as the set of all data retrieved from the repository that validate the qCG.

A knowledge oriented specification of an information source is composed by (i) a Conceptual Graph that visually describes what we expect to know from that source, (ii) an interpretation for the support on which the graph is built, (iii) a repository for the graph (that contains all the data tuples), and (iv) a wrapper that ensures the communication between the user queries and the repository.

A CG Mixer depicts the integrated view, by the means of a Conceptual Graph, and provides the rules to allow for the translation of user queries to the appropriate data sources. The rules are defined by the relation vertices from the integrated view. For each relation in the integrated view, the proper translation is provided. This translation has to preserve the order of nodes in the initial relation, for corectness.

4.1 Data Sources

A Knowledge Oriented Specification (KOS) for a source is, informally, a Conceptual Graph that syntactically describes the data along with the data source itself. The specification does not try to exhaustively describe the sources, but provides a description of the data we have access to. More formally, if we issue a query over this specification, we should have access to the answer by the means of a wrapper. The main purpose of a KOS is to simply inform us how materialized views over the data sources can be obtained.

Usually, CGs are given semantics by translating them to existential first order logic formulae (see, for example, [3]). We propose a semantics based on model theory, adapted for our integration purposes. In order to do this we define what the interpretation of a support is, and how to assign that interpretation to the simple Conceptual Graph defined on that support. This leads to the notion of a repository of a CG on a model. In the following definitions we are building upon the work of [2].

An **interpretation** (or model) for a support is a structure that assigns appropriate values from a domain (universe) to each concept type, relation type and marker. This assignment respects the way the relation /concept types are defined and also preserves their hierarchical order.

Definition 1. (Interpretation)
An interpretation or model $\mathcal{M}$ for the support $S = (T_C, T_R, \mathcal{I}, *)$ *is a pair* $\mathcal{M} = (D, F)$ *where*
- D is a set of objects called the domain *or* universe *of $\mathcal{M}$,*
- F is a function defined on $T_C \cup T_R \cup \mathcal{I}$ such that $F(\mathcal{I}) \subseteq D$, $F(T_C) \subseteq \mathcal{P}(D)$, $F(T_R^i) \subseteq \mathcal{P}(D^i)$ for each $i \in \{1, \ldots, k\}$ (k is the maximum arity of a relation type in T_R) satisfying:
- $\forall t_c, t_c' \in T_C, t_c \leq t_c' \Rightarrow F(t_c) \subseteq F(t_c')$,
- $\forall t_r, t_r' \in T_R^i, t_r \leq t_r' \Rightarrow F(t_r) \subseteq F(t_r')$.

An **assignment** allows to link the concepts of a CG to the domain (universe) of the model defined over its support.

Definition 2. (Assignment)
Let $\mathcal{M} = (D, F)$ be a model for the support $S = (T_C, T_R, \mathcal{I}, *)$, and $SG = [S, G, \lambda]$ be a CG, with $G = (V_C, V_R, N_G)$.
An assignment for SG in $\mathcal{M}$ is a function $f : V_C \rightarrow D$ such that
-$\forall c \in V_C$, if $\lambda(c) = (t_c, ref_c)$ then $f(c) \in F(t_c)$, and if $ref_c \in \mathcal{I}$ then $f(c) = F(ref_c)$;
-$\forall r \in V_R$, if $deg_G(r) = i$ then $(f(N_G^1(r)), \ldots, f(N_G^i(r))) \in F(\lambda(r))$.
The set of all assignments for SG in the model $\mathcal{M}$ is denoted $\mathcal{A}(SG, \mathcal{M})$. If $\mathcal{A}(SG, \mathcal{M}) \neq \emptyset$ then SG holds in $\mathcal{M}$ and is denoted $\mathcal{M} \Vdash SG$.

The soundness of projection now follows as a simple observation. Indeed, let SG and SH be two Conceptual Graphs on the same support S such that $SG \geq SH$ and let $\mathcal{M}$ be a model for S.

Each assignment f for SH in $\mathcal{M}$ can be used to construct an assignment f' for SG in $\mathcal{M}$, by defining $f'(c) = f(\pi(c))$, where π is some projection from SG to SH.

Hence, we have obtained that *if $SG \geq SH$ and $\mathcal{M} \Vdash SH$ then $\mathcal{M} \Vdash SG$.*

Given a data source, we need to be able to link the information (set of tuples) contained therein with the associated Conceptual Graph and its model. To do this we introduce the notion of a **repository**. A repository is a set of tuples, each of which makes the Conceptual Graph true in a given model.

The repository is intentional (as opposed to extensional); one needs to go through the data source to be able to build it. There is no need to materialize the repository in order to make use of it (in the manner of materialized views for databases). Intuitively, a repository contains all possible interpretations for the generic (marked with "*") concepts in the graph.

Definition 3. (Repository)
Let $\mathcal{M} = (D, F)$ be a model for the support $S = (T_C, T_R, \mathcal{I}, *)$, and $SG = [S, G, \lambda]$ be a CG, with $G = (V_C, V_R, N_G)$.
We set $V_C := V_C(*) \cup V_C(\mathcal{I})$, where for each $c \in V_C$ with $\lambda(c) = (t_c, ref_c)$ we have $c \in V_C(*)$ if $ref_c = *$, and $c \in V_C(\mathcal{I})$ if $ref_c \in \mathcal{I}$. We also suppose that $V_C(*) \neq \emptyset$ and that an ordering $V_C(*) = \{c_1, \ldots, c_p\}$ is fixed.
The repository for SG in the model $\mathcal{M}$, is the set $\mathcal{R}(SG, \mathcal{M}) \subseteq D^p$ of all tuples $(d_1, \ldots, d_p) \in D^p$ with the property that the mapping $f : V_C \to D$, defined by

$$f(c_i) := d_i, \text{ for } c_i \in V_C(*), \quad \text{and } f(c) := F(ref_c), \text{ for } c \in V_C(\mathcal{I}),$$

is an assignment for SG in $\mathcal{M}$.

4.2   Querying the Data Sources

Once the data sources are defined, we need to be able to query and integrate them with other sources. In this section we define the main querying mechanisms for our model and how the results are retrieved. We also formally introduce the notion of a knowledge oriented specification, and we present our integration methodology.

In order to be able to query the data sources, we introduce a structure called a **query Conceptual Graph** (qCG). This structure is similar to that introduced in [5], but in this paper we define it in a new, graph theory oriented, light.
A query Conceptual Graph allows one to represent a query over the sources in a Conceptual Graph like notation. Basically, to find all the information about a generic concept, we mark it by "?". The "?" symbol stands for all the instances of a given type in the repository, which make the graph hold. The qCG has an associated Simple Conceptual Graph, whose intuitive purpose is to represent the query graph without any "?". Later on, when defining an answer to a qCG, this graph is important because it helps validate answers.

Definition 4. (Query Conceptual Graph)
A query Conceptual Graph (abbreviated qCG) is quadruple $\mathbb{Q} = [SQ, arity, X, \lambda'_Q]$, where
- $SQ = [S, Q, \lambda_Q]$ is a CG with $Q = (V_C, V_R, N_Q)$,
- arity is a positive integer,
- $X \subseteq V_C(*)$, and
- $\lambda'_Q : X \to \{?^1, ?^2, \ldots, ?^{arity}\}$ is a surjective labelling (with query marks).
SQ is the Conceptual Graph associated to qCG $\mathbb{Q}$, arity is the arity of $\mathbb{Q}$, and X are the query concept vertices of $\mathbb{Q}$.

Let SG be a CG and $\mathbb{Q}$ be a qCG both defined on the same support S. We could define the answer to $\mathbb{Q}$ over SG as the union of the repositories of all spanned subgraph of SG on which SQ (the the Conceptual Graph associated to $\mathbb{Q}$) can be projected. However, if SQ has a concept vertex labelled with an individual marker then this vertex can be projected only in a concept vertex of SG labelled with the same individual marker, by the definition of a projection (if $i_1, i_2 \in \mathcal{I}$

then $i_1 \geq i_2$ if and only if $i_1 = i_2$; however, $* \geq i$ for all $i \in \mathcal{I}$). This works if the source represented by SG is a collection of Conceptual Graphs, which is not feasible in an integration scenario. Therefore we introduce a specific querying mechanism, considering *legal queries* defined as follows.

Definition 5. (Legal Query)
Let SG be a CG and $\mathbb{Q}$ be a qCG both defined on the same support S. Let $SQ = [S, Q, \lambda_Q]$ be the Conceptual Graph associated to $\mathbb{Q}$ with $Q = (V_C, V_R, N_Q)$. We denote by SQ^* the CG obtained from SQ by replacing the individual markers with $*$, for all concept nodes belonging to the set $A = V_C \setminus V_C(*)$.

*We say that $\mathbb{Q}$ is a **legal query** for SG if the set, $Occ(\mathbb{Q}, SG)$, of the occurrences of $\mathbb{Q}$ in SG is nonempty, where $Occ(\mathbb{Q}, SG) = \{\pi(SG^*) | \pi \in \Pi_{SG^* \to SG}$ and for each $v \in A$, if $\lambda_G(\pi(v)) = (type, i)$, then $\lambda_Q(v) = (type', i)\}$.*

In words, $\mathbb{Q}$ is a legal query for SG if there is a spanned subgraph of SG $(\pi(SQ^*))$ into which SQ can be projected or into which SQ^* (the CG obtained from SQ by transforming all nodes in generic conceptual nodes) can be projected. In the second case, if the spanned subgraph of SG has individual concept nodes, these must be "compatible" with the corresponding individual concept nodes from SQ.

Therefore if $SH \in Occ(\mathbb{Q}, SG)$ then $SH \geq SG$ and either $SQ \geq SH$ or $SQ \not\geq SH$. In the last case however, we have chances to find an assignment in the repository of SG for which SQ holds.

By the above definition we have oriented the role of the SG to the description of querying capability of the source, rather then a "schematically" description of it. This is the first step to our "knowledge oriented specification" of a source. We further need to define the answering mechanism.

An **answer** to a qCG is the set of all data retrieved from the repository that validate the qCG. Intuitively, by taking all the instances from the repository that make the graph associated to the qCG true, one obtain its answer. This notion is very important because it helps us define a knowledge oriented specification for a given source.

Definition 6. (Answer to a qCG)
Let $SG = [S, G, \lambda]$ be a CG with $G = (V_C^1, V_R^1, N_G)$, $V_C^1(*) = \{c_1, \ldots, c_p\}$, $\mathcal{M} = (D, F)$ be a model for the support S, and $\mathcal{R}(SG, \mathcal{M})$ be a repository for SG in the model $\mathcal{M}$.
Let $\mathbb{Q} = [SQ, arity, X, \lambda_Q']$ be a legal qCG for the CG SG $(Occ(\mathbb{Q}, SG) \neq \emptyset)$. We define the answer to $\mathbb{Q}$ over $\mathcal{R}(SG, \mathcal{M})$ as being the set $Ans(\mathbb{Q}, \mathcal{R}(SG, \mathcal{M}))$ obtained with the following algorithm:

$Ans(\mathbb{Q}, \mathcal{R}(SG, \mathcal{M})) \leftarrow \emptyset$
for *each* $\pi(SQ^*) \in Occ(\mathbb{Q}, SG)$ **do**
    for *each* $(d_1^1, \ldots, d_p^1) \in \mathcal{R}(SG, \mathcal{M})$ **do** {
        compatible $\leftarrow$ *true*
        for $j = \overline{1, p}$ **do**
            if *compatible* **then**
                if $\exists c \in V_C(SQ^*)$ *s.t.* $[\pi(c) = c_j$ *and* $\lambda_Q(c) = (type, i)]$ **then**

$$\textbf{if } F(i) \neq d_j^1 \textbf{ then } compatible \leftarrow false$$
$$\textbf{if } compatible \textbf{ then } \{$$
$$\textbf{for } i = \overline{1, arity} \textbf{ do } \{$$
$$find \ c \in \lambda'^{-1}(i) \ and \ j \ such \ that \ \pi(c) = c_j$$
$$d_i \leftarrow d_j^1$$
$$\}$$
$$add \ (d_1, \dots, d^{arity}) \ to \ Ans(\mathbb{Q}, \mathcal{R}(SG, \mathcal{M}))$$
$$\}$$
$$\}$$

Theorem 1. *Let* $SG = [S, G, \lambda]$ *be a CG with* $G = (V_C^1, V_R^1, N_G)$, $V_C^1(*) = \{c_1, \dots, c_p\}$, $\mathcal{M} = (D, F)$ *be a model for the support* S, *and* $\mathcal{R}(SG, \mathcal{M})$ *be a repository for* SG *in the model* $\mathcal{M}$. *If* $\mathbb{Q} = [SQ, arity, X, \lambda'_Q]$ *is a legal qCG for the CG SG,then*
(i)if $SQ \geq SG$ *then* $Ans(\mathbb{Q}, \mathcal{R}(SG, \mathcal{M})) \neq \emptyset$.
(ii) if $SQ \ngeq SG$ *but in the* $\mathcal{R}(SG, \mathcal{M})$ *there is a tuple which gives rise to an assignment of a CG SG' obtained from SG by replacing some generic concept nodes by individual concept nodes and having the property that* $SQ \geq SG'$, *then* $Ans(\mathbb{Q}, \mathcal{R}(SG, \mathcal{M})) \neq \emptyset$.

Proof. Part (i) is a trivial corollary of the soundness of projection and part (ii) follows from the algorithm for the construction of the set $Ans(\mathbb{Q}, \mathcal{R}(SG, \mathcal{M}))$.

All notions introduced above lead now to the formal definition of a **knowledge oriented specification**.

A knowledge oriented specification of an information source is composed by
(i) a Conceptual Graph that visually describes what we expect to know from that source,
(ii) an interpretation for the support on which the graph is built,
(iii) a repository for the graph (that contains all the data tuples), and
(iv) a wrapper that ensures the communication between the user queries and the repository.

Definition 7. (Knowledge Oriented Specification of an Information Source)
Let IS *be an information source. A knowledge oriented specification of* IS *is a quadruple* $KOS(IS) = (SG, \mathcal{M}, \mathcal{R}(SG, \mathcal{M}), W)$, *where*
-SG $= [S, G, \lambda]$ *is a CG on the support* S, *the* source support,
- $\mathcal{M} = (D, F)$ *is a model for the support* S, *the* source model,
- $\mathcal{R}(SG, \mathcal{M})$ *is a repository for* SG *in the model* $\mathcal{M}$, *and*
-W is a wrapper, that is a software tool which, for each legal qCG $\mathbb{Q}$ *for* SG, *returns the answer set* $Ans(\mathbb{Q}, \mathcal{R}(SG, \mathcal{M}))$.

Figure 4 integrates all the above definitions and sketches a Knowledge Oriented Specification.

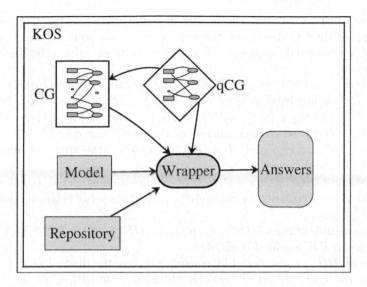

Fig. 4. Knowledge Oriented Specification

4.3   Data Integration

A CG Mixer depicts the integrated view, by the means of a Conceptual Graph, and provides the rules to allow for the translation of user queries to the appropriate data sources. The rules are defined by the relation vertices from the integrated view.

For each relation in the integrated view, the proper translation is provided. This translation has to preserve the order of nodes in the initial relation, hence the extra labelling of concepts (as depicted in greyed out rectangles).

Definition 8. (CG Mixer)
Let $IS^1, \ldots, IS^n$ be a set of information sources, and their knowledge oriented specifications $KOS(IS^i) = (SG^i, \mathcal{M}^i, \mathcal{R}^i(SG^i, \mathcal{M}^i), W^i), i = 1, n$.

A CG Mixer over the information sources $IS^1, \ldots, IS^n$ is a pair $\mathbb{M}(IS^1, \ldots, IS^n)$ $:= (SG^0, \mathbb{R})$, where
- $SG^0 = [S^0, G^0, \lambda^0]$ is a CG with $G^0 = (V_C^0, V_R^0, N_{G^0})$, and
- $\mathbb{R}$ is a mapping which specifies for each $r^0 \in V_R^0$ a set $\mathbb{R}(r^0)$ of rules providing descriptions of the relation vertex r^0 in (some of) information sources. Each rule in $\mathbb{R}(r^0)$ is a triple (IS^k, A, w), where
- *IS^k is an information source specified by $KOS(IS^k)$*
- *$A \subseteq V_R^k$ (the relation vertices set of SG^k)*
- *$w \in V_C^+([A]_{G^k})$ is a sequence of $d_{G^0}(r^0)$ concept vertices of the subgraph $[A]_{G^k}$ spanned in G^k by the relation vertices in A.*

A rule $(IS^i, A, w) \in \mathbb{R}(r^0)$ means that the star graph $G^0[r^0]$, is translated in the source IS^i as $[A]_{G^k}$ and if $w = w_1 \ldots w_k$ $(k = d_{G^0}(r^0))$, then w_j corresponds to $N_{G^0}^j(r^0)$ $(j = 1, k)$.

In other words, a rule interprets each relation vertex in the CG Mixer via a subgraph of the CG describing the appropriate local source. This is done by means of an ordered sequence of concept vertices (the relations' vertex neighbors).

Let $M(IS^1,\ldots,IS^n) = (SG^0, \mathbb{R})$ be a CG Mixer. A *legal query over* M $(IS^1,\ldots,IS^n)$ is any legal qCG for SG^0. Let $\mathbb{Q} = [SQ, arity, X, \lambda'_Q]$, be a legal qCG for SG^0, with $SQ = [S, Q, \lambda_Q]$, $Q = (V_C, V_R, N_Q)$, and $X \subseteq V_C(*)$. Consider also $Occ(\mathbb{Q}, SG)$, the set of the occurrences of $\mathbb{Q}$ in SG (see definition 5).

Let $V_R = \{r_1^0,\ldots,r_m^0\}$ and $H = [\{r_1^0,\ldots,r_m^0\}]_{G^0}$ (the spanned subgraph of G^0 from which is obtained SQ by specialization).

From SQ and $(SG^0, \mathbb{R})$ a set $\mathbb{R}(SQ)$ of graphs is constructed as follows.

For each $H \in Occ(\mathbb{Q}, SG)$ consider $\{r_1^0,\ldots,r_m^0\}$ its set of relation nodes ($H = [\{r_1^0,\ldots,r_m^0\}]_{G^0}$).

For each m-uple of rules $((IS^{k_1}, A^1, w^1),\ldots,(IS^{k_m}, A^m, w^m)) \in \mathbb{R}(r_1^0) \times \ldots \times \mathbb{R}(r_m^0)$ a graph RH is added to $\mathbb{R}(SQ)$.

The graph RH is constructed by considering first the union $RH = F^1 \cup \ldots \cup F^m$. Here, the graph F^i, $(i = 1, m)$, is obtained from $[A^i]_{G^{k_i}}$ in the following way: if the concept vertex w_j^i, $(j = 1, d_{SQ}(r_i^0))$, has a generic marker in SG^{k_i} and in SQ the j-neighbor of r_i^0 has been replaced by an individual marker, then the generic marker of w_j^i is replaced by this individual marker. Note that in the above union, the subgraphs coming from distinct sources are disjoint.

The final graph RH is obtained by adding, to the above obtained graph, a special set of new relation vertices in order to describe the neighborhood structure of the original graph H. All these vertices have the special label (name) $"="$ and have exactly two neighbors (with the meaning that the corresponding concept vertices represent the same object).
More precisely, for each $i, j \in \{1,\ldots,m\}$ such that $N_H^t(r_i^0) = N_H^s(r_j^0)$ (in H the t-neighbor of r_i^0 is the same concept vertex as the s-neighbor of r_j^0), and $IS^{k_i} \neq IS^{k_j}$ (the two subgraphs in which r_i and r_j are coming from distinct sources), a new equality relation vertex is added to the graph already constructed, with the 1-neighbor the vertex w_t^i of F^i and the 2-neighbor the vertex w_s^j of F^j.

The graphs from the set $\mathbb{R}(SQ)$ can be considered as the set of all possible query rewriting of $\mathbb{Q}$. Clearly, each of the $|\mathbb{R}(r_1^0)| \times \ldots \times |\mathbb{R}(r_m^0)|$ graphs added in $\mathbb{R}(SQ)$ for $H \in Occ(\mathbb{Q}, SG)$, $H = [\{r_1^0,\ldots,r_m^0\}]_{G^0}$, is constructed with the above algorithm in polynomial time (with respect to the orders of the subgraphs involved).

By the above construction, each graph $RH \in \mathbb{R}(SQ)$ can be expressed as a disjoint union of source subgraphs, interconnected (as described above) by the equality relation vertices.

Let us denote by RH^j be the (nonempty) subgraph of RH which is also a subgraph of the graph SG^j associated to the source IS^j. If we assign appropriate query marks to the concept vertices corresponding to the vertices of SQ having query marks, we obtain a legal qCG for the source SG^j. This can be obtained with the following algorithm:

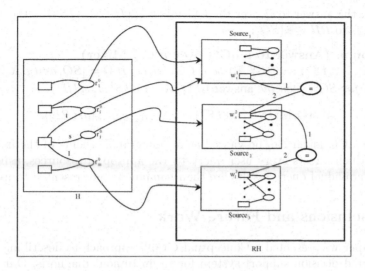

Fig. 5. Query Rewriting

For each concept vertex $c \in V_C(RH^j)$, if $c = w_k^j$, for some k (which means that there is r_i^0 for which a rule (IS^j, A^j, w^j) has been used in the construction of RH), and if $N_{SQ}^k(r_i^0)$ has a query marker, then assign a query marker to w_k^j.

The superscripts of these new query markers can be given such that they form a set $\{1, \ldots, arity'\}$ and also respect the meaning in $\mathbb{Q}$ (that is, if two original vertices in SQ have the same query mark, then their surrogates in RH^j have the same new query mark).

Clearly, the above algorithm constructs a legal qCG $\mathbb{Q}_{RH}^j$ for SG^j.

Therefore to each graph $RH \in \mathbb{R}(SQ)$ we have associated a set of legal qCG's for the graphs describing the sources involved in the construction of RH.

If $J \subseteq \{1, \ldots, n\}$ is the index set of the sources IS^j involved in the construction of RH, then this set of legal qCG's is $\mathbb{Q}(RH) := \{\mathbb{Q}_{RH}^j | j \in J\}$.

The answer, $Ans(RH)$, to RH over $\mathbb{Q}(RH)$ is constructed with the following algorithm:

- For each $j \in J$ find the set $Ans(\mathbb{Q}_{RH}^j, \mathcal{R}^j(SG^j, \mathcal{M}^j))$ (using the wrapper W^j);
- $Ans(RH) := \emptyset$;
- For each element of the set $\times_{j \in J} Ans(\mathbb{Q}_{RH}^j, \mathcal{R}^j(SG^j, \mathcal{M}^j))$ verify if the values corresponding to the concept vertices which are the two neighbors of some equality relation vertex in RH are equal. If all these tests are successfully add the tuple obtained by concatenating the components of this element, to the set $Ans(RH)$.

The above test depends on the number of equality relation vertices of the graph RH and, clearly, can be implemented in linear time. Also, some optimization of the construction can be considered; for example, if some set $Ans(\mathbb{Q}_{RH}^j, \mathcal{R}^j(SG^j, \mathcal{M}^j))$ is empty, then $Ans(RH)$ is also empty.

Finally, the answer to $\mathbb{Q}$ over the CG mixer $\mathbb{M}(IS^1, \ldots, IS^n)$ is the union of the answers to $RH \in \mathbb{R}(SQ)$:

Definition 9. (Answer to a qCG over a CG Mixer)
Let $\mathbb{M}(IS^1, \ldots, IS^n) := (SG^0, \mathbb{R})$ be a CG Mixer. If $\mathbb{Q} = [SQ, arity, X, \lambda'_Q]$ is a legal qCG for SG^0, then the answer to $\mathbb{Q}$ over $\mathbb{M}(IS^1, \ldots, IS^n)$ is

$$\mathcal{A}ns(\mathbb{Q}, \mathbb{M}(IS^1, \ldots, IS^n)) := \cup_{RH \in \mathbb{R}(SQ)} \mathcal{A}ns(RH).$$

Therefore a CG mixer can be viewed as an integrated schema of the individual sources, which directs every user query to the appropriate sources, using a set of rules. Individual query results are then combined and presented to users.

5   Conclusions and Future Work

In this paper we presented a Conceptual Graph approach to describing data in a distributed decision support system for brain tumour diagnosis. Our work is theoretical and it explains how Conceptual Graphs expressivity and easy plug in capabilities benefit such system.

At the moment we do not explicitly represent knowledge regarding problem solving methods. That is to say, our approach captures only the static model rather than the inference procedures. Typical examples of the former are "patient", "particular type of tumour", "MRS scans with their parameters", etc. while examples of the latter are "due to the fact that ... the tumour is malignant" or "peak areas with ... characters suggest ...". Future work will address extending the KOS (and subsequently the CG Mixer) with rules to address this problem.

On the other hand, a medical diagnosis is normally a complicated process with ambiguity and uncertainty which cannot be entirely and precisely formalised in an inference model good for taxonomic knowledge. This, however, does not deny the merit of building a reasoning system on top of HADOM to provide moderate suggestions and warnings to clinicians instead of replacing them. We are also investigating a Conceptual Graph based case base reasoning approach for HealthAgents.

References

1. Arús, C., Celda, B., Dasmahapatra, S., Dupplaw, D., González-Vélez, H., van Huffel, S., Lewis, P., Lluch, M., i Ariet, M.L., Mier, M., Peet, A., Robles, M.: On the design of a web-based decision support system for brain tumour diagnosis using distributed agents. In: WI-IATW'06. 2006 IEEE/WIC/ACM Int Conf on Web Intelligence & Intelligent Agent Technology, Hong Kong, pp. 208–211. IEEE, Los Alamitos (2006)
2. Chein, M., Mugnier, M.-L.: Conceptual graphs: Fundamental notions. Revue d'Intelligence Artificielle 6(4), 365–406 (1992)
3. Chein, M., Mugnier, M.-L., Simonet, G.: Nested graphs: A graph-based knowledge representation model with FOL semantics. In: Proc. of the 6th Int'l Conf. on the Principles of Knowl. Repres. and Reasoning (KR'98), pp. 524–535. Morgan Kaufmann, San Francisco (1998)

4. Croitoru, M., Compatangelo, E.: Conceptual graph projection: a tree decomposition-based approach. In: Doherty, P., Mylopuolos, Welty, C. (eds.) Proc. of the 10th Int'l Conf. on the Principles of Knowledge Representation and Reasoning (KR'2006), pp. 271–276. AAAI, Stanford, California, USA (2006)
5. Dau, F.: Query Graphs with Cuts: Mathematical Foundations. In: Blackwell, A.F., Marriott, K., Shimojima, A. (eds.) Diagrams 2004. LNCS (LNAI), vol. 2980, pp. 32–50. Springer, Heidelberg (2004)
6. Sowa, J.F.: Conceptual Structures: Information Processing in Mind and Machine. Addison-Wesley, London, UK (1984)
7. Tate, A.R., Underwood, J., Acosta, D.M., Julia-Sape, M., Majos, C., Moreno-Torres, A., Howe, F.A., van der Graaf, M., Lefournier, M.M., Murphy, F., Loosemore, A., Ladroue, C., Wesseling, P., Bosson, J.L., Simonetti, A.W., Gajewicz, W., Calvar, J., Capdevila, A., Wilkins, P., Bell, A.C., Remy, C., Heerschap, A., Watson, D., Griffiths, J.R., Arus, C.: Development of a decision support system for diagnosis and grading of brain tumours using in vivo magnetic resonance single voxel spectra. NMR Biomed 19, 411–434 (2006)

A Conceptual Graph Based Approach to Ontology Similarity Measure

Madalina Croitoru, Bo Hu, Srinandan Dashmapatra, Paul Lewis,
David Dupplaw, and Liang Xiao

University of Southampton

Abstract. This paper presents a combinatorial, structure based approach to the problem of finding a (di)similarity measure between two Conceptual Graphs. With a growing number of ontologies and an increasing need for quick, on the fly knowledge integration and querying, ontology similarity measures are essential for building the foundations of the Semantic Web. Conceptual Graphs benefit from a graph based representation that can be exploited in versatile optimisation techniques. We propose a disimilarity measure based on the content and the structure of two graphs. This disimilarity measure is based on the clique number of the matching graph, a combinatorial structure which encodes the two graphs projection information.

1 Motivations and Rationale

In this paper we present a structural, Conceptual Graph [11] based approach to the problem of finding ontology (di)simmilarity measures. Conceptual Graphs are a visual, logic based knowledge representation formalism. The ontological knowledge is represented in the support which is a poset of concept and relation hierarchies. The factual knowledge is represented in a bipartite graph where the two classes of partition contain concept and relation nodes from the support. We propose a (di)similarity measure in between two Conceptual Graphs which considerers the inherent structural properties of the two graphs. More precisely, by considering both relation adjacency in the bipartite graph and the relation hierarchy in the support, we devise a combinatorial structure, the matching graph, which can then be used to deduce an interesting (di)similarity value.

Finding a (di)similarity measure between two Conceptual Graphs is a important problem in an information era where more and more ontologies are employed for powerful applications (e.g. The Semantic Web [2]). Conceptual Graphs benefit from an easy plug in capabilities making it easy to employ and extend existing ontologies. For intelligent knowledge based applications it is important to be able to compare the represented knowledge by providing a "meaningful" (di)similarity measure. The aim of this paper is to theoretically lay the foundations for such (di)similarity measure between two Conceptual Graphs. Future work aims at translating RDF [12] / OWL DL [13] ontologies into Conceptual Graphs and showing how the structural properties of this (di)similarity measure add extra benefit.

U. Priss, S. Polovina, and R. Hill (Eds.): ICCS 2007, LNAI 4604, pp. 154–164, 2007.

Since a large benchmark of Conceptual Graphs is still under development, our work is theoretical and evaluated by its own novel approach to a reasoning founded (di)similarity measure. Indeed, existing work on Conceptual Graph comparison (for example [8], [10], [9]) does not address the interesting structural properties that arise from two neighborhood relation nodes and the inherent combinatorial properties the projection raises on such structural features.

This paper is structured as follows. In Section 2 we formally introduce Conceptual Graphs and the projection operator. The aim of this section is to present rigurous definitions that allow future sections to rigurously present our approach to projection checking and subsequently (di)simmilarity measures. Section 3 presents the Matching Graph, a combinatorial structure for projection checking. This structure exploits both the structural interdependencies between the relation nodes in the bipartite graph and the relation type hierarchy in the support. This, and the fact that projection as a reasoning mechanism in itself aims at knowledge comparison is the motivation of this work. Indeed, our claim is that not only one should consider semantically sound transformation for Conceptual Graph comparison but also the graph structure of the factual knowledge. We believe that the Matching Graph, by its definition, is an effective tool to address this claim. The section finishes by presenting further optimizations which are exploited to construct the Reduced Matching Graph. As mentioned before, we employ projection checking as the foundation of our approach since focusing on reasoning tools for Conceptual Graphs implicitly addresses knowledge comparison problems. This claim will allow us to define a structurally rigorous (di)similarity measure in Section 4 based on the Reduced Matching Graphs clique number of the two graphs to be compared. Section 5 concludes the paper.

2   Conceptual Graphs

Some of the definitions in this section follow the work of [3]. Background knowledge for Conceptual Graphs is encoded in a structure called support which consists of a concept type hierarchy, a relation type hierarchy, a set of individual markers that refer to specific concepts and a generic marker, denoted by $*$, which refers to an unspecified concept.

Definition 1 (Support). *A support is a 4-tuple* $S = (T_C, T_R, \mathcal{I}, *)$ *where:*

- T_C *is a finite, partially ordered set (poset) of* **concept types** $(T_C, \leq)$ *that defines a type hierarchy where* $\forall x, y \in T_C$, $x \leq y$ *means that x is a subtype of y. The top element of this hierarchy is the universal type* $\top_C$.
- T_R *is a finite set of relation types partitioned into k posets* $(T_R^i, \leq)_{i=1,k}$ *of relation types of arity i $(1 \leq i \leq k)$, where k is the maximum arity of a relation type in T_R. Each relation type of arity i, namely $r \in T_R^i$, has an associated* **signature** $\sigma(r) \in \underbrace{T_C \times \ldots \times T_C}_{i \ times}$, *which specifies the maximum concept type of each of its arguments. This means that if we use $r(x_1, \ldots, x_i)$, then x_j is a concept of type$(x_j) \leq \sigma(r)_j$ $(1 \leq j \leq i)$. The partial orders*

*on relation types of the same arity must be **signature-compatible**, i.e.*
$\forall r_1, r_2 \in T_R^i \; r_1 \leq r_2 \Rightarrow \sigma(r_1) \leq \sigma(r_2)$.

- $\mathcal{I}$ *is a countable set of* ***individual markers***.
- $*$ *is the* ***generic marker*** *that refers to an unspecified concept.*
- *The sets T_C, T_R, $\mathcal{I}$ and $\{*\}$ are mutually disjoint.*
- $\mathcal{I} \cup \{*\}$ *is partially ordered by $x \leq y$ if and only if $x = y$ or $y = *$.*

A conceptual graph is a structure that depicts factual information about the background knowledge contained in its support. This information is presented in a visual manner as an ordered bipartite graph, whose nodes have been labelled with elements from the support.

Definition 2. *A **simple conceptual graph** (SCG) is a 3-tuple $SG = [S, G, \lambda]$, where:*

- $S = (T_C, T_R, \mathcal{I}, *)$ *is a support;*
- $G = (V_C, V_R; E_G, l)$ *is an ordered bipartite graph;*
- λ *is a labelling of the nodes of G with elements from the support S:*
  $\forall r \in V_R, \; \lambda(r) \in T_R^{d_G(r)}; \quad \forall c \in V_C, \; \lambda(c) \in T_C \times (\mathcal{I} \cup \{*\})$ *such that*
  if $c = N_G^i(r)$, $\lambda(r) = t_r$ and $\lambda(c) = (t_c, ref_c)$ then $t_c \leq \sigma_i(r)$.

Conceptual graphs represent knowledge at a syntactic level. Projection (subsumption) - a labelled graph homomorphism - is the main tool for reasoning with SCGs. This is done by preserving the order of the neighbors in the two graphs and comparing the types and labels of the nodes / relations. Projection corresponds to deduction for the existential conjunctive and positive fragment of first order logic ([4]). In the following, when the support is implicit we will just use a tuple $(G, \lambda_G$ for denoting a simple conceptual graph SG.

Definition 3 (Projection). *If $SG = (G, \lambda_G)$ and $SH = (H, \lambda_H)$ are two simple conceptual graphs defined on the same support S, then a **projection** from SG to SH is a mapping*
$\pi : V_C(G) \cup V_R(G) \rightarrow V_C(H) \cup V_R(H)$, *such that:*

- $\pi(V_C(G)) \subseteq V_C(H)$ *and* $\pi(V_R(G)) \subseteq V_R(H)$;
- $\forall c \in V_C(G)$ *and* $\forall r \in V_R(G)$, *if* $c = N_G^i(r)$ *then* $\pi(c) = N_H^i(\pi(r))$;
- $\forall v \in V_C(G) \cup V_R(G)$, $\lambda_G(v) \geq \lambda_H(\pi(v))$.

The order on λ in the above definition preserves the order on T_C (T_R) and considers the elements of $\mathcal{I}$ mutually incomparable (as previously defined). If there is a projection from SG to SH (that is $\Pi_{G \rightarrow H} \neq \emptyset$), then SG **subsumes** SH (denoted as $SG \geq SH$). The **subsumption relation** is a pre-order on the set of all Simple Conceptual Graphs defined on the same support. This is the starting point of our research that made us look at projection optimisation techniques for implicitly finding Conceptual Graphs (di)similarity measures.

3   Matching Graph

Let us consider two simple conceptual graphs, G and H, for which we want to test if $G \geq H$ holds. For each relation node r of the graph G let us consider the set $Canditates_0(r)$. This is the set of all relation nodes of H in which r can be individually projected. The only criteria for the nodes in $Canditates_0(r)$ is to be type compatible with r. However, the projections candidates for two relation nodes r and r' have to be compatible also from the shared neighbor concept nodes view point. More precisely, if $s \in Canditates_0(r)$ then the pair (r, s) is an individual projection. Two individual projections (r, s) and (r', s') are compatible if the common neighbors of r and r' in G are preserved (in the right order) by s and s' in H. If we define a graph having individual projection as nodes and the edges given by the nodes compatibility, then a clique in this graph will ensure the overall compatibility of its members. Therefore we translate the projection checking of the two SCGs into finding a clique whose cardinality equals the number of relation nodes of the first graph [5].

Definition 4. *Let $SG = (G, \lambda_G)$ and $SH = (H, \lambda_H)$ be two SCGs with no isolated concept vertices defined on the same support S. The matching graph of SG and SH is defined as the graph $\mathbb{M}_{G \to H} = (V, E)$ where:*

- *$V \subseteq V_R(G) \times V_R(H)$ is the set of all pairs (r, s) such that $r \in V_R(G), s \in V_R(H), \lambda_G(r) \geq \lambda_H(s); \forall i \in \{1, \ldots, d_G(r)\}, \lambda_G(N_G^i(r)) \geq \lambda_H(N_H^i(s))$ and $\forall i, j \in \{1, \ldots, d_G(r)\}$ if $N_G^i(r) = N_G^j(r)$ then $N_G^i(s) = N_G^j(s)$.*
- *E is the set of all 2-sets $\{(r, s), (r', s')\}$, where $r \neq r'$, $(r, s), (r', s') \in V$ and $N_H^i(s) = N_H^j(s') \forall i \in \{1, \ldots, d_G(r)\}, \forall j \in \{1, \ldots, d_G(r')\}$ such that $N_G^i(r) = N_G^j(r')$.*

A clique in a graph F is a set of mutually adjacent vertices. The maximum cardinality of a clique in F is denoted as $\omega(F)$.

The theorem below shows that if SG and SH are two simple conceptual graphs without isolated nodes, then the problem of finding a projection from SG to SH is equivalent to finding a maximum cardinality clique in $\mathbb{M}_{G \to H}$.

Theorem 1. *Let $SG = (G, \lambda_G)$ and $SH = (H, \lambda_H)$ be two simple conceptual graphs without isolated concept vertices defined on the same support S and let $\mathbb{M}_{G \to H} = (V, E)$ be their matching graph. There is a projection from SG to SH if and only if $\omega(\mathbb{M}_{G \to H}) = |V_R(G)|$.*

For a proof of this Theorem see [6].

Let us consider the two Simple Conceptual Graphs depicted in Figure 1 and let us assume that the relation nodes types r and s are comparable: $r > s$.

Figure 2 depicts the matching graph associated to the graphs G and H. The columns associated to the relation nodes x and y represent their Candidates sets. Each candidate is a node in the matching graph. An edge is drawn if the two nodes are compatible from a neighbor concept nodes view point.

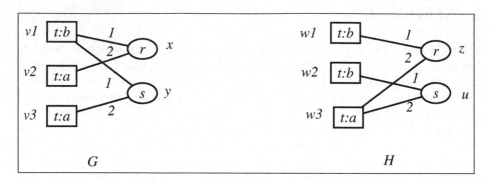

Fig. 1. $G \geq H$ iff $r \geq s$

In the previous section we saw that $Candidates_0(x) = \{z, u\}$ and $Candidates_0$ $(y) = \{u\}$. Therefore the three nodes of the matching graph associated to G and H are the nodes corresponding to $\{x, z\}$, $\{x, u\}$ and $\{y, u\}$. The edges of the matching graph are drawn according to its nodes compatibility. If we try and project x over z and y over u the two projections are not compatible because the first concept neighbor node of the node x does not correspond to the first concept neighbor node of y. However if we project x over u and y over u then the two projections are compatible also from a concept neighbor node view point. Hence, the only edge of the graph will link the nodes $\{x, u\}$ and $\{y, u\}$.

Let us consider another example depicted in Figure 3. The vertices of the graph $\mathbb{M}_{G \to H}$ shown in are represented by black circles labelled using the pair obtained by taking the corresponding column and row relation vertices. For example, vertices (x_1, y_1) and (x_1, y_4) mean that the relation node x_1 of G can be projected to relation node y_1 or y_4 of H. The only drawn edges of $\mathbb{M}_{G \to H}$ are those of the complete subgraph induced by $\{(x_1, y_1), (x_2, y_2), (x_3, y_2), (x_4, y_1),$ $(x_5, y_2)\}$. For example, the edge $\{(x_1, y_1)(x_4, y_1)\}$ means that the projection of x_1 and x_4 to y_1 preserves the (ordered) adjacency in G, since both $v_4 = N_G^2(x_1) = N_G^2(x_4)$ and $v_5 = N_G^3(x_1) = N_G^3(x_4)$ are satisfied by their common projection y_1 in H.

It follows that by projecting x_1 to y_1, x_2 to y_2, x_3 to y_2, x_4 to y_1, x_5 to y_2, as well as $v_1 = N_G^1(x_1) = N_G^1(x_2) = N_G^1(x_3) = N_G^1(x_5)$ to $u_1 = N_H^1(y_1) = N_H^1(y_2)$, $v_2 = N_G^2(x_2)$ to $u_2 = N_H^2(y_2)$, $v_3 = N_G^1(x_4)$ to $u_1 = N_H^1(y_1)$, $v_4 = N_G^2(x_1) = N_G^2(x_3) = N_G^2(x_4) = N_G^2(x_5)$ to $u_2 = N_H^2(y_1) = N_H^2(y_2)$, we obtain a projection Π from SG to SH.

The matching graph described above computes a number of unnecessary comparisons. This process can be optimized by the Reduced Matching Graph which performs a successive step of deletions to avoid unneeded computation [6]. This corresponds to the *arc consistency* processing intensively discussed in the Constraint Satisfaction Problem field [7].

More specifically if $(r, s) \in V(\mathbb{M}_{G \to H})$ is a node such that there is an $r' \in V_R(G)$, $r' \neq r$ with $N_{\mathbb{M}_{G \to H}}((r, s)) \cap V_{r'} = \emptyset$, then (r, s) belongs to no $|V_R(G)|$-clique and can be thus deleted from $\mathbb{M}_{G \to H}$.

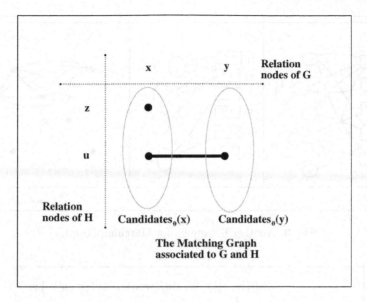

Fig. 2. The Matching Graph

Hence a node in $\mathbb{M}_{G \rightarrow H}$ having no neighbor in some V_r class cannot belong to a clique with $|V_R(G)|$ nodes. Therefore this node can be deleted from the matching graph. However if the deleted node is the unique neighbor in its class to some other node, then the latter node can be deleted as well.

This gives rise to a cascading sequence of deletions which allow us to simplify the matching graph.

The algorithm devised to construct the reduced matching graph is as follows.

1. **for** each $r \in V_R(G)$ **do** {
    $V_r \leftarrow \emptyset$;
    for each $s \in V_R(H)$ **do**
      if $\lambda_G(r) \geq \lambda_H(s)$
        and $\lambda_G(N_G^i(r)) \geq \lambda_H(N_H^i(s))_{i=1,d_G(r)}$
      then $V_r \leftarrow V_r \cup \{(r,s)\}$
                      }

2. **for** each $r \in V_R(G)$ **do**
    for each $(r,s) \in V_r$ **do**
      for each $r' \in V_R(G), r' \neq r$ **do** {
        $d_{\mathbb{M}_{G \rightarrow H}}((r,s))_{r'} \leftarrow 0;\ Adj((r,s))_{r'} \leftarrow \emptyset$;
        for each $(r',s') \in V_{r'}$ **do**
          if $\forall i\ N_G^i(r) = N_G^i(r') \Rightarrow N_H^i(s) = N_H^i(s')$
          then { $d_{\mathbb{M}_{G \rightarrow H}}((r,s))_{r'} + +$;

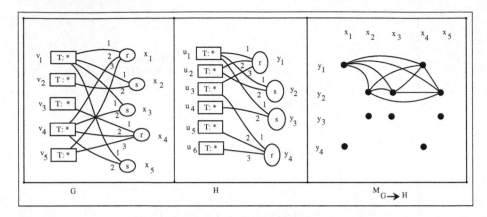

Fig. 3. Another Example of a Matching Graph

$$Adj((r,s))_{r'} \leftarrow Adj((r,s))_{r'} \cup \{(r',s')\ \};$$
$$\}$$
$$\}$$

3. Initialize as empty a stack S of vertices of $\mathbb{M}_{G \rightarrow H}$;
   for each $r \in V_R(G)$ **do**
       for each $(r,s) \in V_r$ **do**
           if $\exists r' \in V_R(G), r' \neq r$ s.t. $d_{\mathbb{M}_{G \rightarrow H}}((r,s))_{r'} = 0$
           then add (r,s) to S ;
   while S is nonempty **do** {
       unstack (r,s), the top element of stack S;
       for each $r' \in V_R(G), r' \neq r$ **do**
           for each $(r',s') \in Adj((r,s))_{r'}$ **do**   {
               delete (r,s) from $Adj((r',s'))_r$;
               $d_{\mathbb{M}_{G \rightarrow H}}((r',s'))_r \leftarrow d_{\mathbb{M}_{G \rightarrow H}}((r',s'))_r - 1$;
               if $d_{\mathbb{M}_{G \rightarrow H}}((r',s'))_r = 0$
               then add (r',s') to S;
                                                              }
       $V_r \leftarrow V_r - \{(r,s)\}$
                                  }

Let us explain in detail the above algorithm. Step 1 of the algorithm constructs the sets V_r. More precisely, for each relation node r in G we test which relation node s in H is a suitable (simply from a type compatibility view point) candidate for projection. We assume, as mentioned before, that type comparison can be done in a constant $O(1)$ time.

Step 2 of the algorithm constructs the edges of the matching graph. For every node of the matching graph, we compute its adjacency sets with respect to every set V_r not containing the node itself. The cardinality of these sets are stored in the node's degree vector $d_{\mathbb{M}_{G \to H}}$.

In step 3 the nodes having a 0 component in the degree vector are collected in a stack S in order to be deleted. While the stack is not empty the top node $\{r, s\}$ of the stack is deleted: if some node has $\{r, s\}$ in the corresponding adjacency set then $\{r, s\}$ is also removed from that set and its corresponding component in the degree vector is decreased.

Let m_G (m_H) be the number of relation vertices of graph G (H) and e_G (e_H) the number of edges of graph G (H). If the comparison of each label is accomplished in constant time, then the time complexity of algorithm step 1 is $\mathcal{O}((m_G + e_G) \cdot m_H)$. This is because every relation node r of G is compared to every relation nodes in H and, if the comparison result is satisfactory, the neighbor concept nodes have to match as well. This last comparison is performed, in the worst case, $d_G(r)$ times. Therefore the overall time spent by Step 1 is $O(m_G + \sum_{r \in V_R(G)} d_G(r)) \cdot m_H = \mathcal{O}((m_G + e_G) \cdot m_H)$.

Moreover, as the number of vertices of graph $\mathbb{M}_{G \to H}$ is $\mathcal{O}(m_G \cdot m_H)$, the time complexity of algorithm step 2 is $\mathcal{O}(m_G^2 \cdot m_H^2)$.
In algorithm step 3, the vertices (r, s) of graph $\mathbb{M}_{G \to H}$ are collected in the stack S in $\mathcal{O}(m_G^2 \cdot m_H)$. These vertices will be deleted, since they have no neighbors in some nonempty stable set $V_{r'}$ ($r' \neq r$).

The node deletion process is controlled in order to capture all the new occurrences of a node that must be deleted in the current graph. In this case, each node occurrence is added to the stack S. The overall complexity of algorithm step 3 is dominated by step 2.

Hence, graph $\mathbb{R}\mathbb{M}_{G \to H}$ can be constructed in $\mathcal{O}(m_G^2 \cdot m_H^2)$ time.

Since the deleted nodes of $M_{G \to H}$ cannot belong to a maximum clique, the result stated by Theorem 1 also holds for the reduced matching graph $\mathbb{R}\mathbb{M}_{G \to H}$. This allows us to derive the following algorithm for projection checking:

---

1. *Construct the reduced matching graph* $\mathbb{M}_{G \to H}$;
2. *Find the clique number* $\omega(\mathbb{M}_{G \to H})$;
3. **If** $\omega(\mathbb{M}_{G \to H}) < |V_R(G)|$
   then return "G $\not\geq$ H"
   else return "G $\geq$ H", Π

The clique number of the reduced matching graph can be considered as a non-trivial measure for comparing conceptual graphs. More precisely, the difference between $|V_R(G)|$ and $\omega(\mathbb{M}_{G \to H})$ can be considered as the "distance" between SG and SH.

In the next section we will exploit the inherent combinatorial, structural properties of the Reduced Matching Graphs clique number in order to retrieve a (di)similarity measure in between two Conceptual Graphs.

4   Comparison Measure

This sections explicitly gives the semantic (di)simmilarity measure intuitively emerged from the reduced matching graph manoeuvre. In this paper we used the term "(di)simmilarity" since we are referring to a measure that does not consider the order of the graphs to be compared. More precisely we want to ensure that the measure of closeness between graph G and H is the same with the one between graph H and graph G.

More importantly since we are founding the proposed (di)simmilarity measure on the reasoning aspect we want to consider both situation in which both graphs project onto each other. Indeed there could be cases in which one graph will project into the other one but the reverse does not hold. These two observations lead to the following (di)simmilarity measure definition.

Let $\mathcal{G}$ the class of all SCGs based on a fixed support S. We want to define a *(di)similarity semantic measure*, that is a function $d : \mathcal{G} \times \mathcal{G} \to \mathbf{R}_+$ such that for every $SCG_1, SCG_2 \in \mathcal{G}$:

(i) $d(SCG_1, SCG_2) = 0$ if and only if $SCG_1 \geq SCG_2$ and $SCG_2 \geq SCG_1$;

(ii) $d(SCG_1, SCG_2) = d(SCG_2, SCG_1)$.

In the previous section we explained how the Reduced Matching Graph is encoding all the projection information between two Conceptual Graphs. The clique number of this structure will give a good indication of how well the relations compare both as types (since the type comparison is done in order to define the RMG nodes) and adjacency (since the RMG edges are defined based on compatible relations). Of course the size of the two graphs to be compared also matters in order to get a good grasp on the problem. Based on the desired properties of the (di)simmilarity function mentioned above, we define the value of the (di)simmilairty function given by:

$$d(SCG_1, SCG_2) = |V_R^{G_1}| + |V_R^{G_2}| - \omega(\mathbb{M}_{G_1 \to G_2}) - \omega(\mathbb{M}_{G_2 \to G_1}).$$

It is easy to see that this function satisfies conditions (i) and (ii) in the definition of the (di)simmilarity measure.

- d is obviously a symmetrical function ($d(SCG_1, SCG_2) = d(SCG_2, SCG_1)$)
- $SCG_1 \geq SCG_2$ and $SCG_2 \geq SCG_1$ hold if and only if $\omega(\mathbb{M}_{G_1 \to G_2}) = |V_R(G_1)|$ and $\omega(\mathbb{M}_{G_2 \to G_1}) = |V_R(G_2)|$ if and only if $d(SCG_1, SCG_2) = 0$ (since the clique number of a matching graph cannot exceed the number of relation nodes).

5 Conclusions and Future Work

In this paper we presented a semantic (di)simmilarity measure between Conceptual Graphs. Our work was motivated by the belief that projection as such performs knowledge comparison, and exploiting he projection implicit combinatorial combinations can give raise to a reasoning oriented (di)simmilarity measure. Our work is theoretical and complexity results are proven. However we have not implemented our work since a large benchmark of Conceptual Graphs is currently unavailable [1]. In future we want to use RDF data as an initial test bed for our work, by translation RDF into Conceptual Graphs and then comparing our (di)simmilarity measure with existing work in the field. We are also investigation the translation of OWL DL ontologies into Conceptual Graph within the same purpose.

Also, from a theoretical point of view the above mentioned technique lays down interesting questions. More precisely let $SCG = (S, G, \lambda_G)$ and $SCH = (S, H, \lambda_H)$ be two simple conceptual graphs defined on the same support S. Suppose that $G \not\geq H$. The explanation of this is that either the *structural parts* (the texture as it has been called in the above problem) of the two graphs are not compatible or the labels order of the two graphs does not match. In the first case, nothing remains to be done for obtaining a positive response. In the second case it is possible to increase the labels of G in order to have a positive response. This could be useful in practical problems in which the factual knowledge expressed by the query G is not so general as it should be. More precisely:

> Given $SCG = (S, G, \lambda_G)$ and $SCH = (S, H, \lambda_H)$ two simple conceptual graphs defined on the same support S, such that $SCG \not\geq SCH$, is there λ^1 a labelling of G such that $\lambda^1|_{V_R^G} = \lambda|_{V_R^G}$ and $SCG^1 = (S, G, \lambda_G^1)$ satisfies $SCG^1 \geq SCH$?

Clearly, if λ^1 is taken $\lambda^1(v_c) = (\top, *)$ for each $v_c \in V_C^G$, and we obtain that $SCG^1 \not\geq SCH$, then the answer is *no*. But, if the answer is *yes* finding a minimal generalization of SCG which subsumes SCH seems to be a difficult algorithmic problem, since it is necessary to implement a search process into the support S.

References

1. Baget, J.-F., Carloni, O., Chein, M., Genest, D., Gutirrez, A., Leclere, M., Mugnier, M.-L., Salvat, E., Thomopuolos, R.: Towards benchmarks for conceptual graph tools. In: de Moor, A., Polovina, S., Delugach, H. (eds.) First Conceptual Structures Tool Interoperability Workshop, pp. 72–87. Aalborg University Press (2006)
2. T. Berners-Lee, J. Hendler, and O. Lassila. The Semantic Web. *Scientific Am.*, pages 28–37, May 2001.
3. Chein, M., Mugnier, M.-L.: Conceptual graphs: Fundamental notions. Revue d'Intelligence Artificielle 6(4), 365–406 (1992)
4. Chein, M., Mugnier, M.-L., Simonet, G.: Nested graphs: A graph-based knowledge representation model with FOL semantics. In: Proc. of the 6th Int'l Conf. on the Principles of Knowl. Repres. and Reasoning (KR'98), pp. 524–535. Morgan Kaufmann, San Francisco (1998)

5. Croitoru, M., Compatangelo, E.: A combinatorial approach to conceptual graph projection checking. In: Webb, G.I., Yu, X. (eds.) AI 2004. LNCS (LNAI), vol. 3339, pp. 130–143. Springer, Heidelberg (2004)
6. Croitoru, M., Compatangelo, E.: Conceptual graph projection: a tree decomposition-based approach. In: Doherty, P., Mylopuolos, Welty, C. (eds.) Proc. of the 10th Int'l Conf. on the Principles of Knowledge Representation and Reasoning (KR'2006), pp. 271–276. AAAI, Stanford, California, USA (2006)
7. Dechter, R.: Constraint Processing. Morgan Kaufmann, San Francisco (2003)
8. Delugach, H.: An Exploration into Semantic Distance. In: Proc. of the 7th Int'l Conf. on W'shop on Conceptual Graphs, pp. 119–124 (1992)
9. Montes-Y-Gomez, M., Gelbukh, A., Lopez-Lopez, A., Baeza-Yates, R., Mayr, C., Lazansky, J., quirchmayr, G., Vogel, P.: Flexible comparison of conceptual graphs. In: Proc of the 12th Intln Conference on Database and Expert Systems Applications. Springer, editor, LNCS, pp. 102–111. Springer, Heidelberg (2001)
10. Myaeng, S.H., Lopez-Lopez, A.: Conceptual graph matching: A flexible algorithm and experiments. Jour. of Exp. and Theor. Comp. Sci. 4, 107–126 (1992)
11. Sowa, J.F.: Conceptual Structures: Information Processing in Mind and Machine. Addison-Wesley, London, UK (1984)
12. W3C RDF Primer. RDF Primer — W3C Recommendation 10/2/2004 (2004)
13. World Wide Web Consortium. Web Ontology language (OWL), current (August 2005)

A Comparison of Different Conceptual Structures Projection Algorithms

Heather D. Pfeiffer and Roger T. Hartley

New Mexico State University
{hdp,rth}@cs.nmsu.edu

Abstract. Knowledge representation (KR) is used to store and retrieve meaningful data. This data is saved using dynamic data structures that are suitable for the style of KR being implemented. The KR allows the system to manipulate the knowledge in the data by using reasoning operations. The data structure, together with the contents of the transformed knowledge, is known as the knowledge base (KB). An algorithm and the associated data structures make up the reasoning operation, and the performance of this operation is dependent on the KB.

In this paper, the basic reasoning operation for a query-answer system, projection, is explored using different theoretical algorithms. Within this discussion, the associated algorithms will be using different KBs for their Conceptual Graph (CG) knowledge representation. The basic projection algorithm defined using the CG representation is looking for a graph morphism of a query graph onto a graph from the KB.

The overall running time for the projection operation is known to be a NP class problem; however, by modifying the algorithm, taking into account the associated KB, the actual time needed for discovering and creating the projection/s can be improved. In fact, a new projection algorithm will be defined that, given a typical query onto a carefully defined KB, presents a running time for the actual projection that only grows with the number of projections present.

1 Introduction

Query-Answer systems are very important in business and industry today. However, these systems need to be able to represent *knowledge* in the computer in order to use reasoning techniques when attempting to answer a query for a problem domain. In the computer, the description of the problem to be solved has become known as *knowledge representation, KR*. This representation must be able to store and retrieve meaningful data so that reasoning operations can be performed. The most common reasoning technique used in query-answer systems is *projection* of the query onto the stored knowledge. Later this work will discuss more about this technique, but first more about storing meaningful data will be presented.

One type of KR is semantic networks. These networks are displayed as a discrete graphical structure of vertices and arcs [1]. Within the graphical structure,

U. Priss, S. Polovina, and R. Hill (Eds.): ICCS 2007, LNAI 4604, pp. 165–178, 2007.

the vertices are called nodes and may be displayed as circles or boxes. The arcs
are called links and are displayed as lines with arrows between the nodes. The
nodes are related to each other through their links where the links are assigned
a one-to-one correspondence to a conceptual meaning [2]. The nodes are sometimes called conceptual units and may be seen as objects within the network.
These objects are of many different types including entities, attributes, events or
even states. On top of the semantic network, abstract hierarchies are organized
according to levels of generalization for the conceptual units. The links of the
network form relational connections between the conceptual units, such that the
valence (or parity) of the connection is the number of units that are associated
with a particular unit. In a semantic network links are usually dyadic (binary)
connecting two conceptual units together.

Even though there are multiple semantic network representations available,
the representation that shows much flexibility is *conceptual structures*. Conceptual Structures (CS) are a logic based representation of C.S. Peirce's existential
graphs [3] developed by John Sowa [4]. Conceptual structures are like a set of
logic building blocks; the definitions for some of the blocks are presented beginning with the *type* block:

Definition 1. *A type is a labeling for an abstract idea which is either a conceptual unit or a relationship. These types are members to a set, T, that may form
several structures including hierarchy trees, lattices, and other related structures.
When the structure is a type hierarchy lattice, the set is labeled T_C, and the function* ctype *maps a conceptual unit to the type label in the structure. When the
structure is a relation hierarchy tree, the set is labeled T_R, and the function* rtype
maps a relationship to the type label in the structure.

A *referent* block would have the following definition:

Definition 2. *A referent is an abstract conceptual unit that has been instantiated with a factual value.*

Graph diagrams that are built out of the blocks of conceptual structures are
conceptual graphs (CG) [4,5]. For this work, a conceptual graph has the following
definition:

Definition 3. *A conceptual graph is a bipartite, connected, directed graph $G =
(V, E)$, such that the set of all vertices (nodes) V is partitioned into two disjoint
sets V_C and V_R. The vertices are labeled, and the set V_C is called the* concept
nodes *and the set V_R is called the* conceptual relations *nodes. $e \in E$ is an ordered
pair that connects an element of V_C to an element of V_R using a directed arc.*

*The label of a concept node is a pair, $c = < type, referent >$. The type is an
element of the set T_C. The referent (if present) contains the individual instantiation for the type field.*

*The label of a conceptual relation node is a pair, $cr = < type, signature >$,
where type is an element of the set T_R, and the signature is a pair, $s = < I, O >$
where I is the arcs that are directed into the conceptual relation and O is*

the arcs that are directed out from the conceptual relation. The signature is further defined by its subset category of either relation *or* actor. *The relation is a tuple, $r =< type, c_1, c_2, ..., c_n >$ where type is defined above and in the signature $I \subseteq V_C$ and $O \in V_C$. The actor is a slightly different tuple, $a =< type, c_1, c_2, ..., \{..., c_{n-1}, c_n\} >$ where type is defined above and in the signature $I \subseteq V_C$ and $O \subseteq V_C$.*

Researchers M. Chein and M.-L. Mugnier [6] from the LIRMM group at the Universite Montpellier and other researchers [7,8] have done research on a subset of conceptual graphs known as *simple conceptual graphs (SCGs)* (see Sowa 3.1.2 [4]). As explained in Baget and Mugnier [7], these SCGs are connected, bipartite graphs where the arcs are labeled and finite but not directed, $SG = ((V_c, V_r), U, \lambda)$.

2 Foundational Projection

In general, the matching part of the projection algorithm is unification [9], and there are known linear unification algorithms for acyclic (tree) graphs [10]. Also, SCGs have been evaluated as both graph homomorphism and graph isomorphism. In their original paper from 1992 [11], Mugnier and Chein looked at general projection running times and injective projection. However, CGs and SCGs are not necessarily trees and only part of the algorithms presented next apply to injective projection, so these linear algorithms give guidance, but do not always directly apply.

As discussed in the Messmer and Bunke paper [12], a naive strategy with forward-checking for establishing a subgraph isomorphism is Ullman's backtracking in search tree algorithm [13]. Since Messmer and Bunke feel that it is a common technique with a good baseline subgraph isomorphism algorithm, the Ullman algorithm and its known complexity (from [13,12]) will be reiterated here for defining a basis for investigating projection algorithms. The basic idea of Ullman's algorithm is to take one vertex of the input vertices (query graph) at a time and map it onto a model (a graph from the KB) such that the resulting mapping represents a subgraph isomorphism for a subgraph of the model (KB graph) projected from the input graph (query graph) (see page 307 and 322 of Messmer and Bunke [12]). If at some point, the mapping being built does not represent a subgraph isomorphism then the algorithm backtracks and tries a different mapping. This process is continued until all vertices, $v_1, ..., v_M$ in V_I of the input graph are successfully mapped onto V of the model. This either produces a subgraph isomorphism from G to G_I or stops when a vertex in V_I can not be mapped to at least one vertex in V. In the second case, the algorithm backtracks to a new v_1 in V or v_{n-1} in V and tries to remap the subgraph isomorphism.

Even though this basic algorithm works well for small model and input graphs, it performs poorly as the graphs become larger. This is because all checks are being done locally. Ullman added a forward-checking procedure to know when it is not possible for v_n to be mapped onto an available vertex in V_I (see page 322 in Messmer and Bunke [12]), so that the algorithm can backtrack immediately and save computational steps. In the best case Ullman's algorithm is bounded by: $O(NIM)$ where $N = \#model$ graphs, $I = \#labeled\,vertices$ in input graph which come from the M set of labels, $M = \#labeled\,vertices$ in model graph that are unique. In the worst case the algorithm is bounded by: $O(NI^M M^2)$ where $N = \#model$ graphs, $I = \#vertices$ *in the input graph* and are unlabeled, $M = \#vertices$ *in the model graph* and are unlabeled.

As can be seen, even with this general algorithm, labeling of vertices greatly improves the efficiency of the algorithm. However, it should be noted, that this algorithm does not take into account any support or hierarchy knowledge information.

2.1   Operator

The *project* operator is defined through a mapping $\pi : u \to v$, where πu is a sub-element of v. When u and v are defined to be conceptual graphs, for graph u to be a subgraph of graph v then all of the nodes and arcs of u are in v [14], and the project operator π holds to the following rules [4,15]:

- Type preserving: For each concept c in u, πc is a concept in πu where $type(\pi c) \leq type(\,c\,)$, and $\leq$ is the subtype relation. If c is an individual, that is an actual instance of an object, then $referent(\,c\,) = referent(\,\pi c)$.
- Structure preserving: For each conceptual relation r in u, πr is a conceptual relation in πu where $type(\pi r) = type(\,r\,)$. If the ith edge of r is linked to a concept c in u, the ith edge of πr must be linked to πc in πu.

Fig. 1. Query Graph

2.2   Operation

A projection operation uses the project operator, which is a matching on a graph morphism, graph data structures with either the support information for SCGs or hierarchies when full CGs, and the actual projection algorithm. Stated in Baget and Mugnier, "the elementary reasoning operation, projection, is a kind of graph homomorphism that preserves the partial order defined on labels" [7]. Not only does projection use a project operator (see its definition in the subsection above), but the support S of the graph be it a SCG or the defined type hierarchy if a CG produces a generalization subgraph during the projection operation.

For the rest of this work, the projection operation evaluation and comparison will be restricted to injective projection. This projection mapping is not necessarily one-to-one; that is, a concept or relation in u may have more than one concept or relation in v that πu is a valid mapping. In this respect, there is more than one valid projection from u to v. When the projection operation is performed using

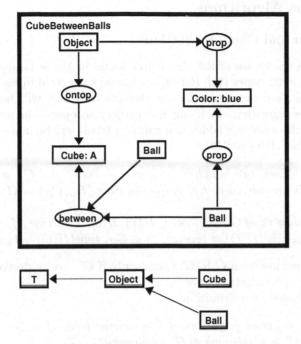

Fig. 2. KB Graph with Type Hierarchy

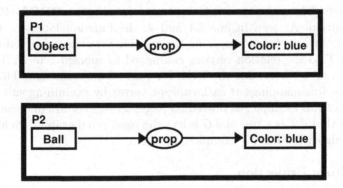

Fig. 3. Projection Results

the query graph from Figure 1 onto the KB graph and hierarchy of Figure 2, the two projections, P1 and P2, discovered are displayed in Figure 3. [1] Using the type hierarchy, both *object* and *ball* are matches; note, if no hierarchy were present, then there would be only one projection. This is a simple injective projection because of the small graphs, however, it can become complex very quickly.

---

[1] The figures in this section were generated by *CharGer* [16].

3   Previous Algorithms

3.1   Mugnier and Chein Projection

Given here is a discussion of the algorithms found in Marie-Laure Mugnier and Michel Chein's 1992 paper [11]. Before discussing the actual injective projection algorithm, given are some added basic definitions which will help the reader understand the algorithm. 1) Using the projection operation provided in the section above, the following additional rules on labels will be added to the graph morphism (from [11] page 240):

Definition 4. *Given two simple conceptual graphs G and G', a projection Π from G to G' is an ordered pair of mappings from (R_G, C_G) to $(R_{G'}, C_{G'})$, such that:*
(i) For all edges rc of G with label i, $\Pi(r)$ $\Pi(c)$ is an edge of G' with label i.
(ii) $\forall r \in R_G$, $type(\Pi(r)) = type(r)$; $\forall c \in C_G$, $type(\Pi(c)) = type(c)$.

There is a projection from G to G' if and only if G' can be derived from G by the elementary specialization rules [4,6].
2) Injective projection definition:

Definition 5. *Injective projection is a restricted form of projection where the image of G in G' is a subgraph of G' isomorphic to G.*

For the algorithm to compute the injective projection from T to G, it is broken into two parts. The first function is used to determine the **PROJ-ROOT** part of the definition. As seen in line 3.1 and 4, the function looks for the actual projection from T to G by comparing the relation vertices connected to concept vertex a in T to the relation vertices connected to concept c in G. The second function is used to determine the **PROJ-r** part of the definition. This function looks for possible mappings at each concept vertex by examining sub-trees. The complexity of this example, as proved on pages 248-249 of Mugnier and Chein92 [11], is such that if T is a tree and G is a cyclic conceptual graph then an injective projection algorithm is a NP-complete problem.

3.2   Croitoru Projection

Madalina Croitoru's projection algorithm is based on SCGs as described in her two 2004 papers [8,17]. This algorithm begins by starting from the foundational injective algorithm given by Mugnier and Chein [11], using SCGs with support which as stated in the Mugnier and Chein 1992 paper [11] is NP-complete. The change applied to this algorithm is to preprocess each graph pair looking for a *matching graph* as defined by the Definition 4.1 on page 8 of Croitoru2004a [8]. The actual matching graph definition is:

Definition 6. *Let $SG = (G, \lambda_G)$ and $SH = (H, \lambda_H)$ be two SCG's without isolated concept vertices defined on the same support S.*
The matching graph of SG and SH is the graph $M_{G \to H} = (V, E)$ where:

- $V \subseteq V_R(G) \times V_R(H)$ *is the set of all pairs* (r, s) *such that* $r \in V_R(G)$, $s \in V_R(H)$, $\lambda_G(r) \geq \lambda_H(s)$ *and for each* $i \in \{1, \ldots, d_G(r)\}$ $\lambda_G(N_G^i(r)) \geq \lambda_H(N_H^i(s))$.

- E *is the set of all 2-sets* $\{(r, s), (r', s')\}$, *where* $r \neq r'$, $(r, s), (r', s') \in V$ *and for each* $i \in \{1, \ldots, d_G(r)\}$ *and* $j \in \{1, \ldots, d_G(r')\}$ *such that* $N_G^i(r) = N_G^j(r')$ *we have* $N_H^i(s) = N_H^j(s')$.

These matching graphs indicate which relation vertices should be used as potential candidates for projection; therefore, reducing the search space. By this addition of preprocessing and then using the matching graphs, this makes the projection of $G \to H$ in its reduced form belong to a class of graphs on which finding the maximum clique can be solved in polynomial time [8].

3.3 Notio Projection

The Notio project is a general conceptual graph implementation with a well-defined API [18]. It is currently being used by several projects [19,20,21] for working with basic reasoning operations with a CG KB. This is the author's derived theoretical algorithm (see Algorithm 1) from the Notio implementation code [18,22] for the injective projection algorithm (note: Southey never wrote any papers or documentation on the actual implemented algorithm). It should be noted for Algorithm 1, all the vertices are all labeled, but the edges are directed.

Also for the analysis of the execution times given above, the following definition of variables hold:

| M_c |= #of concepts in the KB graph
| M_r |= #of relations in the KB graph
| Q_c |= #of concepts in the query graph
| Q_r |= #of relations in the query graph
| Q_e |= #of edges in the query graph
| N |= #of graphs in the KB
| KB_c |= #of concepts in the whole KB

As can be seen in the stated algorithm, in step 1: Notio collects all the concept and relation vertices from both the KB graph and query graph. This takes $O(|M_c| + |M_r| + |Q_c| + |Q_r|)$. In step 2: Notio attempts to see if any of the concept vertices from the KB graph maps to a concept vertex in the query graph. In this way attempting to see if there is any possible subgraph isomorphism of the KB graph to the query graph. In the worst case this step is bounded by: $O(|M_c||Q_c||KB_c|)$.

In step 13: Notio (if a possible mapping was indicated from step 2) will attempt to match all the relation vertices from the KB graph (along with their neighboring concepts along their edges) to query graph vertices with the same edge relationships. As a match is found for relation vertices in the query graph; those relation vertices are now only examined. At end of this step, it is checked that all relation vertices for the query graph were mapped. In the worst case this step is bounded by: $O(|M_r||Q_r||M_c||Q_c||KB_c| + |Q_e|^{|Q_r|}) O(|M_r||Q_r||$

Algorithm 1. Notio Projection

1: Get all concept and relation vertices from the KB and Query graphs
2: **for** $i \leftarrow 0, numfirstconcepts$ **do**   ▷ all concepts in KB graph
3:    **for** $j \leftarrow 0, numsecondconcepts$ **do**   ▷ all concepts in Query graph
4:       $foundmatch \leftarrow false$
5:       **if** $(\text{type}(c_i) == \text{type}(c_j)) \;||\; (\text{supertype}(c_i) == \text{type}(c_j))$ **then**
6:          **if** $(\text{individ}(c_i) == \text{individ}(c_j) \;||\; (\text{individ}(c_j) == \emptyset)$ **then**
7:             $foundmatch \leftarrow true$ ▷ match all concepts in query graph
8:          **end if**
9:       **end if**
10:
11:    **end for**
12: **end for**
13: **if** foundmatch $== true$ **then**
14:    **for** $i \leftarrow 0, numfirstrelations$ **do**   ▷ all relations in KB graph
15:       **for** $j \leftarrow 0, numsecondrelations$ **do**   ▷ all relations in Query graph
16:          **if** $(!\text{relation[j].mapped}) \;\&\&\; (\text{type}(r_i) == \text{type}(r_j))$ **then**
17:             **if** match from r_j to match to each of its concepts **then**
18:                relation[j].mapped = true   ▷ repeat line 2 for all
19:             **end if**
20:          **end if**
21:
22:       **end for**
23:    **end for**
24:    $foundmatch \leftarrow true$
25:    **for** $j \leftarrow 0, numsecondrelations$ **do**
26:       **if** !relation[j].mapped **then**
27:          $foundmatch \leftarrow false$
28:       **end if**
29:    **end for**
30: **end if**
31: **if** foundmatch $== true$ **then**
32:    $P \leftarrow$ build new subgraph projection
33:    **return** P ▷ return new projection
34: **else**
35:    **return** $\emptyset$ ▷ no projection returned
36: **end if**

$M_c \;||\; Q_c \;|\; + \;|\; Q_e \;|)$. In step 31: if a projection is found, it is returned. Therefore the leading step is line 13 for the over all running time, so the worst case bound for the whole KB is very close to the worst case bound given for Ullman's algorithm above: $O((|\; M_r \;||\; Q_r \;||\; M_c \;||\; Q_c \;||\; KB_c \;|\; + \;|\; Q_e \;|^{|Q_r|}))(|\; N \;|)$.

4   New Algorithm

After examining the above algorithms it was discovered that even though the running times were acceptable, the actual projection algorithms were not general.

That is, the user was confined by what parts of a valid conceptual graph could be present in the data or could only have one projection even if more than one was present. The desire to allow the user to use a directed, connected, bipartite conceptual graph (see Definition 3) that was cyclic for both the query and KB graphs prompted a new projection algorithm to be designed.

4.1   Supporting Information

In order to produce a new algorithm, new data structures and supporting routines were needed. Because in the KB the connection between the algorithm and data structures is critical, the new data structures and variables need to be designed around the actual algorithms.

Variables and Given values. Evaluating all the past projection algorithms, and looking at the data structures used for each knowledge base, the authors discovered that handling conceptual graphs as triples as oppose to vectors or linked lists makes the operation of projection much easier and cleaner to process. These authors are not the first researcher to think about using triples. Kabbaj and Moulin in 2001 [23] looked at CG operations using a bootstrapping step. It was at this time that they also looked at defining the join operation using triples as part of the matching data structure. However, they did not look at exploiting the triples in the actual algorithm of the operation.

All conceptual graphs in the KB and the query graph are stored not only with the general conceptual graph information, but also with a C-R-C list in a cs-triple format. Their definitions are given below:

- **cs-triple** is a 3-tuple, $T = < c_i, b, c_j >$, where c_i, c_j are concept nodes, and i and j are not equal. b is a conceptual relation (either a relation or actor node), and $(c_i, b) \in E$ and $(b, c_j) \in E$, and c_i and c_j are members in the signature of b.
- **c-r-c list** is a concept-relation-concept list that holds cs-triple information in which the 'b' in the 3-tuple is a relation node
- **c-a-c list** is a concept-actor-concept list that holds cs-triple information in which the 'b' in the 3-tuple is an actor node
- **defining labels** are all elements in a data structure hold a unique label; that includes concepts, relations, actors, and triples

During the performance of the projection operation two added data structures are used. One data structure holds the matching possibilities of the query concepts with the KB graph concepts, called the *match list*, and the second structure holds the matching triples from the KB graph for each concept in the query graph, called the *anchor list*. These data structures improve performance by making available preprocessed information at the time of the actual projection during which the projection graph or graphs are created.

Actual Supporting Routines. Because the conceptual information is the structural foundation of a conceptual graph and because the relationships between the concepts define the meaning of the graph, the new supporting routines

Algorithm 2. Projection

```
1: function NEWPROJECTION(Q, KB)                    ▷ Query and KB graphs
2:     P = ∅
3:     for each G ∈ KB do                           ▷ All graphs in KB
4:         W ← A list from Q                         ▷ Preprocessing
5:         for each qᵢ ∈ W, where i = 1 to c(W) do
6:             if ((M ← MatchConcepts(qᵢ, G)) > ∅) then
7:                 for each nⱼ ∈ M, where j = 1 to M do
8:                     match = false
9:                     for each tₐ ∈ Q do
10:                                  ▷ where a = 1 to the # of cs-triples in crc list for qᵢ
11:                         for each sᵦ ∈ G do
12:                                  ▷ where b = 1 to the # of cs-triples in crc list for nⱼ
13:                             if MatchTriple(tₐ, sᵦ, true) == true then
14:                                 add (nⱼ, (sᵦ, tₐ)) to qᵢ ∈ W
15:                                 match = true
16:                             end if
17:                         end for
18:                     end for
19:                     if match == false then
20:                         break out of loop and start next graph in KB
21:                     end if
22:                 end for
23:             else
24:                 break out of loop and start next graph in KB
25:             end if
26:         end for
27:         Pset = ∅                                 ▷ Projection processing
28:         for each qᵢ ∈ W, where i = 1 to c(W) do
29:             Pset = Projection(i, W, G, Pset)
30:         end for
31:         P ← P ∪ Pset
32:     end for
33:     return P                         ▷ Set of projections from query onto KB
34: end function
```

have been defined around the *triple* relationship of the *C-R-C*. The routines for *MatchConcept, MatchHierarchy, MatchTriple,* and *MatchConcepts* are examples of the most important matching support routines.

4.2   Actual Algorithm

The overall algorithm (see Algorithm 2) for the projection of the query graph onto the KB is based on looking at all triples that are in the query graph and checking for a complete subgraph match of the query graph onto the KB graph during preprocessing. Because each triple in the query graph is unique, even if the node *type* is not, all projections can be found in the KB graph. Then after all matches of conceptual units and triples are found, the actual projection graphs are built. However, because the temporary data structures are saved from the preprocessing, matching does not have to happen again at build time. The actual projection just uses the match list and anchor list already created to build up or create new the projection graphs. Because the anchor list contains all available projections, both injective and non-injective or homomorphism projections are found.

4.3 Execution Time

Now that the algorithm is split into two sections, there is a running time for answering the decision question of whether or not there is a projection, it will be called the *matching algorithm*, and a running time for the *actual projection*. For the new algorithms, three modifications have been made that affect the execution time of the projection operation: 1) all nodes and triples are uniquely labeled, 2) the edges are not labeled, but do have implied labeling through their directionality within the triples, and 3) the triples are not only part of the data structure of the KB, but also directly effect the actual projection algorithm. The *labeling* drives the execution time of the matching algorithm when doing an injective projection toward the running time for a subgraph 'labeled' isomorphism problem which can be solved in polynomial time as opposed to a straight subgraph isomorphism problem which is known to be NP-complete. The triples allow the matching algorithm to stop sooner when no projection is possible.

For the actual projection creation, the number of triples in the query graph drives the amount of time needed for the actual projection. The size of the graphs in the KB affects the base of the execution time, but the number of times the **Projection** function is executed is based on the number of triples in the query graph. In a typical query-answer scenario where the query graph would potentially contain normally two to four triples compared to possibly a thousand in the KB graph, this algorithm takes into account that the query graph is small. Because of that, the time to do thousands of graphs in a KB is only multiplied by a constant based on the maximum number of triples in a KB graph that the small query graph is projected onto. Therefore the execution time is only multiplicative in the number of projection available with this query graph. Since in the most common case there is only one projection, the actual projection creation algorithm becomes polynomial. Through this shift in problem class, the running time for the projection operation for a typical scenario within a query-answer application shows improvement.[2]

5 Comparison and Conclusion

Four different, yet related, projection algorithms have been described. Examining Table 1 comparisons between basic units, type of graphs, number of possible projections found, overall operation algorithm execution time and just actual projection creation execution time will be evaluated.

---

[2] The complete algorithms and running times were presented as part of the author's PhD dissertation [24].

Table 1. Comparison of four algorithms

	M&C	Croitoru	Notio	New Algorithm
basic unit	relations	relations	relations	concepts
works over	SCGs	SCGs	CGs	CGs
projections found	all	# relations	1	all
overall operation	NP-Complete	NP-Complete	NP-Hard	NP-Hard
actual projection	NP-Complete	NP-Complete	NP-Complete	NP

The Mugnier and Chein and Croitoru algorithms use SCGs, and Notio and the new algorithm work over full CGs. Looking back at the example shown when discussing the projection operation, Notio would only find one projection because it was only designed to look for a single projection graph. Croitoru's algorithm assumes that the total number of relations in the query graph equals the number of possible projections; therefore, with this example it will only find a single projection. However, if the KB graph has multiple projections to a single relation in the query graph; part of the projections would be missed.

It is not clear from the Mugnier and Chein 1992 paper if they can handle two concept pairs with the same relationship between them in a projection operation. However, from later work [25], it is indicated that the same relationship between different concepts can be found and multiple projections are possible between two CGs, but the execution time is at best NP-complete and only works on SCGs (no actors or directed graphs). Mugnier and Chein algorithm is also based on the *relations* found within the graph and must traverse all of their signatures to discover if there is a subgraph morphism. The new algorithm is based on the conceptual units, or *concepts*, within the graph and can stop searching as soon as there is no match for a concept or concept triple in the KB graph for one of query graph's conceptual unit.

Mugnier and Chein's algorithm does the whole projection operation as a single injective projection algorithm, where Croitoru, Notio and the new algorithm all use some form of preprocessing. Notio and the new algorithm have a complete separation between the preprocessing algorithm and projection; where, Croitoru uses the preprocessing algorithm inside of the actual projection, therefore, giving the same running time for both the overall algorithm and the actual projection. Notio does preprocessing at storage time that helps in constructing the projection. However, the actual projection algorithm after the preprocessing is still NP-Complete.

The new algorithm splits the overall projection algorithm into two parts, matching and projection construction. Then data structures are used between these two algorithms to use the structure of the graphs to help in the projection process. In the most common case the matching algorithm is the longest running part of the overall algorithm because the actual projection execution is polynomial. Therefore, in a typical scenario where the query graph is small, the new algorithm is not only able to find all projections for full conceptual graphs, but can use the data structures of the KB to do it faster. Future work is to determine if the actual projection algorithm for all injective projections can be performed in polynomial time by experimental results [24].

Acknowledgment

The authors would like to thank Dr. Desh Ranjan for reviewing the definitions and mathematical results that appear within this paper.

References

1. Lehmann, F. (ed.): Semantics Networks. Pergamon Press, Oxford, England (1992)
2. Schubert, L.: Extending the expressive power of semantic networks. Artifical Intelligence 7, 163–198 (1976)
3. Peirce, C.: Manuscripts on existential graphs. Peirce 4, 320–410 (1960)
4. Sowa, J.: Conceptual Structures: Information Processing in Mind and Machine. Addison-Wesley, Reading, MA (1984)
5. Sowa, J.: Knowledge Representation: Logical, Philosophical, and Computational Foundations. Brooks/Cole (2000)
6. Chein, M., Mugnier, M.L.: Conceptual graphs: Fundamental notions. Revue d'Intelligence Artificielle 6(4), 365–406 (1992)
7. Baget, J.F., Mugnier, M.L.: Extensions of simple conceptual graphs: the complexity of rules and constraints. Journal of Artificial Intelligence Research (JAIR) 16, 425–465 (2002)
8. Croitoru, M., Compatangelo, E.: On conceptual graph projection. Technical Report AUCS/TR0403, University of Aberdeen, UK, Department of Computing Science (2004)
9. Corbett, D.: Reasoning and Unification over Conceptual Graphs. Kluwer Academic/Plenum Plublishers, New York (2003)
10. Paterson, M., Wegman, M.: Linear unification. J. Comput. Syst. Sci. 16, 158–167 (1978)
11. Mugnier, M.L., Chein, M.: Polynomial algorithms for projection and matching. In: Pfeiffer, H.D., Nagle, T.E. (eds.) Conceptual Structures: Theory and Implementation. LNCS, vol. 754, pp. 239–251. Springer, Heidelberg (1993)
12. Messmer, B., Bunke, H.: Efficient subgraph isomorphism detection: A decomposition approach. IEEE Transactions on Knowledge and Data Engineering 12, 307–323 (2000)
13. Ullman, J.: An algorithm for subgraph isomorphism. J. of the Assoc. for Computing Machinery 23(1), 31–42 (1976)
14. Harary, F.: Graph Theory. Addison-Wesley, Reading, MA (1969)
15. Willems, M.: Projection and unification for conceptual graphs. In: Ellis, G., Rich, W., Levinson, R., Sowa, J.F. (eds.) ICCS 1995. LNCS, vol. 954, pp. 278–282. Springer, Heidelberg (1995)
16. Delugach, H.: CharGer 3.3 - A Conceptual Graph Editor, University of Alabama in Huntsville, Alabama, USA (2004), http://www.cs.uah.edu/delugach/CharGer
17. Croitoru, M., Compatangelo, E.: A combinatorial approach to conceptual graph projection checking. In: Webb, G.I., Yu, X. (eds.) AI 2004. LNCS (LNAI), vol. 3339, Springer, Heidelberg (2004)
18. Southey, F., Linders, J.: NOTIO - a Java API for developing CG tools. In: Tepfenhart, W.M. (ed.) ICCS 1999. LNCS, vol. 1640, pp. 262–271. Springer, Heidelberg (1999)
19. Delugach, H.: CharGer: A graphical Conceptual Graph editor. In: Delugach, H.S., Stumme, G. (eds.) ICCS 2001. LNCS (LNAI), vol. 2120, Springer, Heidelberg (2001), http://www.cs.nmsu.edu/hdp/CGTOOLS/proceedings/index.html

20. Benn, D., Corbett, D.: pCG: An implementation of the process mechanism and an extensible CG programming language. In: Delugach, H.S., Stumme, G. (eds.) ICCS 2001. LNCS (LNAI), vol. 2120, Springer, Heidelberg (2001), http://www.cs.nmsu.edu/hdp/CGTOOLS/proceedings/index.html
21. Polovina, S., Hill, R.: Enhancing the initial requirements capture of multi-agent systems through conceptual graphs. In: Dau, F., Mugnier, M.-L., Stumme, G. (eds.) ICCS 2005. LNCS (LNAI), vol. 3596, pp. 439–452. Springer, Heidelberg (2005)
22. Southey, F.: Notio and Ossa. In: Delugach, H.S., Stumme, G. (eds.) ICCS 2001. LNCS (LNAI), vol. 2120, Springer, Heidelberg (2001), http://www.cs.nmsu.edu/hdp/CGTOOLS/proceedings/index.html
23. Kabbaj, A., Moulin, B.: An algorithmic definition of cg operations based on a bootstrap step. In: Delugach, H.S., Stumme, G. (eds.) ICCS 2001. LNCS (LNAI), vol. 2120, Springer, Heidelberg (2001)
24. Pfeiffer, H.D.: The Effect of Data Structures Modifications On Algorithms for Reasoning Operations. PhD thesis, New Mexico State University, Las Cruces, NM (2007)
25. Mugnier, M.L., Leclere, M.: On querying simple conceptual graphs with negation. In: Data and Knowledge Engineering, DKE, Elsevier, Revised version of R.R. LIRMM 05-051 (2006)

A Conceptual Graph Approach to Feature Modeling

Randall C. Bachmeyer and Harry S. Delugach

Department of Computer Science
N300 Technology Hall
University of Alabama in Huntsville
Huntsville, AL 35899 USA
{rbachmey,delugach}@cs.uah.edu

Abstract. A software product-line is a set of products built from a core set of software components. Although software engineers develop software product-lines for various application types, they are most commonly used for embedded systems development, where the variability of hardware features requires variability in the supporting firmware. Feature models are used to represent the variability in these software product-lines. Various feature modeling approaches have been proposed, including feature diagrams, domain specific languages, constraint languages, and the semantic web language OWL. This paper explores a conceptual graph approach to feature modeling in an effort to produce feature models that have a more natural, and more easily expressed mapping to the problem domain. It demonstrates the approach using a standard Graph Product-line problem that has been discussed in various software product-line papers. A conceptual graph feature model is developed for the graph product-line and it is compared to other feature models for this product-line.

1 Introduction

The fact that software is complex, expensive to develop, and virtually impossible to rid of all defects has been widely reported in industry publications as well as the popular press [4, 13]. One of the key techniques to reduce the magnitude of these problems is to increase software reuse, and therefore decrease costs and improve quality. This is critically important for embedded systems where the cost of fixing software defects might entail hardware recalls or complex firmware upgrades. The study of software product-lines is an active area of research that focuses on a particular class of reuse important to the embedded systems community.

Product-lines consist of a set of products that exhibit similar functionality and share a common base of software assets. The capabilities of a particular member of the product-line are described by a set of features. An example of a product-line with a variety of features is a digital camera product-line. Many of the core software functions are common across all cameras, but each model has a unique set of features. By building a formal model of the features, automated configuration tools can be developed that assemble the software assets of the product-line to produce the firmware for a specific camera model. These formal models are referred to as feature models.

U. Priss, S. Polovina, and R. Hill (Eds.): ICCS 2007, LNAI 4604, pp. 179–191, 2007.

A key research area within software product-lines is how to best represent a feature model [1, 2, 3, 5, 12, 14, 16]. In this paper, we review three feature modeling approaches and examine how each one handles feature constraints. We then propose a conceptual graph representation that we believe offers distinct advantages over the other techniques. We also demonstrate how to use conceptual graphs to concisely specify a configuration for a product-line member. Although we do not discuss the generation of the final software product, we do show a procedure that expands a product configuration into a full specification that would be processed by a product generation tool.

2 Feature Modeling

In the software product-line research area, a *feature* is often described as an increment of product functionality [17]; *feature models* are the aggregate set of all features of a product-line along with their usage constraints [1]; and *feature modeling* is the activity of building feature models. The usage constraints in a feature model restrict the set of all possible product-line members to only those that have a valid non-conflicting set of features. Much of the model development and model debugging time is spent establishing an accurate and complete set of usage constraints.

Before discussing some of the more common approaches to feature modeling, we will introduce the graph product-line problem. It will be used throughout the remainder of this paper to compare the different feature modeling approaches.

2.1 Graph Product-Line Problem

Lopez-Herrejon and Batory proposed the graph product-line (GPL) problem for evaluating product-line methodologies [18]. In a later paper, Batory used the GPL problem that showed the connections between feature models, grammars, and propositional formulas [1]. The GPL problem was also used by Wang, et al. to demonstrate a semantic web approach to feature modeling [16]. These papers provide the feature models that we use for comparison.

GPL is a product-line that builds driver applications for exercising graph algorithms.[1] The features in this problem consist of a driver, graph algorithms, graph characteristics, and graph search types. The graph characteristics allow for weighted and unweighted graphs as well as directed and undirected graphs. The search types allowed are breadth-first and depth-first. Not all of the graph algorithms require a search type and many of the algorithms only operate on graphs with specific characteristics. Figure 1 from Lopez-Herrejon detail the feature constraints for this problem [18]. The reader is referred to the original paper for additional details regarding the GPL problem.

---

[1] In this paper, *graph algorithms* do not refer to conceptual graph algorithms, but instead to the classical graph algorithms called out in the graph product-line problem. A graph algorithm example in a conceptual graph paper is unfortunate but unavoidable since GPL is a standard for comparing feature modeling approaches.

Algorithm	Required Graph Type	Required Weight	Required Search
Vertex Numbering	Directed, Undirected	Weighted, Unweighted	BFS, DFS
Connected Components	Undirected	Weighted, Unweighted	BFS, DFS
Strongly Connected Components	Directed	Weighted, Unweighted	DFS
Cycle Checking	Directed, Undirected	Weighted, Unweighted	DFS
Minimum Spanning Tree	Undirected	Weighted	None
Single-Source Shortest Path	Directed	Weighted	None

Fig. 1. GPL Constraints [18]

3 Current Approaches

3.1 Feature Diagrams

The most commonly published feature model notation is the feature diagram, a hierarchical decomposition and annotation of features denoting whether features are optional, mandatory, or alternatives [1, 5, 9, 10, 12]. The feature diagram node notation is shown in Figure 3, and the feature diagram for GPL is shown in Figure 2. The feature diagram is drawn in a manner to convey the commonness and variance of a product-line. To the extent possible, usage constraints are captured in the shape of the graph and in the node notations. However, there is no way to represent usage constraints involving non-adjacent features. Some authors suggest using a separate constraint language [2, 7] to augment the feature diagram.

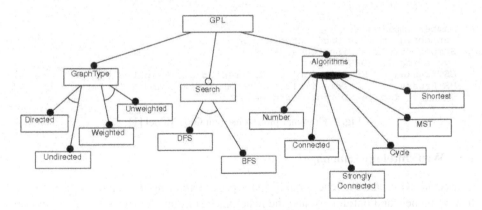

Fig. 2. GPL Feature Diagram [16]

Mandatory	Optional	Alternative	Or
C F	C F	C F1 F2	C F1 F2

Fig. 3. Feature Modeling Node Notation [16]

3.2   Grammars Plus Constraints

The relationship between feature diagrams and grammars was shown originally by deJonge and Visser [11]. Batory later elaborated on the relationship with his excellent paper that showed the relationship between feature diagrams, grammars, and propositional formulas [1].

A grammar alone does not allow an engineer to model all of the usage constraints, similar to the problem with feature diagrams. However, Batory's paper had also shown the relationship of grammars to propositional formulas, therefore it was possible to demonstrate how the grammar and explicit logical constraints could be brought together into a complete set of logical statements. These statements could then be used by a logical truth maintenance system (LTMS) to assist the user in specifying valid product configurations. The grammar and constraints developed by Batory for the GPL problem are shown in Figure 4.

```
// grammar
GPL : Driver Alg+ [Src] [Wgt] Gtp :: MainGpl ;
Gtp : Directed | Undirected ;
Wgt : Weighted | Unweighted ;
Src : BFS | DFS ;
Alg : Number | Connected | Transpose StronglyConnected :: StrongC
              | Cycle | MSTPrim | MSTKruskal | Shortest ;
Driver : Prog Benchmark :: DriverProg ;

%% // constraints

Number implies Src ;
Connected implies Undirected and Src ;
StrongC implies Directed and DFS ;
Cycle implies DFS ;
MSTKruskal or MSTPrim implies Undirected and Weighted ;
MSTKruskal or MSTPrim implies not (MSTKruskal and MSTPrim) ;
Shortest implies Directed and Weighted ;
```

Fig. 4. Grammar plus Constraint GPL Model [1]

3.3   Web Ontology Language

Czarnecki, Kim, and Kalleberg published a paper exploring the relationship between feature models and ontologies; and the idea that feature models were simply views on ontologies [8]. Although they examined several sample domains, GPL was not one of those.

Wang, et al. took a direct approach to studying the relationship between feature models and ontology. They used OWL DL, a semantic web ontology language, to model the GPL problem. In their paper, they postulated that the lack of formal semantics had hampered advancements in feature modeling and they believed that a web ontology approach could provide those semantics [16]. They represented features with OWL classes and relationships with OWL properties. An example of a configuration for a specific product is shown in Figure 5.

$E \sqsubseteq GPLRule$
$E \equiv ((\exists \, hasConnected.Connected) \sqcap (\exists \, hasSearch.Search)\sqcap$
$(\exists \, hasAlgorithms.Algorithms) \sqcap (\exists \, hasDFS.DFS)\sqcap$
$(\exists \, hasGraphType.GraphType) \sqcap (\exists \, hasNumber.Number)\sqcap$
$(\exists \, hasWeighted.Weighted) \sqcap (\exists \, hasDirected.Directed)\sqcap$
$(\exists \, hasStronglyConnected.StronglyConnected)\sqcap$
$(= 0 \, hasUndirected) \sqcap (= 0 \, hasMST) \sqcap (= 0 \, hasShortest)\sqcap$
$(= 0 \, hasUnweighted) \sqcap (= 0 \, hasBFS) \sqcap (= 0 \, hasCycle))$

Fig. 5. OWL DL Product Specification [16]

Although Wang, et al. were successful in their modeling effort, they conceded "modeling at such a low level would be too difficult for most engineers" [16]. They subsequently developed a graphical tool that allowed an engineer to draw the more traditional feature diagram as shown in Figure 2. This tool was then used to generate the underlying OWL DL models.

3.4 Analysis of Current Approaches

Although feature diagrams are very easy to read, the diagram alone cannot capture all usage constraints. The reader must reference separate constraint documentation. Modifying a feature diagram can also be difficult since changing the hierarchy can cause constraints to be added or removed from the constraint documentation.

The grammar plus constraints approach is more complete than the feature diagram, but it requires two distinctly different languages to fully describe the feature model. Like the feature diagram, modifications in the grammar may entail modifications of the constraints.

While the web ontology approach [16] uses a single language to represent features and usage constraints, we believe the authors are correct in their assessment that it would be too difficult to use directly. The escape to a feature diagramming tool that generates OWL does not solve this problem since it has the same problems as noted with the feature diagrams above, with the exception that the constraint language is now fully specified.

Although the above issues are of a concern, the authors of this paper believe the key issue with the current approaches lies in the limitations of the relationships between features that are being captured. As Batory shows [1], the relationships in a feature diagram, and therefore in the grammar, are either implication or choose-n-of-m. We believe that an ontological approach, with a richer set of types and relationships, is a better approach. We also believe that conceptual graphs provide a modeling and configuration notation that is much more approachable than OWL for this problem. Since conceptual graphs are based on first order logic [15], they offer the same level of semantic formality that Wang, et al. were searching for.

4   Conceptual Graph Feature Models

In this section, we present a feature modeling approach based on John Sowa's conceptual graphs [15]. We first define a meta-model. We then discuss the processes of building a feature model and defining product configurations. We conclude the section with the presentation of a procedure for converting a configuration into a complete specification.

4.1   Meta-model for Conceptual Graph Feature Models

The concepts and relations defined in this section of the paper are generic and can be used in any conceptual graph feature model. The three key definitions for features are:

> **Type** Feature(x) **is** [Entity:*x]→(Characteristic)→[Semantic: ##]
>
> **Type** RootFeature(x) **is** [Feature:*x]→(attribute)→[T: #root]
>
> **Type** OptionalFeature(x) **is**
>   [Feature:*x]→(state)→[FeatureState: {#include | #exclude}]

In keeping with normal conceptual graphs usage, this paper uses a '#' followed by an identifier in a concept referent field to represent an individual marker. An identifier in the referent field that does not have a '#' prefix is a name contraction [15].

The *Feature* type requires each feature to have unique semantic (this is shown using the marker '##' that this paper defines as a marker which is unique for each type t ≤ *Feature*). As is the case with all feature modeling techniques, the precise semantic is not defined in the feature model. Tools accepting a specification derived from a feature model are required to know the semantics by some other means.

The *RootFeature* type is used to classify root features of a product-line model. Application engineers can only specify a configuration for features that are a subtype of RootFeature. The root feature is referred to as a *concept* in product-line literature, but the word concept has a much broader, deeper ingrained and not easily changed meaning for conceptual graphs; hence the change.

The types *EntityFeature* and *CharacteristicFeature* are also a part of the meta-model. They do not have supporting differentia, but are provided to simplify the classifications of features.

4.2   Creating the Feature Model

The feature model is defined by the domain engineer; a person who is both knowledgeable about the product domain and skilled at developing knowledge bases. Each product feature is mapped to a unique concept in the conceptual graph feature model. Since there is a one-to-one mapping between concepts and features, this paper uses the words feature and concept interchangeably when discussing conceptual graph feature models.

When designing conceptual graph feature models, the engineer should first focus on classifying the features and relations. The relations should not focus on logical implications or model variance but rather the natural relationships between features (e.g. part-of, characteristic-of, is-a, etc.).

The specification generation procedure described in this paper also supports conceptual graph schemas. Their use is restricted to defining statements that are possible – but not always required. This allows a feature to either make an assertion about the existence of a related feature, or make no statement if it can function with or without the related feature. This will be illustrated in the GPL example in this paper.

Many of the usage constraints will be implicit from type, relation, and schema definitions. Additional usage constraints that do not follow from the natural definition of the problem domain can be specified with global constraints. These constraints are also defined using conceptual graphs.

The conceptual graph feature model described in this section could be used with a variety of specification methods including manually constructing and validating a specification using the types and relations of the feature model. The following three sections describe a procedure that allows the application engineer to specify a minimal configuration and then expand that configuration to generate a complete and valid specification suitable for input to a system generator.

4.2.1 Create the Configuration

An application engineer specifies a member of the product-line by constructing a set of conceptual graph statements. These configuration conceptual graphs use only the concept and relation types defined in the feature model. The configuration process is similar to filling in the blanks of a configuration form. The configuration statements bind to a prescribed location in the specification conceptual graph and restrict concept markers, select an item from a disjunctive set, specify the members of a set or restrict the type of a concept – all of which are permissible operations under the canonical transformation rules [15].

4.2.2 Generating the Specification

The procedure specified in this section was used to manually generate the complete specifications for the GPL conceptual graph feature model developed in Section 5. It is not presented as a general algorithm suitable for all possible conceptual graph based feature models, but is instead a starting point for our later research into automatic specification generation algorithms and requisite limitations to conceptual graph based feature models.

The procedure only uses canonical graph transformations, guaranteeing the resulting specification will be grammatically valid and will satisfy all the implicit usage constraints. The word "graph" in the following procedure always refers to a conceptual graph. The procedure itself is independent of the graph product-line problem.

- Let C be the set of all graphs in a configuration with root feature rf.
- Let T be a set of graphs initialized to the empty set.
- For each graph g in C.
    - Make a copy h of graph g.
    - For each concept c in h
        - If $referent(c)$ is not a set,
            - Perform a maximal type expansion of c in h.

- Perform a maximal schema expansion of *c* in *h*.
  - ▪ Otherwise, for each individual *j* in the set *referent(c)*,
    - Add the expansion of individual *c(j)* to *C*.
- o For each concept *c* in *h* where *referent(c)* is the generic marker '*',
  - ▪ Replace the generic marker with the marker *#type(c)* causing all generic features of the same type to be joined.
- o For each concept *c* in *h* where *type(c) ≤ OptionalFeature* and it is not the case that *[c] → (state) → [FeatureState:#exclude]*.
  - ▪ Assert *[c] → (state) → [FeatureState:#include]*
- o Join *h* to the graphs in T to produce T'. If T' can be generated from T using only the canonical graph transformations then let T=T'; otherwise, a usage constraint has been violated and this procedure should terminate with a failure condition.

The set *T* now contains a grammatically valid specification. To verify that the specification conceptual graph *S* does not violate the global constraints, *S* must be joined to the set of global constraint conceptual graphs. If no contradiction is detected, the specification *S* is a complete and valid specification for the product-line member defined by the configuration conceptual graphs. At this point in our research, we are not prepared to address the details of this more complex join and testing of global constraints.

Manually applying the above procedures to the GPL problem in Section 5 produced the expected results, although the human intuition in this process made the joins and detection of contradictions obvious. The major challenge for the next phase of this research is to convert the above procedure into an efficient and implementable algorithm that works on a well-defined type of feature model and does not rely on human intuition.

5 Graph Product-Line Problem

In this section, the authors develop the feature model for the graph product-line problem. Three configurations are then developed and processed with the procedure from the previous sections. Where possible, the names of the features (eg. Alg, Gtp) are the same as those presented in the grammar presented by Batory [18].

5.1 Designing the GPL Feature Model

In Figure 6, we present the type hierarchy for the GPL feature model. The types for *Graph*, *Src*, *Gtp*, *Wgt* and their subtypes are presented with no additional differentia.

Schema for Graph(x) **is** [EntityFeature:*x]- 　　　　　　　　　　(characteristic)→[Gtp] 　　　　　　　　　　(characteristic)→[Wgt]
type Alg(x) **is** 　　　　[EntityFeature:*x]→(parameter)→[Graph]
schema for Alg(x) **is** 　　　　[EntityFeature:*x]→(parameter)→[Src]
type Driver(x) **is** [RootFeature:*x]→(test)→[Alg:{*}]

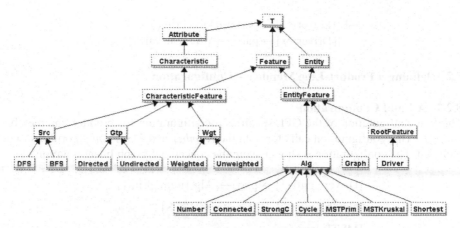

Fig. 6. GPL Feature Model Type Hierarchy

The GPL graph algorithms can now be described using the types and schemas presented above. Note the similarity of the type definitions for the algorithm to the corresponding row in the table of constraints for the GPL. The definitions illustrate the capability of conceptual graphs to use the lexicon and grammar of the feature model to express usage constraints in a language that is very close to the natural language of the domain.

type Number(x) **is** [Alg:*x]→(parameter)→[Src]

type Connected(x) **is** [Alg:*x]-
               (parameter)→[Graph]→(characteristic)→[Undirected]
               (parameter)→[Src]

type StrongC(x) **is** [Alg:*x]-
               (parameter)→[Graph]→(characteristic)→[Directed]
               (parameter)→[DFS]

type Cycle(x) **is** [Alg:*x]→(parameter)→[DFS]

type MSTPrim(x) **is** [Alg:*x]→(parameter)→[Graph]-
                       (characteristic)→[Undirected]
                       (characteristic)→[Weighted]

type MSTKruskal(x) **is** [Alg:*x]→(parameter)→[Graph]-
                       (characteristic)→[Undirected]
                       (characteristic)→[Weighted]

type Shortest(x) **is** [Alg:*x]→(parameter)→[Graph]-
                       (characteristic)→[Directed]
                       (characteristic)→[Weighted]

The only global constraint in this feature model is the requirement to allow either Kruskal's or Prim's algorithm for the minimal spanning tree, but not both [18]. This constraint can be translated directly to the conceptual graph:

¬[[[Driver:*x]→(part)→[MSTPrim]]
  [[Driver:*x]→(part)→[MSTKruskal]]]

5.2 Defining a Product-Line Member – Configuration

5.2.1 A Valid Configuration

The first configuration of the GPL specifies a configuration for product-line member *config1*. This configuration calls for both the *Number* and *MSTPrim* algorithms to be included. The configuration is shown using the conceptual graph linear notation.

[Driver:config1]→(part)→[Alg:{num, prim}]
[Number:num]→(parameter)→[BFS]
[MSTPrim:prim]

The conceptual graph in Figure 7 is the specification conceptual graph generated from the above configuration using the procedure in section 4.3.3. Note that we have only shown the individual markers for concepts where we wanted the reader to observe the implicit joining specified in the procedure.

The driver supports only one graph, therefore the two *Graph* concepts have the same identifier and are connected with a coreferent line. Since the characteristics of the graph for *MSTPrim* are type restrictions of the more general characteristics of the graphs in the *Number* algorithm, the two *Graphs* are compatible. The global constraint is also satisfied in this configuration.

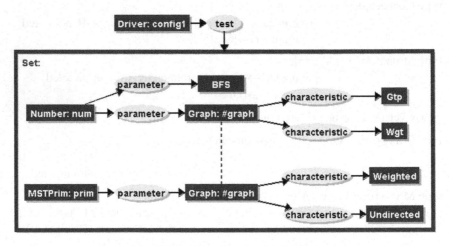

Fig. 7. Final Specification for Config1

5.2.2 A Configuration that Violates a Selectional Constraint

The second configuration *config2* explores the use of type definitions to reduce the number of required constraint equations. This configuration specifies the *MSTPrim* and *Shortest* algorithms are to both be included in the configuration. We use the conceptual graph referent mechanism to include the *prim* conceptual graph from the *config1* example. This assumes that both configurations are part of the same configuration model.

[Driver:config2]→(part)→[Alg:{sp, prim}]

[Shortest:sp]

The final specification, generated by the procedure in section 4.3.3, is shown in Figure 8. When the *Graph* parameters from each algorithm are joined (due to identical referents), a conflict is discovered. A *Graph* cannot be both directed and undirected; therefore *config2* is an invalid configuration.

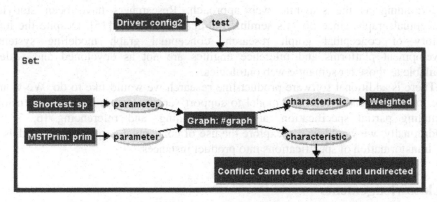

Fig. 8. Invalid Specification for Config2 Showing Conflict

5.2.3 A Configuration That Violates a Global Constraint
The final example configuration specifies both the *MSTKruskal* and *MSTPrim* algorithms.

[Driver:config3]→(part)→[Alg:{prim, kruskal}]

[MSTKruskal:kruskal]

The transformation procedure from Section 4.3.3 would process this configuration and produce a specification without detecting any grammar conflicts. However, the specification would fail the global constraint test since the constraint declares that there is no driver that exercises both the Kruskal and Prim minimum spanning tree algorithms.

6 Conclusions and Future Work

Our preliminary research has shown that a conceptual graph approach to feature modeling is feasible. We are particularly encouraged by the strong mapping between the GPL problem requirements and the statements in the conceptual graph feature models. The description of the GPL features including the driver, algorithm constraints, and the additional global constraint were all easily mapped into conceptual graphs. This simple mapping is not surprising given the close ties between natural language and conceptual graphs [15]. We believe a domain engineer will be able to provide many of the usage constraints simply by defining types that describe the natural relationship between features. This strong mapping between natural language and a conceptual graph feature model speaks directly to both the writability and readability of conceptual graph feature models.

The approach to product configuration and the specification generation procedure presented worked well for the manual generation of product specificiations. The next phase of our research will automate this procedure and determine the bounds on the types of conceptual graph models that can be processed efficiently.

We also like the fact that conceptual graphs have the same strong formal semantic underpinning as the semantic web approach. Researchers have been studying conceptual graphs since Sowa's seminal publication in 1984 [15]. Despite the long history of conceptual graph research, conceptual graph modeling systems, development platforms and inference engines are not as developed and widely available as those for semantic web ontologies.

There is additional software product-line research we would like to do. We would like to validate and extend our model to support various feature diagram extensions including partial specification, attribution, cloning, and referencing [6, 7, 8]. Additionally, we would like to explore the use of conceptual graph feature models in the transformation of specifications into product instances.

Acknowledgements

The authors of this paper are grateful to the reviewers of this paper for their comments and suggestions. The guidance they have provided has given us insight into how we might restrict our graphs in order to develop efficient, automatic algorithms for the generation of complete and valid product specifications.

References

1. Batory, D.: Feature models, grammars, and propositional formulas. In: Proceedings of the Software Product Line Conference (2005)
2. Beuche, D.: Composition and Construction of Embedded Software Families, Ph.D. thesis, Otto-von-Guericke-Universittät, Magdeburg, Germany (2003)
3. Bühne, S., Lauenroth, K., Pohl, K.: Why is it not Sufficient to Model Requirements Variability with Feature Models? In: Proceedings Workshop on Automotive Requirements Engineering (AURE04), Nanzan University, Nagoya, Japan (2004)
4. Brooks, Jr., F.P.: No Silver Bullet – Essence and Accidents of Software Engineering. IEEE Computer 20(4), 10–19 (1987)
5. Czarnecki, K.: Overview of Generative Software Development. In: Banâtre, J.-P., Fradet, P., Giavitto, J.-L., Michel, O. (eds.) UPP 2004. LNCS, vol. 3566, pp. 313–328. Springer, Heidelberg (2005)

6. Czarnecki, K., Helsen, S., Eisenecker, U.: Staged Configuration Through Specialization and Multi-Level Configuration of Feature Models, Software Process Improvement and Practice, 10(2) (2005)
7. Czarnecki, K., Kim, C.H.P.: Cardinality-based feature modeling and constraints: a progress report. In: International Workshop on Software Factories, San Diego, California (October 2005)
8. Czarnecki, K., Kim, C.H.P., Kalleberg, K.: Feature Models Are Views on Ontologies. In: Proceedings of 10th International Software Product Line Conference (SPLC 2006), pp. 41–51. IEEE, Los Alamitos (2006)
9. Czarnecki, K., Eisenecker, U.W.: Generative Programming: Methods Tools and Applications. Addison-Wesley, Reading Mass (2000)
10. Greenfield, J., Short, K.: Software Factories: Assembling Applications with Patterns, Models, Frameworks, and Tools. Wiley, Indianapolis, IN (2004)
11. de Jonge, M., Visser, J.: Grammars as feature diagrams. In: ICSR7 Workshop on Generative Programming, pp. 23–24 (2002)
12. Kang, K., Cohen, S., Hess, J., Nowak, W., Peterson, S.: Feature-Oriented Domain Analysis (FODA) Feasibility Study, Technical Report, CMU/SEI-90TR-21 (November 1990)
13. Levenson, N.G.: Systemic Factors in Software-Related Spacecraft Accidents, AIAA Space 2001 Conference and Exhibition, New Mexico (August 2001)
14. Peng, X., Zhao, W., Xue, Y., Wu, Y.: Ontology-Based Feature Modeling and Application-Oriented Tailoring. In: Morisio, M. (ed.) ICSR 2006. LNCS, vol. 4039, pp. 87–100. Springer, Heidelberg (2006)
15. Sowa, J.F.: Conceptual Structures: Information Processing in Mind and Machine. Addison-Wesley, Reading Mass (1984)
16. Wang, H., Li, Y., Sun, J., Zhang, H., Pan, J.: A semantic web approach to feature modeling and verification. In: Workshop on Semantic Web Enabled Software Engineering (SWESE'05) (November 2005)
17. Zave, P.: FAQ Sheet on Feature Interactions, www.research.att.com/ pamela/faq.html
18. Lopez-Herrejon, R.E., Batory, D.: A Standard Problem for Evaluating Product-Line Methodologies. In: Bosch, J. (ed.) GCSE 2001. LNCS, vol. 2186, pp. 9–13. Springer, Heidelberg (2001)

From Conceptual Structures to Semantic Interoperability of Content

Pavlin Dobrev[1,2], Ognian Kalaydjiev[1], and Galia Angelova[1]

[1] Institute for Parallel Processing, Bulgarian Academy of Sciences
25A Acad. G. Bonchev Str., 1113 Sofia, Bulgaria
[2] ProSyst Labs EOOD, 48 Vladaiska Str., 1606 Sofia, Bulgaria
p.dobrev@prosyst.com, {ogi,galia}@lml.bas.bg

Abstract. Smart applications behave intelligently because they understand at least partially the context where they operate. To do this, they need not only a formal domain model but also formal descriptions of the data they process and their own operational behaviour. Interoperability of smart applications is based on formalised definitions of all their data and processes. This paper studies the semantic interoperability of data in the case of eLearning and describes an experiment and its assessment. New content is imported into a knowledge-based learning environment without real updates of the original domain model, which is encoded as a knowledge base of conceptual graphs. A component called mediator enables the import by assigning dummy metadata annotations for the imported items. However, some functionality of the original system is lost, when processing the imported content, due to the lack of proper metadata annotation which cannot be associated fully automatically. So the paper presents an interoperability scenario when appropriate content items are viewed from the perspective of the original world and can be (partially) reused there.

1 Introduction

The "Levels of Conceptual Interoperability Model" (LCIM) [1] defines different layers of interoperation and how they are related to the ideas of integratability, interoperability, and composability. This paper adopts the model in order to take a closer look at the semantic interoperability of smart applications. We focus on semantic interoperability, which is always implemented via import, exchange or reuse of data, developed for another application. Based on our experience in ontology development and LCIM as a formal metric, we evaluate the problem of ontology reuse in the case of eLearning and present our vision about achieving higher levels of interoperability through reuse of semantically annotated learning content.

The paper is structured as follows. Section 2 considers the background: the issue of semantic interoperability and the state of the art in ontology development. Section 3 discusses the notion of intelligent content in eLearning. Section 4 presents our experiment in semantic interoperability in case of eLearning, which is based on an adaptive strategy for sequencing of learning objects, and its evaluation. Section 5 contains the conclusion.

U. Priss, S. Polovina, and R. Hill (Eds.): ICCS 2007, LNAI 4604, pp. 192–205, 2007.
© Springer-Verlag Berlin Heidelberg 2007

2 Semantic Interoperability

In general, the recent efforts in interoperability and reusability approach the issue from two perspectives: *(i)* interoperability and reusability of content, to be achieved mainly by metadata annotations, and *(ii)* interoperability of software systems. In this section we argue that the two views are strongly interrelated although they reflect two different focuses – on data and on programs.

2.1 Interoperability of Content by Semantic Annotations

Semantic annotation of natural language content started in the 80-ties of last century with the so-called "Text Encoding Initiative", where units of texts were collected with associated metadata annotations in SGML-format. These activities were driven by the needs to exchange linguistic data among the various groups creating very large language resources. Due to the specific features of the formal models, applied in the computational linguistics, the initiative did not influence substantially other application domains which also focus on content creation.

Some 10 years ago the Semantic Web ideas attracted much more attention and the *interoperability view to data* arose. Today it is widely accepted that content creation is very expensive and therefore content should be developed and kept in a way which enables its later use and re-use by other applications, beyond the original settings where it was created. At the same time it becomes clear that the electronic content is a treasury, perhaps more valuable than the software itself, and that metadata should be designed very carefully to ensure the content life-cycle. Much efforts are invested in proper content creation, for instance the European Commission funds the programme "eContent Plus" which supports the content enrichment by semantic annotation.

Today content is (manually) annotated with diverse metadata in many areas:

- html-pages are annotated with ontology labels in the Semantic Web,
- images, video, movies and all kinds of sound records are annotated with metadata in publishing archives, TV and music archives as well as by companies recording e.g. customers complains,
- eLearning content is actively developed and extensively annotated according to widely accepted annotation standards,
- archives in cultural heritage, esp. museums and art galleries are annotated according to application-specific metadata schemes, to enable better storing and reuse of museum and art artefacts, etc.

Semantic annotation according to certain ontology labels seems to be the best way to preserve content meaning and to provide reusability and interoperability of content. Normally the content objects or items - atomic pieces in content archives – are juxtaposed metadata in some formalised format. The tendency is to design special entries for storing references to the underlying conceptual hierarchies of important domain notions. Unfortunately, the lack of standardised nomenclatures and classifications in many areas is a major obstacle for the advances of interoperable content and its practical application.

2.2 Interoperability of Software Applications and LCIM

LCIM is a popular interoperability standard, part of SEI Guide to Interoperability - http://www.sei.cmu.edu/isis/guide/isis-guide.htm. Tolk et al. introduced the model in [1, 2, 3, 4]. Initially LCIM contained four interoperability levels but later they were increased to six levels. Based on previous research, we define a slightly modified LCIM similar to those used in [5, 6]:

- Level 0: Stand-alone systems have **No interoperability**.
- Level 1: On the level of **Technical interoperability**, a communication protocol exists for exchanging data between participating systems. On this level, a communication infrastructure is established which allows the exchange of bits and bytes. The underlying networks and protocols are unambiguously defined.
- Level 2: The **Syntactic interoperability** level introduces a common structure to exchange information, i.e., a common data format is applied and/or API exists. On this level, a common protocol to structure the data is used; the format of the information exchange is unambiguously defined.
- Level 3: If a common information exchange reference model is used, the level of **Semantic interoperability** is reached. On this level, the meaning of the data is shared; the content of the information exchange requests are unambiguously defined.
- Level 4: **Pragmatic interoperability** is reached when both interoperating systems are aware of the methods and procedures that are employed by the other system. In other words, each participating system understands the use of data and the context of its application within the other system. The context of information exchange is unambiguously defined.
- Level 5: As a system operates on data over time, the state of that system will change, and this includes the assumptions and constraints that affect its data interchange. If two systems attain **Dynamic interoperability**, then each system is able to comprehend the state changes that occur in the assumptions and constraints that the other system is making over time, and both systems are able to take advantage of those changes. In particular, this becomes increasingly important regarding the effects of the operations. The effect of the information exchange within the participating systems is unambiguously defined.
- Level 6: Finally, if the conceptual models – i.e. the assumptions and constraints of the meaningful abstractions of the reality – are aligned, the highest level of interoperability is reached: **Conceptual interoperability**. This requires that the conceptual models will be documented based on engineering methods enabling their interpretation and evaluation by other engineers. In other words, on this level we need a "fully specified but implementation independent model" (not just a text describing the conceptual idea).

The key point of LCIM is that, while unambiguous interpretation of shared data between systems is necessary for interoperation, it is not sufficient. We have enough *interoperability* standards for shared data. Almost every W3C specification like XML, RDF, OWL states its objectives to *enhance the functionality and interoperability of the Web*. But the data encoded using these standards are not necessarily interoperable. For example concepts in ontologies that have the same labels, and even the same meaning, can be used completely differently in different applications, so the interoperability of certain atomic items or other fragments does not guarantee the interoperability of the whole data set.

Levels 0 to 3 focus on data interoperability. But what it really means when we say that two systems are *semantically interoperable*? First it is important for both systems to model their data using the same formalism – e.g. ontologies. Second this formalism must have the same data representation format, understandable for both applications (e.g. RDF, OWL, KIF, CL). Is the *semantic interoperability* achievable through ontologies and conceptual structures? In [5] we conclude that some semantic interoperability is possible but not at a very deep level. At present only limited *syntactic interoperability* can be achieved (data can be exchanged in standardised formats). In the next sections we discuss these issues again, when we describe our experiment in semantic content interoperability.

Level 4 and 5 are related to the interoperability of the manner how applications process their data. Is it possible to have *semantic interoperability* without achieving higher levels of interoperability? Most probably not, at least not fully, because there is always one additional factor that is not included in the current definition of LCIM – the human understanding. When people develop software applications today, they encode lots of hidden knowledge in the implementation of the program algorithms. In this way full semantic interoperability needs at least limited dynamic or pragmatic interoperability as well.

Conceptual interoperability requires a formalism that fully specifies the interoperable modules of the systems. A good candidate for this is UML. Latest versions of the UML are formal enough to be automatically processed. A mapping from and to UML to OWL is defined in [7]. Note that the mapping is not full in both directions. And again, without human understanding of both systems, they can not interoperate with each other.

At present the interoperability between two applications is always partial – it is implemented on the common data and modules understandable for both systems. In the next sections we present an approach to partial content interoperability in the eLearning context, based on semantic metadata annotations.

2.3 Ontologies as Interoperability Backbones

Ontologies define the common terms and concepts (meaning) used to describe and encode an area of knowledge. Ontologies can be presented in a very informal or highly formal and explicit way: *thesauruses* (words and synonyms), *taxonomies* (minimal hierarchy or a parent/child structure), *knowledge bases* (with more complex facts), *logical theories* (with very rich, complex, consistent and axiomatic knowledge). To reflect this variety, the ontology community distinguishes between *lightweight* and *heavyweight* ontologies. Lightweight ontologies include concept taxonomies, properties, and few relationships. Heavyweight ontologies, on the other hand, entail a deep, detailed description of concepts and relations in addition to axioms, as a formal knowledge representation paradigm. Heavyweight ontologies are in fact the knowledge bases, used in AI for automatic inference. There are dozens of lightweight ontologies developed for particular applications. A heavyweight ontology is CyC; it is manually developed for a wide range of automatic knowledge processing.

For us, ontologies are the natural decision to enable semantic interoperability, as long as they enable interoperable domain models. However, they are too many, too diverse and describe different cuts of the reality at different depth and granularity.

Although the knowledge representation discipline arose in AI some 35-40 years ago, established international achievements in the standardisation, reusability, interoperability and scalability issues are hardly seen at present. Several advanced groups have convergent views to build an unified upper model; some of them are influenced by linguistic considerations (e.g. the SENSUS ontology) while others take completely formal direction (e.g. CyC). There is a recent proposal for suggested upper merged ontology [8] and a recently accepted international standard for a knowledge representation language called 'common logic' [9].

One of the main obstacles to elaborate satisfactory conceptual models stems from the fact that the reality as such has no any formal model, therefore there are no better or worse domain models and in general, every domain model might be considered a good formalisation only from the perspective of a specific successful application. The helpful metaphor of *knowledge soup* [10] emphasizes on the lack of *per se* granularity of concepts and relations and the lack of any intrinsic or "natural" conceptual structure. Instead, the segmentation into individuals and types is imposed upon the world by our words and the conceptual system associated to them. Therefore, for practical reasons, researchers and developers in knowledge acquisition accept the following postulates:

- Agreement on standard upper model is a practical requirement and needs to be elaborated as soon as possible, to support the international efforts for the collection of formalised conceptual resources;
- Standardisation of the representation languages – e.g. common logic – is needed but it should not impose restrictions on the content;
- There are many alternative domain models that may be considered even as disjoint formalisations of the reality although they deal with the same labels (words and terms);
- Multiple inheritances (due to distinct classification perspectives) are the natural status of conceptual classifications and there should be means to cope with it.

In fact, the availability of underlying ontologies is an issue at the core of smart applications interoperability in particular and the Semantic Web enterprise in general. There are no large and widely accepted ontologies, available today, which can be directly used as annotation standards for intelligent content. Numerous different ontologies over the Web require effective techniques for (semi-automatic) ontology matching in order to provide semantic interoperability. There are no convincing results in the automatic ontology alignment despite the international efforts involved in the task. However, content annotation is progressing meanwhile. The result is that large archives of carefully elaborated content are systematically annotated according to different classifications and nomenclatures without any clear view how the content will be made interoperable in the future. Practical experiments of automatic content reuse are rare and always start by alignment of the semantic meta-annotations schemes, which includes manual human intervention to ensure proper quality.

3 Digital Content in eLearning

Activities in eLearning are related to content development, on the one hand, and content structuring in learning sequences, on the other hand. Content development is guided by popular standards like SCORM [11] and IEEE Learning Object Metadata LOM [12] which are based on XML-representation. But the approaches for building learning sequences – *learning design* - are still a hot research field. Usually the content is defined by the so-called learning objects, entities that may be (re-)used for learning, education or training. The granularity of a given learning object corresponds to a stand-alone learning objective. In other words, learning objects are educational units with coherent semantic meaning in the corresponding subject area. Building high-quality learning content is extremely expensive task and the minimal aim is to provide its portability, i.e. to ensure content loading into various learning management system.

Similarly to the interoperability levels defined in LCIM for software applications, we can talk about *technical, syntactic, semantic, pragmatic/dynamic* and *conceptual* levels of content interoperability. Current learning management systems ensure *syntactic* (often XML-based) reuse and *certain* level of semantic reuse, as the learning object are normally annotated by unambiguous keywords belonging to certain unambiguously defined subject area. However, to achieve the higher level of pragmatic/dynamic reuse would require embedding of imported learning objects into a relevant learning sequence of another learning system, to enable reuse in the proper context. In this way the issue of content interoperability is closely related to the learning design approaches, to the pedagogical strategy for assessment of learning performance and other features of the importing system. So the level of pragmatic/dynamic content interoperability is much more difficult to achieve.

Usually the learning sequence is defined by explicit reference to the next learning object, i.e. by explicit citation of its identification number within a learning object repository. This reference is entered either at the stage of learning object creation in a manual annotation process, or at the stage of explicit design of a learning sequence. The recent versions of SCORM annotation model contain a special metadata field, where the learning sequence is explicated. In this way, importing new (additional) learning content X into a running learning management system, using learning repository Y, would require global change of the learning sequences and manual update of the sequences of training courses. In other words, achieving semantic interoperability of learning content is too expensive and needs manual efforts.

In the next section we present a first attempt to implement a semi-automatic approach to content importing, by minimal human-expert intervention and automatic construction of dummy meta-labels for the imported content.

4 Towards Semantic Interoperability of eLearning Content

In an earlier project, we have developed the learning environment STyLE [13, 14, 15], which supports our recent experiments in semantic interoperability. STyLE is an intelligent tutoring system for self-tuition, which provides readings and tests in foreign language learning, especially in English financial terminology. It compliments the classroom activities and is implemented as a knowledge-based application, which stores separately the domain knowledge in finances and the pedagogical resource of readings and drills to be shown to students. STyLE knowledge base consists of a type hierarchy of concepts, which represents domain notions (most often labelled by English financial terms) and conceptual graphs, encoding domain facts. In this paper we are interested in the type hierarchy as it provides the semantic labels of the drills' annotations. Every drill, when created as a learning object, is manually juxtaposed a domain concept (English financial term) which is tested by the exercise. In this way every learning object, related to concept X, corresponds to the learning objective: "test the student knowledge about concept X". A planner – the so-called Pedagogical Agent, tracks the student performance and decides what is to be shown next to the student. This component provides experiments in learning design, because STyLE has no manually predefined sequence of showing learning content to the students; instead, the assumption is that the planner will display every drill sooner or later, at certain learning situation. Readings are shown to the student in case or errors, according to the text relevance to the currently unknown terms. Every drill (a learning object) is activated together with some software module, which enables the assessment of the students' answers, so we are not interested in the training and assessment process itself. Below we consider the problems and solutions in semantic interoperability of content imported from outside to STyLE environment.

4.1 Adaptive Sequencing of Learning Content

Effective planning is implemented in environments where each action done by plan is largely predictable but needs not to be completely deterministic. The Pedagogical Agent in STyLE shows to the student a sequence of readings and tests which maximises the coverage of learner's knowledge (i.e. the agent searches for concepts, which are not tested, and displays relevant readings and drills to the student). This agent has two strategies. *The local strategy* plans the system' moves between:

- Displaying drills, testing different characteristics of one concept/term and
- Displaying readings chosen to increase the learner's knowledge about this concept/term.

The goal of the local strategy is to show exercises testing all aspects of the learner's knowledge regarding each concept. This strategy operates on the learning objects metadata, which are linked to the terms from the concept hierarchy and to IDs of conceptual graphs, encoding domain facts (aspects) of the tested concepts.

The *global strategy* plans movements to different concepts, i.e. a kind of attention shifts, taking into account the concepts' position in the financial ontology. It constructs a testing route trough all notions available in the pedagogical resource.

It is important to note that there is no predefined sequence of showing learning objects (readings and drills) to the student; rather, the teacher specifies the sequence of general topics only and the planner aligns the route through all learning objects to the main course topics. This intelligent, adaptive approach is feasible for self-tuition because the planner chooses the sub-topics taking into consideration the learners' errors. The gain is that the ontology and the pedagogical resource of exercises become independent to large extent; new drills can be added without changing the annotation of the old learning objects and also, new concepts in the ontology can be appended together with relevant readings and exercises to extend the learners' knowledge about the domain.

4.2 Import of Semantically-Annotated Content

Our experiment was planned when new relevant texts in the financial domain were collected from Internet and the STyLE training expert considered them as suitable additional readings, which are worth to import as learning content into the original pedagogical resource. Another motivation is that static reading archives need updates and upgrades, not to bore the learner with repetitive information. Most of the texts were relevant to the Financial Ontology of the Sigma knowledge engineering system (Sigma is a system for developing, viewing and debugging theories in first order logic, see [16]; it works with the Knowledge Interchange Format KIF and is optimized for SUMO [8]). Therefore we decided to simulate the meta-annotation of the new training materials using Sigma and to update the pedagogical resource with readings in additional topics, which were not originally included in STyLE. The formal models of *Financial Instruments* and *Financial Markets* in STyLE and Sigma are designed with comparable granularity and depth but Sigma contains more concepts and covers more broadly the financial domain, so it is a convenient annotation framework. In addition Sigma has larger upper model. Table 1 presents numbers of concepts and relations in the ontologies in question.

Table 1. Short feature comparison of ontologies

Number of	STyLE Ontology	FinancialOntology.kif	SUMO Merge.kif
Concepts	131	226	696
Relations (incl. hierarchy *subclasses*)	227	219	770

When planning experiments in interoperability of learning content, one quickly realises that the semantically-annotated content reflects the pedagogical philosophy of its developers. For instance, the teaching expert of STyLE distinguished some 18 kinds of bonds, which are important for training of students-economists in financial English (see the left-side concept hierarchy at Figure 1). This teaching expert organised the hierarchy in a special way – in fact, for her own courses and seminars, to reflect the desired granularity of semantic meta-labels for the readings and exercises as well as the general course topics – financial instruments, financial markets, etc. So the 300 learning objects in STyLE were developed as a holistic pedagogical resource, to compliment the class-room activities in a particular discipline, particular faculty, academic year etc. In principle all learning objects are interoperable and reusable in another learning environment, but they can be integrated into a holistic sequence of training materials only if the original concept hierarchy is available in the re-using system as well. Similarly, import of content to STyLE would require at least aligning the learning objects metadata to those applied in STyLE,

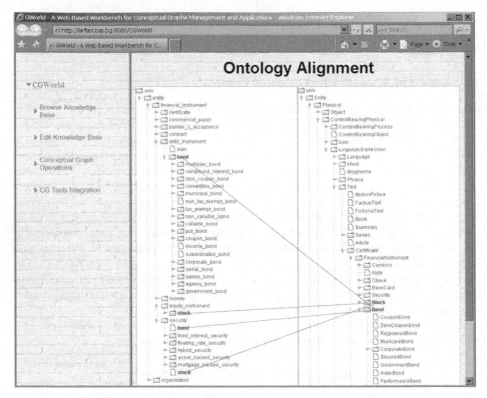

Fig. 1. Alignment of domain concepts used for annotation of learning content

including the semantic labels. So there are constrains regarding the semantic interoperability and eventual lack of functionality (or quality of services), which may stem from unsuccessful alignment of semantic labels in the case of reusable learning content. The metadata concerns only "superficial issues" like semantic markers of learning objects. However, there is much more than that in a pedagogical resource: pedagogical goals and individual manner of expressing them when constructing drills with specific granularity, individual teacher styles, etc. So the content re-usability concerns higher level pedagogical issues which go beyond the subject of this paper.

We show below that new content can be imported to STyLE indeed, but it does not function in the same way and some system tasks cannot be performed for the new content. In our case, this is mainly due to non-consistent assignment of semantic meta-labels to the original and imported learning objects.

As content import and reusability require (semi-)automatic alignment of the semantic metadata annotation, our first care is to consider the concepts of both ontologies. A fragment of Sigma Financial Ontology is shown at the right-hand hierarchy at Figure 1. It is generated from the *subclass* relation of Sigma ontologies – from SUMO (Merge.kif) and the Financial Ontology (FinancialOntology.kif).

When talking about ontology evaluation, alignment and merging, we initially try to find equivalent concepts. If the concepts are identical by name, we can suppose that they have the same meaning. Obviously, some lexical conversion of labels is helpful in this case, by simple transliteration rules (like: *if a concept's label of FinancialOntology.kif has length of more than one word, then add "_" between the words*; and *convert all names to lover case*). After performing these simple conversions, we found about 30 identical concept labels in the STyLE ontology and the Sigma Financial Ontology. Such concepts are:

```
blue_chip_stock, bond, call_option, callable_bond, cash,
certificate_of_deposit, common_stock, contract,
corporate_bond, coupon_bond, financial_instrument,
government, government_bond, growth_stock, index_option,
junk_bond, loan, money_market, ...
```

This means that the labels of some 23% of the concepts in STyLE and Sigma Financial Ontology can be aligned almost directly, as a first guess for identical meanings in the two ontologies. These common concepts are used to draw the lines connecting the two type hierarchies, please see the lines shown at Figure 1. However, there are formal definitions encoded in both ontologies: CG type definitions and other statements in STyLE and KIF statements in the Financial Ontology (Table 2). We see that the two formal models approach the identically-labelled concepts from different perspective and focus on different factual information related to these concepts. For instance, BOND in STyLE is DEBT-INSTRUMENT and SECIRITY, while in Sigma it is INVESTMENT and FINANCIAL-INSTRUMENT. Therefore, only automatic alignment of "key concepts" labels is possible and reasonable, to enable the import of the new learning content into the old pedagogical resource.

Table 2. Different approaches to formal modelling of the concept BOND

Concept	Formal statements
BOND in STyLE	`isa(bond,debt_instrument). isa(bond,security).`

Definition
 	
and
 	
Attributes
	`(def [bond] [Situation: (represent [security`
	`:lambda] [debt *x1])(of ?x1 [corporation])])`
	`/* Bond is a security which represents debt of`
	`corporation */`
	`(attr [bond :lambda *x0] [coupon *x1])(goal`
	`?x1 [represent *x2])(obj ?x2 [interest`
	`*x3])(attr ?x0 ?x3)(att ?x3 [semi-annual]) /*`
	`Coupon bond is a bond with coupons represent-`
	`ting semi-annual interest payments attached */`

BOND in Sigma	`(subclass Bond Investment)` `(subclass Bond FinancialInstrument)`

To support the merging of semantic annotations via concept hierarchy labels, we have implemented the interface shown at Figure 1, where a domain expert can:

- Accept proposed *line of identity* between concept types, based on automatic identification of closer names (see identity lines at Fig. 1; they establish *1:n* mappings from the new, right-side hierarchy 2 to the left-side hierarchy 1);
- Remove some proposed identity lines;

- Add additional *is-a* lines *concept-superconcept,* to define how concept X from hierarchy 2 should be considered as a subconcept of concept Y from hierarchy 1 (by pointing if the automatic identification fails).

This semi-automatic alignment and merging enables to produce automatically meta-annotation of the imported content according to the hierarchy of the old pedagogical resource. In this way we can produce some dummy meta-labels of the imported content, which make them visible in the original pedagogical setting.

Please note that the tasks described above do not add new concept to the original hierarchy 1. The extension is achieved by using dummy concepts only. An example is *RegisteredBond* in Sigma, which should be considered as a subclass of the original concept of *Bond* in STyLE. The expert using the workbench at Fig. 1 actually says: "import the content annotated by *RegisteredBond* in the new environment and consider it as annotated by *Bond* in the original system". Another expert may have a different opinion about integration of content into third system and then another expert-specific dummy concept will be produced and supported. Our experiment required the implementation of two additional modules:

- the interface providing the superficial alignment "hints", as shown at Figure 1, adopted from the environment CGWorld [17] and implemented as Java applet;
- the component producing dummy-concepts and relevant meta-annotation, which mediates between the STyLE software and the resource of interoperable content. Fig. 2 shows its role to align imported content metadata to the original resource.

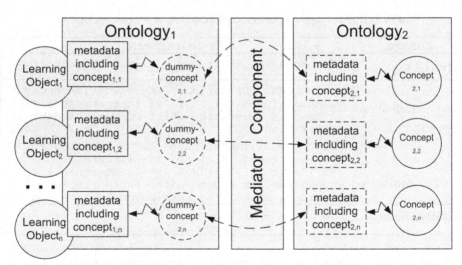

Fig. 2. Dummy annotations supporting views to new content from the original world

To conclude the implementation comments, we notice that the mediator is an additional component which can be easily integrated to any eLearning system. It does not affect the system functionality as such but only prepares data (content) to upgrade the pedagogical resource. Form implementation point of view, as shown at Figure 2, we did not change the algorithms that deal with metadata. Instead we rely on the added dummy-concepts which represent a real concept (or set of concepts) in the second ontology. Both ontologies are kept in CGWorld. After finishing the alignment process all new metadata are re-generated according to the newly created relations between the two ontologies.

4.3 Experiment in Content Reuse

Briefly, the experiment looks encouraging but at the same time it makes evident the limitations of the semantic interoperability scenario. After completing the alignment of the hierarchies as discussed in section 4.2, the content to be re-used is juxtaposed meta-labels which are compliant to the original system's initial annotation. Perhaps the multiplication of annotations is the most interesting idea in our experiment, because the content might have:

- annotation according to the original settings,
- annotation according to scheme X (and be reusable within system X),
- annotation according to scheme Y (and be reusable within system Y) etc.

In our case, the new content becomes visible and accessible for the Pedagogical Agent in STyLE. Every imported learning object is associated to certain concept of the type hierarchy shown at the left column at Figure 1. For the imported learning objects, the Pedagogical Agent cannot start the local planning strategy, because it depends *(i)* on the aspects (conceptual graphs encoding simple facts about the tested concept – which are missing in the imported object as none assigns them manually) and *(ii)* on the complexity ranking (which is also not available in the imported learning objects). So the imported content is only treated by the global planning strategy and shown as a *miscellaneous* pool of readings, presenting general knowledge about the concept to where the readings are "linked". Every imported learning object will be shown to the student in certain learning situation, in case that the student makes errors and needs new additional readings, but the strategy of adaptive sequencing will treat the imported objects differently.

This experiment is evaluated by students and training experts and these were the results seen. The easy extension of the annotated readings' archive is a positive feature which pleases the teachers who always need new reading materials and new samples of terminology usages in concrete texts (including bilingual and parallel ones). In principle more readings are good for the students too, but – as we said above – the imported readings might look slightly less focused to certain learning situations (although the financial texts typically discuss several domain notions together).

Evaluating the experiment, we also see the limitations of the interoperability we achieve by import of new learning content (it is level 3 of LCIM, partial semantic interoperability). Fully interoperable content in eLearning requires development of systems, which may exchange formal models of their pedagogical goals and tasks – including assessment of students' performance, approach to content sequencing etc. But all these formal models are far from the current eLearning practice. So today we can exchange new readings mostly, without much problems and without risks to import incoherent training materials and general pedagogical inconsistencies.

5 Conclusion

In eLearning, "intelligent content" is defined as content, which is aware of itself, as it enables maintenance and exchange of large resources of training materials. Today semantic meta-annotations are viewed as the ultimate, universal recipe of how to make the content "machine understandable". We have considered in this paper an experiment of content interoperability and we see that content by itself cannot provide full reuse and interoperability, as they always depend on the software that processes the content. So our claim is that experiments in content interoperability should be organised much more often, as they reveal the next generation of problems we may face when we create millions of content objects with limited capacity of reuse. As it was demonstrated in the paper, semantic interoperability is possible only between applications with similar functionality in a particular domain. It is also evident that full semantic interoperability of data can be achieved only if certain pragmatic, dynamic and conceptual interoperability are implemented. The idea of a mediating module, which ensures the consistent dynamic view of a software application to new data, seems to be a good decision especially because such a module supports this interface without principal changes of the system and the new data, so they keep their original shape and format.

Looking at interoperability of smart applications from a more general perspective, we notice that usually the enterprises develop their systems independently, with low consideration for the collaboration, i.e. there are to investments to ensure that systems can interoperate with other systems. Lots of systems are built from scratch and there is a lot of redundancy in the company software applications. Increasing interoperability between these systems would reduce the time of the customers and end user, needed for learning new systems and would increase the maintainability of the software environments as a whole. However, interoperability of programs is difficult to achieve. At the same time, regarding interoperability of data, one notices that increasing archives of semantically annotated content are labelled by metadata derived from different nomenclatures, classifications, and ontologies, without a clear view how this content will be re-used outside the original settings where it is created.

In conclusion, interoperability is not a closed concept for which a line can be drawn. Instead, interoperability is a means to achieve a goal, to advance the effective delivery of information and services to the end user. But interoperability requires careful design of the software application as well as its data and content into a holistic approach of modelling and implementation.

Acknowledgements. The work reported in this paper is partially supported by the project BIS-21++ funded by the European Commission in FP6 INCO via contract no.: INCO-CT-2005-016639.

References

[1] Tolk, A., Muguira, J.: The Levels of Conceptual Interoperability Model (LCIM). In: Fall Simulation Interoperability Workshop, Orlando, FL (September 2003)
[2] Tolk, A.: What Comes After the Semantic Web - PADS Implications for the Dynamic Web, PADS. In: 20th Workshop on Principles of Advanced and Distributed Simulation (PADS'06), pp. 55–62 (2006)

[3] Tolk, A.: Composable Mission Spaces and M&S Repositories - Applicability of Open Standards. In: Spring Simulation Interoperability Workshop, Washington, D.C. (April 2004)

[4] Tolk, A., Turnitsa, C.D., Diallo, S.Y.: Ontological Implications of the Levels of Conceptual Interoperability Model. In: WMSCI2006, Orlando (July 2006)

[5] Dobrev, P.: CG Tools Interoperability and the Semantic Web Challenges. In: Hitzler, P. et al (eds.) Conceptual Structures: Inspiration and Application, Contributions to ICCS-06 - 14th Int. Conference on Conceptual Structures, Aalborg University Press, pp. 42–59 (July 2006)

[6] Dobrev, P.: Knowledge Interoperability for the Semantic Web Applications. In: Angelova, G., et al. (eds.) John Vincent Atanasoff Information Days, Proceedings of the Young Researchers Session, 4-6 October 2006, Sofia, Bulgaria, pp. 20–26 (2006) ISBN - 954-91743-5-2

[7] Ontology Definition Metamodel, Sixth Revised Submission to OMG/RFP ad/2003-03-40, http://www.omg.org/cgi-bin/doc?ad/06-05-01.pdf

[8] Suggested Upper Merged Ontology SUMO: http://www.ontologyportal.org/

[9] Information technology — Common Logic (CL) - A framework for a family of logic-based languages, ISO/IEC FDIS 24707 (2006), http://common-logic.org/docs/cl/24707-31-Dec-2006.pdf

[10] Sowa, J.F.: Conceptual Structures: Information Processing in Mind and Machine. Addison-Wesley, Reading, MA (1984)

[11] SCORM® 2004 3rd Edition Documentation (2004), http://www.adlnet.gov/scorm/20043ED/Documentation.cfm

[12] Learning Object Metadata (2002), http://ltsc.ieee.org/wg12/files/LOM_1484_12_1_v1_Final_Draft.pdf

[13] Angelova, G., Nenkova, A., Boycheva, S., Nikolov, T.: Conceptual graphs as a knowledge representation core in a complex language learning environment. In: Contributions to ICCS-2000, Darmstadt, Germany, Shaker Verlag, pp. 45–58 (August 2000)

[14] Angelova, G., Boytcheva, Sv., Kalaydjiev, O., Trausan-Matu, St., Nakov, P., Strupchanska, A.: Adaptivity in Web-Based Computer-Aided Language Learning. In: Proc. ECAI- 2002, pp. 445–449 (2002)

[15] Angelova, G., Kalaydjiev, O., Strupchanska, A.: Domain Ontology as a Resource Providing Adaptivity in eLearning. In: Meersman, R., Tari, Z., Corsaro, A. (eds.) On the Move to Meaningful Internet Systems 2004: OTM 2004 Workshops. LNCS, vol. 3292, pp. 700–712. Springer, Heidelberg (2004)

[16] Pease, A.: The Sigma Ontology Development Environment. In: Working Notes of the IJCAI-2003 Workshop on Ontology and Distributed Systems, Acapulco Mexico (2003), see also http://sourceforge.net/project/showfiles.php?group_id=102489 http://sigmakee.cvs.sourceforge.net/sigmakee/KBs/

[17] Dobrev, P., Toutanova, K.: CGWorld - Architecture and Features. In: Priss, U., Corbett, D.R., Angelova, G. (eds.) ICCS 2002. LNCS (LNAI), vol. 2393, pp. 261–270. Springer, Heidelberg (2002)

Faster Concept Analysis

Adam D. Troy[1], Guo-Qiang Zhang[1,*], and Ye Tian[2]

[1] Department of EECS, Case Western Reserve University
Cleveland, Ohio 44022, U.S.A.
gq@case.edu
[2] Information Technology Division
Cleveland Clinic Health Systems, Ohio, U.S.A.

Abstract. We introduce a simple but efficient, multistage algorithm for constructing concept lattices (MCA). A concept lattice can be obtained as the closure system generated from attribute concepts (dually, object concepts). There are two strategies to use this as the basis of an algorithm: (a) forming intersections by joining one attribute concept at a time, and (b) repeatedly forming pairwise intersections starting from the attribute concepts. A straightforward translation of (b) to an algorithm suggests that pairwise intersection be performed among all cumulated concepts. MCA is parsimonious in forming the pairwise intersections: it only performs such operations among the newly formed concepts from the previous stage, instead of cumulatively. We show that this parsimonious multistage strategy is complete: it generates all concepts. To make this strategy really work, one must overcome the need to eliminate duplicates (and potentially save time even further), since concepts generated at a later stage may have already appeared in one of the earlier stages. As considered in several other algorithms in the literature [5], we achieve this by an auxiliary search tree which keeps all existing concepts as paths from the root to a flagged node or a leaf. The depth of the search tree is bounded by the total number of attributes, and hence the time complexity for concept lookup is bounded by the logarithm of the total number of concepts. For constructing lattice diagrams, we adapt a sub-quadratic algorithm of Pritchard [9] for computing subset partial orders to constructing the Hasse diagrams. Instead of the standard expected quadratic complexity, the Pritchard approach achieves a worst-case time $O(N^2/log N)$. Our experimental results showed significant improvements in speed for a variety of input profiles against three leading algorithms considered in the comprehensive comparative study [5]: Bordat, Chein, and Norris.

1 Introduction

The expanding roles of Formal Concept Analysis (FCA) in many areas make the development of efficient algorithms an important component in any application involving contexts with size beyond small examples. In [5], Kuznetsov and Obiedkov provide an extensive survey and comparative experimental evaluation of algorithms for FCA in the literature.

---

* Corresponding author.

U. Priss, S. Polovina, and R. Hill (Eds.): ICCS 2007, LNAI 4604, pp. 206–219, 2007.

In this paper we introduce a simple but efficient, multistage algorithm for constructing concept lattices (MCA). Given a formal context, a concept lattice can be obtained as the closure system generated from attribute concepts (dually, object concepts). There are two strategies in the literature to use this as the basis of an algorithm:

(a) forming intersections by joining one attribute concept at a time – this falls into the class of algorithms often called incremental concept analysis, and
(b) repeatedly forming pairwise intersections starting from the attribute concepts.

Bordat [1], Norris [8], and CbO [5] use strategy (a), while Chein [2] and our approach use strategy (b).

A straightforward translation of (b) to an algorithm suggests that pairwise intersection be performed among all cumulated concepts. MCA is parsimonious in forming the pairwise intersections: it only performs such operations among the newly formed concepts from the previous stage, instead of cumulatively. We show that this parsimonious multistage strategy is correct: it generates all concepts. We further demonstrate the speed improvements through a set of experimental evaluations.

In comparative evaluation of algorithms for FCA, several important issues must be carefully considered, in no particular order [5]:

1. Whether or not the computation of order relation (i.e. diagram graph) is separated as a different phase than the construction of concept set.
2. Whether or not the computed concept set contains redundant or repeated concepts.
3. Whether or not the intent and extent of concepts are both maintained throughout an algorithm.
4. Whether the concept set is formed by joining one attribute concept (dually, object concept) at a time iteratively, or alternatively, the concept set is formed by setting the initial concept set to include all attribute concepts (dually, object concepts).

Against these features, our MCA is unique in that it

1. separates the computation of diagram graph as a different phase so we can take advantage of sophisticated sub-quadratic partial ordering algorithms proposed by Pritchard [9];
2. maintains a non-redundant set of concepts and uses an auxiliary search tree for quick concept lookup;
3. keeps only the attribute (dually, object) set, or the intent of concepts throughout the algorithm to eliminate the overhead of extent maintenance (the extent can be looked up afterwards in an efficient way if needed, but it is neither necessary for determining the concepts nor for diagram construction);
4. forms the concept system by a multistage intersection operation from the initial concept set consisting of all object concepts.

Of all the algorithms in the literature (see [5]), our algorithm is the closest in spirit to Chein [2], with key distinctions described in items 1, 3, and 4 above (note that Pritchard's algorithm appeared much later). Particularly, each of our stages is independent of the previous ones in that we do not need to modify any concept sets in the previous stage, as is done in Chein's algorithm. The need to mark off concepts in the previous stage and keep it in the current stage is dictated by a simple theoretical justification (Prop. 3, Section4). By developing a more elaborate combinatorial argument, the need to mark off concepts in the previous stage is eliminated, with substantial saving in computational time. Correctness of the latter strategy is non-trival (Prop. 4, Section 4).

To demonstrate that the theoretically advantage of MCA translates to tangible improvement in practice, we performed comparative experimental evaluations against the leading performers from the Kuznetsov-Obiedkov survey [5]: Bordat, Chein, and Norris. The experimental results demonstrated significant improvements in the construction of concept set. They also demonstrated interesting improvements in the diagram graph construction when concept lattice sizes are low-degree polynomials (in terms of context size), by employing Pritchard's approach.

In performing the comparative experimental evaluation, we have been careful in taking into account a number of factors that may influence the result, as suggested in [5]:

– Using a common computational environment for all algorithms under consideration. We implemented Bordat, Chein, MCA, and Norris all in Python, with optimization strategy applied as long as we see fit to do it.
– Using a variety of input context, with varying parameters testing different aspects of each algorithm evaluated.
– Validating the algorithms for both manual and test cases to make sure that the results from all the algorithms agree with each other. We do this by visually inspecting the concept lattice diagrams rendered using both ConExp [13] and Graphviz [4].

In the end, we are surprised that significant improvements in concept analysis algorithm can still be made, particularly by using a simple idea (with a more demanding theoretical justification). Maybe this is exactly Occam's Razor at work.

2 Preliminaries

We follow the notation of [3] in this paper. Readers are referred to [3] and [14] for further details. For any set A, let $\mathcal{P}(A)$ denote the powerset of A. A subset C of the powerset $\mathcal{P}(A)$ is called a *closure system* on A if C is closed under arbitrary intersections, i.e., for every $X \subseteq C, \bigcap X \in C$. By convention, the whole space A is always a member of a closure system C. A *closure operator* on A is a (self-map) function $\varphi : \mathcal{P}(A) \to \mathcal{P}(A)$ which is inflationary ($X \subseteq \varphi(X)$), monotonic ($X \subseteq Y \Rightarrow \varphi(X) \subseteq \varphi(Y)$), and idempotent ($\varphi(\varphi(X)) = \varphi(X)$).

Proposition 1. *Define a* closed set *with respect to a closure operator* $\varphi : \mathcal{P}(A) \to \mathcal{P}(A)$ *to be a fixed point of* φ. *Then closed sets of* φ *are precisely sets of the form* $\varphi(X)$. *The collection of closed sets* $\{\varphi(X)|X \in \mathcal{P}(A)\}$ *forms a closure system on* A.

For closure systems C_1 and C_2 over A, let $C := C_1 \cap C_2$. One can check that C is again a closure system over A. In general, arbitrary intersections of closure systems remain a closure system.

Lemma 1. *Let* A *be a set. Then the set of all closure systems over* A *forms a (meta) closure system over* $\mathcal{P}(A)$. *Every subset of* $\mathcal{P}(A)$ *generates a closure system over* A, *which is the smallest closure system containing the starting subset.*

Concept lattices can be viewed as a closure system generated from a subset of a certain powerset.

Proposition 2. *Let* **K** = (G, M, I) *be a formal context. Then its concept lattice* $\mathcal{B}K$ *is isomorphic to the closure system generated by the set* $\{\{g\}'|g \in G\}$ *and dually,* $\mathcal{B}K$ *is inverse-isomorphic to the closure system generated by the set* $\{\{m\}'|m \in M\}$.

This brings flexibility for procedures for constructing concept lattices. For example, one can partition G into $A \cup B = G$ with $A \cap B = \emptyset$, find the closure system generated by $\{\{a\}'|a \in A\}$ and $\{\{b\}'|b \in B\}$, respectively, and then find the closure system generated by the union of the two closure systems. This view provides an easy-to-understand, straightforward way to justify the correctness of the class of "divide-and-conquer" algorithms in the literature (for which correctness proofs are often omitted).

3 MCA and Example

For terminology, we call object concepts or attribute concepts in a neutral way primitive concepts. In each particular setting, "primitive concept" will refer to either object concept or attribute concept, but not both.

Because concept sets can be viewed as closure systems generated by primitive concepts, an immediate idea is to start from these singleton-generated concepts and repeatedly perform intersections until no new concepts can be formed. Although correct, this naïve approach may involve redundant computations in two ways. One is that intersections of different primitive concepts may give the same resulting set, and hence removing redundancy can reduce the number of potential intersections needed. The second is that pairwise intersections may need only be performed on a subset of existing

```
Input: context (G, M, I)
Output: concept set C
1  Insert M in SearchTree;
2  Insert each member of C_1 in SearchTree if it is not already in, where C_1 := {{g}'|g ∈ G};
3  i = 2;
4  while |C_{i-1}| > 1 do
5      C_i = {};
6      for each pair of (distinct) concepts c_j, c_k in C_{i-1} do
7          candidate = c_j ∩ c_k;
8          if candidate not in SearchTree then
9              C_i = C_i ∪ {candidate};
10         end
11         i + +;
12     end
13 end
14 C = ⋃_i C_i;
```

Algorithm 1. The Multistage Concept Analysis Algorithm

concepts, instead of all existing ones cumulated, even though no redundancy exists. Being parsimonious in both results in a multistage algorithm, as in Algorithm 1.

To illustrate how MCA works, we applied it to the "Living Beings" context from [3]. The context is redisplayed here for easy reference:

	a	b	c	d	e	f	g	h	i
1	×	×					×		
2	×	×					×	×	
3	×	×	×				×	×	
4	×		×				×	×	×
5	×	×		×		×			
6	×	×	×	×		×			
7	×		×	×	×				
8	×		×	×		×			

We have the following concepts, generated in stages:

Stage 0		
	0	abcdefghi
	1	abg
	2	abgh
	3	abcgh
	4	acghi
	5	abdf
	6	abcdf
	7	acde
	8	acdf

Stage 1		
1	9	abg
	10	ag
	11	ab
	12	a
2	13	agh
3	14	acgh
	15	abc
	16	ac
4		
5	17	ad
	18	adf
6	19	acd

The second column in the table for Stage 1 indicates the concept (number) in the previous stage that has been used to obtain potential new pairwise intersections. Each new concept is given a consecutive number, which can be used for reference and book-keeping for the next stage, particularly for manual examples.

Several remarks are in order. First, we did not display Stage 2 which produces no new concepts although this step is needed to ensure the proper termination of the algorithm. Second, the total number of generated concepts is 19, the same as illustrated in [3]. Third, during the "execution" MCA we referred neither to the original context, nor to the extent of any concept. One can easily incorporate a data structure for looking up the extent of a concept, *after* the concepts are already determined by MCA. In contrast, many algorithms in the literature need to have both extent and intent to work properly.

4   Correctness

The correctness of MCA can be shown by induction on the number of primitive concepts used in an intersection. The following simple observation is the theoretical basis of

Chein's algorithm. We mention it here because the justification for our MCA algorithm is based on a similar idea, but a more elaborate combinatorial argument is needed.

For notational preparation, define $S \cap\!\!\!\!\cap\, T := \{s \cap t | s \in S, t \in T\}$, where S and T are collections of subsets. With respect to a given formal context (G, M, I), define $L_0 := \{G\}$, $L_1 := \{\{g\}' | g \in G\}$, and for $i > 1$, $L_i := L_{i-1} \cap\!\!\!\!\cap\, L_{i-1}$.

Proposition 3 (Chein). *With respect to a given formal context (G, M, I),*

$$L = \bigcup_{0 \le i \le |G|} L_i,$$

where L is the set of concepts of (G, M, I).

The proof amounts to an easy induction and we briefly highlight the inductive step. Suppose L_i contains all concepts determined by intersections of $i \ge 0$ primitive concepts. Then, L_{i+1}, formed by pairwise intersections of concepts in L_i, contains all concepts obtained by intersections of $i + 1$ primitive concepts. This can be seen by rewriting an intersection A of $i + 1$ primitive concepts as the pairwise intersection of two concepts B and C in L_i: $A = B \cap C$, where B is the intersection of the first i primitive concepts used for A, and C is the intersection of the primitive concepts omitting the first one, as the following equation shows:

$$\bigcap_{1 \le k \le i+1} \{g_k\}' = \bigcap_{1 \le k \le i} \{g_k\}' \cap \bigcap_{2 \le k \le i+1} \{g_k\}'.$$

Note that L_{i+1} may contain concepts obtained by intersections of other than $i+1$ primitive concepts, because of potential degenerations as well as concepts using $2i$ primitives. There may also be redundancies in that we cannot ensure that L_i and L_{i+1} are non-overlapping. Non-overlapping can be checked by looking up a search tree which contains all existing concepts, as the implementations of several existing algorithms in the literature [5].

Our MCA algorithm uses a different sequence of sets, defined as follows:

$$S_0 := \{G\},$$
$$S_1 := \{\{g\}' | g \in G\}, \text{ and}$$
$$S_{i+1} := (S_i \cap\!\!\!\!\cap\, S_i) - \bigcup_{1 \le k \le i} S_k$$

for $i \ge 1$. The key distinction from Chein lies in the removal of all existing concepts $\bigcup_{1 \le k \le i} S_k$ when forming S_{i+1}, for each stage. This way, only newly generated (and necessary) concepts are kept for subsequent stages, resulting in potentially huge savings in computational cost.

Proposition 4 (Correctness of MCA). *With respect to a given formal context (G, M, I),*

$$L = \bigcup_{0 \le i \le |G|} S_i.$$

Before providing a proof, we need a helper equivalence relation.

Definition 1. *Let (G, M, I) be a formal context. Two sets $X, Y \subseteq G$ are equivalent if*

$$\bigcap \{\{g\}' | g \in X\} = \bigcap \{\{g\}' | g \in Y\}.$$

When X and Y are equivalent, we write $X \equiv Y$. We call a subset of G irreducible if it is not equivalent to any of its proper subsets.

Proof (Proposition 4). It suffices to show by (course of value) induction in i that each concept of the form $\bigcap \{\{g\}' | g \in X\}$ with X irreducible and $|X| = i$ belongs to S_t, where $t = 1 + \lceil \log_2 i \rceil$. In other words, t is an integer such that $2^{t-2} < i \leq 2^{t-1}$ for $i \geq 2$ (we fix $t = 0$ for $i = 0$ and $t = 1$ for $i = 1$).

The base cases ($i = 0, 1$) follow from the definition of S_i. For the induction step, assume that for some $k > 0$, all irreducible subsets X with $|X| = j \leq k$ determine concepts $\bigcap \{\{g\}' | g \in X\}$ belonging to $S_{1 + \lceil \log_2 j \rceil}$. We show that for an irreducible set Y with $|Y| = k + 1$, $\bigcap \{\{g\}' | g \in Y\}$ belongs to $S_{1 + \lceil \log_2 (k+1) \rceil}$.

Let $Y \subseteq G$ be an irreducible set with $|Y| = k + 1$. We have

$$\bigcap \{\{g\}' | g \in Y\} = \bigcap \{\{g\}' | g \in Y_1\} \cap \bigcap \{\{g\}' | g \in Y_2\},$$

where Y_1, Y_2 are subsets of Y with sizes equal to $\lceil (k+1)/2 \rceil$, $|Y_1 \cap Y_2| \leq 1$, and $Y = Y_1 \cup Y_2$. In other words, Y_1, Y_2 is a partition of Y into two equal-sized subsets when $k + 1$ is even, and Y_1, Y_2 is *almost* an equal-sized partition of Y when $k + 1$ is odd – they share a single common element.

When $k + 1$ is even, Y_1 and Y_2 are themselves irreducible. We have, by induction hypothesis, $\{\{g\}' | g \in Y_i\} \in S_{1 + \lceil \log_2 (k+1)/2 \rceil}$ for $i = 1, 2$. Therefore, $\bigcap \{\{g\}' | g \in Y\} \in S_{1 + \lceil \log_2 (k+1) \rceil}$ by the definition of S_i.

When $k + 1$ is odd, Y_1 and Y_2 must also be irreducible since any subset of an irreducible set is irreducible. By induction hypothesis, we have $\{\{g\}' | g \in Y_i\} \in S_{1 + \lceil \log_2 (k+2)/2 \rceil}$ for $i = 1, 2$. However, $\lceil \log_2 (k + 2) \rceil = \lceil \log_2 (k + 1) \rceil$ when k is even. Therefore, $\bigcap \{\{g\}' | g \in Y\} \in S_{1 + \lceil \log_2 (k+1) \rceil}$ again.   □

From the above proof we can see that, incidentally,

$$L = \bigcup_{0 \leq i \leq 1 + \lceil \log_2 |G| \rceil} S_i.$$

5   Computing Diagram Graph Using Pritchard's Sub-quadratic Algorithm

In [9,10], Pritchard addresses the following problem: given a collection

$$\mathcal{F} = \{S_1, S_2, \cdots, S_k\},$$

where $S_i \subseteq D$ for a fixed set D, compute the *complete subset graph*, where the vertices are members of $\mathcal{F}$, and there is an edge from S to S' iff $S' \subseteq S$.

Some notations are needed first. For any subset y of D, let

$$\mathcal{F}.y := \{x \in \mathcal{F} | y \subseteq x\}.$$

Then $\mathcal{F}.\{d\}$ stands for the sub-collection of all subsets in $\mathcal{F}$ containing d, and $x \in \mathcal{F}.y$ iff $y \subseteq x$ iff $x \in \bigcap_{d \in y} \mathcal{F}.\{d\}$. Therefore,

Lemma 2 (Pritchard). *For any $y \in \mathcal{F}$, $\mathcal{F}.y = \bigcap_{d \in y} \mathcal{F}.\{d\}$.*

Hence, the intersection $\bigcap_{d \in y} \mathcal{F}.\{d\}$ contains all supersets of y. Now suppose $\mathcal{F}$ is the collection of all concepts (intents, say). To find all the parents of x in the concept lattice algorithmically, one first finds $\mathcal{F}.\{d\}$ for each $d \in x$, and then computes the intersection $\bigcap_{d \in x} \mathcal{F}.\{d\}$, and then remove y from it. To find all upper neighbors of y, one can further remove all sets in $\bigcup\{\mathcal{F}.z | z \in \bigcap_{d \in x} \mathcal{F}.\{d\} - \{y\}\}$, from $\bigcap_{d \in x} \mathcal{F}.\{d\} - \{y\}$. This gives us the following algorithm, which achieves the optimal bound $O(N^2 / \log N)$ as shown in [9], where N is the sum of the cardinalities of all the sets in the collection $\mathcal{F}$.

The lattice graph construction algorithm is given in Algorithm 2. It is quite elegant. Inputs to this algorithms are the attribute set M and the concept set C computed by some other algorithm, e.g. MCA. Lines 1–3 construct a list, S_m, for each attribute $m \in M$ which contains the index of each concept that contains the particular attribute. Lines 4–6, the linking phase, identify the supersets P_c, of each concept by computing the intersection of the concept list, S_m (computed in the previous step), for each attribute m belonging to concept, c. The final phase, lines 7–9, removes the links to concepts other than direct parents by removing those concepts that are the parents of the parents of the current concept. This is done by removing the union of the parental list P_i for each parent concept i with concept c in P_c.

The second phase, lines 4–6, is the workhorse step of this algorithm. The speed advantage of this algorithm stems from only having to compute $|c|$ number of (the number of attributes in a concept, generally a small number) chained intersections to find the parents for each concept. The intersections are fast to compute because the sets over which they operate are usually short, particularly after one or more intersections in the chain is already computed. The first step can also be completed cheaply as part of the concept construction algorithm. The final step largely consists of union operations, which are relatively inexpensive.

6 Comparative Experimental Study

This section compares the performances of various FCA algorithms in the literature with our algorithms for MCA (Algorithm 1) and diagram graph construction (Algorithm 2). In particular we evaluate the concept construction algorithms of Chein [2] and Norris [8], the diagram graph construction algorithm of Valtchev [12], and the concept and lattice construction algorithm of Bordat [1], against the corresponding algorithms using Algorithm 1 and Algorithm 2, respectively. We also implemented Lindig's algorithm [6] but the result was not interesting enough to be included. All algorithms were implemented in Python on a MacBook Pro 2.33 GhZ and 2 GB RAM. The algorithms are expected to uniformly perform better using C and a desktop computer.

Contexts used in our experiments were randomly generated according to these parameters: $|G|$ for the number of objects, $|M|$ for the number of elements, and $|g'|$ for the number of attributes per object.

Input: attribute set M, concept set C
Output: parent sets P_i
1 **for** $m \in M$ **do**
2 $\quad|\quad S_m = \{c | c \in C \wedge m \in c\};$
3 **end**
4 **for** $c \in C$ **do**
5 $\quad|\quad P_c = \bigcap \{S_m | m \in c\};$
6 **end**
7 **for** *each P_c* **do**
8 $\quad|\quad P_c = P_c / \bigcup \{P_i | i \in P_c\}$
9 **end**

Algorithm 2. Diagram Graph Construction

6.1 Concept Set Only

Here we compare the performances of algorithms that just compute a concept set. In applications such as data mining, concept sets (as opposed to the complete lattice diagram) are of primary interest. We compare Algorithm 1 with the algorithms of Chein [2] and Norris [8]. These are the best performing algorithms reported in [5]. We timed each algorithm on contexts with various sizes and densities where density refers to the number of attributes belonging to an object. Varying these parameters gives a more complete picture of the performance of each algorithm.

As can be seen from Figs. 1a–1c in the Appendix, MCA outperformed the other algorithms in all experiments. Chein was only slightly slower than MCA on sparse contexts (Fig. 1a), but Norris was significantly slower. On medium density contexts (Fig. 1b), Chein and Norris were similar but much slower than MCA. On denser contexts (Fig. 1c), Chein lagged far behind Norris, which was closely behind MCA.

6.2 Lattice Diagram Only

The Pritchard lattice diagram construction algorithm (Algorithm 2) was compared with the algorithm of Valtchev, Missaoui and Lebrun [12]. As discussed earlier, Algorithm 2 computes all the upper-neighbors (supersets) of each node and then removes those which are not direct parents. Valtchev-Missaoui-Lebrun begins at the top of the lattice and then recursively identifies the lower neighbors of each concept. In comparing the two algorithms we computed the concept set using MCA and then timed the construction of lattices using both algorithms. Experiments were run using the same parameters as given in the previous subsection. Figs. 2–4 show the superior performance of the Pritchard in all cases. This may be attributed to the computing of supersets rather than all lower neighbors.

6.3 Concept Set and Lattice Diagram Together

Here we compare the performance of MCA+Pritchard with Bordat [1]. In [5], Bordat was featured as the best algorithm that constructed the concepts and lattice diagrams simultaneously. Our experiments revealed mixed results. For sparse contexts (Fig. 5), MCA+Pritchard outperformed Bordat by a fair amount. For medium density contexts (Fig. 6), MCA+Pritchard outperformed Bordat until $|G| = 75$, and then a reversal occurred. For denser contexts (Fig. 7), Bordat outperformed MCA when the number of objects exceeded 30.

7   Conclusion

The introduced Multistage Concept Analysis (MCA) Algorithm is simple but also efficient, demonstrated through a rigorous theoretical analysis and an experimental evaluation. It is the fastest algorithm we have seen so far for constructing concept set.

This paper also brings Pritchard's Algorithm to concept analysis. When concept generation and diagram generation are combined, MCA+Pritchard outperforms Valtchev-Missaoui-Lebrun for contexts with lattice sizes bounded by a low-degree polynomial in $|G|$ or $|M|$. For contexts whose lattice sizes grow faster than low-degree polynomials, incremental, neighborhood approaches are expected to perform better, since the time complexity of these algorithms is bounded by a product of the lattice size and a low-degree polynomial in $|G|$ and $|M|$, rather than a sub-qudratic function of the lattice size. It would be interesting to bring into the picture more recent incremental algorithms, such as those proposed in [7,11].

References

1. Bordat, J.P.: Calcul pratique du treillis de Galois d'une correspondance. Math. Sci. Hum. 96, 31–47 (1986)
2. Chein, M.: Algorithme de recherche des sous-matrices premieres d'une matrice. Bull. Math. Soc. Sci. Math. R.S. Roumanie 13, 21–25 (1969)
3. Ganter, B., Wille, R.: Formal Concept Analysis. Springer, Heidelberg (1999)
4. http://www.graphviz.org/
5. Kuznetsov, S.O., Obiedkov, S.A.: Comparing performance of algorithms for generating concept lattices. J. Exp. Theor. Artif. Intell. 14(2-3), 189–216 (2002)
6. Lindig, C.: Fast Concept Analysis. In: Ganter, B., Mineau, G.W. (eds.) ICCS 2000. LNCS, vol. 1867, Springer, Heidelberg (2000)
7. van der Merwe, D., Obiedkov, S., Kourie, D.: AddIntent: A new incremental algorithm for constructing concept lattices. In: Eklund, P.W. (ed.) ICFCA 2004. LNCS (LNAI), vol. 2961, pp. 372–385. Springer, Heidelberg (2004)
8. Norris, E.M.: An algorithm for computing the maximal rectangles in a binary relation. Rev. Roumaine Math. Pures et Appl. 23(2), 243–250 (1978)
9. Pritchard, P.: On computing the subset graph of a collection of sets. J. Algorithms 33(2), 187–203 (1999)
10. Pritchard, P.: A fast bit-parallel algorithm for computing the subset partial order. Algorithmica 24(1), 76–86 (1999)
11. Valtchev, P., Missaoui, R., Godin, R., Meridji, M.: Generating frequent itemsets incrementally: two novel approaches based on Galois lattice theory. Journal of Experimental & Theoretical Artificial Intelligence 14(2/3), 115–142 (2002)
12. Valtchev, P., Missaoui, R., Lebrun, P.: A fast algorithm for building the Hasse diagram of a Galois lattice. In: dans Actes du Colloque LaCIM 2000, pp. 293–306, Montreal (2000)

13. Yevtushenko, S.: ConExp. http://sourceforge.net/projects/conexp
14. Zhang, G.-Q.: Chu spaces, formal concepts, and domains. Electronic Notes in Computer Science 83, 16 (2003)
15. Zhang, G.-Q., Shen, G.: Approximable Concepts, Chu spaces, and information systems. In: De Paiva, V., Pratt, V. (eds.) Theory and Applications of Categoiries, Special Volume on Chu Spaces: Theory and Applications, vol. 17(5), pp. 80–102 (2006)

A   Figures

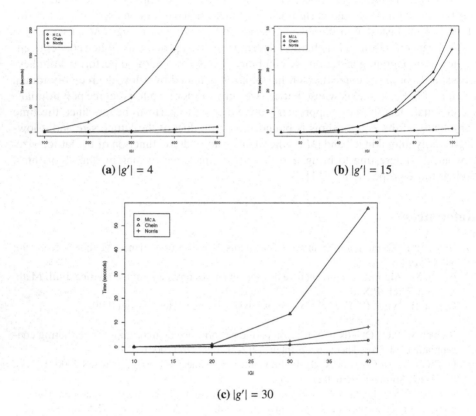

(a) $|g'| = 4$

(b) $|g'| = 15$

(c) $|g'| = 30$

Fig. 1. Comparison of time to compute concept set only – MCA vs. Chein and Norris for contexts with $|M| = 100$, $|g'| = 4$ (sub-figure 1a), $|g'| = 15$ (sub-figure 1b), $|g'| = 30$ (sub-figure 1c), and $|G|$ between 10 and 50.

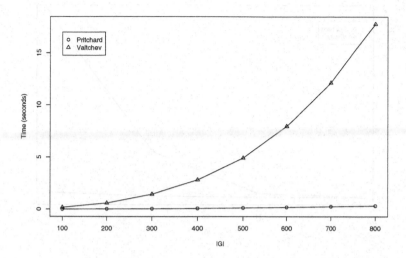

Fig. 2. Comparison of time to construct lattice diagrams only – Pritchard vs. Valtchev for concept sets from contexts with $|M| = 100$, $|g'| = 4$ and $|G|$ between 100–800

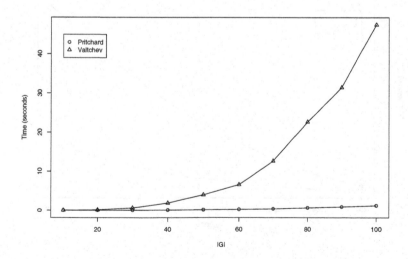

Fig. 3. Comparison of time to construct lattice diagrams only – Pritchard vs. Valtchev for concept sets from contexts with $|M| = 100$, $|g'| = 15$ and $|G|$ between 10–100

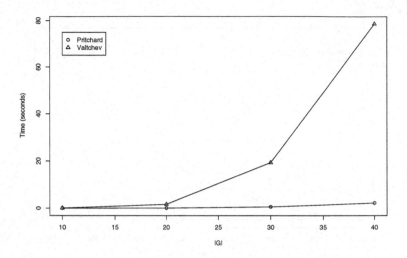

Fig. 4. Comparison of time to construct lattice diagram only – Pritchard vs. Valtchev for concept sets from contexts with $|M| = 100$, $|g'| = 30$ and $|G|$ between 10–40

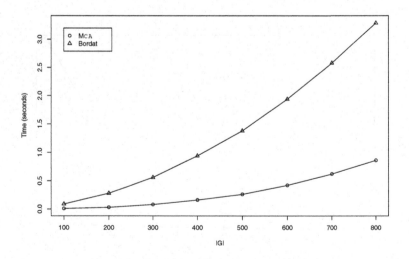

Fig. 5. Comparison of time to compute concept set and lattice diagram – MCA+Pritchard vs. Bordat for contexts with $|M| = 100$, $|g'| = 4$ and $|G|$ between 100–800

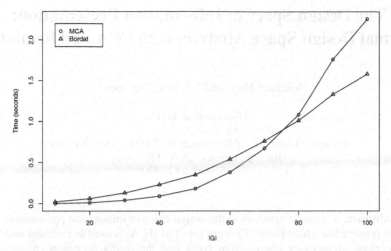

Fig. 6. Comparison of time to compute concept set and lattice diagram – MCA+Pritchard vs. Bordat for contexts with $|M| = 100$, $|g'| = 15$ and $|G|$ between 10–100

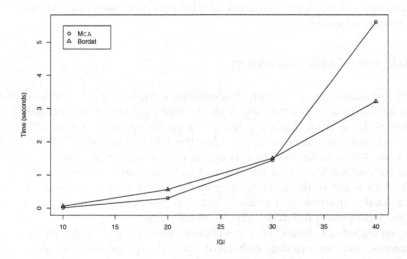

Fig. 7. Comparison of time to compute concept set and lattice diagram – MCA+Pritchard vs. Bordat for contexts with $|M| = 100$, $|g'| = 30$ and $|G|$ between 10–40

The Design Space of Information Presentation: Formal Design Space Analysis with FCA and Semiotics

Michael May[1] and Johannes Petersen[2]

[1] LearningLab DTU
mma@dtv.dk
Technical University of Denmark, DK-2800 Kongens Lyngby
[2] Rambøll Denmark A/S - Oil & Gas
JOAP@ramboll.dk

Abstract. A semiotic approach to the design space of information presentation is presented in which Formal Concept Analysis (FCA) is used to represent and explore attributes of abstract sign types and the media (graphical, haptic, acoustic, gestic) through which they are presented as specific representational forms. Early taxonomies in design have typically been incomplete (in only considering graphics) and inconsistent (in the absence of separation between media and sign types). With digital multimedia and the future "semantic web", we need a consistent taxonomy to support component-based flexible (adaptive, tailorable) presentations with a clear separation between (a) the content forms of data, (b) the representational forms through which data is expressed, (c) the combination of media of presentation, and (d) the specific layout within the constraints of the presentation devices and the ergonomic and aesthetic choices of designers and users.

1 Background and Motivation

With the increasing diversification of electronic devices (e.g. PCs, PDAs, mobile phones) and software platforms, and with the rapid growth of internet-based web-technologies, there is an increasing interest in providing services that are not only device- and platform-independent, but also flexible in interfaces and information presentation. At the same time there is an increasing interest in providing access to information independent of user's physical abilities and preferences. *Design for flexibility* and *reuse* in the form of support for *tailoring* or *automated adaptation* promote ideal requirements for user interfaces and components in terms of a systematic separation of different layers of organization.

This separation of layers of organization was initiated with object-oriented programming, but was further elaborated with the progressive development of internet-based programming from HTML to XML. It is clearly not enough to have a separation of "form" and "content" in the sense of the early specification of HTML and DHTML, where (graphical) *layout* was specified by the mark-up language. In the XML-based multimedia specification language SMIL (Synchronized Multimedia Integration Language) the specification of the "form" of presentations have been extended from graphical layout to spatial and temporal layout of components in different media channels [4]. What is still missing in SMIL 2.1 (2005) as well as in

U. Priss, S. Polovina, and R. Hill (Eds.): ICCS 2007, LNAI 4604, pp. 220–240, 2007.

the current MPEG-7/21 "Content Description Interface" for multimedia is a separation of *representational forms* from *media types*. The need for such a separation is becoming evident in the quest for the "semantic web", i.e. with the need to have conceptual descriptions of information content available on the web to support intelligent search driven by semantic specifications. Even with ontologies to support conceptual descriptions of content, we will however still need an analysis and description of the representational forms that can adequately express any given content, and users should be able to request information expressed in a particular form (e.g. in diagrammatic form or in linguistic form) comparable to the option given today of searching for content presented in a certain type of media as specified by the file extension (e.g. searching for video or static graphics). The "semantic web" of the future will thus necessarily involve some kind of *applied semiotics* in order to be able to handle representational forms district form media.

The ambition of providing *universal access* to information independent of user's physical abilities and preferences have focused on media issues in the attempt to find alternative media for information content (e.g. audio alternatives for visually impaired users), but since different *types of content* cannot be expressed equally well in all media, the universal access will necessarily move in the same direction as the semantic web, i.e. towards some kind of applied semiotics to handle representational forms as district from media. Representational forms should in fact be defined as *invariant across media*, and the universal access to information content should look for a foundation in the link between *forms of content* and the media-independent forms of representation "matching" a given content. Flexible user interface components could therefore be a common goal for universal access and for usability in general, and the development of flexible components could be an alternative to the development of specific assistive technologies. One of the conceptual tools used in the work on universal access is the construction of *abstract user interface descriptions* in languages like UIML (User Interface Markup Language) or XIML (Extensible Interface Markup Language). IBM introduced their concept of "transcoding" as a framework for adaptation of different "modalities" such as text, images, video and audio to individual pervasive devices through "on-the-fly" *content summarization, translation* and *conversion* [30]. A comparison of four abstract UI description languages and their support for accessibility is given in [36]. A major challenge for the future progress of the field have been identified as finding techniques for representing semantics in ways that are scalable, extensible, and "modality-independent", i.e. independent of specific media types and sensory modalities [34].

The problem however has been that there are no consistent taxonomies available for classifying multimedia interfaces, components and presentations. Even within the less complex field of graphical design the problems of representation and layout of illustrated static documents have not yet been fully solved [2], [22]. There have been attempts to classify graphics such as [6], [14], [15], [37], but they all fail in different ways [22]. They have usually both been *incomplete* (from the point of view of multimedia) by dealing only with graphic media (and mainly with static graphics) and *inconsistent* in confusing issues of the media of presentation with the representation of information content.

2 Abstract Media Types

The problems of classification in graphical design are inherited to multimedia design. The best way to look more closely at this extended problem is to start with the concept of media types. First we need to consider terminology. Media types are sometimes called "modalities", because *sensory modalities* (e.g. visual, tactile, auditory, kinesthetic) contribute to the concept of media (e.g. "audio-visual media"). We do however need to add distinctions pertaining to different *channels of communication* in order, for example, to distinguish graphic and gestic communication. Graphic and gestic presentations are both perceived visually, but whereas graphics communicates through properties and relations of presented physical objects, gestic communication is based on embodied movement. In order to abstract media types based on sensory modalities and communication channels, we will not use the sensory terms (i.e. visual, tactile, auditory, and kinesthetic) to describe media, but "objectified" terms: graphic, haptic, acoustic and gestic. These are *abstract* media types and should not be confused with *technologies* of presentation like television or multimedia systems. Classification of abstract media types is thus quite different from classifications of multimedia *systems*.

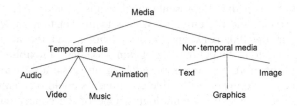

Fig. 1. Media types according to Gibbs & Tsichritzis (modified to make non-temporal media explicit) [9]

Media types (fig.1) were introduced by the object-oriented approach to multimedia programming [9]. The object-oriented approach to "media types" was a big step forward in providing a template for the embedding of operational possibilities along with the representational properties supported by each type of media. This was described by a *feature structure* for multimedia objects called a *media type template*. The types identified were however a mix of media types and representational forms based mainly on which presentations had their own data format at the time. "Audio" is for instance identified as a type separate form "text", whereas the auditory equivalent of text (i.e. spoken language) is not identified. The description of a data format for speech synthesis (i.e. voiceXML) came much later (voiceXML 2.0 adopted by W3C in 2004). But if voice was not identified as separate for audio in general, why was music given its own media type [9] in 1994? Probably because the MIDI (Musical Instrument Digital Interface) data type had been introduced as early as in 1983.

The technology-driven approach to classification leads to arbitrary choices, and there is no systematic differentiation between media types and sign types. The distinction between temporal and non-temporal media types is claimed to establish a correspondence between the media types described (fig.1). This *appears* to be the case for audio versus text, video versus image, and animation versus graphics, but what about music? The temporal form of music does not have any corresponding non-temporal form according to this classification (in our classification music is a complex *diagrammatic* form in sound). A further analysis will break down these apparent analogies. Audio is not the temporal form of text, because audio includes non-linguistic sound. Another problem is that text and graphics cannot be considered as different "media types". As any graphical designer knows text is necessarily a form of graphics, i.e. it is language expressed as graphical text with graphical properties like shape, size and orientation. "Graphic" does indeed indicate a media type, but it has to cover many forms of representation and it has to cross the boundary between temporal and non-temporal. The whole range of representational forms from images to language can in fact be presented as *static*, as *repetitive*, as *sequentially* animated or as *dynamically* controlled presentations (within a movie for instance). Graphic drawings do not cease to be graphic just because they are animated, i.e. animation should be seen as a potential within graphic media, and it therefore cuts across different forms of graphic representation. "Text" is not inherently non-temporal, but can be presented as animated letters, blinking text, ticker-tape text, auto-scrolling text or "fluid" hypertext [3].

In summary the abstract media types suggested here are the *graphic*, the *haptic*, the *acoustic* and the *gestic* type. Other media types could in fact be defined, i.e. the types derived from the chemical senses of smell and taste, but since presentations communicated through chemical channels of communication are not only transient, but also difficult to control, at least with the present state of technology, they are not included. The approach here is to describe media types through their significant properties as media of presentation and communication. In the simplified example below (fig. 2) only five properties have been selected (1) the *immediacy* of interpretation characterizing graphic and acoustic media, (2) the *visual* channel of communication characterizing the graphic and gestic media, (3) the *transient* nature of presentations in the acoustic and the gestic media, (4) the *serialization* of communication in acoustic and the haptic media, and (5) the *movement*-based communication characterizing gestic and haptic media. Although this list of properties is a somewhat arbitrary selection from the properties characterizing media, the lattice shown in fig. 2 demonstrates that this list is enough to differentiate the abstract media types. For practical purposes we will need a more elaborate specification of media properties to determine the specific forms of *cognitive support* [39] provided by different choices of media and representational forms.

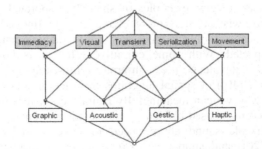

Fig. 2. A media type lattice showing the specification of media types through a small number of properties

The fundamental problem of classification is not solved with clarifying media types. A systematic distinction is needed between media types and *representational forms* derived from *abstract sign types*, i.e. syntactically and semantically different ways of representing information. In order to understand the nature of representational forms as independent from media types, it is useful first to consider the case of natural language.

3 Abstract Sign Types

3.1 Core Semantics and Emergent Semantics

Natural language can be understood as an *abstract sign type* with which we can present information content across different *media* in specific *representational forms* such as graphical language (i.e. text), acoustic language (i.e. speech), haptic language (e.g. Braille embossed text), or embodied gestures (i.e. "body language", sign languages). We should in fact consider representational forms as being *invariant across different media* [33]. The difference between diagrams and natural language, for instance, is invariant across tactile and graphic presentations of diagrams or language. Where the media types are derived from *how the content of a presentation is communicated and perceived*, the abstract sign types as well as their expression in particular media (i.e. the specific representational forms) are derived from *how this content is interpreted* according to an intended interpretation. Braille text as well as graphical text is interpreted as language, whereas graphic and tactile diagrams are interpreted as diagrammatic forms. A semiotic and cognitive theory of information presentation is needed to explain this semantic invariance of interpretation, by which representational forms are constituted. A cognitive foundation for the major representational forms that can be characterized as image-like forms, diagrammatic forms and language-like forms might be found in *modular cognitive structures* of the human brain, i.e. in the modular structures of *perception*, *spatial reasoning* and *language* [38].

It could be claimed that language is a unique representational system in providing invariance across the phonological and gestic system of expression as well as across different graphical writing systems [33]. Graphical representation systems might be different, because they rely on a less abstract mapping of relations in the represented domain and relations in the representation within a presentation media. This could explain why it is easy to give examples of spatially elaborated graphic and tactile *maps* and *diagrams*, whereas it is more challenging to find acoustic equivalents, because transient acoustic objects does not allow diagrammatic relations to be presented and inspected. With some cognitive effort it is however possible to reconstruct mental equivalents (i.e. "internal representations" in the sense of *distributed cognition* [40]) of spatial diagrams from temporally presented acoustic features, which is why we can meaningfully claim the existence of simple *acoustic maps* (e.g. the sound maps used to identify the position of enemy aircrafts in fighter cockpits, or localizable sound alarms used to direct occupants in building to emergency exits [20]) and simple *acoustic diagrams*. Music in fact exemplifies complex diagrammatic structures in sound organized at many levels (tonal, melodic, harmonic etc.)

The important point is that it is possible to give a modified account of the *invariant properties* that are *constitutive* of abstract sign types. We will stipulate that *only a reduced set of core properties are invariant across media* and *this invariant core is what constitutes the abstract sign type*. These invariant properties are then inherited to the media-specific representational forms. Each specific *media type furthermore affords and provides cognitive support for an extended set of properties and operations* that are meaningful within the representational form, although they are not supported in all media [22]. The addition of media-specific properties gives rise to an "emergent semantics".

XML- and web-based multimedia specifications like SMIL implement some level of abstraction in the form of a separation of layout and media from content, but there is no conceptual description of the content and no semantic description of sign types or representational forms independent of media. This is why *semiotics* (i.e. a theory of signs and signification) and techniques for representing and manipulating conceptual structures (FCA, CG) are necessary for the semantic web - as well as for information presentation in general. A first step is to represent and explore a conceptual taxonomy of *abstract sign types* based on their differential attributes (i.e. their semantic properties) and then to investigate their inherited and emergent attributes when they are expressed as *representational forms* within specific physical media of presentation.

3.2 Iconicity and the Iconic-Symbolic Dimension

A point of departure has been the differentiation of forms along the iconic-symbolic dimension described by Peirce [16],[18]. In multimedia semiotics *concrete iconic forms*, *abstract iconic forms* and *symbolic forms* have been suggested [29], whereas a more detailed differentiation includes *Image, Map, Graph, Diagram, Symbol* and *Language* [22]. In the application domain of "technical graphics" an early suggestion of a *feature-based semantics* was given by Alan Manning [15]. Each graphical type in this simple semantic system was described by a set of *feature structures* and it have later been shown that this taxonomy can be described as a formal context [22]. Manning stipulated four types of technical graphics abstracting from their terminological variants and their apparent similarities and differences: chart, diagram, graph, and table. According to his analysis these types can be distinguished by logical combination of two features called "units" and properties: display of one unit (written as –u) or several "units" (+u) and the representation of one property (-p) or several properties (+p). Each graphical type in this simple semantic system can thus be described by a feature structure:

chart: [-p, -u]   graph: [-p, +u]

diagram: [+p, -u]   table: [+p, +u]

A simple *pie chart* only displays one unit (some totality) and it only represents one property (fractions of the total), whereas a graph like a *bar graph* has several units (represented by individual bars) and again only one property (the amount represented by the height of each bar). A *diagram* (as analyzed by Manning) only represents one unit (some object), but display several properties of the represented object, whereas a table usually represent many units and display several properties of each represented object. As described in [22] the feature structures can be described as formal concepts and represented as a lattice, cf. fig. below.

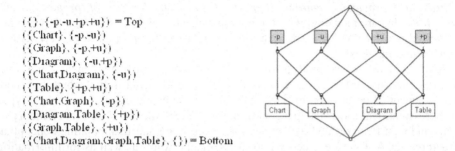

$\{\{\}, \{-p,-u,+p,+u\}\} = \text{Top}$
$\{\{\text{Chart}\}, \{-p,-u\}\}$
$\{\{\text{Graph}\}, \{-p,+u\}\}$
$\{\{\text{Diagram}\}, \{-u,+p\}\}$
$\{\{\text{Chart,Diagram}\}, \{-u\}\}$
$\{\{\text{Table}\}, \{+p,+u\}\}$
$\{\{\text{Chart,Graph}\}, \{-p\}\}$
$\{\{\text{Diagram,Table}\}, \{+p\}\}$
$\{\{\text{Graph,Table}\}, \{+u\}\}$
$\{\{\text{Chart,Diagram,Graph,Table}\}, \{\}\} = \text{Bottom}$

Fig. 3. Formal concepts and lattice representation of Manning's feature structures for technical graphics [22]

Taxonomies of multimedia often reproduce the difficulties encountered in taxonomies of graphics. An example is the classification of media types into text, sound, graphics, motion, and multimedia [10], [11]. The haptic and gestic media are not considered, and the relation between text and graphics is unanalyzed. The model however introduces a *dimension of expression* of media types covering "elaboration", "representation" and "abstraction". In the case of graphics this expression dimension is exemplified by photographs and images at the "elaboration" end, blueprints and schematics as the intermediate form of "representation", and icons at the "abstraction" end. The expression dimension of [10] is close to the dimension of the sign in the so called TOMUS model [29], with its subdivision of sign types into *concrete-iconic, abstract-iconic, and symbolic* (fig. 4). The TOMUS model introduces a three-dimensional model of multimedia objects, based on a differentiation between *sign types*, *syntactic structures* and *sensory modalities* (i.e. media types).

The iconic-symbolic dimension of sign types can be derived from the semiotic analysis of signs according to C.S. Peirce. Where Saussure and later Hjelmslev [12] conceived "semiology" as an extension of linguistics that would study the quasi-linguistic properties of other "sign systems", Peirce developed his more general conception of a "semiotic" from his analysis of logic. For Peirce the sign is not a dyadic relation between expression and content, but a triadic relation between a physical representation (the "representamen") and an interpretation (the "interpretant") of this representation as referring to an object in some respect. Within this triadic relation, it is the representation-object aspect which is categorized as being iconic, indexical or symbolic by Peirce.

The causal or contextual relation implied by the indexical category can however be considered as a separate issue from the dimension of iconic – symbolic. A subdivision of the iconic – symbolic dimension was therefore used as a foundation of the sign

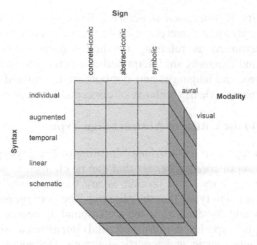

Fig. 4. The TOMUS model of sign types, syntax and (sensory) modalities (i.e. media types) according to [29]

typology presented in [16], [17], [18], [19], [21], with the main sign categories being *Image, Map, Graph, Diagram, Symbol* and *Language*.

The Image and Map categories here correspond to the concrete-iconic sign type in the taxonomy of [29], the diagrammatic forms of representation, Graph and Diagram, correspond to the abstract-iconic sign type, and Symbol and Language to the symbolic type. The TOMUS model only distinguishes the aural and the visual senses supporting the acoustic and the graphic media types, but it is an important feature of the model that all combinations of media types and abstract sign types can be considered in a systematic way.

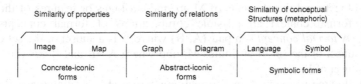

Fig. 5. The major sign types at two levels of detail and their foundation in similarity metrics

"Iconicity" will always be present in some form in order for signs to be informative about some aspect of a real or imagined world. Iconicity is not necessarily "pictorial" or even image-like in a generalized sense (image-like objects in non-visual media). Iconicity can be more or less abstract from the display of object properties in images to the schematization of situations and events in language[16]. From the semiotic point of view we cannot have a single *similarity measure*. The main representational forms correspond to different underlying similarity measures as implied by C. S. Peirce. Concrete-iconic forms rely on a *similarity of properties*, abstract-iconic forms rely on a *similarity of relations*, and symbolic forms rely on an "induced" metaphoric *similarity of conceptual structures*. These types of similarity correspond to systematic differences in the interpretation of the main iconic forms (fig. 5): images and maps are interpreted as referring to their objects through a similarity of properties, graphs and diagrams are interpreted as referring to their objects through a similarity of relations, and languages and symbols are interpreted as referring to their objects through a (metaphorical) similarity of conceptual structures.

3.3   Introduction to the Lattice of Abstract Sign Types

The challenge with regard to the *attributes* defining or "generating" the abstract sign types is to define a set of *core properties* that can be claimed to be invariant over the media of presentation. These core properties are then the attributes of the formal context of the abstract sign types (i.e. the objects). The core properties constitutes the abstract sign types and by derivation the corresponding representational forms. A diagram as an abstract type has certain (potential) attributes as signs that are shared among graphical, haptic, gestic, and acoustic diagrams. Defined in this way the rest of the *actual properties* of *concrete diagrams* follow by inheritance of media attributes and from subsequent combinations of presented components giving rise to an "emergent semantics". Diagrams known from specific work domains are always complex combinations of multiple representational forms – sometimes even within multiple media. The perspective of having an underlying taxonomy describing the attributes of these forms as well as their combinations is that we should be able to support more flexible components – not just in terms of layout, but in terms of transformations between media (e.g. from graphics to sound) or between representational forms (e.g. from diagrams to language).

Feature structures can be represented as formal concept in the sense of Formal Concept Analysis [8]. Features ("attributes") and the concepts they specify (the "objects") are related in a matrix called a *formal context*. A formal context C:= (G, M, I) is defined as two sets G and M with a relation I between G and M. The elements of G are the objects and the elements of M are the attributes or features of the context. From the formal context all possible combinations of formal concepts can be generated. A *formal concept* is a pair (A, B) where A is a subset of the set of objects G, B is a subset of the set of attributes M, and where A´=B and B´=A (i.e. AxB is a maximal subset of I).

In the following we will set up a formal context for abstract sign types. The formal context can be defined at many *levels of detail* dependant on its purpose. We can constructively move between different levels of detail by abstracting from some of the attributes or from some of the sign types. This can be seen as a simple case of *conceptual scaling* [5] [8], i.e. the trivial case where we select a part of the full lattice based on a *sub-context*. The lattice representation of the formal concepts, i.e. the abstract sign types, can be used to explore the attributes defining different selections of objects. Below is shown three example of lattices representing sub-contexts for the *concrete-iconic types*, Image and Map (fig. 6 left), for their extension with the Map type (fig. 6 right), and for the *abstract-iconic types* (Graph and Diagram), but in latter case (fig. 7) specified further into four types, (Numerical) Graph, Categorical graph, Network Chart, and Diagram. For simplicity we will initially show these lattices without their defining attributes (only the objects are shown).

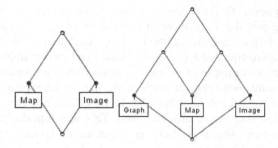

Fig. 6. Two simple lattices without attribute annotation corresponding to two sub-contexts. Left the lattice of concrete-iconic sign types (Image and Map), and right the extension of this lattice with the Graph type.

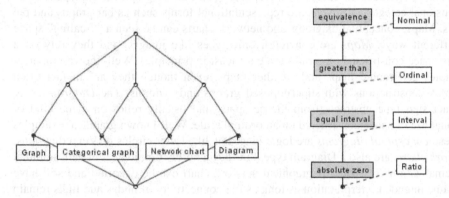

Fig. 7. (Left) The lattice of the abstract-iconic sign types, but at a further level of specification using scale types

Fig. 8. (Right) The chain lattice of scale types according to the classification of Stevens [35]

The further specification of the Graph and Diagram types in fig. 4 can be understood as the result of a specification of the abstract-iconic types using properties of the concept of *scale types* (fig. 8) as defined by Stevens [35][40],[25],[26], i.e. the ratio, interval, ordinal and nominal scale types. By the term "categorical graphs" we here refer to graphs of nominal scale, i.e. categorical data. We could construct a larger lattice as a product of two lattices containing both the attributes generating the abstract sign types and the attributes generating scale types, and then select the relevant sub-context to specify the abstract-iconic sign types specified with scales type attributes. The scale types are themselves ordered in a lattice "chain" [7] with the nominal scale concept as the "weakest" top concept and the ratio scale as the "strongest" concept at the bottom, i.e. the concept that has inherited all the attributes of the weaker scales above it. A richer taxonomy of scale types is utilized in [8].

Nominal scales require only the attribute of *equivalence*, meaning that assignment of content to a nominal scale only constructs data corresponding to equivalence classes, i.e. we can classify content as belonging to a particular class as we do in linguistic descriptions. *Ordinal scales* require equivalence as well as an *order relation*, meaning that the assignment of content to an ordinal scale constructs data with some kind of "greater than" relation among its elements. *Interval scales* (inter-ordinal scales) require equivalence, order relations, and further imposes a requirement of *equal intervals*. An example of an interval scale with equal intervals is the Celsius scale for temperature measurements, where we can meaningfully add and subtract temperatures, but we cannot multiply (20 degrees C is not twice as warm as 10 degrees C), because there is no absolute zero. This was introduced with the Kelvin scale for temperature which is then an example of a *ratio scale* adding the requirement of an *absolute zero*.

The scale types can be used to differentiate abstract sign types. Numerical *graphs* are *ratio scale graphs* (1, 2, 3 or n-dimensional graphs) whereas *categorical graphs* introduce one or more dimensions with data on a *nominal scale* (ordinal scale data can also be included in categorical graphs). Well-known graphical examples of 2-dimensional categorical graphs are representational forms such as bar graphs and pie charts. Maps, conceptual diagrams and network charts can be seen as "scaling" space in different ways. *Maps* are concrete-iconic types like images, and they rely on a metric space constructed on a ratio scale (at least in principle). Well-known graphical examples geographic maps and weather maps, even though they are in fact often complex constructions with superimposed graphs and symbols. The *Diagram* as an abstract sign type abstracts from metric space and usually relies on some kind of regional use of space constructed on an ordinal scale. Well-known graphical examples of these *conceptual diagrams* are logic diagrams like Euler circles or Venn diagrams. *Network charts* are also a Diagram type, but they a purely topological in their nominal "schematization" of space. A graphical network chart can be distorted and still have the same intended interpretation as long as the connectivity of nodes and links remain the same. A well-known graphical example is process diagrams used in e.g. chemical industrial plants and in power plants. Process diagrams are however often made more "familiar" to operators by aligning their abstract parts with some kind of *superimposed* map of the layout of the components that the process diagrams refer to (they are often called "mimic diagrams") and abstract symbols representing physical components can be *substituted* with more "iconic" images that "looks like" the components they represent (reactors, pumps, pipes etc.). *Superimposition* and *substitution* are examples of general *constructive operations* on presentations in the sense of [13].

3.4   Specification: The Formal Context of Abstract Sign Types

A formal context that could generate the abstract sign types required will now have to be introduced. We will here stick with the extended version of the taxonomy with two types of graphs and two types of diagrams, even though we could specify abstract sign types with greater detail (for instance by using the full formal context of scale types). Below (fig. 9) is shown a semi-lattice of the suggested attributes that we will use the conceptual scaling of abstract sign types.

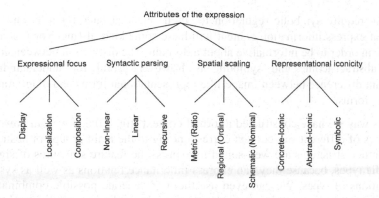

Fig. 9. Semi-lattice of the attributes of the expression for conceptual scaling of the combined media type – sign type lattice (the ordinal scale of "regional space" here includes inter-ordinal scales, i.e. intervals)

As shown above these attributes have two levels and could therefore be used to define a multi-valued context with the four first-level attributes as variable attributes, and with the second-level attributes as their values. We will however only use this ordering of attributes to provide an overview, whereas we will "flatten" the attributes in the formal context for the abstract sign types to enhance simplicity. The first-level attributes that distinguish different aspects of expressions of information content are (1) *expressional focus*, (2) *syntactic parsing*, (3) *spatial scaling*, and (4) *representational iconicity*.

(1) Sign types impose a *focus* on the expression of content. The focus can be on display-like continuous properties of the presentation, on localization of significant parts within a whole, or on its compositional structure. These attributes called *display, localization,* and *composition* are not entirely independent, but they define three types of main focus for an intended interpretation of sign types.

(2) Another set of attributes concerns the reading or syntactic parsing of presentations. The second-level attributes are here *non-linear, linear and recursive*. Images, maps and diagrams all exemplify non-linear reading, whereas graphs and symbols primarily rely on a linear reading although they also exhibit global features that can be accessed in a non-linear way (e.g. the shape of graphs). Network charts and (formal and natural) languages are the main examples of sign types with recursive syntax.

(3) The third set of attributes of the expression concerns the "scaling" of space as *metric space, regional space* or *schematic space*. These "uses" of space according to the intended interpretation of sign types can be seen as related to the concept of scale types. Images, maps and graphs rely primarily on a metric space for their expression, and (conceptual) diagrams primarily on a regional space. The symbolic types and network charts rely mainly on an abstract schematizations of space.

(4) The fourth set of attributes concerns the iconic-symbolic differentiation of "representational iconicity" with the image-like types at one end and the symbolic types at the other end. These are again not entirely independent types, since iconic

form also require symbolic regulation for their intended interpretation, and all sign types that express information content will have some kind of (more or less abstract) iconicity in order to be informative about a domain. The distinction between concrete-iconic, abstract-iconic and symbolic is however useful for understanding the significant difference between image-like representational forms and diagrammatic or linguistic forms.

In this way we have constructed a formal context (fig. 10) and we can now explore the lattices of different sub-context corresponding to the main groups of abstract sign types. Lattice structures are well-suited to express the feature structures of sign types and media types, because they can express inheritance relations as well as systematic combinations of types. We can even use them to generate possible combinations of sign types and media types in cases where we might not know a prototypical example in advance, i.e. we can use lattices to *explore the design space* of all possible type *combinations* and their *expression* in different media.

	display	localisation	composition	non-linear	linear	recursive	metric	regional	schematic	concrete	abstract	symbolic
Image	×			×			×			×		
Map		×		×			×			×		
Graph		×			×		×				×	
Categorical graph		×			×				×		×	
Diagram			×	×				×			×	
Network chart		×	×			×			×		×	
Symbol	×				×				×			×
Language			×		×	×			×			×

Fig. 10. The formal context of abstract sign types with 8 "objects" (sign types) and 12 attributes

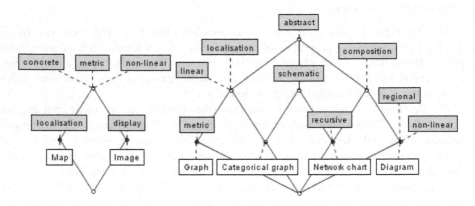

Fig. 11. (Left) The lattice of concrete-iconic types with attributes   **Fig. 12.** (Right) The lattice of abstract-iconic types with attributes

Important relations between types can be explored in the lattice representation by exploring different *sub-context* (a primitive form of *conceptual scaling*), for instance by including the diagrammatic forms in their relation to the symbolic forms (fig.13).

Having constructed different parts of the lattice it is now easier to grasp the overall structure of the full lattice representation (not shown here) for all the sign types at the specified level of detail. A useful graphical operation on the lattice of abstract sign

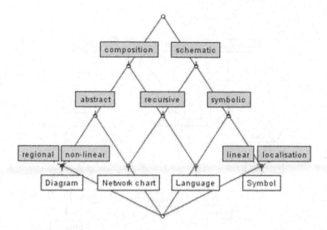

Fig. 13. The lattice of symbolic sign types (Symbol and Language) extended with the main diagram types, (conceptual) Diagram and Network chart ("abstract" is short for abstract-iconic forms)

types is to highlight parts of the lattice corresponding to the formal operation of a *filter*, i.e. showing the objects (if any) and attributes above a selected sign type.

3.5 Articulation: The Product Lattice of Representational Forms

What we have constructed so far is however only (an exemplification of) a lattice of *abstract sign types*. A lattice representation of *actual representational forms* within specific media, will require a *lattice product* given a formal context defining the attributes that will generate the *media types* as formal concepts. Media attributes will include attributes of sensory modalities and communication channels, for instance the transient or non-transient nature of physical presentations. We will call this further specification of signs through media types for an *articulation* of the abstract sign types in representational forms, whereas any concrete *presentation* carrying specific *information content* about a domain will be understood as an *expression* of this content through representational forms. Beyond these unit representational forms that constitutes the *lexical level* of meaning in the expression of content, a semiotic analysis will have to focus on the combination of representational forms into sentence-like units (phrases). At this *phrastic level* of expression other methods of formal analysis will be more adequate than FCA, i.e. specifically the syntactic structures expressed in Conceptual Graphs (CG) [31] [32]. A full semiotic analysis will reveal further levels of expression, i.e. the narrative, the discursive and the pragmatic level, but this is beyond the scope of the present paper [23].

Turning again to media types we should expand the simplified lattice of media types (fig. 2) with temporal distinctions in order to differentiate static graphics, sequential graphics (i.e. animation) and dynamic graphics (i.e. video). An example of such a lattice is shown below with attributes (fig. 14).

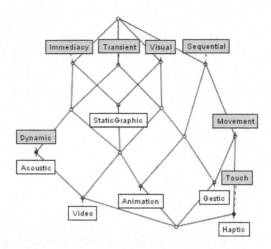

Fig. 14. An extended lattice of media types (including sequential and dynamic graphics, i.e. animation and video)

The usefulness of *abstract sign types* relies in the fundamental idea that each type has a *core semantics* that can be described as invariant across media types. It is necessary to distinguish the core semantics of abstract sign types (Image, Map, Graph, Diagram, Symbol and Language at the top level of detail) from the *emergent semantics* of representational forms when sign types are *articulated* in particular media of presentation (graphic, acoustic, haptic, gestic) and furthermore utilized to express information content within a Sign types are abstractions from concrete presentations. For most practical purposes concrete presentations are never pure (unimodal) signs, but are constructed as complex *combinations* of different sign types *articulated* through a single media like graphics or through a combination of media, i.e. multimedia. Abstract sign types are thus useful as an analytical tool to understand the complex articulations, combinations, and expression of meaning in information presentations.

In fig. 15 below we have exemplified the articulation of three abstract sign types (Image, Map, and Graph) within two different media (Graphic and Acoustic). This is again to simplify the idea of forming a product lattice in which we can easily track the inheritance of core properties of sign types and the emergent properties arising from articulation within specific media. In the lattice we have shown the articulation by introducing a new set of objects, i.e. the set {G-Image, G-Map, G-Graph, A-Image, A-Map, A-Graph}, where G stands for Graphic and A stands for Acoustic.

We want to emphasize that alternative conventions of representation could be used here. We could utilize the method of *conceptual scaling* in FCA [5][8] and define media, signs and scales as three dimensions of a multi-valued context. The convention of representation preferred in this paper is only meant to *introduce* the approach of semiotics and FCA in the domain of information presentation, i.e. for the future elaboration of the approach we should probably switch to conceptual scaling of multi-valued contexts and to a full specification of scale types (rather than the simplification of Stevens [35]) as described in [8]. Similarly we have also simplified the semiotic analysis, since we focus entirely on the *lexical level of meaning*, i.e. on the form of expression constituted by the unit representational forms, whereas any future elaboration of the approach should include the *phrastic level of meaning* constituted by *combinations* of representational forms (where Conceptual Graphs will become important), as well as higher levels of meaning (narrative, discursive, pragmatic) and the relation between these *forms of expression* (of content) and the property dimensions of the presentation that are not used to express specific information content, i.e. the "free channels" that are therefore available for modification and manipulation according to other communicative, aesthetic, and ergonomic purposes.

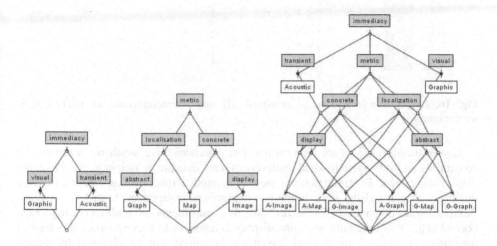

Fig. 15. A lattice product construction specifying representational forms (lattice on the far right) from abstract media types (Graphic, Acoustic) and abstract sign types (Image, Map, Graph)

4   Formal Design Space Analysis (FDSA) of Smart Instrument Components

In order to present a short illustrative case study utilizing the approach of semiotics and FCA to the domain of information presentation, let us take the simple case of a temperature measurement. The practical use context of this could be the work domain of process control, where many measurements are recorded, combined and utilized in calculations   and further inferences [27] about the dynamic states of controlled processes within a production plant (e.g. in a chemical process plant or a power plant). The simple measurement of temperature on an interval scale (Celcius) could for instance be the specific information content presented within the *media* of graphics in one of the well-known *display component* showed below (fig. 16). The four different components shown are all derived from the abstract sign type of the *graph* and they all use the *interval scale* as the scale type *matching* the scale type of data (intervals of distance or angle in graphic space corresponding to intervals in temperature measurements) [40][26][28]. The graph representation is adequate because graphs can express *content forms* such as single variables and relations between variables (constraints). The four examples show are however also different, but only in the specific *layout* of the graphical components, e.g. the graphical shape of the pointer (bar graph, pointer or arrow). They are *semantically equivalent* and one component can be obtained by *layout transformations* from the other.

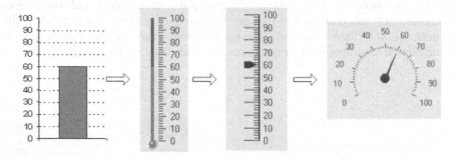

Fig. 16. Layout transformations of graphical 1D graph representations of interval scale temperature

Layout transformations are relevant because communicative, aesthetic or ergonomic constraints as well as individual preferences can indicate a rationale for changing display component. From the point of view of semiotics, it is however clear that we are dealing with one and the same graph-based graphical component. In order to support flexible display components for "smart" adaptive instruments and multimedia in process control [1][27][28], we only need one display component to accommodate this type of measurement, because the related layout modifications can be obtained by simple transformations (as automated adaptation or user-assisted tailoring).

Temperature measurements in process control will typically be recorded in order for assist comparisons and assessment of trends etc. We can therefore imagine our 1D *smart instrument component* extended with the capability to abstract from the dimension of time, i.e. as in the examples of fig. 16 displaying an *implicit time* index (they always show the temperature at the present time) or to reintroduce time by transforming the sub-type of the graphical graph to a 2D history graph (fig. 17).

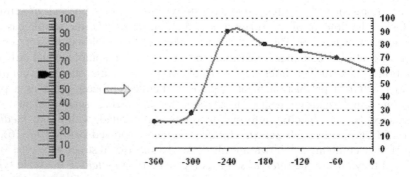

Fig. 17. Sub-type transformation of a 1-D graph from implicit to explicit time representation (2-D history graph)

A more advanced form of transformation that can still be built into our smart component is the ability to *change scales according to changes in the preferred scale type of data*. In fig. 18 we see the same temperature measurement displayed on a ratio scale (Kelvin), on an interval scale (Celcius), on an interval scale *superimposed* by an ordinal scale to support pragmatic evaluations of the reading (e.g. normal, high or critically high temperature according to constraints of the work domain), and *substituted* with the ordinal scale (presented by color coding: green below, yellow middle, red above).

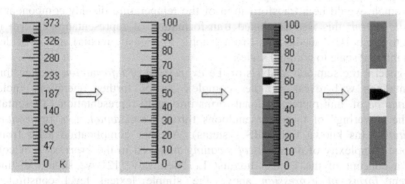

Fig. 18. Scale transformations of a ratio scale graph (left) for temperature (Kelvin scale) to an interval scale graph (Celcius scale), and transformations of the latter by superimposition (middle right) and substitution (far right)

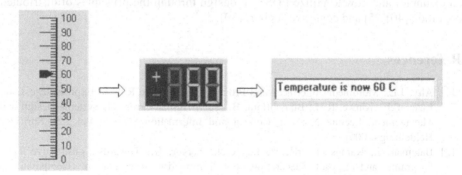

Fig. 19. Transformations of representational form: from graph to numerical symbol to language

In fig. 19 we have shown the ability of the smart instrument display component to be flexible with regard to its representational form. The underlying sign type for representation of the displayed data is changed from a 1-D *graph* to a *numerical symbol* and finally to *language*. According to the semiotic theory presented here the display component based on the graph type is already specified on a phrastic level of meaning by being an instrument reading: the graph component literally says the same thing as the sentence through its pointer and scale [17][18][21].

The semiotic approach to design advocated here could be called *Formal Design Space Analysis* (FDSA). The point of the short illustrative case given above is to suggest how a systematic *analysis and synthesis of the design space* of instrument display components could be supported by the use of semiotics and formal concept analysis. Since this description of components will include a feature structure representation of the significant properties of the different parts and their variations, we have a basis for transforming the components (in *layout, sign type, subtype, scale type, media type*) according to e.g. human factors criteria or individual preferences [21].

Complex operations are supported in the form of combinations of transformations. An example would be a transformation of the temperature display component to an *acoustic alarm*: this is a combined transformation of representational form (from graph to symbol), of media type (from graphic to acoustic media), and of scale type (from interval scale to nominal scale).

In general the semiotic analysis of the *expression of information content* through presentations will reveal a rather complex set of further operations including *combination* of unit representational forms into multi-representational presentations, and the "layering" of these presentations through e.g. *annotation, projection* and *substitution* (as known from GIS systems). Another complication arise from the semiotic complexity of the *levels of meaning* involved in the *expression* of content. With a concept of the Danish linguist L. Hjelmslev [12] we should distinguish different *forms of expression* above the simple lexical level constituted by representational forms. Further research will be carried out to explore the relation between different levels of meaning in the design and use of information presentation and interface components (of which flexible instrument display components are just an example) and how to utilize FDSA in design through the principles of distributed cognition [40][25] and cognitive support [39].

References

[1] Alty, J.L.: Multimedia Interfaces and Process Control: The Role of Expressiveness. In: Elzer, P.F., Kluwe, R.H., Boussoffara, B. (eds.) Human Error and Systems Design and Management. Lecture Notes in Control and Information Science, vol. 253, Springer, Heidelberg (2000)

[2] Bateman, J., Kamps, T., Kleinz, J., Reichenberger, K.: Towards Constructive Text, Diagram, and Layout Generation for Information Presentation. Computational Linguistics 27(3), 409–449 (2001)

[3] Bouvin, N.O., Zellweger, P.T., Grønbæk, K., Mackinlay, J.D.: Fluid annotations through open hypermedia: using and augmenting emerging Web standards. In: Proceedings of 11th International World Wide Web Conference, Honolulu, pp. 160–171. ACM, New York (2002)

[4] Bulterman, D.: SMIL 2.0. Part 1: Overview, Concepts, and Structure. IEEE Multimedia 8(4), 82–88 (2001)

[5] Cole, R., Eklund, P.: Scalability in formal context analysis. Computational Intelligence 15(1), 11–27 (1999)

[6] Cox, R.: Representation construction, externalized cognition and individual differences. Learning and Instruction 9, 343–363 (1999)

[7] Davey, B.A., Priestley, H.A.: Introduction to Lattices and Order. Cambridge U.P., Cambridge (1990)

[8] Ganter, B., Wille, R.: Formal Concept Analysis. Mathematical Foundations. Springer, Heidelberg (1999)

[9] Gibbs, S., Tsichritzis, D.C.: Multimedia Programming: Objects, Environments and Frameworks. Addison-Wesley, Workingham Reading Menlo Park (1995)

[10] Heller, R.S., Martin, C.D.: A Media Taxonomy. IEEE Multimedia 2(4), 36–45 (1995)

[11] Heller, R.S., Martin, C.D., Haneef, N., Gievska-Krliu, S.: Using a Theoretical Multimedia Taxonomy Framework. Journal of Educational Resources in Computing 1(1), 1–22 (2001)

[12] Hjelmslev, L.: Omkring Sprogteories Grundlæggelse. Travaux du Cercle Linguistique du Copenhague, XXV. 1943, Reissued 1993, Reitzel. English translation: Hjelmslev, L.: Prolegomena to a Theory of Language. Indiana University Press, Bloomington (1961)

[13] Kamps, T.: Diagram Design. A Constructive Theory. Springer, Heidelberg (1999)

[14] Lohse, G.L., Walker, N., Biolsi, K., Rueter, H.: Classifying graphical information. Behaviour & Information Technology 10(5), 419–436 (1991)

[15] Manning, A.D.: The semantics of technical graphics. Journal of Technical Writing and Communication 19(1), 31–51 (1989)

[16] May, M.: Diagrammatic reasoning and levels of schematization. In: Johansson, T.D., Skov, M., Brogaard, B. (eds.) A Fundamental Problem in Semiotics. NSU Press, Copenhagen (1999)

[17] May, M.: Semantics for Instrument Semiotics. In: Lind, M. (ed.): Proceeding of the 20th European Annual Conference on Human Decision Making and Manual Control, EAM-2001 Ørsted DTU, pp. 29–38 (2001)

[18] May, M., Andersen, P.B.: Instrument Semiotics. In: Liu, K., Clarke, R.J., Andersen, P.B., Stamper, R.K. (eds.) Information, Organisation and Technology. Studies in Organisational Semiotics, Kluwer Academic Publishers, Boston, Dordrecht & London (2001)

[19] May, M.: Instrument Semiotics: A Semiotic Approach to Interface Components. Knowledge-Based Systems 14, 431–435 (2001)

[20] May, M.: Wayfinding, Ships, and Augmented Reality. In: Andersen, P.B., Qvortrup, L. (eds.) Virtual Applications: Applications with Virtual Inhabited 3D Worlds, pp. 212–233. Springer, Heidelberg (2004)

[21] May, M., Petersen, J.: Media, Signs and Scales for the Design Space of Instrument Displays. In: Bust, P.D. (ed.) Contemporary Ergonomics 2006 (Proceedings of the Annual Conference of the Ergonomics Society), pp. 93–97. Taylor & Francis, Abington (2006)

[22] May, M.: Feature-based Multimedia Semantics: Representational Forms for Instructional Multimedia Design. In: George, G., Sherry, Y.C. (eds.) Digital Multimedia Perception and Design, Idea Group Publishing, Hershey London Melbourne Singapore (2006)

[23] May, M.: Handbook of Multimedia Semiotics and Digital Multimedia Design. IGI Global, Hershey London Melbourne Singapore, (to appear)

[24] Nack, F., Hardman, L.: Towards a Syntax for Multimedia Semantics, Report INS-RO204, Information Systems (INS) at CWI, Amsterdam (2002)

[25] Petersen, J., May, M.: Scale Transformations and Information Presentation in Supervisory Control. In: Lützhöft, M. (ed.) Proceedings of the 22nd Conference on Human Decision Making and Control. University of Linköping, June 2-4, 2003, pp. 75–85 (2003)

[26] Petersen, J., May, M.: Scale Transformations and Information Presentation in Supervisory Control. International Journal of Human-Computer Studies 64, 405–419 (2006)

[27] Petersen, J.: Model-Based Integration and Interpretation of Data. In: Thissen, W., Wieringa, P., Pantic, M., Ludema, M. (eds.) IEEE International Conference on Systems, Man and Cybernetics, October 10-13, 2004, pp. 815–820 (2004)

[28] Petersen, J., May, M.: Extending Interface Design Principles based on Scale Types. In: Proc. of the 25th European Annual Conference on Human Decision-Making and Manual Control, Valenciennes, France, September 27-29, 2006 (2006)

[29] Purchase, H.C., Naumann, D.: A Semiotic Model of Multimedia: Theory and Evaluation. In: Rahman, S.M. (ed.) Design and Management of Multimedia Information Systems: Opportunities and Challenges, pp. 1–21. Idea Group Publishing, Hershey London Melbourne Singapore (2001)

[30] Smith, J.R., Mohan, R., Li, C.-S.: Transcoding Internet Content for Heterogenous Client Devices. In: Proc. IEEE Inter. Symp. on Circuits and Systems (ISCAS), June 1998. http://www.research.ibm.com/networked_data_systems/transcoding/Publications/iscas98.pdf

[31] Sowa, J.F.: Conceptual Structures: Information Processing in Mind and Machine. Addison-Wesley, Reading MA (1984)

[32] Sowa, J.F.: Knowledge Representation: Logical, Philosophical, and Computational Foundations. Brooks/Cole Publishing Co. Pasific Grove, CA (2000)

[33] Stenning, K., Inder, R., Neilson, I.: Applying Semantic Concepts to Analyzing Media and Modalities. In: Glasgow, J., Narayanan, N.H., Chandrasekaran, B. (eds.) Diagrammatic Reasoning. Cognitive and Computational Perspectives, AAAI Press & MIT Press, Menlo Park CA (1995)

[34] Stephanidis, C.: Adaptive Techniques for Universal Access. User Modelling and User-Adapted Interaction 11, 159–179 (2001)

[35] Stevens, S.S.: On the theory of scales and measurement. Science 103(2684), 677–680 (1946)

[36] Trevin, S., Zimmermann, G., Vanderheiden, G.: Abstract User Interface Representations: How Well do they Support Universal Access? ACM Sigcaph 73, 77–84 (2003)

[37] Twyman, M.: A schema for the study of graphic language. In: Kolers, P.A., Wrolstad, M.E., Bouma, H. (eds.) Processing of visible language, vol. 1, pp. 117–150. Plenum Press, New York (1979)

[38] Van der Zee, E., Nikanne, U. (eds.): Cognitive Interfaces. Constraints on Linking Cognitive Information. Oxford U.P., Oxford (2000)

[39] Walenstein, A.: Foundations of Cognitive Support: Towards abstract patterns of usefulness. In: Forbrig, P., Limbourg, Q., Urban, B., Vanderdonckt, J. (eds.) DSV-IS 2002. LNCS, vol. 2545, pp. 133–147. Springer, Heidelberg (2002)

[40] Zhang, J.: A representational analysis of relational information displays. International Journal of Human Computer Studies 45, 59–74 (1996)

Reducing the Representation Complexity of Lattice-Based Taxonomies

Sergei Kuznetsov[1], Sergei Obiedkov[1,2], and Camille Roth[3,4]

[1] Department of Applied Mathematics, Higher School of Economics, Moscow, Russia
[2] Moscow Institute of Physics and Technology, Moscow, Russia
[3] European Center for Living Technology, Venice, Italy
[4] Department of Sociology, University of Surrey, Guildford, UK
skuznetsov@yandex.ru, sergei.obj@gmail.com, camille.roth@polytechnique.edu

Abstract. Representing concept lattices constructed from large contexts often results in heavy, complex diagrams that can be impractical to handle and, eventually, to make sense of. In this respect, many concepts could allegedly be dropped from the lattice without impairing its relevance towards a taxonomy description task at a certain level of detail. We propose a method where the notion of stability is introduced to select potentially more pertinent concepts. We present some theoretical properties of stability and discuss several use cases where taxonomy building is an issue.

1 Introduction

Formal Concept Analysis (FCA) is generally an appropriate framework for building categories defined as object sets sharing some attributes, irrespectively of a particular domain of application. In this framework, categories are called "formal concepts" each concept being a pair of an object set and an attribute set such that every attribute holds for every object. This presents a convincing formal model of the philosophical notion of a "concept" characterized extensionally by the set of entities it covers and intensionally by the set of properties they have in common [1]. Formal concepts, in turn, can be gathered in a lattice structure, thus providing an overlapping taxonomy for the underlying categories.

Besides, traditional lattice operations translate properly in taxonomical and categorical terms: on the one hand, the meet of two categories is a sub-category holding objects belonging to both categories, along with their associated shared attributes; on the other hand, the join of two categories is the super-category defined by attributes shared by both categories, and the associated objects.

While formal concept lattices are theoretically robust, in practice, one often has to face huge structures containing a prohibitive number of categories, even for rather small datasets. Even if navigation in the structure is possible despite large sizes [2], readability generally remains a problem as, computational issues set apart, "even carefully constructed line diagrams lose their readability from a certain size up" [3] (p. 75).

U. Priss, S. Polovina, and R. Hill (Eds.): ICCS 2007, LNAI 4604, pp. 241–254, 2007.

Solutions may consist in representing only the most meaningful portions of the lattice by assuming that some concepts are likely to be less relevant than others from the standpoint of taxonomy description [4,5,6]. To this end, we need to filter out nodes that do not satisfy specified constraints of a certain kind. In this paper, we develop one such pruning technique. In particular, the notion of stability introduced in [7,8] to discriminate irrelevant nodes seems to be particularly appropriate and was fruitfully used in a previous attempt to prune concept lattices in the practical case of epistemic community representation [5].

Here, we apply the method to larger datasets and other domains—other kinds of epistemic communities, but also other kinds of data—as well as address the dynamic description of the resulting reduced structures. While stability was satisfactorily applied to a small sub-context consisting of agents using particular notions, thus yielding meaningful taxonomies, it was unclear whether it could be possible to go further in other domains and with much larger contexts.

2 Formal Framework

Before proceeding, we briefly recall the FCA terminology [3]. Given a *(formal) context* $\mathbb{K} = (G, M, I)$, where G is called a set of *objects*, M is called a set of *attributes*, and the binary relation $I \subseteq G \times M$ specifies which objects have which attributes, the derivation operators $(\cdot)^I$ are defined for $A \subseteq G$ and $B \subseteq M$ as follows:

$$A^I = \{m \in M \mid \forall g \in A : gIm\};$$

$$B^I = \{g \in G \mid \forall m \in B : gIm\}.$$

Put differently, A^I is the set of attributes common to all objects of A and B^I is the set of objects sharing all attributes of B.

If this does not result in ambiguity, $(\cdot)'$ is used instead of $(\cdot)^I$. The double application of $(\cdot)'$ is a closure operator, i.e., $(\cdot)''$ is extensive, idempotent, and monotonous. Therefore, sets A'' and B'' are said to be *closed*.

A *(formal) concept* of the context (G, M, I) is a pair (A, B), where $A \subseteq G$, $B \subseteq M$, $A = B'$, and $B = A'$. In this case, we also have $A = A''$ and $B = B''$. The set A is called the *extent* and B is called the *intent* of the concept (A, B). In categorical terms, (A, B) is equivalently defined by its objects A or its attributes B.

A concept (A, B) is a *subconcept* of (C, D) if $A \subseteq C$ (equivalently, $D \subseteq B$). In this case, (C, D) is called a *superconcept* of (A, B). We write $(A, B) \leq (C, D)$ and define the relations $\geq$, $<$, and $>$ as usual. If $(A, B) < (C, D)$ and there is no (E, F) such that $(A, B) < (E, F) < (C, D)$, then (A, B) is a *lower neighbor* of (C, D) and (C, D) is an *upper neighbor* of (A, B); notation: $(A, B) \prec (C, D)$ and $(C, D) \succ (A, B)$.

The set of all concepts ordered by $\leq$ forms a lattice, which is denoted by $\underline{\mathfrak{B}}(\mathbb{K})$ and called the *concept lattice* of the context $\mathbb{K}$. The relation $\prec$ defines edges in the *covering graph* of $\underline{\mathfrak{B}}(\mathbb{K})$.

3   Stability

3.1   Rationale

An obvious solution to reducing the number of groups of individuals defined as concept extents by selecting "most interesting groups" is to compute only an upper part of the concept lattice: concepts with extents comprising at least $n\%$ of all objects. This approach produces an order filter of a concept lattice often called nowadays an "iceberg lattice". There are several well-known top-down lattice construction algorithms (see a review in [9]) and algorithms for computing frequent itemsets [4,10] suitable for building such iceberg lattices. The reduction in the number of concepts, as compared to the number of concepts in the whole lattice, can be considerable. However, one should be careful not to overlook small but interesting groups, for example, "exotic" or "emergent" groups not yet represented by a large number of objects, or, groups that contain objects who are not members of any other group.

Undoubtedly, the size of the concept lattice is not only a computational problem. The lattice may contain nodes that are just too similar to each other because of noise in data or real minor differences yet irrelevant to a given purpose. In this case, taking an upper part of the lattice does not solve the problem, since this part may well contain such similar nodes.

To tackle the problem of selecting "meaningful" concept intents, the notion of concept stability was proposed in [7,8] and developed in [5]. The general idea of stability is as follows: A concept is stable if its intent does not depend much on each particular object of the extent.

3.2   Definition

In this section, we define the notion of stability of a formal concept introduced in [7] and [8] in a slightly different form than the one we use here. The definition given below is the one from [5].

Definition 1. *Let* $\mathbb{K} = (G, M, I)$ *be a formal context and* (A, B) *be a formal concept of* $\mathbb{K}$. *The stability index,* σ, *of* (A, B) *is defined as follows:*

$$\sigma(A, B) = \frac{|\{C \subseteq A \mid C' = B\}|}{2^{|A|}}.$$

In [5], it is shown that the following proposition holds:

Proposition 1. *Let* $\mathbb{K} = (G, M, I)$ *be a formal context and* (A, B) *be a formal concept of* $\mathbb{K}$. *For a set* $H \subseteq G$, *let* $I_H = I \cap (H \times M)$ *and* $\mathbb{K}_H = (H, M, I_H)$. *Then,*

$$\sigma(A, B) = \frac{|\{\mathbb{K}_H \mid H \subseteq G \text{ and } B = B^{I_H I_H}\}|}{2^{|G|}}.$$

Thus, the stability index of a concept is the probability of the intent B if all subcontexts of $\mathbb{K}$ over the attribute set M are equally probable. Stability indicates how much the concept intent depends on particular objects of the extent: a stable intent is less sensitive to noise in object descriptions. Besides, the extent of a stable concept is not "very close" to extents of its lower neighbors.

3.3   Properties

In [5], an algorithm for the computation of stability indices is given. In general, computing stability is a #P-complete problem [8]; hence, simple heuristics (easily computable sufficient conditions) to discard concepts with low stability would be useful. The following two propositions give conditions of this sort.

Proposition 2. *Given a concept (A, B) of a context (G, M, I), if there is a set $A_1 \subset A$ such that $A_1' \neq B$, then $\sigma(A, B) \leq 1 - 1/2^{|A \setminus A_1|}$.*

Proof. Since $A_1 \subset A$, $A' = B$, and $A_1' \neq B$, we have $B \subset A_1'$ and $B \subset A_2'$ for all $A_2 \subseteq A_1$. Therefore, $|\{A_2 \subseteq A \mid A_2' = B\}| \leq 2^{|A|} - |\{A_2 \mid A_2 \subseteq A_1\}| = 2^{|A|} - 2^{|A_1|} = 2^{|A \setminus A_1|}$ and $\sigma(A, B) \leq \frac{2^{|A|} - 2^{|A_1|}}{2^{|A|}} = 1 - 1/2^{|A \setminus A_1|}$.   □

In particular, if $|A_1| = |A| - 1$, one has $\sigma(A, B) \leq 1/2$ and the concept (A, B) has fairly low stability: usually, one retains concepts with stability very close to 1. Testing if $A_1' \neq B$ for some A_1 with $|A_1| = |A| - 1$ takes $O(|A|^2 \cdot |M|)$ time.

Proposition 3. *Given a concept (A, B) of a context (G, M, I), if there are two sets $A_1, A_2 \subset A$ such that $|A_1| = |A_2|$, $A_1 \neq A_2$, and $A_1', A_2' \neq B$, then $\sigma(A, B) \leq 1 - \frac{3}{2^{|A \setminus A_1| + 1}}$.*

Proof. By the condition, $A_1 \subset A$, $A_2 \subset A$, $A' = B$, and $A_1' \neq B$, $A_2' \neq B$. To obtain an upper bound of $\sigma(A, B)$ we need to consider the situation where the size of the set $\mathfrak{P}(A_1) \cup \mathfrak{P}(A_2)$ (by $\mathfrak{P}(X)$ we denote the powerset of X) is as small as possible. This is attained when A_1 and A_2 are as close as possible, i.e., $|A_1 \setminus A_2| = |A_2 \setminus A_1| = 1$. In this case (since $|A_1| = |A_2|$) we have $|\mathfrak{P}(A_1) \cup \mathfrak{P}(A_2)| = 2^{|A_1|} + 1/2 \cdot 2^{|A_1|} = 2^{|A_1|} \cdot 3/2$. Therefore, $\sigma(A, B) \leq \frac{2^{|A|} - 2^{|A_1|} \cdot 3/2}{2^{|A|}} = 1 - \frac{3}{2^{|A \setminus A_1| + 1}}$.   □

In the same way one may obtain a condition for three, four, and so on extent subsets that do not give rise to intent B. However, it seems hard to get a useful general statement for an arbitrary number of such extent subsets, which is related to the #P-completeness of computing stability.

   Now we describe how stability of concepts changes with the growth of the data sample. Consider the situation when a context (G, M, I) is updated with a new object g to form context $(G \cup \{g\}, M, J)$ such that $(G \times M) \cap J = I$. Then, according to [11], we distinguish three possible types of concepts of the context $(G \cup \{g\}, M, J)$: an *old* concept that is equal to a concept in the old context, a *modified* concept of the form $(A \cup \{g\}, B)$ such that (A, B) is the concept of (G, M, I), and a *new* concept of the form $((A \cup \{g\})'', B \cap \{g\}')$ such that (A, B) is a concept of (G, M, I) and $B \cap \{g\}'$ is not an intent of (G, M, I). We denote stabilities in contexts (G, M, I) and $(G \cup \{g\}, M, J)$ by σ_I and σ_J, respectively.

Proposition 4. *Given a concept (A, B) of a context (G, M, I), if a new object g is added to the set of objects to form context $(G \cup \{g\}, M, J)$ (such that $(G \times M) \cap J = I$), then for the stability of concepts of the new context $(G \cup \{g\}, M, J)$ there can be the following three possibilities:*

1. For an old concept (A, B), we have $\sigma_J(A, B) = \sigma_I(A, B)$.
2. For a modified concept $(A \cup \{g\}, B)$, we have

$$\sigma_I(A, B) \leq \sigma_J(A \cup \{g\}, B) \leq 1/2 + \sigma_I(A, B)/2.$$

3. For a new concept (A, B), we have

$$\sigma_J(A, B) \begin{cases} = 1/2, \text{ if } B = \{g\}'; \\ < 1/2, \text{ otherwise.} \end{cases}$$

Proof. (1) The extents of an old concept and all its subconcepts do not change, neither does the stability.

(2) First, we prove that $\sigma_I(A, B) \leq \sigma_J(A \cup \{g\}, B)$. Indeed, by definition of $\sigma_I(A, B)$, there are $\sigma_I(A, B) \cdot 2^{|A|}$ subsets $A_1 \subseteq A$ ("old" subsets) such that $A_1' = B$, and, since $g' \cap A_1' = B$ for every A_1 such that $A_1' = B$, we also have $\sigma_I(A, B) \cdot 2^{|A|}$ subsets A_2 such that $g \in A_2 \subseteq A \cup \{g\}$ and $A_2' = B$. Since all such A_1 and A_2 are different, we have

$$\sigma_J(A \cup \{g\}, B) \geq (2^{|A|} \cdot \sigma_I(A, B) + 2^{|A|} \cdot \sigma_I(A, B))/2^{|A|+1} = \sigma_I(A, B).$$

To prove $\sigma_J(A \cup \{g\}, B) \leq 1/2 + \sigma_I(A, B)/2$, note that the largest stability of a modified concept $(A \cup \{g\}, B)$ is attained when $A_2' \cap \{g\}' = B$ for each of the $2^{|A|}$ subsets $A_2 \subseteq A$. Taking into account $2^{|A|} \cdot \sigma_I(A, B)$ subsets $A_1 \subseteq A$ with $A_1' = B$, we have

$$\sigma_J(A \cup \{g\}, B) \leq (2^{|A|} \cdot \sigma_I(A, B) + 2^{|A|})/2^{|A|+1} = \sigma_I(A, B)/2 + 1/2.$$

(3) In the case of a new concept (A, B), we can have $A_1' = B$ for $A_1 \subseteq A$ only if $g \in A_1$. The largest stability will be attained if $A_1' = B$ for all such $A_1 \subseteq A$ with $g \in A_1$. Formally,

$$\sigma_J(A, B) = \frac{|\{A_1 \subseteq A \mid A_1' = B\}|}{|2^A|} \leq \frac{|\{A_1 \subseteq A \mid g \in A_1\}|}{|2^A|} = \frac{|2^{|A|-1}|}{2^{|A|}} = \frac{1}{2}.$$

It is easy to see that the equality $\sigma_J(A, B) = 1/2$ holds only for the object concept of g, and only this concept has stability $1/2$. All other new concepts are less stable.   □

4   Applying Intensional Stability

As such, stability measures how much a community depends on some of its individual members. This may be useful in building attribute-based taxonomies representing intensional categories. In particular, this notion is likely to be relevant when investigating taxonomies of epistemic communities, i.e., groups of agents jointly interested in identical topics, sharing the same notions [12,6]. In this respect, contexts where scientists are objects and the topics on which they work are attributes are particularly adequate: here, formal concepts represent epistemic communities as groups of topics along with corresponding agents. Removing a few scientists from the context should not change the topics of an epistemic community—"real" epistemic communities ought to be stable in spite of noisy data. Apart from noise-resistance, a stable field does not collapse (e.g., merge with a different field or split into several independent subfields) when a few members stop being active or switch to another topic.

We illustrate this criterion using two case studies featuring scientists attending a particular conference and biologists working on a particular model animal. In both cases, while stability-based lattice reduction is significant—from thousands of concepts to less than 30—we are still able to tell a meaningful "story" with respect to what field experts may describe.

4.1 "European Complex Systems Conference"

Using the database of all papers submitted to the second European Conference on Complex Systems in 2006[1], we build a context made of authors and terms mentioned in article titles and abstracts. The resulting context contains 401 authors and 109 terms, which yields a lattice of 6011 concepts. The reduced substructure featuring the 25 most stable concepts is presented in Fig. 1. Note that the set of all stable concepts (for an arbitrary threshold) does not have to be a lattice, even if it is in the examples used in this paper.

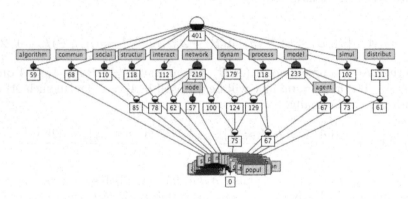

Fig. 1. The 25 most stable concepts in the ECCS dataset. Figures in squares show the sizes of concept extents.

From this lattice, it is possible to provide the following desription of the community attending the ECCS:

- The notion of "network" is obviously a central issue: in addition to being a large community, it is also a parent for several associated subtopics: "social network" (agent-based networks), "structure network" (topological issues), "interact network" (networks as representation of interactions), "node network" (a node being a basic unit), "dynamics network" (evolution of networks) and "model network" (modeling of networks).

- "model" is an important topic too and is related to "agents" and "simulation", as well as "dynamics" (dynamical models in general), in addition to "networks"—it is also worth noting that there exists a sizeable community around "network dynamics model" which refers to scientists interested in the modeling of network dynamics (morphogenesis). The use of models to reconstruct distributions of any kind is represented by the "model distribut" community. Finally, the modeling of dynamical processes ("model dynam process"), although sensibly less significant, is an interesting field as well in this framework.
- Some topics are more isolated as they do not form any joint epistemic community in the stabilized lattice—such as "algorithm" and "community" (to the left). These concepts are likely to refer to minor fields focused on particular issues: community and cluster detection, or introduction and use of novel and general algorithms to achieve empirical measurements in a variety of cases.

In this lattice of 25 concepts it is not possible to see some fields which are actually representative of minor yet active subcommunities — such as "network distrib", "algorithm network" and "algorithm model", which are respectively the 40th, 50th and 75th most stable concepts. Nonetheless, on the whole and given a certain (high) level of epistemological description, the above story appears to be fairly consistent with what experts of the field would perceive as the main topics of complex systems science at that time.

4.2   Embryologists Working on the "zebrafish"

In [5], we have applied stability-based pruning to data obtained from the bibliographical database of MedLine abstracts coming from a well-bounded community of embryologists working on the zebrafish during the period 1998–2003, the goal of the application being to build a taxonomy of this research field.[2] Since the purpose of that paper was to illustrate the proposed technique, we used a small random sample context consisting of 25 authors and 18 words. The incidence relation of the context indicated which authors used which words in their papers on the subject. The lattice of this context consisted of 69 concepts, of which we selected the 17 most stable ones (with stability ≥ 0.52). They constitute a lattice shown in Fig. 2 (some of the 18 attributes are not contained in any stable intent except for the intent of the bottom concept; they are not shown on the diagram).

Taking a larger data sample, 250 authors using the same 18 words, we get 1146 concepts. Of course, in the larger structure, stability indices are also larger

---

[2] Data is obtained from a query on article abstracts containing the term *"zebrafish"* at http://www.pubmed.com.

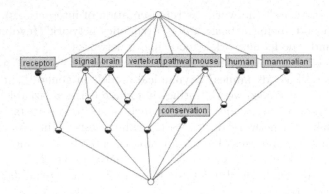

Fig. 2. The lattice of the 17 most stable concepts of a context built from 25 zebrafish researchers and 18 words they used in their papers (taken from [5])

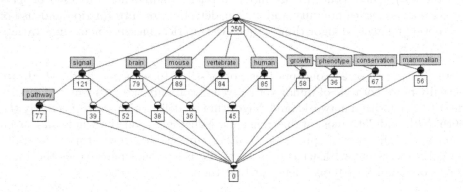

Fig. 3. The lattice of the 17 most stable concepts of a context built from 250 zebrafish researchers and 18 words they used in their papers

(see Section 3), which makes it impractical to use the same stability thresholds for pruning in both cases. Hence, we simply take the same number of the most stable concepts. Fig. 3 shows the lattice formed by the 17 most stable concepts of this context (figures in boxes indicate the extent size of corresponding concepts).

As should be expected, the lattices in Figs. 2 and 3 are not identical, but still share a lot of features in common. One interesting difference is that, in the structure based on the larger data sample, "pathway" occurs as a subconcept of "signal", which certainly makes sense from the domain point of view ("pathway" on its own is still a concept intent, but it is not sufficiently stable in the larger context). Some less important communities, like "mouse, conservation" or "signal, pathway, mouse" are missing from Fig. 3. Instead, the taxonomy resulting from a larger number of authors focuses on more solid associations ignoring some particularities, which can be reintroduced by increasing the number of stable concepts included in the taxonomy.

4.3 Improving the Quality of Taxonomies: Linguistic Processing

In our analysis of the ECCS and zebrafish data, we described authors in terms of words they used in their papers. We removed stop words (such as "and" or "was"), common words with no special meaning for the domain (such as "size" or "function"), as well as "paradigmatic" words, i.e., those relevant to all members of the entire community even if they are not explicitly used by all members (the obvious examples are "complex" and "system"). The remaining words were stemmed using the Porter algorithm [13].

In the process, it has become clear that these techniques are certainly not sufficient: the resulting author–word tables contain a lot of noise coming, in particular, from homonymy and synonymy. We approached the latter by manually combining synonyms—or other semantically related words that could be considered equivalent for our purposes—into one attribute. Of course, this requires some expert knowledge of the domain and cannot be done simply using a general-purpose English thesaurus: words that are synonymous in everyday language can be used differently in the domain to be described or, on the contrary, there may be domain-specific associations between otherwise unrelated words. Homonymy is even more difficult to deal with: words used on their own, without taking the context into consideration, are not very informative; it seems more appropriate to use word phrases.

That we still get rather meaningful taxonomies from formal contexts obtained with such poor means suggests that the methods we use further on may, in general, be valid, but we believe that better linguistic preprocessing will have a significant effect on the quality of the resulting taxonomies.

5 Applying Extensional Stability

The stability index discussed so far relates actually to intensional stability. In a dual manner, it is possible to define an *extensional stability index*, which indicates how a concept extent depends on particular attributes: would the objects of a given concept still belong to the same category if they stop sharing some attributes? A stable extent is thus likely to indicate a group of objects which do not depend on particular attributes.

Definition 2. *The* extensional stability index σ^e *of a concept* (A, B) *is defined as follows:*

$$\sigma^e(A, B) = \frac{|\{C \subseteq B \mid C' = A\}|}{2^{|B|}}.$$

Like intensional stability, the relevance of this index depends on the domain and the aim of the lattice-based taxonomy. For instance, affiliation data, in social science, defines people related to some organizations or events; in this case, formal concept lattices represent taxonomies of agents who share identical affiliations. Extensional stability may be helpful in this situation in measuring how durable links between people within a community are. In this respect, it principally relates to the social aspect of the group: if some people are together because they have a given activity, one may wonder whether they will still be together if they stop doing this activity. People who are also doing something else together are more likely to belong to stable extents. Here, extensional stability tests how much the community as a group of people depends on particular activities. In other words, if one of the activities that unites them becomes less appropriate, will they still survive as a separate community?

To illustrate this, we focus on data stemming from a celebrated case study by Davis, Gardner and Gardner (DGG) [14] which features ladies attending particular events in a small Mississippi town in the 1930s. Using a context where objects are people and attributes are attendance to social events, it is possible to build a concept lattice representing groups of women attending jointly some sets of events [15]. However, even in this simple case the resulting lattice is already rather sizeable with 65 concepts; finding cohesive subgroups in such a structure could be uneasy. By contrast, the lattice corresponding to extensionally stable concepts (stability index strictly above .5) contains only three concepts, in addition to top and bottom nodes: their extents are {g14}, {g12, g13, g14} and {g1, g3}. The stabilized lattice is shown on Fig. 4.

The identification, in this data, of subcommunities together with core and peripheral members has already been the focus of several studies in social science. While interview-based identification in the original DGG study suggests that {g1, g2, g3, g4} and {g13, g14, g15} are respectively core members of two distinct groups, a comprehensive review given in [16] reveals a collection of remarkably diverse results, depending on whether subgroups were identified using, *inter alia*, principal component analysis, matrix algebra, information theory, as well as concept lattices—by means, in this latter case, of a relatively manual approach [15].

Most interestingly, a study by Doreian [17] agrees particularly well with our results: it yields the same core members as those found in our stabilized lattice, i.e., {g1, g3} and {g12, g13, g14}. His approach relies on Q-analysis [18], whose principles are unsurprisingly analogous to FCA: for a given context, each object

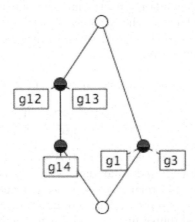

Fig. 4. The lattice of concepts with extensional stability above .5 for the DGG data

is defined as a "polyhedron" where attributes are edges. In this framework, a formal concept can thus be seen as an intersection of polyhedra—extents and intents are respectively defined by polyhedra-objects and edges-attributes participating in a given intersection [19]. Additionally, Q-analysis introduces the notion of connected paths between polyhedra, which are plausibly useful for dealing with connection patterns between objects, yet irrespective of the actual attributes underlying these connected paths. As Freeman underlines, "by considering subsets of women who were connected at higher levels, Doreian was able to specify degrees of co-attendance ranging from the core to the periphery according of each group" [16]. The idea of strong relationships between objects independently of particular attributes is not dissimilar to our notion of extensional stability and could perhaps account for our identical results.

More broadly, while extensional stability appears to yield a satisfying outcome in this small case study, it is nonetheless the matter of further research to check its adequacy on larger datasets and different domains of application.

6 Dynamic Mappings

Let $\mathbb{K}_1 = (G, M, I)$ and $\mathbb{K}_2 = (H, N, J)$ be two contexts describing the same domain in two different time points (or periods). How has the domain changed between these time points? In particular, if $(A, B) \in \mathfrak{B}(\mathbb{K}_1)$ is a concept of $\mathbb{K}_1$, what has happened to it in $\mathbb{K}_2$?

Consider a concept $(C, D) \in \mathfrak{B}(\mathbb{K}_2)$. If the closure of $B \cap D$ equals B in $\mathbb{K}_1$ and D in $\mathbb{K}_2$, we may say that (A, B) and (C, D) are *intensionally related*. In the case of the ECCS data, concepts intensionally related to (A, B) represent the evolution of the field B between the two periods.

Figure 5 shows two diagrams corresponding to the ECCS conferences in 2005 and 2006 (assuming that the words are the same, i.e., $M = N$). In both cases, we have selected the 15 most stable concepts. The differences are as follows: the diagram for 2005 contains concepts with intents {network, dynamics}, {dynamics, model, process}, {dynamics, process}, and {information}—all missing from the 2006 diagram, which contains its own unique intents: {interaction}, {network, social}, {model, agent}, and {simulation, model}. The only 2006 concept intensionally related to {network, dynamics} is the one with intent {network, dynamics, model}. This suggests that the 2005 topic described by {network, dynamics} has merged with the topic described by {network, dynamics, model}; at least, the difference between the two is no longer important at the given level of detail. The other three 2005-specific communities are intensionally related only to the bottom node of the 2006 diagram, which means that they have disappeared or become less important. On the other hand, the bottom node of the 2005 diagram is intensionally related to the four 2006-specific concepts, suggesting that they correspond to new subareas of research.[3] Even though {model, agent} has

---

[3] Again, as we deal with "stabilized" lattices, these new areas are such only at the chosen level of detail. It would be more accurate to say that their importance has increased.

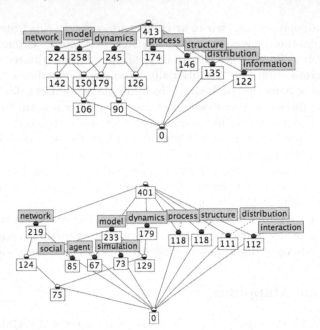

Fig. 5. Stabilized 15 concept lattice for ECCS 2005 and 2006

a parent topic, {model}, already present in the 2005 taxonomy, the "agent" aspect is new, thus, defining a new subfield, that has no corresponding nodes in 2005, which is indicated by the fact that it is intensionally related to the 2005 bottom node.

In the above discussion, the social aspect of the communities has been completely ignored. In some contexts, it is more appropriate to describe the history of a community in terms of what happens to its members. In this case, if all authors dealing with topic A in 2005 switch to topic B in 2006, B should be considered as the 2006 equivalent of the 2005 A-community, even if A is still an active topic in 2006 (supported by newcomers, for example). Such population moves can be captured by *extensional relations* between nodes defined dually to the intensional relations.

It is worth noting that extensional and intensional relations defined in this section originate from the mappings in nested line diagrams [3]. In the case of intensional relations, we assume that that $G \cap H = \emptyset$ (if this is not so, we can always time-tag the objects) and define a context $\mathbb{K}_3 = (G \cup H, M \cup N, I \cup J)$. If a nested line diagram of $\mathbb{K}_3$ is constructed so that G is used as the object set for the outer diagram and H is used as the object set for the inner diagram, then the nodes intensionally related to an outer node are the "realized" nodes of the inner diagram inside this outer node.

Another approach to dynamic mappings could be based on the theory of multicontexts [20], which however has to be adapted for our reduced lattice-based structures.

7   Conclusion

We extend a previous approach based on the notion of stability to build arbitrarily small concept lattices from sizeable contexts. After presenting theoretical properties of stability, introducing in particular several propositions useful for incremental computation of stability in evolving contexts, we distinguish intensional stability from extensional stability and illustrate them through selected case studies, where one or the other could be suitable. In particular, intensional stability appears to be useful for epistemic community taxonomy building, while extensional stability seems to be more effective for finding cohesive subgroups in communities of agents involved in common activities. These examples also demonstrate how different expectations regarding what makes a formal concept relevant for a given taxonomical description task may call for distinct usages of stability, extensional or intensional, which admittedly might not apply in all domains. We have also shown how it is possible to track taxonomy evolution using dynamic mappings between stability-reduced lattices.

Acknowledgements. The authors wish to thank Bertrand Chardon and Marc Schoenauer for providing us with the ECCS 2005 and 2006 data, as well as Nadine Peyriéras for her kind feedback regarding the zebrafish community. Diagrams have been produced with ConExp (http://sourceforge.net/projects/conexp), whose primary developer is Serhiy Yevtushenko. The first author was supported by the project COMO (Concepts and Models) of Deutsche Forschungsgemeinschaft (DFG) and Russian Foundation for Basic Research (RFBR).

References

1. Wille, R.: Concept lattices and conceptual knowledge systems. Computers & Mathematics with Applications 23, 493–515 (1992)
2. Ferré, S., Ridoux, O.: A file system based on concept analysis. In: Palamidessi, C., Moniz Pereira, L., Lloyd, J.W., Dahl, V., Furbach, U., Kerber, M., Lau, K.-K., Sagiv, Y., Stuckey, P.J. (eds.) CL 2000. LNCS (LNAI), vol. 1861, pp. 1033–1047. Springer, Heidelberg (2000)
3. Ganter, B., Wille, R.: Formal Concept Analysis: Mathematical Foundations. Springer, Heidelberg (1999)
4. Stumme, G., Taouil, R., Bastide, Y., Pasquier, N., Lakhal, L.: Computing iceberg concept lattices with TITANIC. Data & Knowledge Engineering 42, 189–222 (2002)
5. Roth, C., Obiedkov, S., Kourie, D.G.: Towards concise representation for taxonomies of epistemic communities. In: Ben Yahia, S., Mephu Nguifo, E. (eds.) CLA 4th International Conference on Concept Lattices and their Applications, Tunis, Faculté des Sciences de Tunis, pp. 205–218 (2006)
6. Roth, C., Bourgine, P.: Lattice-based dynamic and overlapping taxonomies: The case of epistemic communities. Scientometrics 69(2), 429–447 (2006)
7. Kuznetsov, S.O.: Stability as an estimate of the degree of substantiation of hypotheses derived on the basis of operational similarity. Nauchn. Tekh. Inf. Ser.2 (Automat. Document. Math. Linguist.) 12, 21–29 (1990)

8. Kuznetsov, S.O.: On stability of a formal concept. In: SanJuan, E. (ed.) JIM, Metz, France (2003)
9. Kuznetsov, S.O., Obiedkov, S.: Comparing performance of algorithms for generating concept lattices. J. Expt. Theor. Artif. Intell. 14(2/3), 189–216 (2002)
10. Bayardo, Jr., R., Goethals, B., Zaki, M. (eds.) In: Proc. of the IEEE ICDM Workshop on Frequent Itemset Mining Implementations (FIMI 2004), CEUR-WS.org (2004)
11. Godin, R., Missaoui, R., Allaoui, H.: Incremental concept formation algorithms based on Galois lattices. Computational Intelligence 11(2), 246–267 (1995)
12. Haas, P.: Introduction: epistemic communities and international policy coordination. International Organization 46(1), 1–35 (1992)
13. Porter, M.F.: An algorithm for suffix stripping. Program 14(3), 130–137 (1980)
14. Davis, A., Gardner, B.B., Gardner, M.R.: Deep South. University of Chicago Press, Chicago (1941)
15. Freeman, L.C., White, D.R.: Using Galois lattices to represent network data. Sociological Methodology 23, 127–146 (1993)
16. Freeman, L.: Finding social groups: A meta-analysis of the southern women data. In: Breiger, R., Carley, K., Pattison, P. (eds.) Dynamic Social Network Modeling and Analysis, pp. 39–97. National Academies Press, Washington, D.C. (2003)
17. Doreian, P.: On the delineation of small group structure. In: Hudson, H.C. (ed.) Classifying Social Data, pp. 215–230. Jossey-Bass, San Francisco (1979)
18. Atkin, R.: Mathematical Structure in Human Affairs. Heinemann Educational Books, London (1974)
19. Johnson, J.H.: Stars, maximal rectangles, lattices: A new perspective on Q-analysis. International Journal of Man-Machine Studies 24(3), 293–299 (1986)
20. Wille, R.: Conceptual structures of multicontexts. In: Eklund, P.W., Mann, G.A., Ellis, G. (eds.) ICCS 1996. LNCS, vol. 1115, pp. 23–29. Springer, Heidelberg (1996)

An FCA Perspective on n-Distributivity

Heiko Reppe

Institut für Algebra
Technische Universität Dresden
D-01062 Dresden
Heiko.Reppe@mailbox.tu-dresden.de

Abstract. Distributive lattices belong to the best studied ordered structures. A. Huhn introduced a generalisation of this lattice property, called *n-distributivity*. We present two new methods to recognise the parameter n of this property for a given structure. For this purpose we use the *arrow relations* in a formal context and *implications with proper premise*. Additionally, we consider subsets of an order relation $\leq$ on a finite set P with an additional property. These subsets will be called *left clearings* of $\leq$. We show that the family of left clearings forms a complete dually l-distributive lattice, where l denotes the length of $(P, \leq)$. Using these results, we determine that parameter n for *Tamari* lattices for the n-distributivity and dually n-distributivity.

1 n-Distributivity and Dimension

Definition 1. *A lattice $\underline{L}$ is* distributive, *if for all $x, y_0, y_1 \in L$:*

$$x \vee (y_0 \wedge y_1) = (x \vee y_0) \wedge (x \vee y_1). \qquad \Diamond$$

This is similar to the distributivity of the arithmetic operations multiplication and addition. However, in the case of lattice operations, the same formula with the two operations interchanged, yields an equivalent condition. A parametric version of this identity was investigated by A. Huhn in [H72] for modular lattices. He called this n-distributivity, where for $n = 1$ we get the usual distributive law.

Definition 2. *Let n be a positive integer. A lattice $\underline{L}$ is called* n-distributive, *if for all $x, y_0, \ldots, y_n \in L$:*

$$x \vee \left(\bigwedge_{i=0}^{n} y_i \right) = \bigwedge_{j=0}^{n} \left(x \vee \bigwedge_{\substack{i=0 \\ i \neq j}}^{n} y_i \right). \qquad (\mathrm{D}_\vee^n)$$

$\mathrm{D}_\wedge^n$ *denotes the dual equation. A lattice $\underline{L}$ is called* dually n-distributive *if for all $x, y_0, \ldots, y_n \in L$ the equation $\mathrm{D}_\wedge^n$ holds.* $\Diamond$

In [H88] non-modular lattices were studied. In this case the identities $\mathrm{D}_\vee^n$ and $\mathrm{D}_\wedge^n$ are not equivalent anymore. However, lattices satisfying the identity $\mathrm{D}_\vee^n$ satisfy the identity $\mathrm{D}_\vee^{n+1}$ as well. In the following three sections we illustrate the close relationship between n-distributivity and different concepts of dimension.

U. Priss, S. Polovina, and R. Hill (Eds.): ICCS 2007, LNAI 4604, pp. 255–268, 2007.

1.1   Order Dimension

By a well-known *Theorem of Ore* the order dimension of a lattice $\underline{L}$, or more general of a poset, is the minimal number of chains such that $\underline{L}$ embeds as an order into their product. L. Libkin observed the following.

Lemma 1. *[L95, Prop. 4.6] If a finite lattice $\underline{L}$ is n-distributive but not $(n-1)$-distributive, then its order dimension is at least n.*

Planar lattices can be drawn into the plane without edge crossing. The order dimension of planar lattices is at most 2. By Lemma 1 we conclude that a finite planar lattice is either distributive or 2-distributive, cf. [L95, Cor. 4.7].

The order dimension of a lattice and the parameter n of the n-distributive law can be far from each other, as finite *Boolean lattices* indicate. They are all 1-distributive, but their order dimension is as large as their number of atoms. We want to illustrate this further by an example taken from the front cover of the Formal Concept Analysis book by B. Ganter and R. Wille [GW99].

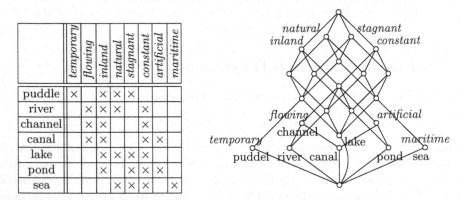

	temporary	flowing	inland	natural	stagnant	constant	artificial	maritime
puddle	×		×	×	×			
river		×	×	×		×		
channel		×	×			×		
canal		×	×			×	×	
lake		×	×	×	×			
pond		×		×	×	×		
sea				×	×	×		×

Fig. 1. This example of waters is taken from the front cover of [GW99], only, we clarified the set of objects. This concept lattice is an example for a 2-distributive and non-planar lattice.

Example 1. Figure 1 shows a formal context with different types of waters as objects and some properties to distinguish them. On the right we see the concept lattice of this context. The lattice is not distributive, as for example:

$$\gamma \text{ canal} = \gamma \text{ canal} \vee (\gamma \text{ river} \wedge \gamma \text{ pond})$$
$$(\emptyset, M) = (\gamma \text{ canal} \vee \gamma \text{ river}) \wedge (\gamma \text{ canal} \vee \gamma \text{ pond}).$$

However, this concept lattice is 2-distributive and dually 2-distributive as we will confirm later on. Obviously, it is not a planar lattice. Its order dimension has to be at least 4, as it contains the product of 4 chains as a sublattice.

1.2 Lattices of Convex Sets of Real Vector Spaces

An interesting class of lattices arises when we study convex sets of real vector spaces. In [H88] and [B05] they were studied in detail.

Theorem 1. *[H88] Let $X \subseteq \mathbb{R}^d$ and $\underline{L} \cong \{X \cap C \mid C \subseteq \mathbb{R}^d, \ C \ convex\}$ ordered by inclusion, then the smallest $n \in \mathbb{N}$ for which $\underline{L}$ is dually n-distributive is at most $d + 1$.*

Example 2. In Figure 2 we selected 5 points of the real plane to illustrate Theorem 1. Considering the points 1, 2, and 3, we see that point 5 is contained in the convex hull of all three of them, but not contained in the convex hull of only two. Whereas the point 4 is contained in the convex hull of all three point 1, 2, and 3, but it is already contained in the hull of 1 and 2. This kind of argumentation can be formalised, as we will see in Section 2.

Figure 2 also shows the standard context for the lattice of all convex sets of $\mathbb{R}^2$ intersected with $\{1, 2, 3, 4, 5\}$. The sets $A, B, \ldots, H$ turn out to be minimal representatives of the structure. The formal context includes the arrow relations. These will be introduced in Definition 3.

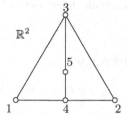

	A	B	C	D	E	F	G	H
1	↗	×	×	×	↙	×	×	↙
2	×	↗	×	↙	×	×	↙	×
3	×	×	↗	×	×	↙	×	×
4	×	×	×	↙	↙	×	↙	↙
5	×	×	×	×	×	↗	↙	↙

Fig. 2. Five selected points of the Euclidean space $\mathbb{R}^2$ and the formal context of the lattice of convex sets

1.3 Modular Lattices and Projective Spaces

If we add the restriction $x \leq y_1$ to the precondition of Definition 1, we receive modular lattices instead of distributive ones. The original definition of n-distributivity of Huhn required the lattice to be modular.

Theorem 2. *[H72, S. 3.1] If $\underline{L}$ is a modular lattice, then $\mathrm{D}_\vee^n$ and $\mathrm{D}_\wedge^n$ are equivalent.*

A nice class of examples is formed by projective spaces. An n-dimensional projective space can be constructed from an $(n + 1)$-dimensional vector space over any field, or more general over a division ring, by identifying all non-zero vectors that are multiples of each other. A 2-dimensional projective space is called projective plane.

Proposition 1. *Projective planes are is 3-distributive. Projective spaces of dimension d are $(d + 1)$-distributive.*

Example 3. The smallest example of a projective plane is given in Figure 3, the so called *Fano plane*. It is the projective space of $GF(2)^3$. The points indicate the seven non-zero vectors of this vector space. The straight lines and the circle indicate projective lines. Each of them contains 3 elements, but any two points specify a line.

The formal context shows the incidence relation of points and lines. The concept lattice has 14 elements, beside points and lines we find the top and the bottom element. The closure of the points 1, 2, 3 contains the point 5, whereas none of the closures of only 2 points of 1, 2, 3 contains this point.

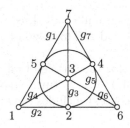

	g_1	g_2	g_3	g_4	g_5	g_6	g_7
1	×	×	╱	×	╱	╱	╱
2	╱	×	×	╱	×	╱	╱
3	╱	╱	×	×	╱	×	╱
4	╱	╱	╱	×	×	╱	×
5	×	╱	╱	╱	×	×	╱
6	╱	×	╱	╱	╱	×	×
7	×	╱	×	╱	╱	╱	×

Fig. 3. The well known Fano plane with labelled points and lines, and its formal context representation

2   Characterisations of n-Distributivity

Definition 3. *For a formal context* (G, M, I) *and* $g \in G$ *and* $m \in M$ *the arrow-relations are defined as:*

$$g \swarrow m :\Longleftrightarrow g \not\!I\, m \text{ and if } g' \subsetneq h' \text{ then } h\, I\, m,$$
$$g \nearrow m :\Longleftrightarrow g \not\!I\, m \text{ and if } m' \subsetneq n' \text{ then } g\, I\, n,$$

If both $g \swarrow m$ *and* $g \nearrow m$ *holds we write* $g \nearrow\!\!\!\!\swarrow m$.   ◇

If $g \swarrow m$ holds in a formal context, then not only $g \not\!I\, m$, but the intent of every lower neighbour of γg contains the attribute m. As we are dealing with complete lattices, this is true if and only if there is exactly one lower neighbour, which is equivalent to γg being join-irreducible.

A set $P \subseteq M$ is called *premise* of an attribute set C if $C \subseteq P''$. These notions are also used for single attributes. We focus on a special kind of premises that were defined in [GW99].

Definition 4. *Let* $\mathbb{K} = (G, M, I)$ *be a formal context and* $A \subseteq M$. *By*

$$A^\bullet := A'' \setminus (A \cup \bigcup_{n \in A} (A \setminus \{n\})'')$$

we denote the set of attributes which are contained in A'' *but neither in* A *nor in the closure of any proper subset of* A. *The set* A *is called proper premise of* $A^\bullet$ *if* $A^\bullet \neq \emptyset$.   ◇

Proposition 2. *[GW99, Prop. 23] Let (G, M, I) be a doubly founded context and $P \subseteq M$. P is a premise of m if and only if*

$$(M \setminus g') \cap P \neq \emptyset$$

holds for all $g \in G$ with $g \nearrow m$. P is a proper premise of m if and only if $m \notin P$ and P is minimal with respect to the property that $(M \setminus g') \cap P \neq \emptyset$ holds for all $g \in G$ with $g \nearrow m$.

If we consider an attribute $m \in M$ that has no proper premise in $M \setminus \{m\}$ at all, then the condition $(M \setminus g') \cap P = \emptyset$ holds for all $P \subseteq M \setminus \{m\}$.

Corollary 1. *Let $\mathbb{K} = (G, M, I)$ be a finite context and $m \in M$. There is a proper premise $P \subseteq M \setminus \{m\}$ of m if and only if $m \in (M \setminus \{m\})''$.*

If $\mathbb{K}$ is clarified, the attributes having no proper premise are known already. They are extremal points of the intent M, cp. [GW99].

We can derive from Proposition 2 further information of proper premises. Considering an attribute $m \in M$, such that μm is meet-irreducible a proper premise P of m cannot be a subset of m''. Furthermore, if we leave out one element $p \in P$, then we find an object $g_p \in (P \setminus \{p\})''$ with $g_p \nearrow m$. A set $P \subseteq M \setminus \{m\}$ satisfying all these conditions is a proper premise of m. This yields to Corollary 2.

For short hand notation we denote by $m^{\nearrow}$ the following set of objects:

$$m^{\nearrow} := \{g \in G \mid g \nearrow m\}.$$

Corollary 2. *Let $\mathbb{K} = (G, M, I)$ be a finite context and $m \in M$ with μm meet-irreducible. Then there is a proper premise of m in $M \setminus \{m\}$ if and only if it exists a set $P \subseteq M$ with $P \not\subseteq m''$ and for all $p \in P$ there is a $g_p \in m^{\nearrow}$ with for all $q \in P$ holds $g_p I q \iff p \neq q$ and $P'' \cap m^{\nearrow} = \emptyset$.*

Theorem 3. *[L95, Thm. 2.1] Let $\underline{L}$ be a finite lattice, then $\mathrm{D}_{\vee}^n$ is equivalent to*

$$x \vee \left(\bigwedge_{t \in T} y_t \right) = \bigwedge_{\substack{K \subseteq T \\ |K| = n}} \left(x \vee \bigwedge_{k \in K} y_k \right),$$

for every index set T and for possible choices of $x, y_t \in L$, $t \in T$.

For distributive lattices there is a nice characterisation within the context by means of arrow-relations and another one with the help of implications with proper premises.

Theorem 4. *[GW99, Thm. 41] Let (G, M, I) be a reduced formal context and $\underline{L} = \underline{\mathfrak{B}}(G, M, I)$ a doubly founded concept lattice then the following statements are equivalent:*

1. *$\underline{L}$ is distributive.*
2. *Every proper premise is a singleton.*

3. $g \nearrow m$ implies $g \nearrow\!\!\!\!\diagup m$, $g \nearrow\!\!\!\!\diagup m$ implies $g \nearrow m$,
   $g \nearrow m$ and $g \nearrow n$ imply $m = n$, and $g \nearrow m$ and $h \nearrow m$ imply $g = h$.

For our characterisation of n-distributivity we restrict to finite concept lattices, but we exclude the property for the formal context being reduced.

Theorem 5. *Let* $\mathbb{K} = (G, M, I)$ *be a finite context then the following are equivalent:*

1. $\mathfrak{B}(\mathbb{K})$ *is* n-*distributive.*
2. *Every implication* $A \to A^{\bullet}$ *with proper premise satisfies:*
   either A *has at most* n *elements or all elements of* $A^{\bullet}$ *are reducible.*
3. *Every attribute* $m \in M$ *with* μm *is meet-irreducible satisfies:*
   if $\{g_0, g_1, \ldots, g_n\} \subseteq m^{\checkmark}$, *and if* $\{m_0, m_1, \ldots, m_n\} \subseteq M$ *with* $\{m_0, m_1, \ldots, m_n\} \nsubseteq m''$ *and for* $i, j \in \{0, 1, \ldots, n\}$ *holds* $g_i \, I \, m_j \iff i \neq j$
   then there is an object $g \in \{m_0, m_1, \ldots, m_n\}' \cap m^{\checkmark}$ *and for all* $n \in M$ *with* $g \nmid n$ *we find a* $h \in \{g_0, g_1, \ldots, g_n\}$ *with* $h \nmid n$.

Figure 4 illustrates Item 3 of this theorem. Suppose μm is meet-irreducible and $\{g_1, g_2, g_3\} = m^{\checkmark}$. Then the subcontext $\mathbb{K} = (\{g_1, g_2, g_3\}, \{m_1, m_2, m_3\}, I)$ describe the prerequisites of Item 3. In the formal context of a 2-distributive lattice we can find this situation only if $\{m_1, m_2, m_3\} \subseteq m''$.

	m	$\ldots$	m_1	m_2	m_3
g_1	$\checkmark$			$\times$	$\times$
g_2	$\checkmark$		$\times$		$\times$
g_3	$\checkmark$		$\times$	$\times$	
$\vdots$					

Fig. 4. The drawing shows a part of a reduced formal context that cannot appear within the formal context of a 2-distributive concept lattice, if μm is meet-irreducible with $m^{\checkmark} = \{g_1, g_2, g_3\}$ and one of the attributes m_1, m_2, m_3 is not contained in m''

Proof. The equivalence $2 \iff 3$ is a reformulation of Corollary 2. We replaced the arbitrary proper premise by an $(n+1)$-elementary set $P = \{m_0, m_1, \ldots, m_n\}$. This set P is not a proper premise of m, neither can it be enlarged to a proper premise of m.

$1 \Rightarrow 2$: Suppose $P \to C$ is an implication with proper premise of the context $\mathbb{K}$ with $|P| > n$. Let $c \in C$. We can apply Theorem 3 to the following equation of an n-distributive lattice:

$$\mu c = \mu c \vee \bigwedge_{p \in P} \mu p = \bigwedge_{\substack{A \subseteq P \\ |A| = n}} (\mu c \vee (A', A'')).$$

As P is a proper premise of c, none of the sets A'' contains c. Hence c'' is the closure of proper subsets of c'' which are closed itself and do not contain the attribute c. In other words μc is meet-irreducible.

$2 \Rightarrow 1$: Let $(A', A''), (B'_0, B''_0), (B'_1, B''_1), \ldots, (B'_n, B''_n) \in \underline{\mathfrak{B}}(\mathbb{K})$. Then

$$(A', A'') \vee \bigwedge_{i=0}^{n}(B'_i, B''_i) = \left((A' \cup \bigcap_{i=0}^{n} B'_i)'', A'' \cap (\bigcup_{i=0}^{n} B_i)'' \right).$$

By Condition 2 a proper premise of an attribute m with μm meet-irreducible has at most size n. Hence, $m \in A'' \cap (\bigcup_{i=0}^{n} B_i)''$ with μm meet-irreducible is already contained in at least one of the sets $A'' \cap (\bigcup_{\substack{i=0 \\ i \neq j}}^{n} B_i)''$ for $j \in \{0, 1, \ldots, n\}$.

Thus

$$\left((A' \cup \bigcap_{i=0}^{n} B'_i)'', A'' \cap (\bigcup_{i=0}^{n} B_i)'' \right) = \bigwedge_{j=0}^{n} \left((A' \cup \bigcap_{\substack{i=0 \\ i \neq j}}^{n} B'_i)'', A'' \cap (\bigcup_{\substack{i=0 \\ i \neq j}}^{n} B_i)'' \right)$$

$$= \bigwedge_{j=0}^{n} \left((A', A'') \vee \bigwedge_{\substack{i=0 \\ i \neq j}}^{n} (B'_i, B''_i) \right)$$

which concludes the proof. □

Using the equivalence $1 \iff 3$ of Theorem 5 we can derive:

Corollary 3. *Let (G, M, I) be a finite context and*

$$n := \max\{|m^{\nearrow}| \mid m \in M, \mu m \text{ meet-irreducible}\},$$

then $\underline{\mathfrak{B}}(G, M, I)$ is n-distributive.

In general this n of Corollary 3 is not the least possible one as the following example illustrates.

Example 4. For the formal context of Example 1 we calculated the arrow relations as it is shown in Figure 5. We see the structure is very rich on $\nearrow$. Therefore we demonstrate the dual of Theorem 5 and consider $\nearrow$ and objects instead. A row does contain four $\nearrow$ at most. The row of *channel* contains three, the row of *lake* contains four, and all the others two or fewer $\nearrow$.

By Corollary 3 the lattice is dually 4-distributive, but we may do better. To check this we must verify if the pattern of Theorem 5 (3) occurs. It does not, since *channel* is not an $\vee$-reducible element and in the case of *lake*, we examine

$$lake^{\nearrow} = \{temporary, \ flowing, \ artificial, \ maritime\}.$$

Here we find the following situation: Two of these columns contain one $\times$ only, there is one with two $\times$, and one with three $\times$. Thus this lattice is at least dually 2-distributive. We already know it is not distributive.

The equivalence $1 \iff 2$ of Theorem 5 leads to:

Corollary 4. *Let (G, M, I) be a finite formal context such that $\underline{L} := \underline{\mathfrak{B}}(G, M, I)$ is n-distributive. If $A_1, \ldots, A_n$ are intents and $m \in (\bigcup_{i=1}^{n} A_i)'' \setminus \bigcup_{i=1}^{n} A_i$ then there is subset $B \subseteq \bigcup_{i=1}^{n} A_i$ with $2 \leq |B| \leq n$ and $m \in B^{\bullet}$.*

	temporary	*flowing*	*inland*	*natural*	*stagnant*	*constant*	*artificial*	*maritime*
puddle	×	↙	×	×	×	↙	↙	↙
river	↙	×	×	×	↗	×	↗	↙
channel		×	×	↗	↗	×	↗	
canal	↙	×	×	↗	↗	×	×	↙
lake	↗	↙	×	×	×	×	↙	↗
pond	↙	↙	×	↗	×	×	×	↙
sea	↙	↙	↗	×	×	×	↙	×

Fig. 5. The chart shows the formal context of the waters example including the arrow relations

With the help of this we can finally prove that the lattice of Example 1 is 2-distributive.

Example 5. The complete list of implications with proper premise of the formal context of Example 1 is the following:

$\{temporary\} \rightarrow \{inland, natural, stagnant\}$,
$\{flowing\} \rightarrow \{inland, constant\}$,
$\{artificial\} \rightarrow \{inland, constant\}$,
$\{maritime\} \rightarrow \{natural, stagnant, constant\}$,
$\{temporary, constant\} \rightarrow \{flowing, artificial, maritime\}$,
$\{temporary, flowing\} \rightarrow \{artificial, maritime\}$,
$\{temporary, artificial\} \rightarrow \{flowing, maritime\}$,
$\{temporary, maritime\} \rightarrow \{flowing, artificial\}$,
$\{flowing, stagnant\} \rightarrow \{natural, temporary, artificial, maritime\}$,
$\{flowing, maritime\} \rightarrow \{temporary, artificial\}$,
$\{artificial, natural\} \rightarrow \{stagnant, temporary, flowing, maritime\}$,
$\{artificial, maritime\} \rightarrow \{temporary, flowing\}$,
$\{inland, maritime\} \rightarrow \{temporary, flowing, artificial\}$.

At the first sight we determine that the maximal size of a premise of an implication with proper premise is 2. By Theorem 5 ($2 \Rightarrow 1$) we have verified that the concept lattice of the waters example forms a 2-distributive lattice.

The stem base of this formal context contains 8 implications only, including the first four of this list. All other implications listed here (and in the stem base as well) are implications referring to the bottom element of the lattice which means the premise is already contradictory.

3   Left Clearings of a Poset and Tamari Lattices

In this section we will analyse the correlation of transitivity and n-distributivity. For this purpose we consider an order relation $\leq$ on a set P, i.e. reflexive, antisymmetric and transitive subset of $P \times P$. The structure $(P, \leq)$ is called partially ordered set, or poset for short.

Definition 5. *Let* $(P, \leq)$ *be a poset. An order relation* $R \subseteq \leq$ *is called* left clearing *of* $\leq$ *if and only if for all* $p, q, r \in P$

$$(p, r) \in R \text{ and } p < q \leq r \text{ implies } (p, q) \in R.$$

The set of all left clearings of $\leq$ *is denoted by* $\mathcal{L}(P, \leq)$. ◊

For any poset $(P, \leq)$ we easily find two order relations which are left clearings of $\leq$, namely $\leq$ itself and the diagonal $\Delta := \{(p, p) \mid p \in P\}$. We will see by means of an example that the family of left clearing can be very large, even if P has only few elements.

Example 6. We consider the poset $(P, \leq)$ shown in Figure 6. Beside $(P, \leq)$ we see 10 order relations which are left clearings of $\leq$ and on the right we see a formal context relating the elements of $\leq \setminus \Delta$ to the 10 left clearings of $\leq$. All left clearings of $\leq$ except $\leq$ itself can be retrieved as intersections of those 10 relations. All together we find 42 left clearings of $\leq$. They form a complete lattice, which is depicted in Figure 7.

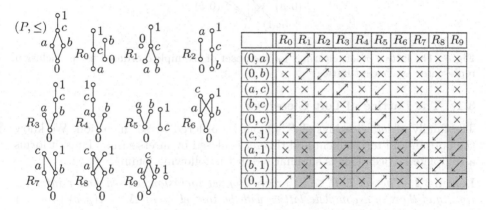

Fig. 6. We see 10 order relations on P which are left clearings of $\leq$. The formal context displays the $\in$-relation. We have left out all elements of Δ as they belong to all order relations. The shaded cells of that tableau show that the concept lattice is not 3-distributive.

Applying Theorem 5 to the formal context we observe four ↗ in column R_9 which is the maximal number. Restricting the context to the rows R_9'' and columns R_1, R_3, R_4, R_6, we recognise exactly the subcontext that we would have to exclude for 3-distributivity. In the formal context of Figure 6 we highlighted those cells.

Observe that R_1 for example is not contained in R_9''. Hence we have found a four-element proper premise of R_9 and we have shown the concept lattice depicted in Figure 7 is 4-distributive.

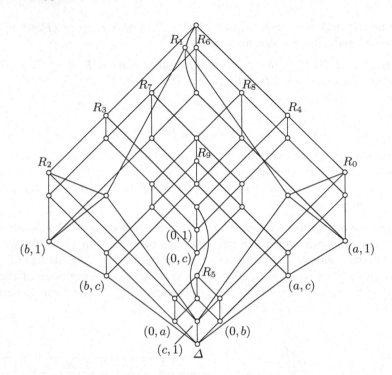

Fig. 7. The lattice of left clearings for the poset of Example 6 ordered by set inclusion represented as concept lattice

3.1 The Lattice of Left Clearings

The family of left clearings $\mathcal{L}(P, \leq)$ can be ordered by set inclusion. We know the bottom and top element and this set is closed by intersection. Thus, it forms a complete lattice. This is summarised in the following lemma.

Lemma 2. *The set* $\mathcal{L}(P, \leq)$ *is ordered by set inclusion, closed under intersection, and thereby a complete lattice with bottom element* $\Delta := \{(p, p) \mid p \in P\}$ *and top element* $\leq$. *The infimum and supremum in* $\mathcal{L}(P, \leq)$ *are given by:*

$$\bigwedge_{t \in T} R_t := \bigcap_{t \in T} R_t,$$

$$\bigvee_{t \in T} R_t := \left(\bigcup_{t \in T} R_t \right)^{\downarrow},$$

where $(\cdot)^{\downarrow}$ *denotes the iteration of the transitive closure and the closure with respect to the property of being a left clearing.*

This lattice is not necessary a sublattice of the lattice of all order relations on the set P, as we already see from the different join-operation. However, it is a meet-subsemilattice. Our aim now is to find the parameter n of the dual n-distributivity for these lattices. To begin with we recall the definition of the length of a poset.

The *length* of a poset $(P, \leq)$ is defined to be the supremum of the cardinalities of the maximal chains in $(P, \leq)$. Coming back to the property we are interested in, we relate the length of the poset $(P, \leq)$ with the smallest $n \in \mathbb{N}$ for which $(\mathcal{L}(P, \leq), \subseteq)$ is dually n-distributive.

Lemma 3. *If $(P, \leq)$ is a poset with finite length l, then $(\mathcal{L}(P, \leq), \subseteq)$ is dually l-distributive but not dually $(l-1)$-distributive.*

Proof. Let $A = \{a_0, a_1, \ldots, a_l\}$ be a maximal chain in $(P, \leq)$ and $a_0 < a_1 < \cdots < a_l$. Then

$$R_1 := \Delta \cup \{(a_0, a_1)\}, \ R_2 := \Delta \cup \{(a_1, a_2)\}, \ldots, \ R_l := \Delta \cup \{(a_{l-1}, a_l)\}$$

are left clearings of $\leq$. The closure of the union of $(l-1)$ of these relations does not contain the pair (a_0, a_l). The closure of all of these relations does contain this element. Hence we have found an implication between objects with proper premise of size l. Thus, $(\mathcal{L}(P, \leq), \subseteq)$ is not dually $(l-1)$-distributive.

If we consider more then l left clearings of $\leq$ and observe the join of them, then every element of the join is already contained in the join of a proper subset of these relation. Hence, the lattice is dually l-distributive. □

Summarising our results for the lattice of Figure 7 we can state that it is 4-distributive and dually 3-distributive.

3.2 Tamari Lattices

Tamari lattices were introduced in [T62]. The Tamari lattices $\mathbb{T}_n$ arise when we consider all possible bracketings of a term with $n+1$ variables. For every consequent triple of variables a bracketing has to state which pair is embraced. On this set a preference relation is defined.

If a term t_1 can be constructed from another term t_2 according to the rule

$$A(BC) \longrightarrow (AB)C$$

and its iterated application to subterms, we prefer t_1 instead of t_2. The least preferable bracketing is given by the term $x_0(x_1(\ldots(x_{n-1}x_n)\ldots))$ and the most preferable one is $(\ldots(x_0x_1)x_2\ldots)x_n$, cp. [GW99, Example 11].

This set of bracketings together with the preference relation forms a complete lattice, as shown by Tamari. The join- and meet-irreducible elements have been characterised by Bennett and Birkhoff in [BB94]. A recursive construction of the formal context of $\mathbb{T}_n$ was developed by Geyer in [G94]. This was modified in [GW99, Example 11] and we use a further modification for our description.

Let $P := \{1, 2, \ldots, n\}$ and $\leq$ the natural order on these numbers. We define a context $\mathbb{K}_n := (<, R_<, \in)$ with

$$R_{(p,q)} := \{(a, b) \in \ \leq \ | \ p \leq a < q \Rightarrow b < q\}$$

and $R_< := \{R_{(p,q)} \mid (p, q) \in \ <\}$. Figure 8 shows the formal context $\mathbb{K}_4$ and the concept lattice $\underline{\mathfrak{B}}(\mathbb{K}_4)$ which is isomorphic to $\mathbb{T}_4$.

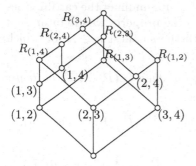

$\in$	$R_{(1,2)}$	$R_{(2,3)}$	$R_{(1,3)}$	$R_{(3,4)}$	$R_{(2,4)}$	$R_{(1,4)}$
$(1,2)$	↗	×	×	×	×	×
$(2,3)$	×	↗	↙	×	×	×
$(1,3)$	↗	×	↗	×	×	×
$(3,4)$	×	×	×	↗	↙	↙
$(2,4)$	×	↙	×	×	↗	↙
$(1,4)$	↗	×	↗	×	×	↗

Fig. 8. We see the lattice $\mathfrak{B}(\mathbb{K}_4)$, which is isomorphic to $\mathbb{T}_4$ and the context $\mathbb{K}_4$ including its arrow-relations, cp. [GW99]

Lemma 4. *If $(P,\le)$ is an m-elementary chain and $(p,q) \in <$, then $R_{(p,q)}$ is a left clearing. If $R \in \mathcal{L}(P,\le)$, then*

$$R = \bigcap_{(p,q) \in \ <\backslash R} R_{(p,q)}.$$

Proof. $R_{(p,q)}$ is reflexive, transitive and antisymmetric. Suppose $(u,w) \in R_{(p,q)}$. Then of course for all $v \in P$ with $u < v \le w$ we derive $(u,v) \in R_{(p,q)}$. Hence, it is a left clearing.

For the second part we observe, $(p,q) \notin R_{(p,q)}$, thus $\bigcap_{(p,q)\in \ <\backslash R} R_{(p,q)} \subseteq R$. Suppose now $(u,v) \in R \backslash (\bigcap_{(p,q)\in \ <\backslash R} R_{(p,q)})$. Then it exists a pair $(r,s) \in < \backslash R$ and $(u,v) \notin R_{(r,s)}$. Hence $r = u$ and $s \le v$. Since $(u,v) \in R$ and $(r,s) \in < \backslash R$ we have a contradiction to R is a left clearing. Thus there is no element like (u,v) which concludes the proof. $\qquad\square$

Lemma 5. *If $(P,\le)$ is an m-elementary chain, then $(\mathcal{L}(P,\le), \subseteq) \cong \mathbb{T}_m$*

Proof. By [GW99, Example 11] and Lemma 4 it remains to show,

$$(u,v) \in R_{(p,q)} \iff \neg(p \le u < q \le v).$$

This is obviously the case. $\qquad\square$

Lemma 6. *The Tamari lattices $\mathbb{T}_{n+1}$, $n \in \mathbb{N}$ are*

- *n-distributive, but not $(n-1)$-distributive and*
- *dually n-distributive, but not dually $(n-1)$-distributive.*

Proof. The second item is a consequence of the result in Lemma 3 and 5. For the first we consider $P := \{0, 1, \ldots, n\}$ ordered in the usual way.

By definition if $p \in P$ and $(p \ne n)$ then $(p,n) \notin R_{(0,n)}$. If $(p,n)' \subsetneq (q,r)'$ then $p = q$ and $r < n$. But $(p,r) \in R_{(0,n)}$. Hence $(p,n) \nearrow R_{(0,n)}$. By the same argument we can prove $|R_{(p,q)}^{\nearrow}| = q - p$. Hence $|R_{(0,n)}^{\nearrow}|$ is maximal.

For $q, p \in P$ and $q < n$ we easily see $(p, n) \in R_{(q, q+1)}$ if and only if $p \neq q$. The relation $R_{(0,1)}$ is not contained in $R_{(0,n)}$ except if they are equal. If they are equal then $n = 1$ and we consider the two-element lattice which is distributive. If they are not equal, we have found a proper premise of $R_{(1,n)}$ of size n, namely $\{R_{(p, p+1)} \mid 0 \leq p \leq n - 1\}$. Applying Theorem 5 we proved the claim.   □

Lemma 7. *If P is an m-elementary set, then $\mathbb{T}_m$ is isomorphic to a sublattice of the lattice of all order relations on P.*

Proof. We already know $\mathbb{T}_m$ is isomorphic to the lattice of left clearings of an m-elementary chain. The latter lattice is a meet-subsemilattice of the lattice of all order relations.

To show that it is actually a sublattice in the case of chains, we have to prove that the transitive closure of the union of left clearings of a total order is a left clearing itself.

As all sets are finite, it suffices to show this for a single case. Suppose R_1, R_2 are left clearings of a total order $\leq$ on P. Let $(a, b) \in R_1$ and $(b, c) \in R_2$, but $(a, c) \in \text{trans}(R_1 \cup R_2) \setminus (R_1 \cup R_2)$. By definition of a left clearing we know for all $d_1, d_2 \in P$

$$a < d_1 \leq b \Rightarrow (a, d_1) \in R_1 \text{ and}$$
$$b < d_2 \leq c \Rightarrow (b, d_2) \in R_2.$$

As there is no element of the interval $[a, c]$ in $(P, \leq)$ incomparable to b we can conclude that by transitivity all elements $d \in P$ with $a < d \leq c$ also satisfy $(a, d) \in \text{trans}(R_1 \cup R_2)$.   □

4   Conclusion

Lattice representations of geometric structures, as projective geometries and convex geometries, inherit the parameter n for the dual n-distributivity from the dimension of the underlying vector space. If on the other hand we look for an representation of a given lattice in either way we have to choose the dimension of the vector space sufficiently high.

We have introduced two new methods for recognising n-distributivity of a lattice. The first one was by excluding a subcontext, that involves the arrow relation $\swarrow$. The second method refers to implications with proper premise of the implicational theory. Here we have to find the size of the largest premise.

Applying these methods to the family of left clearings of an order relation of finite length l, we have shown that this lattice dually l-distributive but not necessarily l-distributive. From this result we could derive that the Tamari lattice $\mathbb{T}_{n+1}$ is n-distributive, but not $(n-1)$-distributive. In this case the dual equality holds as well. For Tamari lattices we have a nice description of the standard context. This we could not achieve for left clearings so far.

References

[BB94]   Bennett, M.K., Birkhoff, G.: Two families of Newmann lattices. Algebra Universalis 32, 115–144 (1994)

[B05]   Bergmann, G.M.: On lattices of convex sets in R^n. Algebra Universalis 53, 357–395 (2005)

[GW99]   Ganter, B., Wille, R.: FCA– Mathematical Foundations. Springer, Heidelberg (1999)

[G94]   Geyer, W.: On Tamari lattices. Discrete Mathematics 133, 99–122 (1994)

[H72]   Huhn, A.P.: Schwach distributive Verbände. I. Acta Sci. Math. (Szeged) 33, 297–305 (1972)

[H88]   Huhn, A.P.: On non-modular n-distributive lattices, I. Lattices of convex sets. Acta Sci. Math. (Szeged) 52, 35–45 (1988)

[L95]   Libkin, L.: n-distributivity, dimension and Carathéodory's theorem. Algebra Universalis 34, 72–95 (1995)

[T62]   Tamari, D.: The algebra of bracketings and their enumeration. Nieuw Arch. Wiskunde III. Ser. 10, 193–206 (1962)

Towards a Semantology of Music

Rudolf Wille and Renate Wille-Henning

Technische Universität Darmstadt, Fachbereich Mathematik
wille@mathematik.tu-darmstadt.de

Abstract. The aim of this paper is to approach a *Semantology of Music* which is understood as the theory and methodology of musical semantic structures. The analysis of music structures is based on a threefold semantics which is performed on the musical level, the abstract philosophic-logical level, and the hypothetical mathematical level. Basic music structures are discussed by examples, in particular: tone systems, chords, harmonies, scales, modulations, musical time flow, and music forms. A specific concern of this paper is to clarify how a Semantology of Music may support the *understanding of music*.

Contents

1. Semantology
2. Music Structures as Semantic Structures
3. Understanding Music

1 Semantology

Understanding music is based on the ability of humans to be affected by music so that it reaches human feelings, emotions, and thought. Humans can even be deeply moved by music, particularly by its musical senses and meanings which may be represented by semantic structures. It is the aim of this paper to develop some first steps towards a theory and methodology of musical semantic structures by using the notion *"Semantology"* as introduced in [GW06] and [EW07].

In general, *Semantology* is understood as the theory and methodology of semantic structures which belong to a declared field of knowledge. The most extensive Semantology (up to now) has been developed for the field of *Conceptual Knowledge Processing* based on the mathematical theory of *Formal Concept Analysis* (see [EW07], [Wi06], [GSW05], [SW00], [GW99], etc.). This Semantology has already supported a large number of applications for a wide range of subjects for which the developed theories and methods have been proven successfully. Many applications made clear that *meaning of semantic structures* in the field of Conceptual Knowledge Processing can be analysed on at least three levels. This shall be briefly scetched (cf. [EW07]):

- First, there is the meaning on the *concrete level* on which the considered conceptual knowledge originates. This is usually the semantics belonging to the scientific fields whose language and understanding are used to describe that knowledge.

U. Priss, S. Polovina, and R. Hill (Eds.): ICCS 2007, LNAI 4604, pp. 269–282, 2007.
© Springer-Verlag Berlin Heidelberg 2007

- Second, there is the meaning on the general *philosophic-logical level* on which the semantics is highly abstracted from the semantics of the concrete level, but is still related to actual realities. It is the semantics of the traditional philosophical logic based on the main functions of human thought: concept, judgment, and conclusion (cf.[Ka88]).
- Third, there is the meaning on the *mathematical level* on which the semantics is strongly restricted to the purely abstract: like numbers, ideal geometric figures and, since the twentieth century, set structures (and their generalizations). This very rigid semantics makes possible the high consensus about the validity of mathematical results, from which the semantic structures may also benefit.

Emphasizing the three levels of semantics for semantic structures has been inspired by *Peirce's classification of sciences* in which *Mathematics* is viewed as the most abstract science studying hypotheses exclusively and dealing only with potential realities, *Philosophy* is considered as the most abstract science dealing with actual realities, while all *other sciences* are more concrete in dealing with special types of actual realities (cf.[GW06]). The sketched threefold semantics shall be demonstrated by an example from music.

Since, in the 16th century, Gioseffo Zarlino took over the new *conception of harmony* from musical practice to the theory of music, many efforts had been made for conceptually penetrating this conception. Jean-Philippe Rameau succeeded in the 18th century in drafting the most lasting theoretical conception of harmony in which he reduced the richness of harmonic phenomena to few basic principles: chord inversion, superposition in thirds, fundamental bass. Rameau's principles are still relevant today which is witnessed by treatises on harmony such as [Wo72] (cf. [Wi85], p.18ff.). Since *tonal music* is grounded on the 7-tone-scale pattern, the most general *forms of tonal chords* (*harmonies*) are representable by selections of numbers out of 1, 2, 3, 4, 5, 6, and 7. For instance, the triad may be represented by 1,3,5, but also by 2,4,6, by 3,5,7, by 4,6,1, by 5,7,2, by 6,1,3, and by 7,2,4 (use Fig. 4 for visualizing (forms of) tonal chords). In the *cross table* of Fig. 1, the chord forms are indicated twice: as headings of the rows and as headings of the columns. The names of the chord forms heading the rows reflect Rameau's principle of superposition in thirds. A cross in the table means that the chord form heading the row of the cross can be represented by a subselection of the numbers heading the column of the cross; in other words: the crosses inform about the *subchord-superchord-relation*.

The cross table gives rise to the *concept hierarchy* presented in Fig. 2 which visualizes the *conceptual logic* of the given chord forms. For instance, a chord form can be extended to a second chord form if and only if there is an ascending path of line segments leading from the circle representing the first chord form to the circle representing the second chord form. An example for another type of logical conclusion is that every chord form extending the triad form and the fifth-seventh-free ninth chord form also extends the third-seventh-free ninth chord form; that follows in this case from the observation that there is a circle

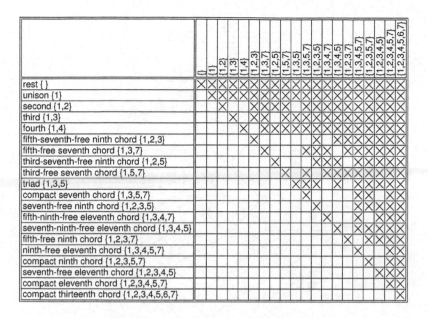

Fig. 1. Cross table describing the subchord-superchord-relation in a 7-tone-scale

which can be reached from each of the three circles, representing the three named chord forms, by two ascending line segments, respectively.

The meanings of the visualized concept hierarchy becomes more transparent if we consider the introduced three levels of semantics:

Mathematically, the labelled line diagram in Fig. 2 represents the *concept lattice* of the formal context represented by the cross table in Fig. 1. The 20 black circles represent the 20 formal object concepts and the 20 formal attribute concepts of the concept lattice, respectively. The *mathematical implications* between the 20 formal object concepts (resp. 20 formal attribute concepts) can be derived from a reduced *Duquenne-Guigues-Basis* (cf. [GW99], p.83) which consists of the 28 object implications (resp. 28 attribute implications) shown in Fig. 3 (dually).

Philosophically, the labelled line diagram in Fig. 2 represents the *concept hierarchy* derived from the context represented by the cross table in Fig. 1. A *concept* of the hierarchy is formed by an *extension* and an *intension* (cf. [Ka88]) where the extension consists of objects whose names are heading rows in the cross table and the intension comprises attributes whose names are heading columns in the cross table; the extension contains exactly those objects which have all attributes of the intension and the intension contains exactly all those attributes applying to all objects of the extension (an object has an attribute in the cross table exactly if there is a cross in the cell belonging to the row headed by the object name and the column headed by the attribute name). The *philosophic-logical semantics* of such a concept is constituted by the meanings of its objects, its attributes, and their relationships. In our example, the philosophic-logical meaning of an object

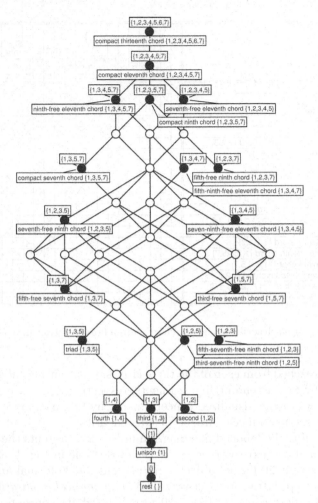

Fig. 2. Concept hierarchy derived from the cross table in Fig. 1

chord form is an abstraction of the object meaning of all chords having that object form, and the philosophic-logical meaning of an attribute chord form is an abstraction of the attribute meaning of all chords having that attribute form. Such philosophic-logical meanings may cause a dynamics which leads to changes and even expansions of the underlying context.

Musicologically, the labelled line diagram in Fig. 2 represents the *concept hierarchy* which offers the most general logical ordering of the chord forms of the 7-tone-scale. The ordering clarifies a rich combinatorics full of symmetries. First of all, the possibilities of *reducing* and *extending* chord forms (even simultaneously) can be easily recognized in the line diagram of the hierarchy. Furthermore, there are many possibilities to *symmetrically interchange chord forms* either order-preserving or order-reversing. For instance, a symmetric form-interchange of the second and the third forces the form-interchange of the triad and the

→{ }; {1,2}→{1}; {1,3}→{1}; {1,4}→{1};
{1,2,3}→{1,2},{1,3}; {1,3,7}→{1,2},{1,3},{1,4}; {1,2,5}→{1,2},{1,4};
{1,5,7}→{1,2},{1,3},{1,4}; {1,3,5}→{1,3},{1,4};
{1,2,3},{1,2,5}→{1,3,5}; {1,2,3},{1,3,5}→{1,2,5}; {1,2,5},{1,3,5}→{1,2,3};
{1,3,5,7}→{1,3,5},{1,3,7},{1,5,7}; {1,2,3},{1,2,5},{1,3,5},{1,3,7}→{1,2,3,5,};
{1,2,3,5}→{1,2,3},{1,2,5},{1,3,5},{1,3,7}; {1,3,4,7}→{1,2,5},{1,3,7},{1,5,7};
{1,2,3},{1,2,5},{1,3,5},{1,5,7}→{1,3,4,5}; {1,3,4,5}→{1,2,3},{1,2,5},{1,3,5},{1,5,7};
{1,2,3,7}→{1,2,3},{1,3,7},{1,5,7}; {1,3,4,5},{1,3,5,7}→{1,3,4,5,7};
{1,3,4,5,7}→{1,3,4,5},{1,3,5,7}; {1,2,3,7},{1,3,5,7}→{1,2,3,5,7};
{1,2,3,5,7}→{1,2,3,7},{1,3,5,7}; {1,2,3,7},{1,3,4,7}→{1,2,3,4,5,7};
{1,2,3,4,5,7}→{1,2,3,7},{1,3,4,7}; {1,2,3,4,5},{1,2,3,5,7},{1,3,4,5,7}→{1,2,3,4,5,7};
{1,2,3,4,5,7}→{1,2,3,4,5},{1,2,3,5,7},{1,3,4,5,7}; {1,2,3,4,5,6,7}→{1,2,3,4,5,7}.

Fig. 3. A reduced Duquenne-Guigue-Basis of implications between the formal object concepts of the concept lattice in Fig. 2; for instance, the implication {1,2,3},{1,2,5},{1,3,5},{1,5,7}→{1,3,4,5} means that each superconcept of the formal object concepts represented by {1,2,3},{1,2,5}, {1,3,5},{1,5,7} is also a superconcept of the formal object concept represented by {1,3,4,5}

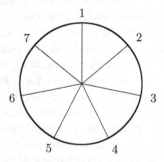

Fig. 4. The twenty chord forms can be recognized by subpolygons in the heptagon diagram where two subpolygons represent the same chord form exactly if one subpolygon can be rotated onto the other

third-seventh-free ninth chord. In total, there are twelve order-preserving symmetry transformations of the presented concept hierarchy which indicates the complex interrelationships of the considered chord forms. Additionally, there are also twelve order-reversing symmetry transformations. They can be constructed by composing the order-preserving transformations with the reflection which assigns to each chord form its complementary chord form; for instance, the triad {1,3,5} has as complementary form the chord form being represented by {2,4,6,7} (and therefore also by {1,3,5,7}). All those phenomena can be visualized in the regular heptagon shown in Fig. 4.

2   Music Structures as Semantic Structures

For musicologists it seems to be uncommon to view *musical structures as semantic structures*. A reason could be that meanings of music are usually accepted as semantical only if they refer to something outside music, like the birdcall imitations of a nightingal, a quail, and a cuckoo in Beethoven's 6th symphony, 2nd movement. Even in his comprehensive book on musical semantics [Ka86], the musicologist V. Karbusicky is very cautious in identifying semantic structures in music which becomes distinct when he uses the term *"semantic enclaves"*; for him, musical structures mostly have more a structuralist sense than a semantic meaning. The musicologist L. M. Zbikowski elaborates a stronger emphasis on meaning in music in his book *"Conceptualizing Music. Cognitive Structure, Theory, and Analysis"* [Zb02]; in summing up his chapter "Categorization, Compositional Strategy, and Musical Syntax", he writes: "... the construction of meaning in music can be achieved through the way composers choose to deploy the elements of musical syntax. Of course, compositional strategy is not the only source of meaning construction in music ... but it is one to which our knowledge of categorization can be profitably applied. As a meeting place for concerns of composers and listeners, categories of musical events are important to both compositional strategy and musical syntax, for they represent a means through which uniquely musical meaning can be created."

In this paper, the *syntax of musical structures* shall be understood as given by syntactical descriptions of music, such as the conventional musical notations, and the *semantics of musical structures* as given by the sounding music in human perception, such as the created music in the mind of composers and the perceived music in the mind of concert listeners. Both representations of musical structures may be activated at the same time, for instance, when a conductor reads musical scores for preparing the next rehearsal with his ochestra. The described understanding of musical structures relates closely to the understanding of cognitive structures in Piaget's structure-genetic approach to developmental psychology (cf. [Pi70]). But our semantic view makes it appropriate to understand musical structures as *semantic structures* and not as cognitive structures in the sense of Piaget's more epistomological view.

In our approach to a Semantology of Music, we assume that musical structures are grounded on *tone sytems* which have been established in some human music culture. An elementary example of such a tone system is a *diatonic scale* structured by its major and minor triads. A data table of such a scale and its corresponding concept hierarchy is shown in Fig. 5 in which the circles represent the concepts and the ascending line segments the subconcept-superconcept-relation. It might surprise that, in the diagram of the concept hierarchy representing the C-major diatonic scale, the vertical symmetry axis of the diagram leads through the circle of the object concept of the tone d which is the second tone of the diatonic scale. This indicates that even elementary musical structures as the diatonic scales contain unexpected relationships.

Our semantological approach with its threefold semantics may help to uncover *coherences* which are musically interesting. On the *music level*, for example, the tones and the major/minor triads of the diatonic scale can be viewed as semantic units which are structurally related as visualized in Fig. 6. The straight line segments of the diagram represent six fifths joined consecutively from the

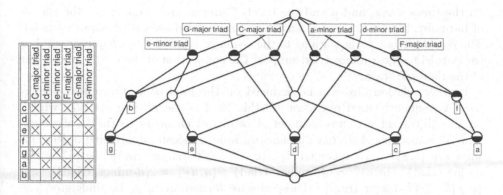

	C-major triad	d-minor triad	e-minor triad	F-major triad	G-major triad	a-minor triad
c	X			X		
d		X			X	
e	X		X			
f		X		X		
g	X		X		X	
a		X		X		X
b			X		X	

Fig. 5. Data table and concept hierachy of a music structure formed by the C-major diatonic scale with its major and minor triads

fifth $f - c$ to the fifth $e - b$. The arc between the two endpoints of each line segment carries a third point representing a tone which forms a major or minor triad together with the fifth of the corresponding line segment. The structure presented in Fig. 6 may help to understand and to create music based on diatonic scales.

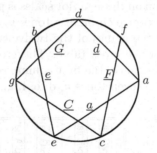

Fig. 6. The structure of the major/minor triads of the C-major diatonic scale

A concept hierarchy as presented in Fig. 5 stimulates to use also the level of the *philosophic-logical semantics* for analysing musical structures. For example, in the diagram of that hierarchy, the meaning of the concepts represented by the non-shaded circles might become interesting. Of course, the extremal concept represented by the circle on the top (bottom) has the meaning that its extension (intension) consists of all objects (attributes) of the context, while its intension (extension) is empty. But, what is the meaning of the three concepts

$(\{e, g\}, \{C\text{-major triad}, e\text{-minor triad}\}),$
$(\{c, e\}, \{C\text{-major triad}, a\text{-minor triad}\}),$
$(\{a, c\}, \{F\text{-major triad}, a\text{-minor triad}\}),$

represented by the three non-shaded circles in the middle of the diagram? An answer is that they form the center of the C-major diatonic scale which is based on the tones a, c, e, and g and the triads C-major and a-minor (in the theory of harmony, the C-major triad and the a-minor triad are called parallel triads). The e-minor triad as the parallel to the G-major triad and the F-major triad as the paralel to the d-minor triad support the connection of the center to the rest of the diatonic scale.

Further relationships can be deduced on the level of *mathematical semantics*. A frequently used operation on this level is the *closure operator* which, on the object set, extends each set A of object elements to the smallest extent A'' containing A. In our diatonic example, musically interesting object closures are $\{b, e\}'' = \{e$-minor triad$\}'$, $\{g, d\}'' = \{G$-major triad$\}'$, $\{g, c\}'' = \{C$-major triad$\}'$, $\{a, e\}'' = \{a$-minor triad$\}'$, $\{a, d\}'' = \{d$-minor triad$\}'$, and $\{c, f\}'' = \{F$-major triad$\}'$. These closure formations may be understood as object implications. The *Duquenne-Guiges-Basis* of all object implications of the C-major diatonic scale reads as follows:

$$b \to g; \ f \to a; \ d, g \to b; \ d, a \to f; \ c, g \to e; \ e, a \to c;$$
$$g, a \to \text{all}; \ d, e \to \text{all}; \ c, d \to \text{all}; \ c, e, g, b \to \text{all}; \ c, e, f, a \to \text{all}.$$

In tonal music it is quite common to change from one diatonic scale to another; such a change is called a *modulation*. For a systematic understanding of such modulations, it is necessary to know the relationships between the different scales. Here we restrict our consideration on the twelve major scales in equal temperament. The most elementary relation on those major scales is presented in the cross table of Fig. 7. The names of the twelve chromatic tones in an octave are heading the rows of the cross table and the names of the twelve corresponding major scales are heading the columns of the cross table. The crosses indicate which tones belong to which scale. For instance, the first column of the table shows that the C-major scale consists of the seven tones c, d, e, f, g, a, and b. If one compares

	C	C sharp/D flat	D	D sharp/E flat	E	F	F sharp/G flat	G	G sharp/A flat	A	A sharp/B flat	B
c	X	X		X		X		X	X		X	
c sharp/d flat		X	X		X		X		X	X		X
d	X		X	X		X		X		X	X	
d sharp/e flat		X		X	X		X		X		X	X
e	X		X		X	X		X		X		X
f	X	X		X		X	X		X		X	
f sharp/g flat		X	X		X		X	X		X		X
g	X		X	X		X		X	X		X	
g sharp/a flat		X		X	X		X		X	X		X
a	X		X		X	X		X		X	X	
a sharp/b flat		X		X		X	X		X		X	X
b	X		X		X		X	X		X		X

Fig. 7. Cross table representing the twelve major diatonic scales in the well-tempered chromatic scale

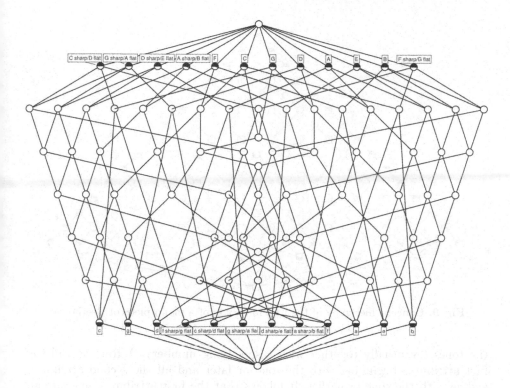

Fig. 8. Concept hierarchy derived from the cross table in Fig. 7

the first column with the third, one learns that the C-major scale coincides with the D-major scale in the five tones d, e, g, a, and b, but differ in the other two. In the *circle of fifths*: C-G-D-A-E-B-(G flat)-(D flat)-(A flat)-(E flat)-(B flat)-F-C, neighbouring scales differ in exactly one tone so that modulations between them are easy to perform. This becomes visible by the grid structures in the line diagram of Fig. 8 which represents the concept hierarchy of the cross table in Fig. 7. Besides the circular grid and the top and bottom circle, suprisingly, there are six further circles which represent concepts having only two objects and two attributes, both in a distance of a *tritone*. This indicates that, for each scale, there exists a direct modulation to its opposite scale and vice versa.

Up to now, the *time flow of music* have not really be considered, although time is essential for the existence of music. The question, of course, is how time can be formalized so that it fits with the already considered structures. Again, an example shall help to imagine an approach for expanding music structures such as tone systems, chords, harmonies, scales, modulation structures, etc. by integrating appropriate time structures. The example in Fig. 9 is taken out of the introduction of *Beethoven's Eroica-Variations* for piano. The musical time flow is modelled by an *interordinal scale* (cf. [GW99], p.42), which has the quarters of the bars twice as its attributes: first to mark the beginning of a time section and second to mark the ending of a time section. The objects of the scale are

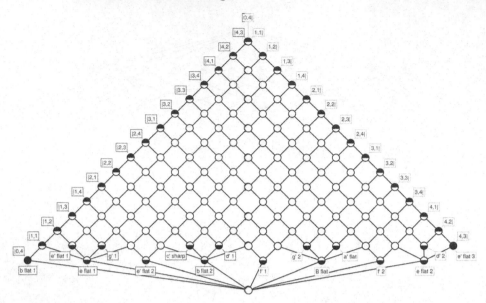

Fig. 9. Concept hierarchy of the first four bars of a piano-piece of Beethoven

the tones (eventually together with a specifying number). A tone has all the first attributes beginning with the tone or later and all the second attributes ending with the tone or earlier. It follows that the first attributes generate an ascending chain of concepts and the second attributes generate a descending chain of concepts.

In the line diagram of Fig. 9, the beginning of a tone can be identified by determining the most left circle with an attribute name which can be reached from the circle of the tone name by a path of ascending line segments; that attribute name indicates the beginning. The ending of a tone can be identified by determining the most right circle with an attribute name which can be reached from the circle of the tone name by a path of ascending line segments; that attribute name indicates the ending. In this way, it can be easily seen that, for instance, the tone "e flat 1" starts with the first quarter of bar 1 and ends with the fourth quarter of bar 1.

The described modelling of musical time flow allows one to represent also an analysis of large music pieces. We demostrate this only by one example: The line diagram in Fig. 10 presents the first step of an analysis of Brahhms' 4th symphony, 1st movement (cf. [Kl81]). On the first level, the movement is arranged in the parts "exposition" [1,1-136,3], "development" [136,4-246,2], "recapitulation" [246,3-393,4], and "movement-coda" [393,4-404,4]. On the second level, the exposition is partitioned into the "1st group" [1,1-53,1], the "2nd group" [53,1-94,4], the "3rd group" [95,1-106,4], and the "epilog" [107,1-136,3]. Subparts of the three groups in the exposition are the "1st main theme" [1,1-18,1], the "2nd main theme a)" [53,1-56,3], the "2nd main theme b)" [56,4-64,4], the "middle

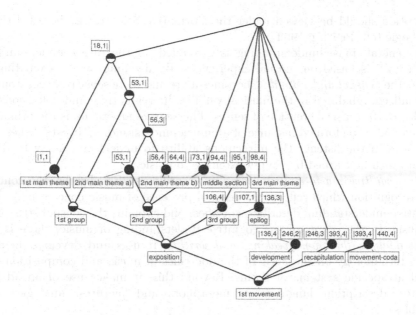

Fig. 10. Concept hierarchy of the first movement of Brahms' Symphony No. 4

section" [73,1-94,4], and the "3rd main theme" [95,1-98,4]. All these parts are densly combined to an impressive work of music.

The examples and the theoretical considerations above may indicate that a Semantology of Music can be developed to a great extend. The few examples of semantic structures in music given in this introductory paper can be multifariously extended. For this, further methods of Conceptual Knowledge Processing can be used, in particular the TOSCANA methodology (cf. [KSVW94], [VW95], [BH05]). For integrating mathematical semantic structures in the semantics of music, developments on mathematical music theory may be adapted as, for instance, ideas and results from [Wi76], [Wi80], [Wi82], [GHW85], [Wi85], [NW90], [Ma90], [Wi00], [Ma02], and [WW06].

3   Understanding Music

If we want to clarify how a Semantology of Music may support the understanding of music, it should be made clear what is meant by *"understanding music"*. For this, the musicologist H. H. Eggebrecht has given a convincing elaboration of the notion *"music-understanding"* in [Eg95]. For Eggebrecht, "music-understanding" means that music affects the listener so that music is accepted by him, gains admission in his feelings and thought, and delights and moves the listener. Eggebrecht distinguishes between the *direct* and the *indirect music-understanding* where the first is of sensual and the second of cognitive nature. On the philosophical level in Peirce's classification of sciences, the sensual perception of music should be located under Phenomenology, while the cognitive perception should be viewed under the Normative Sciences: Esthetics, Ethics, and Logic (cf. [Pe98], p.196ff).

In general, music-understanding is grounded on the direct understanding which itself is based on sensual experiences. By listening and co-executing of music, the *sensual understanding* becomes more and more sensitive, accustomed, internalized, inhabited in the musical validity, its regulatives and definitions, in its play of the musical creation of senses. The *sensual experience* is the primarily formed and to be formed instance of all music-understanding. It establishes and cultivates in the listener the miraculous ability to understand many kinds of music which he learned by simply listening to music.

The *cognitive music-understanding* is, according to Eggebrecht, the conceptual recognition which reflects the sensual presence of music conceptually. The cognitive understanding describes, specifies, and explains the sensual structures of music. In the center of the cognitive understanding of music, there is the *music-analytic thinking, speaking, and writing*. It uses and develops an adequate terminology on the basis of the doctrines of music and composition oriented at specific systems of norms. Beyond this, it makes use of an ad hoc selected description language in metaphors and pictures, analogies and parables.

Semantology becomes supportive on the level of cognitive music-understanding. There the semantic structures of music are of conceptual nature and can hence be analysed and assessed by the *threefold semantics* discussed in Section 1. On the *musical level*, the semantic structures of music can be explained and evaluated by methods based on the cognitive understanding of music. On the *philosophic-logical level*, the semantic structures may be disclosed by using, in general, the main functions of human thought: concept, judgment, and conclusion; in particular, the distinction between the extension and intension of concepts is meaningful. Finally, on the *mathematical level*, the semantic structures of music should be abstracted to mathematical structures which appropriately cover the considered forms of music.

References

[BH05]   Becker, P., Correia, J.H.: The ToscanaJ Suite for implementing conceptual information systems. In: Ganter, B., Stumme, G., Wille, R. (eds.) Formal Concept Analysis. LNCS (LNAI), vol. 3626, pp. 324–348. Springer, Heidelberg (2005)

[Eg95]   Eggebrecht, H.H.: Musikverstehen. Piper, München (1995)

[EW07]   Eklund, P., Wille, R.: Semantology as basis for Conceptual Knowledge Processing. In: Kuznetsov, S., Schmidt, St. (eds.) ICFCA 2007. LNCS (LNAI), vol. 4390, Springer, Heidelberg (2007)

[GHW85]   Ganter, B., Henkel, H., Wille, R.: MUTABOR - Ein rechnergesteuertes Musikinstrument zur Untersuchung von Stimmungen. Acustica 57, 284–289 (1985)

[GSW05]   Ganter, B., Stumme, G., Wille, R. (eds.): Formal Concept Analysis: foundations and applications. In: Ganter, B., Stumme, G., Wille, R. (eds.) Formal Concept Analysis. LNCS (LNAI), vol. 3626, Springer, Heidelberg (2005)

[GW99]   Ganter, B., Wille, R.: Formal Concept Analysis: mathematical foundations. Springer, Heidelberg (1999)

[GW06]   Gehring, P., Wille, R.: Semantology: basic methods for knowledge representations. In: Schärfe, H., Hitzler, P., Øhrstrøm, P. (eds.) ICCS 2006. LNCS (LNAI), vol. 4068, pp. 215–228. Springer, Heidelberg (2006)

[Ka88]   Kant, I.: Logic. Dover, Mineola (1988)

[Ka86]   Karbusicky, V.: Grundriß der musikalischen Semantik. Wissenschaftliche Buchgesellschaft, Darmstadt (1986)

[Kl81]   Kloiber, R.: Handbuch der klassischen und romantischen Symphonie. Breitkopf & Härtel, Wiesbaden (1981)

[KSVW94] Kollewe, W., Skorsky, M., Vogt, F., Wille, R.: TOSCANA - ein Werkzeug zur begrifflichen Analyse und Erkundung von Daten. In: Wille, R., Zickwolff, M. (Hrsg.): Begriffliche Wissensverarbeitung – Grundfragen und Aufgaben. B.I.-Wissenschaftsverlag, Mannheim, pp. 267–288 (1994)

[Ma90]   Mazzola, G.: Geometrie der Töne. Elemente der Mathematischen Musiktheorie. Birkhäuser, Basel (1990)

[Ma02]   Mazzola, G.: The topos of music. Geometric logic of concepts, theory, and performance. Birkhäuser, Basel (2002)

[NW90]   Neumaier, W., Wille, R.: Extensionale Standardsprache der Musiktheorie: eine Schnittstelle zwischen Musik und Informatik. In: Hesse, H.-P. (Hrsg.): Mikrotöne III. (edn.) Helbing, Innsbruck, pp. 149–167 (1990)

[Pe98]   Peirce, Ch., S.: The three normative sciences. In: The Essential Peirce. Selected philosophical writings, edited by the Peirce Edition Project. vol. 2, Indiana University Press, Bloomington (1998)

[Pi70]   Piaget, J.: Genetic Epistomology. Columbia University Press, New York (1970)

[SW00]   Stumme, G., Wille, R. (Hrsg.): Begriffliche Wissensverarbeitung: Methoden und Anwendungen. Springer, Heidelberg (2000)

[VW95]   Vogt, F., Wille, R.: TOSCANA – A graphical tool for analyzing and exploring data. In: Tamassia, R., Tollis, I.G. (eds.) GD 1994. LNCS, vol. 894, pp. 226–233. Springer, Heidelberg (1995)

[Wi76]   Wille, R.: Mathematik und Musiktheorie. In: Schnitzler, G. (Hrsg.): Musik und Zahl. Verlag für systematische Musikwissenschaft. Bonn-Bad Godesberg, pp. 233–264 (1976)

[Wi80]   Wille, R.: Mathematische Sprache in der Musiktheorie. In: Fuchssteiner, B., Kulisch, U., Laugwitz, D., Liedl, R. (Hrsg.): Jahrbuch Überblicke Mathematik, pp. 167–184. B.I.-Wissenschaftsverlag, Mannheim (1980)

[Wi82]   Wille, R.: Symmetrien in der Musik - für ein Zusammenspiel von Musik und Mathematik. Neue Zeitschrift für Musik 143, 12–19 (1982)

[Wi85]   Wille, R.: Musiktheorie und Mathematik. In: Götze, H., Wille, R. (Hrsg.): Musik und Mathematik - Salzburger Musikgespräch 1984 unter Vorsitz von Herbert von Karajan, pp. 4–31. Springer, Heidelberg (1985)

[Wi00]   Wille, R.: Eulers Speculum Musicum und das Instrument MUTABOR. DMV-Mitteilungen, Heft 4/2000, pp. 9–12

[Wi06]   Wille, R.: Methods of Conceptual Knowledge Processing. In: Missaoui, R., Schmid, J. (eds.) Formal Concept Analysis. ICFCA 2006, vol. LNAI, pp. 1–29. Springer, Heidelberg (2006)

[WW06]   Wille, R., Wille-Henning, R.: Beurteilung von Musikstücken durch Adjektive. In: Proost, K., Richter, E. (Hrsg.): Von Intenionalität zur Bedeutung konventionalisierter Zeichen. Festschrift für Gisela Harras zum 65. Geburtstag. Narr, Tübingen, pp. 453–475 (2006)

[Wo72]   Wolpert, F.A.: Neue Harmonik. Heinrichshofen, Wilhelmshaven (1972)

[Zb02]   Zbikowski, L.M.: Conceptualizing music. Cognitive structure, theory, and analysis. Oxford University Press, New York (2002)

Analysis of the Publication Sharing Behaviour in BibSonomy

Robert Jäschke[1,2], Andreas Hotho[1], Christoph Schmitz[1], and Gerd Stumme[1,2]

[1] Knowledge & Data Engineering Group,
Department of Mathematics and Computer Science,
University of Kassel, Wilhelmshher Allee 73, 34121 Kassel, Germany
http://www.kde.cs.uni-kassel.de
[2] Research Center L3S, Appelstr. 9a, 30167 Hannover, Germany
http://www.l3s.de

Abstract. BibSonomy is a web-based social resource sharing system which allows users to organise and share bookmarks and publications in a collaborative manner. In this paper we present the system, followed by a description of the insights in the structure of its bibliographic data that we gained by applying techniques we developed in the area of Formal Concept Analysis.

1 Introduction

Social resource sharing systems provide new means for organising and sharing information on the web. These tools, such as Flickr[1] or del.icio.us,[2] have acquired large numbers of users within a very short time after their introduction. They all use the same kind of lightweight knowledge representation, called *folksonomy*. The word "folksonomy" is a blend of the words "taxonomy" and "folk", and stands for conceptual structures created by the people. Folksonomies provide an intuitive structure for navigating the data by following the link structure.

Until now, however, there is no specific support for a systematic analysis of the content of a folksonomy. One way towards this aim is the application of *Triadic Concept Analysis* [18], which is an extension of Formal Concept Analysis [8] that fits to the structure of folksonomies. Formal Concept Analysis can be considered as a conceptual hierarchical co-clustering technique.

After some remarks on folksonomies in Section 2, we will present in Section 3 our own system for sharing bookmarks and bibliographic references, called BibSonomy. We will then recall in Section 4 the basics of Formal Concept Analysis and of its triadic extension, together with our clustering algorithm TRIAS [15]. Section 5 contains the main contribution of this paper, an analysis of the bibliographic data of BibSonomy.

2 Social Resource Sharing and Folksonomies

Social resource sharing systems are web-based systems used to manage resources on the web in a collaborative way. Users can describe the resources with arbitrary words,

---

[1] http://www.flickr.com
[2] http://del.icio.us

U. Priss, S. Polovina, and R. Hill (Eds.): ICCS 2007, LNAI 4604, pp. 283–295, 2007.

so-called *tags*. The systems can be distinguished according to what kind of resources are supported. Flickr, for instance, allows the sharing of photos, del.icio.us the sharing of bookmarks, CiteULike[3] and Connotea[4] the sharing of bibliographic references, and 43Things[5] even the sharing of personal goals and resolutions. Our own system, *BibSonomy*,[6] can be used for sharing bookmarks and BibTEX entries simultaneously. In their core, these systems are all very similar. Once a user is logged in, he can add a resource to the system and assign arbitrary tags to it.

The collection of all users' tag assignments is called a *folksonomy*. A typical user interface allows for exploration of the folksonomy in all dimensions: for a given user one can see all resources he has uploaded, together with the tags he has assigned to them (see Figure 1 on the facing page); when clicking on a resource one sees which other users have uploaded this resource and how they tagged it; and when clicking on a tag one sees who assigned it to which resources.

Current systems provide additional functionality. For instance, one can copy a resource from another user, and label it with one's own tags. Overall, these systems provide a very intuitive navigation through the data.

2.1 A Formal Model for Folksonomies

A folksonomy describes the users, resources, and tags, and the user-based assignment of tags to resources. The following definition is underlying our BibSonomy system.

A *folksonomy* is a tuple $\mathbb{F} := (U, T, R, Y, \prec)$ where

- U, T, and R are finite sets, whose elements are called *users*, *tags* and *resources*, resp.,
- Y is a ternary relation between them, i.e., $Y \subseteq U \times T \times R$, whose elements are called tag assignments (*tas* for short), and
- $\prec$ is a user-specific subtag/supertag-relation, i.e., $\prec \subseteq U \times T \times T$, called *is-a relation*.

Users are described by a user ID, and tags may be arbitrary strings. What is considered as a resource depends on the type of system. In BibSonomy they are either URLs or publication entries. In this paper, we will disregard the user-specific tag hierarchy.

For convenience we also define the set P of all *posts* as $P := \{(u, S, r) \mid u \in U, r \in R, S = \text{tags}(u, r)\}$ where, for all $u \in U$ and $r \in R$, $\text{tags}(u, r) := \{t \in T \mid (u, t, r) \in Y\}$ is the set of all tags user u has assigned to resource r. Thus, a *post* consists of a user, a resource and all tags that this user has assigned to that resource.

2.2 Related Work

General overviews on the rather young area of folksonomy systems and their strengths and weaknesses are given in [11,19,20]. Recently, work on more specialized topics such as structure mining on folksonomies – e.g. to visualize trends [5] and patterns [25] in

---

[3] http://www.citeulike.org
[4] http://www.connotea.org
[5] http://www.43things.com
[6] http://www.bibsonomy.org

users' tagging behavior – as well as ranking of folksonomy contents [14], analyzing the semiotic dynamics of the tagging vocabulary [3], or the dynamics and semantics [10] have been presented.

Another upcoming field of research is the learning of more formal, usually hierarchical conceptual structures (i.e. taxonomies, ontologies) from folksonomies, which has been approached using different mining techniques [22,26,12,16].

3   Sharing Bookmarks and Publications with BibSonomy

This section briefly describes the BibSonomy system developed by our group[7]. After an introduction to the user interface and architecture of BibSonomy, we give an overview about some of its advanced features. BibSonomy allows to share bookmarks (i.e., URLs) as well as publication references. The data model of the publication part is based on BibTeX [23], a popular literature management system for LaTeX [17].

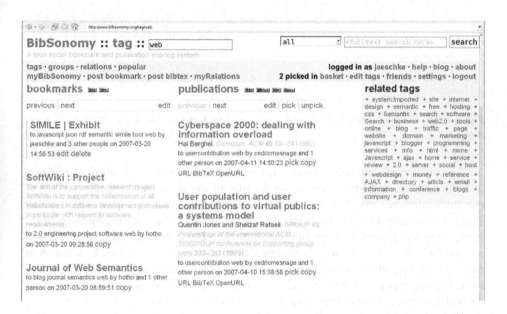

Fig. 1. BibSonomy displays bookmarks and BibTeX based bibliographic references simultaneously

3.1   User Interface

A typical list of posts is depicted in Figure 1 which shows bookmark and publication posts containing the tag *web*. The page is divided into four parts: the header (showing information such as the current page and path, navigation links and search boxes), two lists of posts – one for bookmarks and one for publications – each sorted by date in

---

[7] This section is a summary of [13].

286   R. Jäschke et al.

descending order, and a list of tags related to the posts. This scheme holds for all pages showing posts and allows for navigation in all dimensions of the folksonomy.

A detailed view of one bookmark post can be seen in Figure 2. The first line shows in bold the title of the bookmark which has the URL of the bookmark as underlying hyperlink. The second line shows an optional description the user can assign to every post. The last two lines belong together and show detailed information: first, all the tags the user has assigned to this post (*web, service, tutorial, guidelines, api* and *rest*), second, the user name of that user (*hotho*) followed by a note, how many users tagged that specific resource. These parts have underlying hyperlinks, leading to the corresponding tag pages of the user (`/user/hotho/web` [8], `/user/hotho/service, ...`), the users page (`/user/hotho`) and a page showing all four posts (i. e., the one of user *hotho* and those of the 3 other people) of this resource (`/url/r`, where r is a hashed representation of the resource). The last part shows the posting date and time followed by links for actions the user can do with this post – depending on if this is his own post (*edit, delete*) or another user's post (*copy*).

<table>
<tr>
<td>

REST web services
Good intro to the REST "architecture"
to web service tutorial guidelines api rest by hotho and 3 other
people on 2006-04-04 16:11:47 copy

</td>
<td>

Semantic Network Analysis of Ontologies
Bettina Hoser and Andreas Hotho and Robert Jäschke and
Christoph Schmitz and Gerd Stumme. *Proceedings of the 3rd
European Semantic Web Conference \emph{{accepted for
publication}} (2006)*
to web 2006 social ontology myown semantic analysis network
sna by hotho and 1 other person on 2006-04-06 21:32:23 pick
copy URL BibTeX

</td>
</tr>
</table>

Fig. 2. detail showing a single bookmark post

Fig. 3. detail showing a single publication post

The structure of a publication post displayed in BibSonomy is very similar, as shown in Figure 3. The first line shows again the title of the post, which equals the title of the publication in BibTeX. It has an underlying link leading to a page which shows detailed information on that post. This line is followed by the authors or editors of the publication, the journal or book title and the year. The following lines show the tags assigned to this post by the user, whose user name comes next, followed by a note how many people tagged this publication. As described for bookmark posts, these parts link to the respective pages. After date and time of the posting follow the actions the user can do, which in this case include picking the entry for later download, copying it, accessing the URL of the entry or viewing the BibTeX source code.

3.2   Additional Features

This section briefly describes some of the features of BibSonomy which distinguish it from similiar systems and ease the everyday work with it.

Tagging gained so much popularity in the past three years because it is simple and no specific skills are needed for it. Nevertheless the longer people use systems like BibSonomy, the more often they ask for options to structure their tags. A user specific *binary relation* $\prec$ between tags as described in our model of a Folksonomy (see Section 2.1) is

---

[8] All paths given in brackets are relative to http://www.bibsonomy.org

an easy way to arrange tags. Therefore we included this possibility in BibSonomy and extended it further to use it for conceptual navigation. For instance, it is possible, given a tag, to show all posts with one of the subtags of the given tag.

In particular for literature references there is the problem of *detecting duplicate entries*, because there are big variations in how users enter fields such as journal name or author. On the one hand it is desirable to allow a user to have several entries which differ only slightly. On the other hand one might want to find other users' entries which refer to the same paper or book even if they are not completely identical. Hence it is necessary to map these entries together to allow such browsing functionality. To fulfill both goals we implemented two hashes to compare publication entries at different levels of granularity.

To encourage users to transition from other systems, we implemented an *import* functionality. For del.icio.us, this functionality also takes into account the del.icio.us bundles. Furthermore it is possible to import bookmark files of the Firefox[9] web browser, where the typical folder hierarchy of the bookmarks can be added to the users $\prec$ relation.

Import of existing BibTeX files is also simple: after uploading the file, the user can tag the entries or automatically assign the tag *imported*. If a BibTeX entry contains a field *keywords* or *tags*, its contents are attached as tags to the resource and added to the system. BibTeX-Fields unknown to BibSonomy are saved in the *misc* field and will not get lost. Besides this an automatic import from miscellaneous digital library systems like ACM[10] is implemented, too.

To support also the import of unstructured literature entries often found at web pages in the form of human readable publication lists we need to mediate between these entries and BibTeX. MALLET [21] is a learnable information extraction system which allows after a training phase to extract references from publications lists automatically. It was integrated into BibSonomy to further support the use of the system.

Exporting BibTeX is accomplished by preceding the path of a URL with the string /bib – this returns all publications shown on the respective page in BibTeX format. For example the page http://www.bibsonomy.org/bib/search/text+clustering returns a BibTeX file containing all literature references which contain the words "text" and "clustering" in their fulltext.

This holds also for other export formats like typical HTML styled publication lists, XML, RSS feeds, RDF according to the SWRC ontology, BibTeX and EndNote. All those export options[11] simplify the interaction of BibSonomy with other systems. The same idea is behind the integration of links to OpenURL[12] resolvers which allows for a close interaction with local libraries.

The last feature we would like to mention briefly is group management. In many situations it is desirable to share resources only among certain people. If the resources can be public, then one could agree to tag them with a special tag and use that tag to find the shared resources. The disadvantage is, that this could be undermined by other users

---

[9] http://www.mozilla.com/firefox/

[10] http://portal.acm.org

[11] For an overview have a look at http://www.bibsonomy.org/export/.

[12] http://www.exlibrisgroup.com/sfx_openurl.htm

(or spammers) by using the same tag. To solve this problem and also to allow resources to be visible only for certain users, we introduced *groups* in BibSonomy which gives users more options to decide with whom they share their resources.

4   Formal Concept Analysis

While a folksonomy with its many crosslinks in all different dimensions provides a good framework for serendipitous discovery by navigating along apparently interesting links, it is more difficult to obtain a *systematic* insight into its contents. We might for instance want to know who are the key users of the system, which interests they share, and how they differ.

A canonical definition to this end is to let a community of interest be the set of users who assign the same set of tags to the same set of resources. This understanding is reflected by the theory of Formal Concept Analysis and its triadic extension, which fits perfectly to the notion of a folksonomy.

4.1   Key Notions of Formal Concept Analysis

We briefly recall the key notions of Formal Concept Analysis [8], as the triadic FCA is built upon them. FCA starts with a *(formal) context* $\mathbb{K} := (G, M, I)$ which consists of a set G of objects [German: Gegenstnde], a set M of attributes [Merkmale], and a binary relation $I \subseteq G \times M$. $(g, m) \in I$ is read as "object g has attribute m". This data structure equals the set of transactions used for association rule mining, if we consider M as the set of items and G as the set of transactions.

We define (following [29]), for $A \subseteq G$, $A' := \{m \in M \mid \forall g \in A \colon (g, m) \in I\}$; and dually, for $B \subseteq M$, $B' := \{g \in G \mid \forall m \in B \colon (g, m) \in I\}$. Now, a *formal concept* is a pair (A, B) with $A \subseteq G$, $B \subseteq M$, $A' = B$ and $B' = A$. A is called *extent* and B is called *intent* of the concept. This definition is equivalent to saying that $A \times B \subseteq I$ such that neither A nor B be can be enlarged without violating this condition.

The set $\mathfrak{B}(\mathbb{K})$ of all concepts of a formal context $\mathbb{K}$ together with the partial order $(A_1, B_1) \le (A_2, B_2) \colon\Leftrightarrow A_1 \subseteq A_2$ (which is equivalent to $B_1 \supseteq B_2$) is a complete lattice, called the *concept lattice* of $\mathbb{K}$ [29].

The concept lattice provides the same information as the context – it is thus a lossless representation – but additionally, it structures the data in a concept hierarchy, which can be used for a systematic analysis. If we are interested in the most general concepts only, we can cut off the hierarchy at a given threshold: For $\tau \in [0, 1]$, let $\mathfrak{B}_\tau(\mathbb{K}) := \{(A, B \in \mathfrak{B}(\mathbb{K}) \mid \frac{\text{card}(A)}{\text{card}(G)} \ge \tau\}$. This construction is known as an *iceberg concept lattice* and has a close relationship to association rule mining [27].

4.2   Triadic Concept Analysis

Since folksonomies have three rather than two dimensions (i. e., users, tags, and resources), one cannot apply FCA directly. Fortunately there exists an extension of the theory to the triadic case.

Inspired by the pragmatic philosophy of Charles S. Peirce with its three universal categories [24], Wille and Lehmann extended Formal Concept Analysis in 1995 with a third category [18]. They defined a *triadic formal context* as a quadruple $\mathbb{K} :=$ (G, M, B, Y) where G, M, and B are sets, and Y is a ternary relation between G, M, and B, i. e., $Y \subseteq G \times M \times B$. The elements of G, M, and B are called *(formal) objects*, *attributes*, and *conditions*, resp, and $(g, m, b) \in Y$ is read "object g has attribute m under condition b". A folksonomy as defined above without considering the user-specific tag hierarchy $\prec$ (i. e., considered as quadruple $\mathbb{F} := (U, T, R, Y)$) matches exactly this definition of a triadic formal context.

A *triadic concept* of $\mathbb{K}$ is a triple (A_1, A_2, A_3) with $A_1 \subseteq G$, $A_2 \subseteq M$, and $A_3 \subseteq B$ with $A_1 \times A_2 \times A_3 \subseteq Y$ such that none of its three components can be enlarged without violating this condition. This is the natural extension of the definition of a formal concept to the triadic case. Alternatively the definition can be described with $\cdot'$ operators similar to the dyadic case, but as there are now three dimensions involved, the notation (which we omit here, cf. [18]) becomes more complex.

With the three dimensions one obtains three quasi-orders $\lesssim_1$, $\lesssim_2$, and $\lesssim_3$ on the set of all tri-concepts: $(A_1, A_2, A_3) \lesssim_i (B_1, B_2, B_3)$ iff $A_i \subseteq B_i$, for $i = 1, 2, 3$. The set of all tri-concepts together with these three quasi-orders is called the *concept tri-lattice* of the triadic context $\mathbb{K}$.

For a first approach to the data, the full concept tri-lattice provides usually far too many details. As in the dyadic case, we can restrict the tri-lattice to the frequent tri-concepts. For given thresholds u-minsup, t-minsup, r-minsup $\in [0, 1]$, we call a tri-concept (A, B, C) *frequent* if $|A| \geq \tau_u$, $|B| \geq \tau_t$, and $|C| \geq \tau_r$, where $\tau_u :=$ $|U| \cdot$ u-minsupp, $\tau_t := |T| \cdot$ t-minsupp, and $\tau_r := |R| \cdot$ r-minsupp. By starting with high thresholds, the concept tri-lattice reveals first a high-level view of the data, and allows step-by-step access to more details by subsequently decreasing the thresholds. See Section 5 for an example.

Our TRIAS algorithm computes, for a given triadic context and for given frequency thresholds, all frequent tri-concepts of a triadic context [15]. It splits the computation of (frequent) tri-concepts into two nested iterations of computing (frequent) concepts of dyadic contexts which are derived from the triadic one. These two iterations are each realized by a modification of the seminal NEXTCLOSURE algorithm[9,8]. The modification additionally considers the frequency constraints.

4.3 Related Work

The amount of publications on Formal Concept Analysis is abundant. A good starting point for the lecture are the textbooks [8,2,7], the proceedings of the Intl. Conference on Formal Concept Analysis[13] and the Intl. Conference on Conceptual Structures[14] series, as well as http://www.bibsonomy.org/tag/fca.

Following the initial paper [18] by Lehmann and Wille, several researchers started to analyse the mathematical properties of tri-lattices, e. g., [1,6,30]. [18] and [4] present several ways to project a triadic context to a dyadic one. [28] presents a model for navigating a triadic context by visualising concept lattices of such projections. In [25], we discussed how to compute association rules from a triadic context. Triadic implications have been discussed in [6].

---

[13] http://www.informatik.uni-trier.de/~ley/db/conf/icfca/
[14] http://www.informatik.uni-trier.de/~ley/db/conf/iccs/

5   Conceptual Analysis of the BibSonomy Publication Data

For our analysis we focused on the publication management part of BibSonomy. We first made a snapshot of BibSonomy's publication entries, including all publication posts made until November 23, 2006 at 13:30 CET. From the snapshot we excluded the publication posts from the DBLP computer science bibliography[15] since they're automatically inserted and all owned by one user and all tagged with the same tag (*dblp*). Therefore they do not provide meaningful information for the analysis. Similarly we excluded all tag assignments with the tag *imported* and hence all publication posts which exclusively have this tag. The reason for this is that this tag is automatically assigned by the system to all posts which were added by one of the import functions. The resulting snapshot contains $|Y| = 44,944$ tag assignments built by $|U| = 262$ users, containing $|R| = 11,101$ publication references tagged with $|T| = 5,954$ distinct tags.

The TRIAS algorithm needed on a 2 GHz AMD Opteron machine 75 minutes to compute all 13,992 tri-concepts of this dataset. Among those there are 12,659 tri-concepts which contain only one user, representing the individual conceptualisations of the users. (These could be used to present personal concept hierarchies by means of dyadic Hasse diagrams.) The remaining 1,333 tri-concepts thus all contain at least two users and therefore represent shared concepts. To further analyse these concepts we next take a closer look on the tri-concepts which contain at least three users, two tags and two publication entries (i. e., with minimal support values $\tau_u = 3$, $\tau_t = 2$, $\tau_r = 2$). Each of these 21 tri-concepts expresses the fact that all of its users tagged all its publications with all its tags.

The diagram in Figure 4 on the next page shows the triadic concept lattice of all 21 tri-concepts described above. The titles of the publications in the figure are substituted by numbers for space reasons. The corresponding titles can be found in Table 1, the full bibliographic information was tagged in BibSonomy (after the evaluation) with the tag *iccs_example*.[16]

The diagram follows the conventions introduced in [18]. The 21 nodes in the center of the triangle represent the 21 frequent tri-concepts. The sets of users, tags, and resources composing a tri-concept can be read off the three sides of the triangle. There, three Hasse diagrams display the three quasi-orders $\lesssim_1$, $\lesssim_2$, and $\lesssim_3$ as introduced in Section 4.2. The arrows guide the reader to the larger elements of each quasi-order. Each node in a hierarchy represents the set containing the labels attached to it plus all labels below. The empty nodes are not part of the quasi-order. They are just used to be able to place each label once only.

For instance, the lower-most node in the triangle represents the tri-concept consisting of the set {*jaeschke, schmitz, stumme*} of users, the set {*fca, triadic*} of tags, and the set {*1, 37*} of resources. Similarly, the node in the user hierarchy labelled *brotkasting* represents not only the user *brotkasting* but also all users in nodes laying below this node. Therefore the users *jaeschke* and – since it is located below both *brotkasting* and *jaeschke* – *stumme* also belong to this node. Note that it fulfills thus the minimal support constraint $\tau_u = 3$ for the users.

---
[15] http://www.informatik.uni-trier.de/~ley/db/
[16] http://www.bibsonomy.org/group/kde/iccs_example?items=50

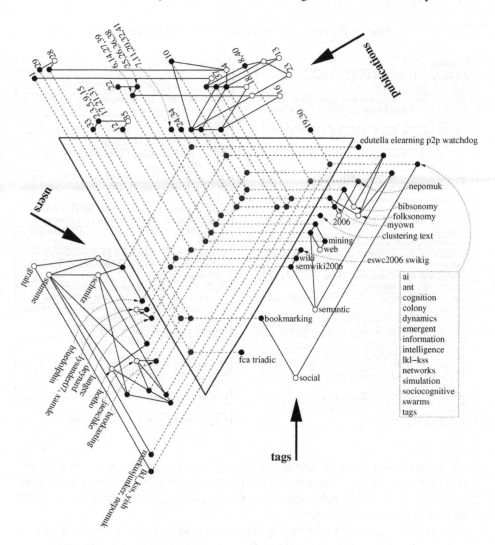

Fig. 4. All frequent tri-concepts for the minimum support thresholds $\tau_u = 3$, $\tau_t = 2$, $\tau_r = 2$

A closer look on the tag hierarchy reveals the content of the most central publications in the system. The tag *social* co-occurs with most of the tags. On the level of generality defined by the τ thresholds, this tag is (together with the tags *ai* [=Artificial Intelligence], ..., *tags*) assigned by the users *lkl_kss* and *yish* to the publications *19* and *30*, (together with the tag *bookmarking*) by the users *hotho, jaeschke, stumme* to the publications *4* and *28*, and (again together with the tag *bookmarking*) by the users *brotkasting, jaeschke, stumme* to the publications *28* and *29*. The tags as well as the corresponding publication titles indicate that the two sets of users {*lkl_kss, yish*} and {*brotkasting, hotho, jaeschke, stumme*} form two sub-communities which both work on social phenomena in the Web 2.0, but from different perspectives.

Table 1. The mapping of publication IDs to publication titles

ID	Publication Title
1	A Finite-State Model for On-Line Analytical Processing in Triadic Contexts
2	Annotation and Navigation in Semantic Wikis
3	A Semantic Wiki for Mathematical Knowledge Management
4	BibSonomy: A Social Bookmark and Publication Sharing System
5	Bringing the "Wiki-Way" to the Semantic Web with Rhizome
6	Building and Using the Semantic Web
7	Conceptual Clustering of Text Clusters
8	Content Aggregation on Knowledge Bases using Graph Clustering
9	Creating and using Semantic Web information with Makna
10	Emergent Semantics in BibSonomy
11	Explaining Text Clustering Results using Semantic Structures
12	Harvesting Wiki Consensus - Using Wikipedia Entries as Ontology Elements
13	Information Retrieval in Folksonomies: Search and Ranking
14	KAON – Towards a Large Scale Semantic Web
15	Kaukolu: Hub of the Semantic Corporate Intranet
16	Kollaboratives Wissensmanagement
17	Learning with Semantic Wikis
18	Mining Association Rules in Folksonomies
19	On Self-Regulated Swarms, Societal Memory, Speed and Dynamics
20	Ontologies improve text document clustering
21	Proceedings of the First Workshop on Semantic Wikis – From Wiki To Semantics

ID	Publication Title
22	Proc. of the European Web Mining Forum 2005
23	Semantic Network Analysis of Ontologies
24	Semantic Resource Management for the Web: An ELearning Application.
25	Semantic Web Mining
26	Semantic Web Mining and the Representation, Analysis, and Evolution of Web Space
27	Semantic Web Mining for Building Information Portals (Position Paper)
28	Social Bookmarking Tools (I): A General Review
29	Social Bookmarking Tools (II). A Case Study – Connotea
30	Social Cognitive Maps, Swarm Collective Perception and Distributed Search on Dynamic Landscapes
31	SweetWiki : Semantic Web Enabled Technologies in Wiki
32	Text Clustering Based on Background Knowledge
33	The ABCDE Format Enabling Semantic Conference Proceedings
34	The Courseware Watchdog: an Ontology-based tool for Finding and Organizing Learning Material
35	Towards a Wiki Interchange Format (WIF) – Opening Semantic Wiki Content and Metadata
36	Towards Semantic Web Mining
37	TRIAS - An Algorithm for Mining Iceberg Tri-Lattices
38	Usage Mining for and on the Semantic Web (Book)
39	Usage Mining for and on the Semantic Web (Workshop)
40	Wege zur Entdeckung von Communities in Folksonomies
41	WordNet improves text document clustering

A second topical group is spanned by the tag *semantic*, which occurs in three different contexts. The first is on semantic wikis, which correlates with the isolated group $\{2, \ldots, 31, 12, 33, 35\}$ of publications, and the – equally isolated – group $\{lysander07, xamde, deynard, langec\}$ of users.

The second context in which the tag *semantic* occurs is on Semantic Web Mining, being connected by the users $\{grahl, hotho, stumme\}$ with different combinations of the additional tags *web* and *mining* to the publications *6, 14, 22, 25, 26, 27, 36, 38*, and *39*. These assignments are witnessed by the three tri-concepts in the very middle of the diagram. On the same line are two more tri-concepts, which indicate that these users are also interested in *text clustering* and in *nepomuk* (the acronym of a European project).

The third context in which the tag *semantic* occurs is in combination with *folksonomy*. This provides a link to the group $\{2006, myown, nepomuk, bibsonomy, folksonomy\}$ of tags which are used by the authors of this paper and by other researchers from the European project Nepomuk[17] to describe their own publications.

Two more topical groups can be found at the top and bottom of the tags quasi-order. One is related to a Peer-to-Peer eLearning application, and the other to triadic Formal Concept Analysis.

---

[17] http://nepomuk.semanticdesktop.org/

Since the diagram shows the frequent tri-concepts only, we cannot deduce from the absence of a relationship that two objects are not related at all. When the thresholds are lowered, links between the topical islands discussed above will show up.

6   Conclusion and Outlook

In this paper we presented the insights that we gained from a qualitative analysis of the bibliographic data of our social resource and publication management system BibSonomy. Our main observations were, on the one hand that tags (like *social*) which act as a bridge between different topics can easily identified by means of triadic formal concept analysis and on the other hand that this method allows to comprehend how this "clustering" into separate (but still connected topics) on the set of tags is reflected on the sets of users and publications, resp. As triadic Formal Concept Analysis respects the full symmetry of a folksonomy (i. e., the three dimensions *users*, *tags*, and *resources* are of equal importance), one can use either of them as a starting point for the analysis.

A possible application of triadic FCA is within a recommender system. Frequent tri-concepts could be used inside BibSonomy for enhancing tag recommendations. For instance one might recommend a tag to a user if it appears together with the resource in a frequent tri-concept. Another application will be to guide users of BibSonomy to consolidate their publication entries.

A challenge on the methodological level is the automatic layout of the diagrams of concept tri-lattices. This is subject of future work, together with the search for more intuitive visualisation metaphors.

At the time being the publication references and topics inside BibSonomy refer often to Web2.0 and related topics. The system is hence a good starting point to analyse this rather new field of research. Our expectation is that the topics will become more diverse in the future – a fact that we observed when analysing data from del.icio.us – since more and more users share their publication entries with BibSonomy. In particular there are currently three EU projects that organize their publication management with BibSonomy.

Acknowledgement. Part of this research was funded by the European Union in the Nepomuk project (FP6-027705) and in the Tagora project (FP6-2005-34721).

References

1. Biedermann, K.: How triadic diagrams represent conceptual structures. In: Delugach, H.S., Keeler, M., Searle, L., Lukose, D., Sowa, J.F. (eds.) ICCS 1997. LNCS, vol. 1257, pp. 304–317. Springer, Heidelberg (1997)
2. Carpineto, C., Romano, G.: Concept Data Analysis. Wiley, Chichester (2004)
3. Cattuto, C., Loreto, V., Pietronero, L.: Collaborative tagging and semiotic dynamics (May 2006), http://arxiv.org/abs/cs/0605015
4. Dau, F., Wille, R.: On the modal unterstanding of triadic contexts. In: Decker, R., Gaul, W. (eds.) Classification and Information Processing at the Turn of the Millenium. Proc. Gesellschaft für Klassifikation (2001)

5. Dubinko, M., Kumar, R., Magnani, J., Novak, J., Raghavan, P., Tomkins, A.: Visualizing tags over time. In: Proc. of the 15th International WWW Conference (2006)
6. Ganter, B., Obiedkov, S.A.: Implications in triadic contexts. In: Wolff, K.E., Pfeiffer, H.D., Delugach, H.S. (eds.) ICCS 2004. LNCS (LNAI), vol. 3127, pp. 186–195. Springer, Heidelberg (2004)
7. Ganter, B., Stumme, G., Wille, R. (eds.): Formal Concept Analysis – Foundations and Applications. In: Ganter, B., Stumme, G., Wille, R. (eds.) Formal Concept Analysis. LNCS (LNAI), vol. 3626, Springer, Heidelberg (2005)
8. Ganter, B., Wille, R.: Formal Concept Analysis: Mathematical Foundations. Springer, Heidelberg (1999)
9. Ganter, B.: Algorithmen zur formalen Begriffsanalyse. In: Ganter, B., Wille, R., Wolff, K.E. (eds.) Beiträge zur Begriffsanalyse, pp. 241–254. B.I.–Wissenschaftsverlag, Mannheim (1987)
10. Halpin, H., Robu, V., Shepard, H.: The dynamics and semantics of collaborative tagging. In: Proceedings of the 1st Semantic Authoring and Annotation Workshop (SAAW'06) (2006)
11. Hammond, T., Hannay, T., Lund, B., Scott, J.: Social Bookmarking Tools (I): A General Review. D-Lib Magazine 11(4) (April 2005)
12. Heymann, P., Garcia-Molina, H.: Collaborative creation of communal hierarchical taxonomies in social tagging systems. Technical Report 2006-10, Computer Science Department, (April 2006)
13. Hotho, A., Jäschke, R., Schmitz, C., Stumme, G.: BibSonomy: A social bookmark and publication sharing system. In: Proceedings of the Conceptual Structures Tool Interoperability Workshop at the 14th International Conference on Conceptual Structures (2006)
14. Hotho, A., Jäschke, R., Schmitz, C., Stumme, G.: Information retrieval in folksonomies: Search and ranking. In: Sure, Y., Domingue, J. (eds.) ESWC 2006. LNCS, vol. 4011, pp. 411–426. Springer, Heidelberg (2006)
15. Jäschke, R., Hotho, A., Schmitz, C., Ganter, B., Stumme, G.: Trias - an algorithm for mining iceberg tri-lattices. In: Perner, P. (ed.) ICDM 2006. LNCS (LNAI), vol. 4065, Springer, Heidelberg (2006)
16. Lambiotte, R., Ausloos, M.: Collaborative tagging as a tripartite network (December 2005) arXiv:cs.DS/0512090
17. Lamport, L.: LaTeX: A Document Preparation System. Addison-Wesley, London, UK (1986)
18. Lehmann, F., Wille, R.: A triadic approach to formal concept analysis. In: Ellis, G., Rich, W., Levinson, R., Sowa, J.F. (eds.) ICCS 1995. LNCS, vol. 954, pp. 32–43. Springer, Heidelberg (1995)
19. Lund, B., Hammond, T., Flack, M., Hannay, T.: Social Bookmarking Tools (II): A Case Study - Connotea. D-Lib Magazine 11(4) (April 2005)
20. Mathes, A.: Folksonomies – Cooperative Classification and Communication Through Shared Metadata (December 2004), http://www.adammathes.com/academic/computer-mediated-communication/folksonomies.html
21. McCallum, A.K.: MALLET: A Machine Learning for Language Toolkit (2002), http://mallet.cs.umass.edu
22. Mika, P.: Ontologies Are Us: A Unified Model of Social Networks and Semantics. In: Gil, Y., Motta, E., Benjamins, V.R., Musen, M.A. (eds.) ISWC 2005. LNCS, vol. 3729, pp. 522–536. Springer, Heidelberg (2005)
23. Patashnik, O.: BibTeXing (1988) (Included in the BIBTEX distribution)
24. Peirce, C.S.: Collected Papers. Harvard Universit Press, Cambridge (1931-1935)
25. Schmitz, C., Hotho, A., Jäschke, R., Stumme, G.: Mining association rules in folksonomies. In: Batagelj, V., Bock, H.-H., Ferligoj, A., Žiberna, A. (eds.) Data Science and Classification: Proc. of the 10th IFCS Conf. Studies in Classification, Data Analysis, and Knowledge Organization, pp. 261–270. Springer, Heidelberg (2006)

26. Schmitz, P.: Inducing ontology from flickr tags. In: Collaborative Web Tagging Workshop at WWW2006, Edinburgh, Scotland (May 2006)
27. Stumme, G., Taouil, R., Bastide, Y., Pasqier, N., Lakhal, L.: Computing iceberg concept lattices with titanic. J. on Knowledge and Data Engineering 42(2), 189–222 (2002)
28. Stumme, G.: A finite state model for on-line analytical processing in triadic contexts. In: Ganter, B., Godin, R. (eds.) ICFCA 2005. LNCS (LNAI), vol. 3403, pp. 315–328. Springer, Heidelberg (2005)
29. Wille, R.: Restructuring lattice theory: An approach based on hierarchies of concepts. In: Rival, I. (ed.) Ordered Sets, pp. 445–470. Reidel, Dordrecht-Boston (1982)
30. Wille, R.: The basic theorem of triadic concept analysis. Order 12, 149–158 (1995)

The MILL – Method for Informal Learning Logistics

Andreas Faatz, Manuel Goertz, Eicke Godehardt, and Robert Lokaiczyk

SAP Research
CEC Darmstadt
64283 Darmstadt, Germany
`firstname.lastname@sap.com`

Abstract. The paper presents the MILL – a system which supports planning of training measures. Its background technique, task-competency modeling, is based on formal concept analysis as an indirect and qualitative way of determining the abilities of learners. In that context, the fulfillment or failure of a work task is the indicator of a set of necessary competencies. The core of the presented approach is a matrix structure – a formal context – which has tasks as labels for its rows and competencies labeling its columns. This matrix is the basis for defining formal concepts which can be ordered in a lattice for navigation and systematic decision support on training measures in an organisation.

1 Introduction

Informal learning means learning activities without a pre-defined support by curricula, textbooks or other classical didactical material imposing a learning path. This paper describes a method for informal learning logistics. It aims at managing training measures in informal learning environments. This particularly means learning environments that are closely embedded into the working place and which are related to the logical and temporal order of tasks (i.e. the workflow) a worker needs and wants to fulfill on the job. Typically, the nature of this work is knowledge-intensive.

The informality of a training measure becomes clear by the fact, that the effect of some or all of the training measures can often only be monitored by the outcome of working tasks. This contrasts with the situation at school or university, where the taught competencies to can be tested by exams. The system described in this paper enables the detection of positive or negative expected consequences of particular training measures in workplace-embedded learning. Additionally, it shows how to plan individual training measures for the worker corresponding to the tasks s/he is responsible for.

The approach to these questions is a mapping of well-ordered and well-structured tasks to a task-competency model. This conceptual structure is queried systematically by the system to exploit the temporal order and conditional interdependency of tasks. The querying creates a feedback on potential positive or negative consequences for the task structure and the progress of the workflow. The

U. Priss, S. Polovina, and R. Hill (Eds.): ICCS 2007, LNAI 4604, pp. 296–309, 2007.

achieved results allow a human resources manager to precisely determine competencies which should be achieved by training measures. We abbreviate the system to be presented with the MILL – Method for Informal Learning Logistics.

The paper is organised as follows. We start with related work and identify the basic requirements on our system. In section 3 we explain the necessary elements of formal concept analysis and Petri nets as the building blocks of the MILL. Section 4 is dedicated to a detailed description of the system, especially regarding the interplay between the task-competency model reflected as a formal concept lattice and the workflow formalised as a Petri net. Section 5 will present our conclusions and a brief outlook on future work.

2   Related Work and Basic Requirements

Task-competency modeling [1] is based on formal concept analysis. It is an indirect and qualitative way of determining the abilities of learners. The fulfillment of a task is the indicator of a set of necessary competencies. The core is a matrix structure, the so-called formal context [2] which has tasks as labels for its rows and competencies labeling its columns. This matrix is the basis for defining so-called formal concepts [2] which can be ordered in a lattice for navigation.

Ley et al. show two application scenarios in the domain of informal learning [3]. First, a task included in a workflow, for example the preparation of a document, is an indicator of the learner's competencies. Example competencies might be language capabilities, ability to abstract, ability to structure a topic and particular domain knowledge. Thus, by judging about the quality of a task output (in this example: the quality of a document) we gain a detailed picture on the competencies which enable a person to fulfill the task. Ley argues, that the resulting conceptual structures (intensions of the concepts are tasks, extension of the concepts are competencies) allow planning the training measures of an organization. In a second scenario the authors establish the lattice from the competencies and taks and present it in a graphical representation for navigation. This enables self-directed learning, where the user can easily determine, which documents match best for his/her training needs.

Although the authors present a flexible formal basis for conceptual structures derived from competencies and taks, the technical progress with respect to the core ideas of formal concept analysis does not become clear – except by the advances made by the approach of the former knowledge space theory due to Korossy [1] which identifies sets of competencies in a non-quantitative way. The authors do not explain systematic browsing or algorithms for browsing through the conceptual structures or typical queries which might support the planning of training measures. The task-competency model is not linked to other parts of a learning environment or workplace. The interpretation of the concept lattice and its mapping on a temporal sequence of tasks or interdependent tasks is an open question we solve by the techniques of the MILL.

From the point of view which enhances Petri nets (as an example for formally modeled workflows) by conceptual structures there is prior work in describing

the elements of Petri nets semantically [4]. Koschmieder et al. express the whole
Petri net as constructs in the web ontology language OWL [5]. Koschmieder's
aim is a modularization and recombination of workflows expressed by Petri nets
across several business units or organizations. The semantic description is not
applied to planning of and reasoning about tasks or developing competencies by
training measures.

With the gaps of the related work in mind we state the following main requirements for a system for planning training measures:

- The system should consider a learning situation outside the classroom and
  thus also apply to training in organisations or during the work process. The
  organisational workflow must be part of the system's underlying model.
- The system should be able to assist planning based on the interdependencies
  of tasks as well as on the interdependencies of the competencies which are
  characteristic for a task.
- The assistance provided by the system should be formalised in a way which
  helps the planner to understand the decisions or suggestions made by the system. Especially the consequences of training measures (or the consequences
  of left-out training measures) should be trackable for the user. The user
  might be a planner as well as a person who gets trained.

3   Existing Background Techniques

The following main sections of the paper will give a survey of the relevant background techniques, namely formal concept analysis capturing the task-competency
model and Petri nets as a universal formalism for descriptions of organisational
workflows. Moreover, we continue explaining the innovative core system (MILL)
and end with an example. Throughout our explanations, we work with the definition of a task as 'an action performed to reach a particular goal' [6].

Notation: the paper uses the following notations for mathematical constructs:

- sets are italic capitals or denoted in the usual way with comma-separated
  elements between braces: { and }.
- **I** is a relation between tasks and competencies (in formal concept analysis
  between general objects and properties)
- small italic Latin characters denote tasks and competencies (in general formal concept analysis between objects and properties)
- small Greek characters denote Petri nets
- ' is the derivation operator for formal contexts
- formal concepts are written as pairs of sets or as bold italic capitals
- if * is attached to a set notation, it denotes a subset of the set originally
  notated
- a set followed by $\sim time$ is partially ordered regarding some time scale,
  similarly $\sim cost$ stands for partial order regarding some cost scale
- *l* is a list of insufficient competencies

3.1   Elements from Formal Concept Analysis

Formal Concept Analysis (FCA) is a theory of conceptual structures (lattices) which result from the simultaneous reasoning about objects and their properties. FCA was founded by Wille [2]. The derivation of the concept lattice is based on tables called formal contexts, where an entry (a cross) indicates, if a property is fulfilled or not. We define along with Wille's theory the notion of a Formal Context.

Definition 1. *Formal Context: let G be a set of objects and M a set of properties. Let* **I** *be a binary relation indicating, which object from G fulfills which property from M. We write g*$\mathbf{I}$*m, if an object g from G fulfills property m from M. G, M and* **I** *together are called formal context* $(G, \mathbf{I}, M)$.

Table 1 shows an example of a formal context named task-competency. Objects a_1 through a_6 denote tasks, where particular documents of a software engineering process must be written or completed. Thus we will simply refer to these tasks as 'documents 1 through 6' in the following explanations. The attributes (denoting the columns) are abbreviations for language (lang), abstraction (abs) and Unified Modeling Language (UML). A checked box means, that for writing the document in the specific row, the marked competencies (e.g. language, abstraction or knowing UML) have to be fulfilled. For instance, for document 1 all these competencies are necessary, but for document 3 only language and UML are required. (The figure was generated with the online tool JaLaBA, see http://maarten.janssenweb.net/jalaba/JaLaBA.pl). Throughout the document we will use Arabic numbers for indexing to temporally order tasks (the task with index number 1 denotes the first one in the workflow). Concurrent tasks would be denoted for example as 1 (concurrency to task 1).

Table 1. Formal context

| | Competencies | | |
Task	lang	abs	UML
a_1	✓	✓	✓
a_2		✓	✓
a_3	✓		✓
a_4	✓	✓	✓
a_5	✓		✓
a_6	✓	✓	

We define a derivative operator $'$ for subsets $X \subseteq G$ and $Y \subseteq M$ mapping the objects (properties respectively) from X (Y, respectively) to those properties from Y (objects from X, respectively) which fulfill the relation **I** for at least all objects (properties respectively) from X (Y, respectively). For the derived sets we write X' and Y', respectively. The construction of a concept lattice is possible by applying several derivations to sets of objects or properties. Note: X

is always a subset of X'' and X' equals X'''. We can define the notion of a *formal concept* of a formal context $(G, \mathbf{I}, M)$ is a pair (A, B) where A is a subset of G and B is a subset of M and $A = B'$. We say that two formal concepts (C, D) and (E, F) fulfill the sub-concept relation $\leq$, if and only if C is a subset of E. Here $(C, D) \leq (E, F)$ reads "(C, D) is a sub-concept of (E, F)". $\geq$ would denote the inverse relation of $\leq$, and $(C, D) \geq (E, F)$ reads "(C, D) is a super-concept of (E, F)". Table 2 shows the concepts resulting from the above task-competency context (computed with JaLaBA). Elements of the first set in the pair forming a formal concept are called extension of the concept, elements of the second set are called intension of the concept.

Table 2. Formal concepts

Formal Concepts of task competency
A < a_1, a_4, {language, abstraction, UML} >
B < a_1, a_2, a_4, {abstraction, UML} >
C < a_1, a_3, a_4, a_5, {language, UML} >
D < a_1, a_4, a_6, {language, abstraction} >
E < a_1, a_2, a_3, a_4, a_5, {UML} >
F < a_1, a_2, a_4, a_6, {abstraction} >
G < a_1, a_3, a_4, a_5, {language} >
H < a_1, a_3, a_4, a_5, a_6, {} >

A visualization generated with ToscanaJ (see ToscanaJ project page at source-forge.net) of the resulting lattice is shown in Figure 1.

Concepts are shown as colored circles, super-sub-conceptual relationships are shown as lines. A line between a darker and a brighter concept means, that

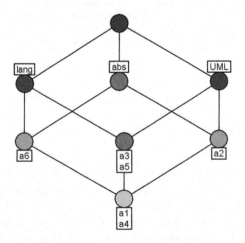

Fig. 1. Concept lattice

the brighter one is a sub-concept of the darker one. The concept at the very top refers to concept H, the concept with the label 'lang' refers to concept G, the one with the label 'abs' refers concept F, the concept with the 'UML' label refers concept E, the concept with the label 'a6' below it refers concept D, the concept with the label 'a3 a5' below it refers concept C, the concept with the label 'a2' below it refers concept B, the concept with the label 'a1 a4' below it refers concept A. The intension of a concept can be read by following all paths, where any step in the path end in a darker concept and collecting all labels in top of a circle (i.e. concept), the extensions can be read by following paths the other way around and collecting all labels below a circle (i.e. concept).

3.2 Elements from Petri Nets

A Petri net consists of tokens, transitions and places as well as arcs which connect places and transitions. The distribution of tokens indicates the state of the Petri net. If a transition is enabled, all places pointing to the transition must be filled with tokens. Tokens can be of different color. This refers to typed events. For example, in a system different colors might refer to different users of a system and the respective state of their workflows. Different colors of tokens might also result from several cases or applications using the same workflow (i.e. here: Petri net). An important characteristic of Petri nets is their openness regarding the way transitions are fired. Tokens at incoming places only enable the transition. There is the possibility of assigning time information to each transition. Time information can be attached to each transition as a real number characterizing the behavior of the transition. The number means e.g. the amount of milliseconds the transaction will need after its enabling to produce its outgoing tokens. A detailed introduction to Petri nets can be found in [7].

Van der Aalst [8] also showed why Petri nets are a good candidate for organisational workflow modeling. First, their graphical expression has a clear formal meaning. Moreover, a Petri net is always a state-based notation in contrast to an event-based notation. The enabling of transitions can be seen at a glance and is not due to any kind of (semi-formal or informal) interpretation. The state-based notation also eases the expression of concurrency. A process leaving the system completely can simply be expressed by removing all its (colored) tokens. Finally, a multitude of mathematical formalisms exist that allow the checking of the properties of a Petri net. The formalisms can directly be applied to Petri nets expressing workflows and for proving the soundness of a workflow. For example it can be shown, that all tasks are free from the danger of falling into a deadlock situation.

Van der Aalst has shown that Petri nets are capable of formalizing current process-aware information systems. Figure 2 shows an example Petri net without tokens. The circles indicate places (for instance 'indirect users found'), the quadratic shapes transitions (for example 'prepare developers list'). Tokens would be indicated by colored circles centered at places. The arcs starting at a place and ending at a transition can be understood as 'is necessary condition for', arcs starting at a transition and ending at a place can be understood as 'triggers state'. Note that this is rather a snapshot of a larger Petri net, as in real-world applications the places at the very left would be connected to a starting transition.

4   Description of the Core System

Prerequisites: the MILL-system has the following prerequisites and components:

- a workflow is modeled as or transformable to a Petri net,
- for the sake of our explanation we assume a single worker in a single workflow,
- a task-competency-matrix exists, thus the conceptual structures like in Figure 1, Table 1 and Figure 2 exist,
- the tasks are the same or a super-set of the tasks corresponding to the transitions in the Petri net.

For example, relating these prerequisites to the examples of Figure 1 and Figure 2 would mean, that besides the modeling expressed by the two figures concepts A through H would correspond to transitions (= tasks) from Figure 2. Thus Figure 2 without the transition 'pool facilitating stakeholders' and without the arcs incoming in and outgoing from it would fulfill the prerequisites, if additionally the rectangular shapes were foreseen with the elements $\{A, \ldots H\}$.

If the tasks of the workflow are a proper subset, the task-competency matrix and the resulting formal context can be restricted to those tasks which are actually contained in the workflow.

The operations of the MILL-system starts at the moment, when a task t of a worker fails (case A) or is judged to be fulfilled in a non-sufficient way (case B). This situation might occur for reasons which originate outside of the organization (e.g. customers refuse the outcome of a work package) or for internal reasons (e.g. internal reviews). The workflow, i.e. the Petri net with all its tokens at the current places is frozen before (case A) or directly after (case B) the transition is fired. Two ratings supporting the identification of competencies to be improved start immediately and (potentially) in parallel: a competency-based rating and a task-based rating triggering a planning procedure. The competency-based rating is a direct one, the task-based rating a more indirect one. Both ratings return a set of tasks D, which we call critical tasks.

Competency-based rating: the competencies which are necessary to fulfill the task t, are checked against the competencies of the worker. The MILL-system either has access to such competency ratings for each worker or prompts an evaluator to judge about the competencies. Up to this point, competency-based rating resembles and formalizes the ideas of [3]. From this point on, all further steps and techniques we introduce (for competency-based rating, task-based rating and beyond) are novel.

The insufficient competencies are returned as an ordered list l (if possible, ordered from the worker's worst competency to best but still insufficient competency). An example for such a list could be: $<$ language, UML $>$. The critical

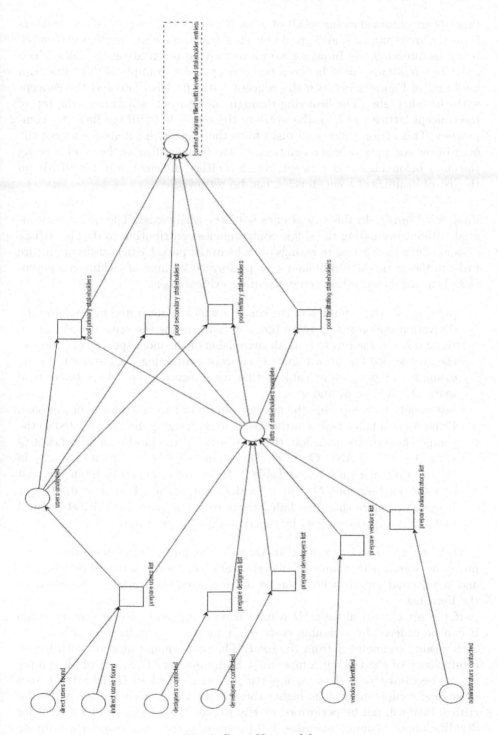

Fig. 2. Petri Net model

tasks D are obtained as the result of browsing down the concept lattice, starting from the most upper (more grey) concepts which have an insufficient competency as intension. By browsing we mean moving down along the lines which indicate a relation $\geq$ or $\leq$ between two concepts. An example of these lines can be found in Figure 1 between the concept with the label 'a6' and the concept with the label 'abs'. The browsing through the concepts starts from the top of the concept lattice and from the worst to the best – but still insufficient – competency. This action collects all tasks from the respective extensions except the ones below concepts, where a competency (to be identified as the label heading the circle belonging to the concept) which is either sufficient or not explicitly on the list of insufficient competencies, can be retrieved.

Task-based rating: In this case, l does not necessarily exist. The task t is evaluated without evaluating the single competencies contributing to it. The MILL-system might have access to ratings of tasks in the past. Critical tasks are future tasks in the workflow which cannot be performed because of lacking competencies. The task-based rating generates other critical tasks:

- promptly as the extension of the concept which is generated by applying the derivation operator to t twice (i.e. t''). This means the critical tasks D are those tasks in the future from the extension of the most special (brightest in the grey-scaled layout of Figure 1) concept containing t. Consider from the example in Figure 1 and Table 2, that for instance $t = a_1$. Then the critical tasks D would be a_1 and a_4.
- historically by comparing the current failure of t to failed tasks in the past. If there was a failed task s in the past and no training measure related to the competencies in its intension, then extension of the most general concept Q with $Q \leq (t'', t')$ AND $Q \leq (s'', s')$. For instance, let $t = a_4$ and $s = a_2$ in the example in Figure 1 and Table 2. Here, s is assumed to be in the past as $s < t$. Furthermore, $t'' = a_1, a_4$ and $s'' = a_1, a_2, a_4$. Then $D = a_1, a_4$.
- If there are more than two failed tasks from the past, the critical (future) tasks D can be determined by repetition of this procedures.

Ordering and selecting critical tasks: the necessary training measures to improve the overall performance of an organization are now prioritized by temporal and conditional aspects which can be derived from the workflow expressed by the Petri net.

If the set of critical tasks D results from competency-based ranking, then it can be ordered by assigning costs (that means: a positive real number) to each lacking competency from the list l. The costs should increase with higher insufficiency of the lacking competency. Summing over the costs of all competencies necessary for performing a particular critical task in D yields total costs of missing competencies. The higher these costs, the more likely the particular critical task will not be performed by the worker. This fact provides a basis for deciding about training measures. For instance, if the costs reflect the training costs, the planner might decide to chose the least expensive training measures first. Another potential cost function is the cost of scheduling another person to fulfill the task.

If D is the outcome of a purely task-based ranking, then there is no similar ordering regarding costs of missing competencies, as this innovative way of obtaining D judges tasks based on competencies. Thus there might be other competencies which are not measurable in a single step task-based rating only. Task-based rating rather provides immediate information on other surely critical tasks.

In both cases an alternative ordering of D by temporal aspects is possible. The tasks (corresponding formally to transitions) in the Petri net are partially ordered in the sense that for some pairs of tasks (transitions) it is possible to state, which task (transition) will be enabled or executed after the other one. In the example based on Figure 2, some of the tasks (transitions) can be ordered temporally as follows: 'prepare users list' comes before 'pool primary stakeholders', 'prepare administrators list' comes before 'pool facilitating stakeholders'. In cases of concurrency, Petri nets with time information are even able to resolve the concurrency and order the tasks along a timeline. From this ordering point of view, a critical future task which approaches earlier in the workflow, could get priority in comparison to a later critical task from the planning point of view.

In the example shown in Figure 2, all tasks described as 'pool ' are concurrent and all tasks described as 'prepare ' are concurrent. If for example 'prepare users list' and 'prepare designers list' were attached with time information, that the firing of the transition needs 3 days for 'prepare users list' and 1 day for 'prepare designers list', then the order created would place 'prepare designers list' before 'prepare users list'.

No matter which alternative is chosen, the ordering step of the MILL-system results in a partially ordered set $D \sim time$ (if it is partially ordered by temporal aspects) or $D \sim cost$ (if it is partially ordered by costs). This set already contains not only the critical tasks, but also a priority, either driven by criticality or by time issues. Before we proceed with the final selection of critical tasks, we remark that D might also result from a mixture between or union of task-based and competency-based rating. In this case, the partial ordering will be created as $D \sim time$.

The last step in determining prioritized training measures is applying van der Aalst's second soundness criterion [8] to simple transformations of the workflow captured by the Petri net. This soundness criterion investigates, if a procedure expressed by a Petri net terminates eventually. This might be generalized to a criterion, that the procedure terminates in time due to a real-time condition. An example of a Petri net which does not terminate, is depicted in Figure 3(a). This Petri net always misses a token to proceed, whereas the Petri net in Figure 3(b) will terminate. Nevertheless, even the Petri net from Figure 3(b) would hurt the real-time criterion, if for instance the firing of the transition dies not take place within a specified time threshold.

Let β be a Petri net capturing the workflow. Then let $\beta(D^*)$ denote a Petri net equal to β except for the fact, that all transitions which correspond to a subset

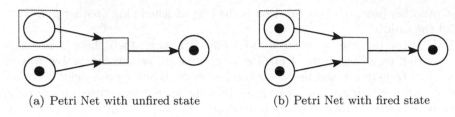

(a) Petri Net with unfired state   (b) Petri Net with fired state

Fig. 3. Petri nets

D^* of D, are absolutely disabled. An example of such a $\beta(D^*)$ is the Petri net resulting from adding incoming absolutely empty places to all transitions corresponding to tasks from D. We call this example of a Petri net with disabled transitions partially disabled. If for example β is the Petri net from Figure 2 and D^* turns out to be the single-element set 'pool primary stakeholders', then $\beta(D^*)$ would be Figure 2 with an additional token-less place which has an arc to the transition 'pool primary stakeholders' as its only connection to the example Petri net β.

If $\beta(D^*)$ obeys the soundness or real-time criterion for all subsets D^* of D, then the MILL suggests to train the tasks from l which cause the endangerment of the earliest (or most costly) critical task from $D \sim time$ (or $D \sim cost$). If l does not exist, but still all $\beta(D^*)$ obey the soundness or real-time criterion, the MILL suggests training of competencies from the intension of the most general concept Q which is a sub-concept to all $|D|$ concepts generated by twofold derivation of a single-element subset of D. If this intension is empty, the derivations and the computing of Q are repeated on the basis of successively removing late tasks (due to $D \sim time$) or expensive tasks (due to $D \sim cost$) from D, until Q is nonempty. The MILL also informs the planner, that the suggested training measure is not necessary for keeping the workflow sound.

In all other cases with non-sound $\beta(D^*)$ for some D^*, the MILL passes a warning to the planners, saying that the workflow as a whole is critical. If there are one or several subsets $D^*(1), \ldots, D^*(n)$ of D, for which the resulting partially disabled Petri nets do not obey the soundness or real-time criterion, then the MILL operates through the following steps:

- determining minimal subsets of each $D^*(1), \ldots, D^*(n)$ in the sense, that for each proper subset of each $D^*(1), \ldots, D^*(n)$ the corresponding partially disabled Petri net would still obey the soundness criterion (real-time or van der Aalst's).
- determining the intension(s) of the most abstract concept(s) which contain a single task from the respective $D^*(1), \ldots, D^*(n)$ in their extension. Determine the latest of these single tasks for $D^*(1), \ldots, D^*(n)$ respectively, if there is more than one formal concept with the same most-like level of abstraction.
- for each set $D^*(1), \ldots, D^*(n)$ the MILL proposes all the intensions of the aforementioned most abstract concepts.

If in addition l exists, the MILL only suggests training measures corresponding to those insufficient competencies. The strategy described for cases with non-sound $\beta(D^*)$ for some D^* corresponds to the idea of reserving enough time in the organization to train competencies which will in the end keep the workflow possible and alive. Following this strategy, the resulting intensions are again ordered by the MILL proceeding through the following steps:

- for each of the retrieved intensions identify the earliest task (regarding $D \sim time$) in their respective extension and denote its time.
- the intension with the time closest to the current failed task is proposed by the MILL as the training measure.

The proposition is based on this order, because it allows for step-by-step safe traversing of competencies (contained in the intensions) to be sure to cover tasks (contained in the extensions) which would stop the future workflow in case of failure. To keep this safety, the MILL is applied after enabling and before firing a transition, as long as there are current or past tasks which failed and still could fail due to the task-competency context and an update of the MILL-system. All time-dependent decisions are then relative to the current task (i.e. transition). We conclude with the description of a potential update mechanisms.

Update: For the update phase, the frozen state of the workflow is removed. That means, that the workflow is continued: the next enabled transition(s) is (are) fired. The update of the MILL-system can be - up to decision of the organization, in which it is implemented - foreseen with one or several paradigms:

- the MILL might work pessimistically with a virtual list l containing all competencies except the ones in the derivation of successful tasks from the past.
- alternatively and potentially complementarily, whenever a competency is taught, the MILL-system triggers a new task-competency matrix, where the relation $\mathbf{I}$ between the competency just taught is removed from the task-competency matrix or removed from the (virtual or real) list l.
- also alternatively and potentially complementarily, an update could also include outdating-functions for competencies - for instance, if no task with an application of language competencies has to be performed after the language competencies were trained. The outdate is implemented as insufficient competency re-appearing on l or as re-appearing relation $\mathbf{I}$ in the task-competency matrix.

Finally, note that the MILL supports planning of training measures. It gives no direct clue to alternative paths through the workflow, planning of working tasks or workflow optimization. We also abstract from the time a training measure would take and keep the worker from participating in the workflow. With the introduction of a more time-based view and a multi-user scenario this could be improved. This is also an additional reason why our proposition is merely related to informal learning, because in informal learning we expect learners to schedule the time needed for learning by themselves. Another time critical issue might be freezing the worklow for applying the steps of the MILL. As the computation of intensions and extensions can be done in advance, there is no runtime endangerment with this respect. We suggest to compute and store potential $\beta(D^*)$ and its soundness properties in advance, too. This would avoid collisions between time requirements of the executed workflow and the MILL.

5   Conclusion

We presented the MILL as a novel approach to a system which connects the task-competency approach to well-structured workflows which are typical for organizations (enterprises, administrations et cetera). The worker's competencies are all abilities or skills which support tasks in the workflow and which in a narrower sense can be taught (e.g. programming skills, language skills etc.) or in a broader sense be developed (e.g. management skills, communication skills etc.). Our approach matches the three central requirements from Section 2 in the following ways:

- Classical knowledge space theory and applied competency-performance structures [1] operate in a classroom situation, thus its applications focus on adaptive e-Learning systems [9] which try to cover a landscape of competencies while the learner interacts with these systems. The interdependencies of tasks (in many cases the tasks are test items) is not in the focus of this (virtual and non-virtual) classroom applications (this means: the order of test items is irrelevant). The MILL is focused on real-world tasks in an organization and its aim is to cover the competencies necessary for concrete future actions of the learner.
- The temporal interdependency of tasks which are past, current or future activities of a worker is systematically exploited by the system. Prior work has not established or exploited any orders and relations of tasks. The MILL distinguishes past and future tasks as well as costs of lacking competencies.
- The consequences of missing or improving competencies for future and repetitive tasks are formalized. This formalization is a novel mapping of task-competency modeling to workflow analysis. Prior work has not coupled training activities with consequences for the actual workflow. The MILL also performs a second check after critical future tasks are identified.
- The MILL uses systematic browsing of a concept lattice resulting from the task-competency structure. Prior work [3] focuses on the task-competency structure and resulting learning paths itself; its traversing by (temporally and conditionally) structured tasks from a workflow was not formalized, yet. The MILL innovates the view on the tasks and competencies as a dyadic one: it is possible to reason from failed tasks and from lacking competencies.
- The approach includes rich diagrammatic structures which can be used as an explanation of the system's decisions. In particular, the MILL gives reason to the priorities of training measures.

Future work will focus on the question, if an extension of the approach considering a fuzzyfication of the task-competency structure might be fruitful. The idea behind this is capturing increasing or decreasing competencies over time. This will be challenging, as there are several approaches to fuzzy formal concept analysis. Another useful extension would be a reasoning mechanism which considers the competencies and workflow participation of a whole team instead of single workers. This extension would be neccessary to treat training measures as time consuming tasks, too. This would enable us to shift from informal to formal learning and to find strategies, how the workflow could be executed by the team while one team member is busy in a formal training. We have shown the selection of training measures, which is abstract enough to apply in enterprises and organisations with an arbitrary catalogue of training measures. However, a piloting evaluation scenario should be applied in an e-learning setting like the one from the APOSDLE project [10], for which we also modeled the process shown in Figure 2 and for which a task-competency model exists. The reason why we envisage the evaluation in an e-learning application is the (in comparison) low organisational threshold to be passed. The challenge in these evaluations will be to detect, if a training measure fails or succeeds because of the models (process and task-competency), the quality of the documents to be composed or the prioritising algorithm, the MILL.

References

1. Korossy, K.: Extending the theory of knowledge spaces: A competence-performance approach. Zeitschrift für Psychologie (1997)
2. Wille, R., Ganter, B.: Formal concept analysis. In: Mathematical Foundations, Springer, Heidelberg (1997)
3. Ley, T., Lindstaedt, S., Albert, D.: Supporting competency development in informal workplace learning. In: Althoff, K.-D., Dengel, A., Bergmann, R., Nick, M., Roth-Berghofer, T.R. (eds.) WM 2005. LNCS (LNAI), vol. 3782, Springer, Heidelberg (2005)
4. Koschmider, A., Oberweis, A.: Ontology based business process description. In: Pastor, Ó., Falcão e Cunha, J. (eds.) CAiSE 2005. LNCS, vol. 3520, Springer, Heidelberg (2005)
5. W3C: Overview and features of owl (2004),
   http://www.w3.org/TR/owl-features/
6. van Welie, M., van der Veer, G., Eliens, A.: An ontology for task world models. In: Proceedings of DSV-IS98, Abingdon, UK, Springer, Heidelberg (1998)
7. Girault, C., Valk, R.: Petri Nets for Systems Engineering: A Guide to Modeling, Verification, and Applications. Springer, Heidelberg (2000)
8. van der Aalst, W.M.P.: The application of petri nets to workflow management. Journal of Circuits Systems and Computers 8(1) (1998)
9. Hockemeyer, C., Held, T., Albert, D.: Rath-a relational adaptive tutoring hypertext www-environment based on knowledge space theory. In: Proceedings of CALISCE (1998)
10. Lindstaedt, S.N., Ley, T., Mayer, H.: Integrating working and learning with aposdle. In: Proceedings of the 11th Business Meeting of Forum Neue Medien, 10–11 November 2005, Vienna, Verlag Forum Neue Medien (2005)

Bilingual Word Association Networks

Uta Priss and L. John Old

Napier University, School of Computing
u.priss@napier.ac.uk, j.old@napier.ac.uk

Abstract. Bilingual word association networks can be beneficial as a tool in foreign language education because they show relationships among cognate words of different languages and correspond to structures in the mental lexicon. This paper discusses possible technologies that can be used to generate and represent word association networks.

1 Introduction

The research for this paper is motivated by a word association "game" which traces the cognates of a word in different languages. This word association game is driven by a fascination for detecting conceptual structures that appear to be inherent in natural languages. The game is played by one player starting with a word and then the other player responding with "doesn't this relate to ...". Together the players will then spin a network of words that are associated as bilingual translations (for example, English/German) or by shared etymological root or lexicographic or semantic similarity.

As an example, figure 1 shows an English/German word association network that started with the word "two". The associations of "two" that come first to mind are words, such as "twelve", "twenty" and their German counterparts; then "twilight" (i.e. two lights), "twill" and "tweed" (made from two threads), "twin", and "twig" (branching into two). The German words "Zwieback" and "Zweifel" which both associate with "zwei" (the German word for "two") are interesting because they enlarge the network by leading to other words for "twoness" which do not start with the letters "tw" or "Zw". The literal translation of "Zwieback" is "two bake" which is a literal translation of "biscuit" (via Italian), although in the modern languages "Zwieback" and "biscuit" are not direct translations of each other. The English translation of "Zweifel" is "doubt", which happens to relate to "duo" - the derivative of the Latin/Greek word for "two". Thus "Zwieback" and "Zweifel" establish a connection from "two" to other words of "twoness". The network appears to have three centres relating to the number words "two/zwei", "duo" and "bi" (the other Latin word for "two") which are connected via a bilingual connection.

Because the game is associational, the result is not deterministic. Humans do not have strict rules about etymology in what Miller et al. (1990) call the "mental lexicon", i.e. the mental representation of lexical knowledge in the mind. Any word association game is influenced by the mental lexicons of its players. The relations in a mental lexicon do not need to correspond to linguistically significant relations, such as etymology, but, instead, a mental lexicon is influenced by social, psychological, and cultural factors and can be idiosyncratic.

U. Priss, S. Polovina, and R. Hill (Eds.): ICCS 2007, LNAI 4604, pp. 310–320, 2007.

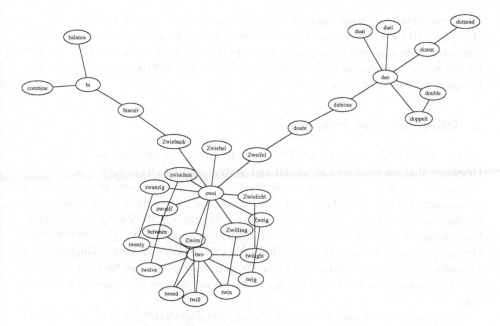

Fig. 1. A bilingual word association network

At some stage the players of a word association game may want to utilise dictionaries and thesauri to add further related words and to compare their intuitively derived associations with established lexical relations, such as etymology. In the example of figure 1, an etymological dictionary will reveal that "Zwiebel" (onion) is not etymologically related to "zwei"; and "duel" is not related to "duo" but instead to "bellum" (Latin for "war"). On the other hand, "bi", "duo" and "two" are related because they are all descendants of the Indo-European root "DWO".

So, is there a more serious background to such kinds of word association networks? Word association networks have been researched in the areas of linguistics and psychology for decades (cf. Nelson et al. (1998)). Word association tests have been used by psychologists and psychiatrists to evaluate patients. Though less popular these days as a projective/interpretive, psychoanalytic technique, they are still used to identify disorders. A stimulus or cue word is given and the response is recorded. The reaction time and type of response (for example concrete or multiword responses) can have diagnostic value. In contrast to the game described above, traditional word association studies are monolingual. But in principal there is no difference between monolingual and bilingual associations if the players have some bilingual linguistic competence.

In addition to the psychological relevance of word associations, Inkpen et al. (2005) explain that the detection of cognate words across different languages has many applications in natural language processing, for example, for sentence alignment and statistical machine translation models. An important application for bilingual word associations is the construction of tools that aid learners of foreign languages. Barrière and St-Jacques (2005) argue that a visual interface that displays word associations (or what they call "semantic context") promotes vocabulary learning. Word association tests show that native speakers store words in their mental lexicons with associations. Thus the use of explicit visualisations of word associations can help language learners to establish structures in their mental lexicons that are similar to those of native speakers. The aim for this paper is to suggest FCA-based[1] technologies for constructing and representing bilingual word association networks.

2   Definitions and Technologies

Inkpen et al. (2005) suggest the following definitions: *cognates* are words from two languages that are perceived as similar and mutual translations of each other, such as "two/zwei". *Partial cognates* are words that have the same meaning in some, but not all contexts. For example, "twig" and "Zweig" are used similarly in some contexts, but in other contexts "Zweig" is better translated as "branch". Both "Zweig" and "branch" have metaphoric meanings ("a branch of a business"), which "twig" does not share. *Genetic cognates* are words that derive from the same word in an ancient language. Due to gradual changes in phonetics and meaning, genetic cognates can have very different forms and meanings in modern languages. The ancestor of genetic cognates is called a *Root*.

Linguists usually distinguish cognates from *false friends* (Inkpen et al, 2005), which have similar spellings but very different meanings. Linguists tend to exclude *lexical borrowings* from being genetic cognates because of their different historical development. Furthermore, linguists will normally distinguish genetic cognates from words that are related via *folk etymologies*, which are words that are lexicographically similar but not actually etymologically related. For example, the vegetable "Jerusalem artichoke" has no relationship with Jerusalem, but, in fact, "Jerusalem" is a folk etymology of the Italian word "girasole" for "sunflower". This means that the vegetable should really be called "sunflower artichoke" instead of "Jerusalem artichoke". For the purposes of constructing bilingual word association networks for language learners, these linguistic distinctions between cognates, false friends, borrowings and folk etymologies are not always important because these are linguistic structures which may not correspond to the conceptual structures in a mental lexicon. For the purposes of learning English, it would not matter if learners associate "Jerusalem artichokes" with Jerusalem instead of sunflowers if that helps them to learn the vocabulary.

The remainder of this section discusses technologies and lexical databases that can be used in the creation of bilingual word association networks. Priss & Old (2004) present a variety of applications of Formal Concept Analysis to lexical databases which are potentially useful for word association networks. In particular, the extraction of *semantic neighbourhoods*, i.e., fields of semantically related words, from lexical databases using *neighbourhood lattices* is relevant. Since *semantic* association networks can easily be derived from WordNet (Miller et al. 1990) and Roget's Thesaurus using neighbourhood lattices and have already been discussed elsewhere (Priss & Old, 2004), the focus of this paper is on *word* association networks. Of special interest are word association networks that establish genetic cognates (such as the one presented in figure 1). Inkpen at al.

---

[1] Formal Concept Analysis, FCA, is a mathematical method for knowledge representation using formal contexts and concept lattices (Ganter & Wille, 1999).

(2005) compare different algorithms that have been used to automatically identify cognates. These phonetic algorithms are usually based on orthographic similarity measures. Inkpen et al. conclude that in order to determine genetic cognates, the Soundex algorithm is most suitable among the phonetic algorithms based on orthographic similarity.

The Soundex algorithm was invented by Russell and Odell (c.f. Knuth, 1998). It codes a word into a letter followed by three numbers, where the letter is the first letter of the word and the numbers encode the remaining consonants. Consonants that are phonetically similar, such as b, f, p, v, are encoded by the same number. The algorithm was originally used by government officials, hospitals and genealogists to detect names that are similar and may be misspelled variants. For example, a patient might tell a hospital staff member that his name is "Rupert", which the staff member might erroneously record as "Robert" - both have the same Soundex code. Because the processes that lead to misspellings of names are similar to historical, phonetic changes, the Soundex algorithm can also be used to detect genetic cognates. For example, the consonant shifts that occurred during the First Germanic Sound Shift (also called Grimm's Law), which separates Proto-Germanic from other Indo-European languages, mostly pertain to consonants that receive the same code in Soundex (for example, German "Vater", Latin "pater", English "father"). More details about the Soundex algorithm are provided in the Appendix.

The usefulness of the Soundex algorithm can be further improved by adapting its rules to the specific characteristics of the languages that are involved instead of using generic rules. Historical linguists might disagree with the use of the Soundex algorithm because it might yield false friends and because it lacks the contextual sensitivity that their linguistic techniques usually display. But, as discussed before, word association networks do not necessarily require the same amount of precision as historical linguistics.

Because of its simplicity, the Soundex algorithm is a powerful tool for detecting possible cognates. But unless it is combined with other technologies, its resulting association networks would be far too large. If one is interested in genetic cognates only, then the easiest approach might be to use an etymological dictionary. Unfortunately, as far as we know, it is quite difficult to obtain comprehensive etymological dictionaries that are of reasonable quality and in the public domain. The second author has complied a database of proto-language Roots and their descendants from a variety of available sources, which is used in some of the examples below (and called "ETYM" in this paper). But this database only provides Roots for a subset of the English and German words. Even in this limited database, the set of all descendants of a Root can be quite large (up to 360 words) and contain many words that are false friends in modern languages.

In order to reduce the size of the resulting networks, lexical databases, such as Roget's Thesaurus[2] or WordNet[3] can be used to identify semantic similarity among sets of phonetically similar words. Both Roget's Thesaurus and WordNet are available on-line, although in the case of Roget's Thesaurus only older versions are in the public domain (for example, from the Project Gutenberg website) because of copyright restrictions. In

---

[2] http://www.roget.org/
[3] http://wordnet.princeton.edu/

this paper, a relational database of Roget's Thesaurus (called "RIT") is used, which is based on Roget (1962) and not in the public domain. Another possibility is the use of bi-lingual dictionaries, which are in the public domain for many language pairs. Any of these could be combined with either the Soundex algorithm or etymological dictionaries to reduce the number of words in the resulting word association networks.

Another possible data source is word association data derived from psychological experiments. Two sets of word association data are easily available: the first set, "USF data", from the University of South Florida (Nelson, et al., 1998), is available on-line for download in the form of multiple ASCII text files. The USF data consists of about 5,000 cue words, or prompts, and about 10,000 target words, or responses. The second set of word association data, from the University of Edinburgh (Kiss et al., 1973), is available online in the form of the Edinburgh Word Association Thesaurus (EAT) and forms part of the MRC Psycholinguistic Database (Wilson, 1988). The EAT and the USF data differ in their distribution of response frequencies because word associations are culturally determined (in this case, UK versus US data). Word association data is loose and probabilistic. Responses to cue words can vary by subject, as can the frequency of the words or targets, given as responses. In contrast to the cue words, the target words are not a fixed set. Individual subjects may give unique (and sometimes puzzling) responses. However, usually the data is aggregated across participants. The associations of higher frequency counts have a reasonable amount of validity in their temporal, cultural contexts. The word association database (WAD) used in this paper consists of a set of 5,018 cue words and was constructed by the second author as an intersection of the British EAT and the US USF word association databases.

3   Examples

This section shows several examples of word association networks relating to the word "two" which demonstrate the use of different lexical databases and algorithms. The example in figure 2 is a neighbourhood lattice of the genetic cognates of "two" in Roget's Thesaurus, although at the category level. This is a standard construction described by Priss & Old (2004). The set of objects consists of the genetic cognates of "two". The attributes are all the categories in Roget's Thesaurus to which these words belong. The SQL statement that was used to select the data is as follows:

```
SELECT word, category FROM RIT WHERE word IN
    (SELECT descendants FROM ETYM WHERE root = 'two')
```

The labels of the attributes are omitted in the diagram in figure 2 because it is sufficient to observe the grouping of the words. This example is monolingual because Roget's Thesaurus is an English database ("zwieback" is an English word borrowed from German). Some of words seem to cluster according to similar phonetics (as in figure 1) because some words have similar meanings in addition to their similar phonetics. In summary, neighbourhood lattices from Roget's Thesaurus by themselves have some similarity with word association networks but do not appear to be sufficient to represent all of the structures that are expressed in figure 1. Furthermore, these lattices can be quite large and may need some manual editing pertaining to the level of detail to be used in Roget's Thesaurus and the manual creation of the graph layout.

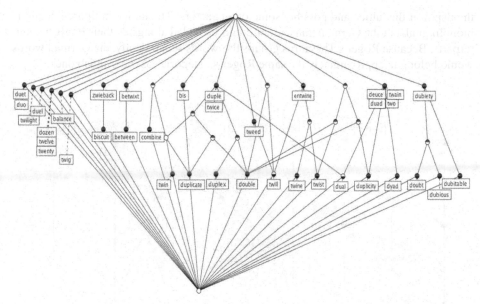

Fig. 2. A neighbourhood lattice for the genetic cognates of "two" from Roget's Thesaurus

In order to reduce the complexity of a neighbourhood lattice, a process of "restriction" can be applied. Figure 3 shows a "restricted neighbourhood lattice" of the lattice in figure 2 which contains only the polysemous cognates of "two" (i.e. the ones with more than one sense in Roget's Thesaurus) and only the corresponding categories from Roget's Thesaurus that contain more than one of the words from the set of formal objects. The following SQL statement was used:

```
SELECT r1.word, r1.sense INTO temptable
FROM RIT r1 WHERE r1.word IN
    (SELECT descendants FROM ETYM WHERE root = 'two')
AND EXISTS
    (SELECT * FROM RIT r2 WHERE r1.word = r2.word
    AND r1.sense != r2.sense));

SELECT t1.word, t1.sense FROM temptable t1
WHERE EXISTS
    (SELECT * FROM temptable t2 WHERE t1.sense = t2.sense
    AND t1.word != t2.word);
```

The restricted lattice contains the core structures from the original lattice in figure 2 and is sufficiently simple that it can be automatically constructed. But the process of restriction is not symmetric with respect to objects and attribute because a different result is obtained depending on whether the objects or the attributes are restricted first. Thus in order to automatically generate such diagrams a variety of heuristics might need to be developed depending on the number of concepts in the lattice, the depth of the lattice and possibly some other factors. The lattice in figure 3 again is monolingual, but the German translations could be added alongside their English counterparts. Because Roget's Thesaurus groups the words semantically, the German words would belong to approximately the same Roget's categories as the English ones.

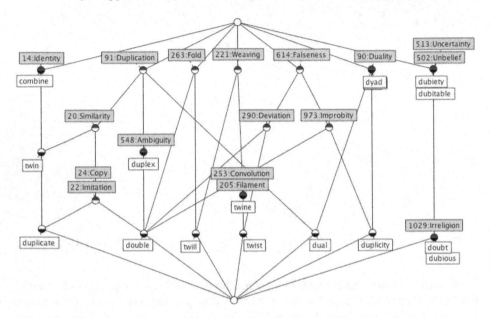

Fig. 3. A restricted neighbourhood lattice for the genetic cognates of "two" from Roget's Thesaurus

Neighbourhoods and the process of restriction can also be applied to graphs in general, not just lattices. Figure 4 shows a restricted neighbourhood of "two" derived from the word association database WAD. Word association data is asymmetric – there is a difference depending on whether a word is given as the "cue" in the word association test, or whether it is the word that participants give in response to a cue (the "target"). This neighbourhood of "two" shows words that invoke "two" as a response, when they are given as cue; or are invoked as a response to "two" when it is given as the cue. The set is "restricted" so that it contains only the words that have other, mutual, associations, in addition to their association with "two". That is, they additionally invoke another, or are invoked by another, of the words associated with "two". Two is excluded for simplicity – it would be connected to every node in the network.

The arrows in figure 5 show the cue/target direction (which may be reciprocal). The strength of the arrows corresponds to the percentage of participants who produced that particular association. This diagram has been included to demonstrate that traditional word associations are quite different both from the bilingual word association networks, as in figure 1, and from the purely semantic neighbourhood lattices in Roget's Thesaurus.

The final example in figure 5 in this section combines data from Roget's Thesaurus, from the etymological database ETYM and from a bilingual dictionary with a very

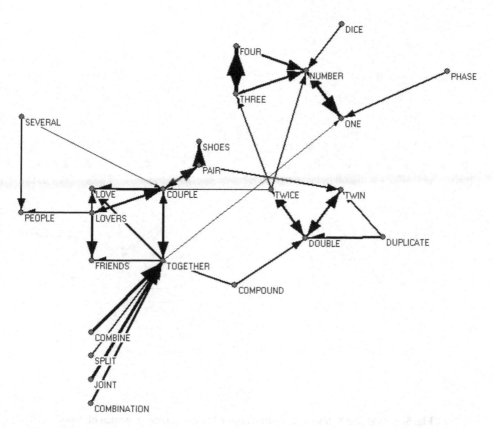

Fig. 4. Words relating to "two" derived from a restricted word association neighbourhood

simple form of the Soundex algorithm. The idea for this example is that a user could manually select a list of words (in this case all the words from figure 1) and a Root, and then automatically generate a diagram. Thus, for the concept lattice in figure 5, the English and German descendants of the Indo-European Root "DWO" were derived from ETYM. The set of descendants was restricted to words where the Root describes the stem of the word instead of a prefix. This is possible because ETYM contains information about whether a Root pertains to the prefix or stem of a word. In addition all the words from figure 1 were included, i.e. the "false" descendants "Zwiebel" and "duel" were also included. Furthermore, translations were derived for most of the words using a bilingual dictionary.

The English descendants of the Indo-European Root are then chosen as formal objects and the German descendants as formal attributes. A simplified Soundex algorithm is used by adding the objects and attributes "tw-", "Zw-", "bi-", and "du-", which facilitate clustering of phonetically similar words. For example, the English words starting with "tw" are assigned the additional attribute "tw-", and so on. In the centre of the figure, there are a few words that appear to be less regular and that correspond to some of the more interesting aspects that were discussed in relation to figure 1. The pairs "Zwieback/biscuit" and "Zweifel/doubt" are cases where the German and English

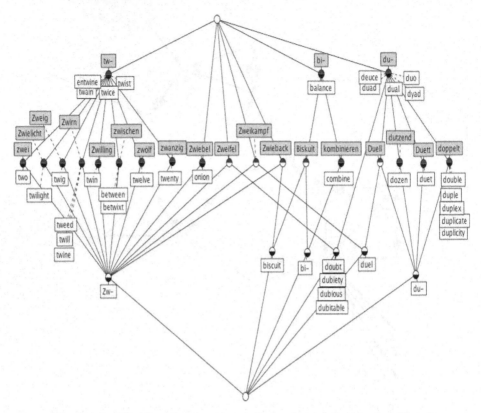

Fig. 5. A modified Soundex algorithm applied to the genetic cognates of "two"

words have a different Soundex encoding. The pair "Zweikampf/duel" is also in this category. As mentioned before, "duel" is not a descendant of "DWO", but "Zweikampf" is a descendant. The lattice does not clarify this fact. On the other hand, the pair "Zwiebel/onion" stands out because, in this case, the English word does not correspond to any of the given Soundex codes. It should be noted that only the words from figure 1 were manually added, but not their relationships. All relationships in figure 5 are automatically derived. Thus overall, the lattice in figure 5 presents a sufficient amount of the structures from figure 1. This example demonstrates that an approach that combines the databases, RIT, ETYM, a bilingual dictionary with a simplified Soundex algorithm appears to be most promising.

4 Conclusion

In order to automatically construct bilingual word association networks that can aid language learners, a variety of tools and technologies can be utilised. Table 1 summarises the different lexical databases and tools that facilitate the representation of genetic and semantic cognates and of conceptual structures in the mental lexicon. The previous section illustrates the use of some of these tools and databases in examples. The example in figure 5, which is based on the use of an etymological dictionary, a modified Soundex algorithm and Roget's Thesaurus, displays structures that are most similar to the target structure in figure 1. Thus an algorithm based on these components seems most appropriate for representing bilingual word association networks. Further testing using other examples will need to be conducted in order to establish whether this strategy is successful in other areas of the lexicon.

Table 1. Possible components in the creation of word association networks

genetic cognates	semantic cognates	mental lexicon
etymological dictionary Soundex	Roget's Thesaurus WordNet bilingual dictionary	word association data

The other examples from the previous section demonstrate that reliance on Roget's Thesaurus or word association data by itself produces results that are quite different from the target in figure 1. In particular, networks derived from traditional word association structures are very different from the bilingual word association network in figure 1. Our intuition is that networks, as in figure 1, are more suitable for vocabulary learning than networks as in figure 4, even though networks as in figure 4 are based on psychologically derived data. The reason for this is that in modern language instruction, each lesson usually focuses on a topic. Thus some of the structures in networks as in figure 4 are implicitly contained in each language lesson anyway. Bilingual networks as in figure 1, however, allow a focus on similar but contrasting language elements. This kind of information is not usually available in modern language teaching resources and would be an additional resource.

References

1. Barrière, C., St-Jacques, C.: Semantic Context Visualization to Promote Vocabulary Learning. In: Association Computers in the Humanities & Association for Literary and Linguistic Computing Conference, pp. 10–12 (2005)
2. Ganter, B., Wille, R.: Formal Concept Analysis. In: Mathematical Foundations, Springer, Heidelberg (1999)
3. Inkpen, D., Frunza, O., Kondrak, G.: Automatic Identification of Cognates and False Friends in French and English. In: Proceedings of the International Conference on Recent Advances in Natural Language Processing (RANLP 2005), pp. 251–257 (2005)
4. Kiss, G.R., Armstrong, C., Milroy, R., Piper, J.: An associative thesaurus of English and its computer analysis. In: Aitken, A.J., Bailey, R.W., Hamilton-Smith, N. (eds.) The Computer and Literary Studies. University Press, Edinburgh (1973)
5. Knuth, D.: The Art of Computer Programming Volume 3: Sorting and Searching, 2nd edn. Addison-Wesley Professional, London, UK (1998)
6. Nelson, D.L., McEvoy, C.L., Schreiber, T.A.: The University of South Florida word association, rhyme, and word fragment norms (1998), Available at
   http://w3.usf.edu/FreeAssociation/
7. Miller, G., Beckwith, R., Fellbaum, C., Gross, D., Miller, K.J.: Introduction to WordNet: an on-line lexical database. International Journal of Lexicography 3(4), 235–244 (1990)

8. Phillips, L.: The Double Metaphone Search Algorithm. C/C++ Users Journal 18(6) (June 2000)
9. Priss, U., Old, L.J.: Modelling Lexical Databases with Formal Concept Analysis. Journal of Universal Computer Science 10(8), 967–984 (2004)
10. Peter Mark, R.: Roget's International Thesaurus, 3rd edn. Thomas Crowell, New York (1962)
11. Wilson, M.D.: The MRC Psycholinguistic Database: Machine Readable Dictionary. Version 2. Behavioural Research Methods, Instruments and Computers 20(1), 6–11 (1988)

Appendix - Soundex Algorithm

The encoding of the Soundex algorithm consists of a letter followed by three numbers: the letter is the first letter of the word, and the numbers encode the remaining consonants (up to three). Vowels and "h, w, y" are ignored unless they are the first letter. The vowels are coded loosely following Grimm's Law. For example, "b, f, p, v" are each assigned the numeric code "1". Duplicates, such as "tt", are assigned only one code or the second consonant is ignored.

The exact Soundex algorithm is as follows
(see http://en.wikipedia.org/wiki/Soundex):

1. Retain the first letter of the string
2. Remove all occurrences of the following letters, unless it is the first letter: a, e, h, i, o, u, w, y
3. Assign numbers to the remaining letters (after the first) as follows:
   b, f, p, v = 1
   c, g, j, k, q, s, x, z = 2
   d, t = 3
   l = 4
   m, n = 5
   r = 6
4. If two or more letters with the same number were adjacent in the original name (before step 1), or adjacent except for any intervening h and w (American census only), then omit all but the first.
5. Return the first four characters, right-padding with zeroes if there are fewer than four.

There exist many soundex variants. The Double Metaphone algorithm (Phillips, 2000) provides significant improvements over the basic Soundex algorithm and is, for example, used to suggest spelling corrections in spell-checkers.

Using FCA for Encoding Closure Operators into Neural Networks*

Sebastian Rudolph

University of Karlsruhe
Institute AIFB
rudolph@aifb.uni-karlsruhe.de

Abstract. After decades of concurrent development of symbolic and connectionist methods, recent years have shown intensifying efforts of integrating those two paradigms. This paper contributes to the development of methods for transferring present symbolic knowledge into connectionist representations. Motivated by basic ideas from formal concept analysis, we propose two ways of directly encoding closure operators on finite sets in a 3-layered feed forward neural network.

1 Introduction

On the scientific quest for enabling machines to fulfill more and more sophisticated cognitive tasks, two basic antagonistic paradigms evolved.

On one side, one tries to capture (in a top-down manner, essentially by introspection and psychological experiments) the basic entities of human thought and their interplay in terms of symbols[1], symbol manipulation systems and formal logic. Also approaches involving conceptual graphs mainly fall into that class.

On the other side, advances in biology and medicine have provided bottom-up insights into the human way of information processing via networks of neurons. So, a contrary approach – started by [2] – tries to employ these findings by simulating neural structures (although this is mostly done in an extremely simplified way).

The interest in the integration of symbolic methods based on computational logic with artificial neural networks (also known as connectionist systems) has grown significantly in the last years. As a motivating goal of those efforts appears to combine the advantages of both approaches: While symbolic systems are superior in dealing (i.e., representing and reasoning) with structured data, connectionist systems show impressive capabilities when it comes to learning on

---

* This is an extended version of a paper presented at NeSy07 – the Third Workshop on Neural-Symbolic-Integration at IJCAI 2007. Sebastian Rudolph is supported by the Deutsche Forschungsgemeinschaft (DFG) under the ReaSem project and by the European Union under the NeOn project (IST-2005-027595).

[1] As opposed to the rich semiotic meaning of the term *symbol*, we use this word in a more syntactic sense following [1].

U. Priss, S. Polovina, and R. Hill (Eds.): ICCS 2007, LNAI 4604, pp. 321–332, 2007.

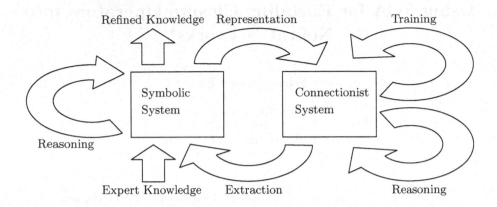

Fig. 1. The neural-symbolic learning cycle

larger datasets and generalizing the results to new input. See [3] for an overview of this prospering research area.

The neural-symbolic learning cycle depicted in Fig. 1 (see also [1]) proposes a general framework for organizing a neural-symbolic integration. In our paper, we focus on the representation subtask, i.e. encoding explicitly prespecified background knowledge. In particular, we investigate ways of canonically encoding *closure operators* into neural networks. Closure operators on attribute or feature sets arise naturally in various domains; whenever the validity of some features enforces the validity of others (as in human associative thinking and logic entailment to name just two extremes of a wide spectrum), this can be described by closure operators. So assume a neural network for some purpose has to be designed, where some rule-like partial information about the network's desired behavior is already known and can be stated in form of implications on the feature set. We now look for a neural network obeying those prescribed rules (which can then be trained on an example set to acquire further behavior). Previous approaches [4,5] tackle this problem by assigning a node of the network to each implication. We propose a contrary approach, where – roughly speaking – network nodes don't take the role of enforcing wanted features but preventing unwanted ones. This approach is motivated by the mathematical area of formal concept analysis.

In Section 2.1 we will introduce the basic notions *closure operator* and *implication* and show their correspondence. Section 2.2 will sketch the elementary ideas of *formal concept analysis*, based on which we will unfold our representation approach. Very briefly, Section 2.3 will recall the notion of a *3-layered feedforward network*. Section 3 then combines the approaches and provides two ways of encoding a formal context's closure operator into a neural network of the specified kind. In Section 4, we show how the approach can be applied to propositional logic programs, where it can be used to compute models. Finally, in Section 5, we conclude and give topics for ongoing research.

2 Preliminaries

2.1 Closure Operators and Implications

In this section, we will introduce two notions – closure operator and implications – and show their tight correspondence.

The following considerations are based on an arbitrary set M. Intuitively, it may be conceived as a set of *features* or *attributes* or *atomic propositions*, depending on the modelled problem. Many of the definitions and theoretical results presented in this and the next section apply to arbitrary sets, however, when it comes to questions of practical realization and computability, finiteness of M has to be presumed.

We will first define the fundamental notion of a closure operator. Roughly speaking, applying such an operator to a set can be understood as a minimal extension of that set in order to fulfill certain properties.

Definition 1. *Let M be an arbitrary set. A function $\varphi : \mathcal{P}(M) \to \mathcal{P}(M)$ (where $\mathcal{P}(M)$ denotes the powerset of M) will be called*

- EXTENSIVE, *if $A \subseteq \varphi(A)$ for all $A \subseteq M$,*
- MONOTONE, *if $A \subseteq B$ implies $\varphi(A) \subseteq \varphi(B)$ for all $A, B \subseteq M$, and*
- IDEMPOTENT, *if $\varphi(\varphi(A)) = \varphi(A)$ for all $A \subseteq M$.*

If φ is extensive, monotone, and idempotent, we will call it a CLOSURE OPERATOR. In this case, we will additionally call

- $\varphi(A)$ *the CLOSURE of A,*
- A CLOSED, *if $A = \varphi(A)$.*

The family of all closed sets is also called CLOSURE SYSTEM. Furthermore, any closure system constitutes a *lattice* with set inclusion as the respective order relation.

Mark that the notion of a closure operator is ubiquitous in both human associative thinking and classical (at least monotonic) logics.[2]

In the sequel, we show in which way closure operators are closely related to implications.

Definition 2. *Let M be an arbitrary set. An IMPLICATION on M is a pair (A, B) with $A, B \subseteq M$. To support intuition, we write $A \to B$ instead of (A, B).[3]*

For $C \subseteq M$ and a set $\mathfrak{I}$ of implications on M, let $C^{\mathfrak{I}}$ denote the smallest set with $C \subseteq C^{\mathfrak{I}}$ that additionally fulfills

$$A \subseteq C^{\mathfrak{I}} \quad implies \quad B \subseteq C^{\mathfrak{I}}$$

for every implication $A \to B$ in $\mathfrak{I}$.[4] If $C = C^{\mathfrak{I}}$, we call C $\mathfrak{I}$-CLOSED.

---

[2] For example, in classical first order logic, taking the set of all consequences $cons(\Phi) := \{\varphi \mid \Phi \models \varphi\}$ of a formula set Φ is a closure operator.

[3] To facilitate reading we will occasionally omit the parentheses, i.e., we will write $a, b \to c$ instead of $\{a, b\} \to \{c\}$.

[4] Note, that this is well-defined, since the mentioned properties are closed wrt. intersection.

It is well known that for a given set $A \subseteq M$ and implication set $\mathfrak{I}$, $A^{\mathfrak{I}}$ can be computed in linear time with respect to $|\mathfrak{I}|$ (see [6] or [7]). As can be easily seen, for any set $\mathfrak{I}$ of implications on any set M, $(.)^{\mathfrak{I}}$ is a closure operator. Moreover, for any closure operator $\varphi : \mathcal{P}(M) \to \mathcal{P}(M)$, there exists (at least) a set $\mathfrak{I}$ of implications on M such that $(.)^{\mathfrak{I}} = \varphi.$[5]

An elementary observation from logic becomes particularly obvious in this setting: a contradiction implies everything. Thus, if, say, two elements $a, b \in M$ are contradictory, this can be expressed by the implication $\{a, b\} \twoheadrightarrow M$. In the sequel, we will use the shorthand $a, b \twoheadrightarrow \bot$ for these special cases.

2.2 Formal Concept Analysis

The mathematical theory of formal concept analysis mainly deals with conceptual hierarchies which are generated from basic data structures encoding object-attribute relationships. Thereby, it provides a rather applied access to lattice theory For a comprehensive introduction into formal concept analysis, see [8].

In this section, we sketch the basic definitions and some results from formal concept analysis, as far as they are needed for this work. We start by defining the central underlying data structure.

Definition 3. *A* (FORMAL) CONTEXT $\mathbb{K}$ *is a triple* (G, M, I) *with*

- *an arbitrary set* G *called* OBJECTS,
- *an arbitrary set* M *called* ATTRIBUTES,
- *a relation* $I \subseteq G \times M$ *called* INCIDENCE RELATION

We read gIm *as: "object* g *has attribute* m".

Intuitively, a formal context is represented by a so-called cross table, where each row is associated to an object, each column to an attribute, and crosses indicate which object has which attributes.

Definition 4. *Let* $\mathbb{K} = (G, M, I)$ *be a formal context. We define a function* $(.)^{I} : \mathcal{P}(G) \to \mathcal{P}(M)$ *with*

$$A^{I} := \{m \mid gIm \ \ for \ all \ g \in A\}$$

for $A \subseteq G$*. Furthermore, we use the same notation to define the function* $(.)^{I} :$ $\mathcal{P}(M) \to \mathcal{P}(G)$ *where*

$$B^{I} := \{g \mid gIm \ \ for \ all \ m \in B\}$$

for $B \subseteq M$*.*

For convenience, we sometimes write g^{I} *instead of* $\{g\}^{I}$ *and* m^{I} *instead of* $\{m\}^{I}$*.*

Applied to an object set, this function yields all attributes common to these objects; by applying it to an attribute set we get the set of all objects having those attributes. The following facts are consequences of the above definitions:

---

[5] A naïve way to achieve this: given φ, let $\mathfrak{I} = \{A \twoheadrightarrow \varphi(A) \mid A \subseteq M\}$.

Proposition 1

- $(.)^{II}$ is a closure operator on G as well as on M.
- For $A \subseteq G$, A^I is a $(.)^{II}$-closed set and dually
- for $B \subseteq M$, B^I is a $(.)^{II}$-closed set.

The next definition shows how a conceptual hierarchy can be built from a formal context.

Definition 5. *Given a formal context* $\mathbb{K} = (G, M, I)$*, a* FORMAL CONCEPT *is a pair* (A, B) *with* $A \subseteq G$*,* $B \subseteq M$*,* $A = B^I$*, and* $B = A^I$*.*
We call the set A EXTENT *and the set* B INTENT *of the concept* (A, B)*.*
Let (A_1, B_1) *and* (A_2, B_2) *be formal concepts of a formal context. We call* (A_1, B_1) *a* SUBCONCEPT *of* (A_2, B_2) *(written:* $(A_1, B_1) \leq (A_2, B_2)$*) if* $A_1 \subseteq A_2$*. Then,* (A_2, B_2) *will be called* SUPERCONCEPT *of* (A_1, B_1)*.*

Proposition 2. *The concept intents of a formal concept are exactly those attribute sets closed with respect to* $(.)^{II}$*.*

It is well known from FCA that the set of all formal concepts of a formal context together with the subconcept-superconcept-order form a complete lattice, the so called *concept lattice*.

2.3   On Neural Networks

In this section, we recall the notion of a particular neural network giving a formal definition that we will build upon in the subsequent sections.

Definition 6. *A* 3-LAYERED FEEDFORWARD NETWORK *is defined as a tuple* $N = (\mathcal{I}, \mathcal{H}, \mathcal{O}, t, w)$ *where*

- $\mathcal{I}, \mathcal{H}, \mathcal{O}$ *are finite disjoint sets called* INPUT NODES, HIDDEN NODES, *and* OUTPUT NODES,
- $t : (\mathcal{I} \cup \mathcal{H} \cup \mathcal{O}) \to \mathbb{R}$ *is the* THRESHOLD FUNCTION, *and*
- $w : (\mathcal{I} \times \mathcal{H}) \cup (\mathcal{H} \times \mathcal{O}) \to \mathbb{R}$ *is the* WEIGHT FUNCTION.

Clearly, neural networks are intended as computational models, i.e. they are designed to calculate something. Hence given a neural network we can define a function capturing its computational behaviour.

Definition 7. *Given a 3-layered feedforward network* N *as specified in Definition 6, the* ASSOCIATED NETWORK FUNCTION $f_N : \mathcal{P}(\mathcal{I}) \to \mathcal{P}(\mathcal{O})$ *is defined in the following way: For a given argument set* S*, we define the set* $\mathcal{A}_S \subseteq \mathcal{I} \cup \mathcal{H} \cup \mathcal{O}$ *of* ACTIVATED NEURONS *as follows (using the shortcut* $\chi_A(a) = |\{a\} \cap A|$*):*

- *for every* $i \in \mathcal{I}$*, we set* $i \in \mathcal{A}_S$ *exactly if* $\chi_S(i) - t(i) > 0$*,*
- *for every* $h \in \mathcal{H}$*, we set* $h \in \mathcal{A}_S$ *exactly if* $\sum_{i \in \mathcal{I}} \chi_A(i) w_{ih} - t(h) > 0$*, and*
- *for every* $o \in \mathcal{O}$*, we set* $h \in \mathcal{A}_S$ *exactly if* $\sum_{h \in \mathcal{H}} \chi_A(i) w_{ho} - t(o) > 0$*.*

Finally, we set $f_N(S) = \mathcal{A}_S \cap \mathcal{O}$*.*

This definition exactly mirrors the usual way of calculating with neural networks, presuming the Heaviside step function as activation function.

In the sequel, we aim at the special case of simulating a closure operator $\varphi : \mathcal{P}(M) \to \mathcal{P}(M)$ with this kind of neural network, i.e., input and output layer correspond to the same set (namely M).

3   Encoding of Closure Operators Inspired by FCA

The basic idea for this paper is to use formal contexts to represent closure operators. In particular (as we have seen in Section 2.2), for a formal context $\mathbb{K} = (G, M, I)$, the function $(.)^{II} : \mathcal{P}(M) \to \mathcal{P}(M)$ is a closure operator on M. Moreover, *any* closure operator on a finite set M can be represented by an appropriate formal context.[6]

So, in this section, we propose two canonical ways to translate a formal context into a 3-layered feedforward network, which – given a set $A \subseteq M$ – computes its closure A^{II}.

The intuition hereby is to identify the hidden layer neurons with the object set of the formal context. We realize the $(.)^{II}$-operator by first applying $(.)^{I}$ to A (which by definition yields an object set represented by the activated neurons in the hidden layer) and, afterwards, applying $(.)^{I}$ to A^{I} thus obtaining the closure of the attribute set at the output layer.

Definition 8. *For a given formal context* $\mathbb{K} = (G, M, I)$, *we define a corresponding 3-layered feedforward network* $N_{\mathbb{K}}$ *in the following way:*

- $\mathcal{I} = \{i_m \mid m \in M\}$
- $\mathcal{O} = \{o_m \mid m \in M\}$
- $\mathcal{H} = \{h_g \mid g \in G\}$
- $t(i) := 0.5$ *for all* $i \in \mathcal{I}$
- $w_{i_m h_g} = w_{h_g o_m} = \begin{cases} 0 & \text{if } gIm \\ -1 & \text{otherwise.} \end{cases}$
- $t(n) := -0.5$ *for all* $n \in \mathcal{H} \cup \mathcal{O}$.

Next, we will prove that indeed the associated network function $f_{N_{\mathbb{K}}}$ corresponds to the closure operator $(.)^{II}$, i.e., for all $A \subseteq M$, we have that $A^{II} = \{m \mid o_m \in f_{N_{\mathbb{K}}}(\{i_{\tilde{m}} \mid \tilde{m} \in A\})\}$

Proposition 3. *Let* $\mathbb{K} = (G, M, I)$ *be a formal context and* $N_{\mathbb{K}}$ *the corresponding neural network. Then*

1. *for every* $A \subseteq M$, *activating (exactly) the set* $\{i_m \mid m \in A\}$ *of input neurons leads to an activation of (exactly) the set* $\{h_g \mid g \in A^{I}\}$ *of hidden neurons and*
2. *for every* $B \subseteq G$, *activating (exactly) the set* $\{h_g \mid g \in B\}$ *of hidden neurons leads to an activation of (exactly) the set* $\{o_m \mid m \in B^{I}\}$ *of output neurons.*

---

[6] One method, how to construct a formal context with this property will be explicated in Section 4.

Proof. Consider the hidden layer neuron h_g representing the object $g \in G$. Now, since $A^I = \{g \mid gIm \text{ for all } m \in A\}$ we have that $g \in A^I$ exactly if g has all attributes from A. Obviously, this is the case if and only if

$$\sum_{m \in M} \chi_{A_A}(i_m)w_{i_m h_g} = \sum_{m \in A} w_{i_m h_g} = 0 > -0.5.$$

The second claim is proved in exactly the same manner.

The next corollary then follows immediately be the definition of $(.)^{II}$ as twofold application of $(.)^I$.

Corollary 1. $N_{\mathbb{K}}$ *computes* $(.)^{II}$.

This approach is quite close to formal concept analysis since the neurons of the hidden layer directly correspond to the object set of the represented formal context. The negative weights are necessary due to the fact that $(.)^I$ is (in both variants) an *antitone* function (i.e. $A \subseteq B$ implies $B^I \subseteq A^I$).

However, this can be overcome by a simple "work around": instead of mirroring the functions $A \mapsto A^I$ and $B \mapsto B^I$ (for $A \subseteq M$ and $B \subset G$), one could use the functions $A \mapsto M \setminus A^I$ and $B \mapsto (M \setminus B)^I$ instead. Both of them are monotone and can hence be modelled with only positive weights, and still their composition yields the wanted operator $(.)^{II}$. In the sequel, we will elaborate this idea.

Definition 9. *For a given formal context* $\mathbb{K} = (G, M, I)$, *we define a corresponding 3-layered feedforward network* $\tilde{N}_{\mathbb{K}}$ *in the following way:*

- $\mathcal{I} = \{i_m \mid m \in M\}$
- $\mathcal{O} = \{o_m \mid m \in M\}$
- $\mathcal{H} = \{h_g \mid g \in G\}$
- $t(i) := 0.5$ *for all* $i \in \mathcal{I}$
- $t(o_m) := -0.5 + |\{g \in G \mid \neg gIm\}|$ *for all* $o_m \in \mathcal{O}$
- $w_{i_m h_g} = w_{h_g o_m} = \begin{cases} 0 \text{ if } gIm \\ 1 \text{ otherwise.} \end{cases}$
- $t(h) := 0.5$ *for all* $h \in \mathcal{H}$

Proposition 4. *Let* $\mathbb{K} = (G, M, I)$ *be a formal context and* $\tilde{N}_{\mathbb{K}}$ *the corresponding neural network. Then*

1. *for every* $A \subseteq M$, *activating (exactly) the set* $\{i_m \mid m \in A\}$ *of input neurons leads to an activation of (exactly) the set* $\{h_g \mid g \in G \setminus A^I\}$ *of hidden neurons and*
2. *for every* $B \subseteq G$, *activating (exactly) the set* $\{h_g \mid g \in B\}$ *of hidden neurons leads to an activation of (exactly) the set* $\{o_m \mid m \in (G \setminus B)^I\}$ *of output neurons.*

Proof. 1. Consider the hidden layer neuron h_g representing the object $g \in G$. Now, since $A^I = \{g \mid gIm \text{ forall } m \in A\}$, we have that $g \in A^I$ exactly if g has all attributes from A. Obviously, this is the case if and only if

$$\sum_{m \in A} w_{mg} = 0 < 0.5.$$

Therefore, any h_g being activated must be in $G \setminus A^I$.

2. Now, consider the output layer neuron o_m representing the attribute $m \in M$. If B is activated in the hidden layer, o_m will be activated exactly if

$$\sum_{g \in B} w_{gm} = |\{g \in B \mid \neg gIm\}| > -0.5 + |\{g \in G \mid \neg gIm\}|$$

Yet, due to $B \subseteq G$, this can only be the case iff $\{g \in B \mid \neg gIm\}| = |\{g \in G \mid \neg gIm\}|$ which is equivalent to the statement that gIm for all $g \in G \setminus B$. Hence, o_m is activated exactly if $m \in (G \setminus B)^I$.

Corollary 2. $\tilde{N}_{\mathbb{K}}$ computes $(.)^{II}$.

Proof. Due to the preceding proposition, applying $\tilde{N}_{\mathbb{K}}$ to an attribute set A will first activate the hidden neurons representing $G \setminus A^I$ and then the output neurons representing $(G \setminus (G \setminus A^I))^I = (A^I)^I = A^{II}$

As already mentioned, using this type of network will activate exactly those hidden layer neurons *not* contained in A^I, if A is entered.

An interesting feature of both presented networks is their symmetry: for all $m \in M$ and $g \in G$, $w_{i_m h_g} = w_{h_g o_m}$. Although this puts structural constraints on the neural network and might therefore hamper the application of learning strategies, it might be useful from a quite different point of view: in cases, where the neural network will be hardwired, input and output layer could be identified and calculation be done in a "back-and-forth manner" using the links twice for every calculation.

4   Application to Propositional Logic Programs

In this section, we will show, how the presented strategy can be applied in a propositional logic programming scenario.

Logic programming is especially suited for this approach, since

- any logic program essentially consists of a set of implications and hence
- entailment can (at least in the negation-free case) therefore be described by a closure operator on the ground facts.

Consequently, one can assign to every logic program an operator T_P which applied to a set of ground facts intuitively calculates the immediate consequences by "applying" each implication once. The entailment closure operator can then be simulated by iteratively applying T_P until a fixed point is reached. [5] presents an approach to encode T_P into a recurrent 3-layered neural network, by assigning every implication to a node of the middle layer. To make this clear, consider the following example.

Imagine, some kind of animal has to be determined via some tests. Let furthermore the only available tests be to indicate whether the animal is a mammal, a bird, a monkey, a donkey, an owl, a fowl or a frog. Hence $M := \{\text{donkey}, \text{monkey}, \text{mammal}, \text{frog}, \text{bird}, \text{owl}, \text{fowl}\}$. Then the implications presented in Fig. 2 characterize the setting:

$$
\begin{aligned}
\text{monkey} &\dashrightarrow \text{mammal} \\
\text{donkey} &\dashrightarrow \text{mammal} \\
\text{owl} &\rightarrow \text{bird} \\
\text{fowl} &\rightarrow \text{bird} \\
\text{monkey, donkey} &\dashrightarrow \bot \\
\text{owl, fowl} &\dashrightarrow \bot \\
\text{mammal, bird} &\dashrightarrow \bot \\
\text{mammal, frog} &\dashrightarrow \bot \\
\text{bird, frog} &\dashrightarrow \bot
\end{aligned}
$$

Fig. 2. Implication representing the knowledge in our example

Following [4], the neural network corresponding to the T_P-operator representing those implications interpreted as a logic program would look like the one represented in Fig. 3.

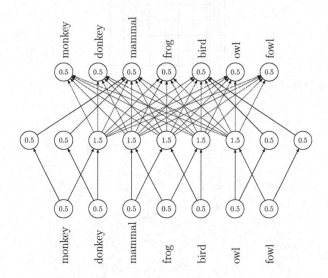

Fig. 3. Neural network corresponding to the T_P-operator of our propositional setting. All weights are set to 1. The dotted lines are those indicating a contradiction.

The set $\{donkey, fowl\}$ demonstrates that, in general, T_P may have to be applied several times to calculate the closure, since

$$
T_P(\{\text{donkey}, \text{fowl}\}) = \{\text{donkey}, \text{mammal}, \text{fowl}, \text{bird}\}
$$

and

$$
T_P(\{\text{donkey}, \text{mammal}, \text{fowl}, \text{bird}\}) = M.
$$

Now we consider how our method would apply. So, we have to find a formal context $\mathbb{K} = (G, M, I)$, where $A^{II} = A^{\mathfrak{I}}$ for all $A \subseteq M$. One possibility to do so is to consider the lattice of all $\mathfrak{I}$-closed sets. Fig. 4 represents this.

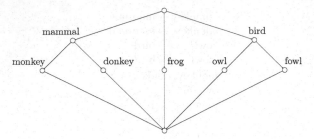

Fig. 4. Lattice of the ℑ-closed sets

	monkey	donkey	mammal	frog	bird	owl	fowl
g_1					×		×
g_2					×	×	
g_3				×			
g_4		×	×				
g_5	×		×				

Fig. 5. The formal context 𝕂 corresponding to the closure operator to describe

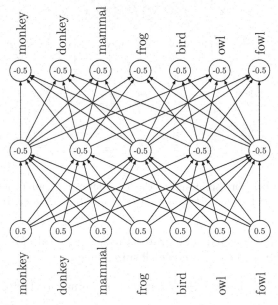

Fig. 6. Neural network corresponding to the consequence operator of our propositional setting. All weights are set to -1

Yet, a well-known result of FCA provides a direct way to find a minimal set of objects for a formal context that is supposed to generate a given lattice. One has to take all supremum-irreducible elements as objects. Looking at the diagram, the supremum-irreducible elements are exactly those having only one lower neighbour. In our particular case, these are exactly all upper neighbours of the bottom element. Hence, we can derive the formal context depicted in Fig. 5.

According to the preceding section, there are two ways of using this kind of formal context to define a neural network that computes the closure of a given set directly (i.e., no manyfold application – likewise no recurrent organisation – of the net would be necessary).

The first one (corresponding to the definition of $N_{\mathbb{K}}$) is shown in Fig. 6. Note that all drawn edges correspond to weights of -1.

The second network (corresponding to the definition of $\tilde{N}_{\mathbb{K}}$) is shown in Fig. 7. Here all drawn edges carry weight of 1.

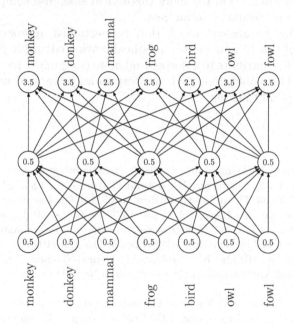

Fig. 7. Neural network $\tilde{N}_{\mathbb{K}}$ corresponding to the consequence operator of our propositional setting. All weights are set to 1.

5   Conclusion and Future Work

In our paper, we presented two new canonical ways for generating neural networks that compute the closure operator of a given finite set. We thereby provide a method to support the representation part of the neural-symbolic learning cycle by presenting an encoding strategy for a kind of background knowledge generically occurring in the area of knowledge processing.

In contrast to other methods, where the closure is approximated iteratively (using a recurrent network), the networks presented in our approach will calculate it directly, i.e., by a single run of the network.

Moreover, as shown by our example, there are cases where this kind of representation is also advantageous in terms of the number of hidden layer neurons needed. In general, this approach seems to be especially beneficial, if the number of implications becomes large.

Naturally, the proposed method requires preprocessing of the implicative information to be encoded. Depending on how this information is given, it has to be transformed into a formal context. The way we presented here – namely generating the whole lattice of the closed sets and identifying the supremum-irreducible elements of it – is certainly not optimal with respect to time costs (in the worst case, the size of the lattice can be $2^{|M|}$). So one important field of future research is to find more efficient methods to convert implicative knowledge into small contexts. On the more theoretical side, also complexity bounds for this kind of task would be of interest.

More generally, we are convinced, that connectionist approaches – if taken into the focus of (up to now more symbolically oriented) fields like conceptual structures – could contribute to a deeper and more comprehensive understanding of either field. This could even pave the way to an integrated neural-symbolic theory of conceptual thinking.

References

1. Bader, S., Hitzler, P.: Dimensions of neural-symbolic integration - a structured survey. In: Artemov, S., Barringer, H., d'Avila Garcez, A.S., Lamb, L.C., Woods, J. (eds.) We Will Show Them: Essays in Honour of Dov Gabbay. International Federation for Computational Logic, vol. 1, pp. 167–194. College Publications (2005)
2. McCulloch, W.S., Pitts, W.: A logical calculus of the ideas immanent in nervous activity. Bulletin of Mathematical Biophysics 5, 115–133 (1943)
3. d'Avila Garcez, A., Broda, K., Gabbay, D.: Neural-Symbolic Learning Systems: Foundations and Applications. In: Perspectives in Neural Computing, Springer, Heidelberg (2002)
4. Hölldobler, S., Kalinke, Y.: Towards a massively parallel computational model for logic programming. In: Proceedings ECAI94 Workshop on Combining Symbolic and Connectionist Processing, ECCAI, pp. 68–77 (1994)
5. Hitzler, P., Hölldobler, S., Seda, A.K.: Logic programs and connectionist networks. J. Applied Logic 2, 245–272 (2004)
6. Dowling, W.F., Gallier, J.H.: Linear-time algorithms for testing the satisfiability of propositional Horn formulae. J. Log. Program. 1, 267–284 (1984)
7. Maier, D.: The Theory of Relational Databases. Computer Science Press (1983)
8. Ganter, B., Wille, R.: Formal Concept Analysis: Mathematical Foundations. Springer-Verlag New York, Inc. Secaucus, NJ, USA, Translator-C. Franzke (1997)

Arc Consistency Projection: A New Generalization Relation for Graphs

Michel Liquiere

LIRMM,
161 Rue ada,
34392 Montpellier cedex 5,
France
Liquiere@lirmm.fr

Abstract. The projection problem (conceptual graph projection, homomorphism, injective morphism, θ-subsumption, OI-subsumption) is crucial to the efficiency of relational learning systems. How to manage this complexity has motivated numerous studies on learning biases, restricting the size and/or the number of hypotheses explored. The approach suggested in this paper advocates a projection operator based on the classical arc consistency algorithm used in constraint satisfaction problems. This projection method has the required properties : polynomiality, local validation, parallelization, structural interpretation. Using the arc consistency projection, we found a generalization operator between labeled graphs. Such an operator gives the structure of the classification space which is a concept lattice.

1 Introduction

The complexity of the computation of the generality relation between two relational descriptions, is a crucial problem. For conceptual graphs [1] this operation is named projection. Such an operation is linked to a classical problem in the graph community: the search for an homomorphism between two graphs. As stated in [2], "the elementary reasoning operation, projection is a kind of graph homomorphism that preserves the partial order defined on labels". The search for an homomorphism between a tree and a graph is polynomial but between general graphs, the problem is NP complete [3]. In conceptual graph community, different algorithms are proposed for the projection problem [4,5,6].

From another point of view, Inductive Logic Programming systems (ILP) commonly used a generality relation, a decidable restriction of logical implication named θ-subsumption. The homomorphism is also directly linked to the θ-subsumption operation [7]. In machine learning, the complexity of this operation has motivated the use of learning biases: syntactic biases (trees [8], specific graph [9,10]), efficient implementation [7] and approximation of θ-subsumption [11].

Finally, the homomorphism is also linked to the classical constraint programming resolution (CSP) [12]. This final link gives an interesting algorithmic point

U. Priss, S. Polovina, and R. Hill (Eds.): ICCS 2007, LNAI 4604, pp. 333–346, 2007.
© Springer-Verlag Berlin Heidelberg 2007

of view since CSP community has many results which improve the resolution algorithm.

In this paper, we propose to use a part of these three domains for a classical unsupervised machine learning problem. We represent each example by a labeled graph. We use a new generality relation named AC-projection based on the arc consistency algorithm [12] and we prove that the search space, for a relational machine learning classification problem, is a concept lattice [13].

2   A New Projection: AC-Projection

Any constraint satisfaction problem can be viewed as a "network" of variables and constraints. In this network each variable is connected to the constraints that involve it and each constraint is connected to the variables it involves. Among backtracking based algorithms for constraint satisfaction problems, algorithm employing constraint propagation, like forward checking and arc consistency [12], have had the most practical impact. These algorithms use constraint propagation (arc consistency) during a search to prune inconsistent values from the domains of the uninstantiated variables.

2.1   AC-Projection and Arc Consistency

In this paragraph we present the arc consistency using a graph notation. For other presentations see the books [3,12].

Notation. For a labelled directed graph, named *digraph* in this paper, G, we note $V(G)$ the set of vertices of G, $A(G)$ the set of arcs of G, $L(G)$ the set of labels of G. For a vertex $x \in V(G)$ we note $l(x) \in L(G)$ the label of x, $N(x)$ the set of all the neighbors of x, $P(x) \subseteq N(x)$ the predecessors of x and $S(x) \subseteq N(x)$ the successors of x.

For a finite set S we note 2^S the set of all subsets of S (power set). In this paragraph we study some important properties of the arc consistency.

Definition 1 (labeling). *Let G_1 and G_2 be two digraphs. We named labeling from G_1 into G_2 a mapping $\mathcal{I}:V(G_1) \rightarrow 2^{V(G_2)} \mid \forall\, x \in V(G_1), \forall\, y \in \mathcal{I}(x), l(x)=l(y)$.*

Thus for a vertex $x \in V(G_1)$, $\mathcal{I}(x)$ is a set of vertices of G_2 with the same label $l(x)$. We can think of $\mathcal{I}(x)$ as the set of "possible images" of the vertex x in G_2. This first labeling is trivial but can be refined using the neighborhood relations between vertices.

Definition 2 ($\sim\!\succ$). *Let G be a digraph, $V_1 \subseteq V(G)$, $V_2 \subseteq V(G)$. We note $V_1 \sim\!\succ V_2$ iff*
1) $\forall x_k \in V_1\ \exists y_p \in V_2 \mid (x_k, y_p) \in A(G)$
2) $\forall y_q \in V_2\ \exists x_m \in V_1 \mid (x_m, y_q) \in A(G)$.

In this definition we give a direct relation between two sets of vertices V_1 and V_2. So for each vertex x_k of V_1 there is at least one vertex y_p of V_2 which is a neighbor of x_k:$((x_k, y_p) \in A(G))$ and all vertices of V_2 are a neighbor of, at least, one vertex of V_1 (oriented condition). This is not a one to one relation like the subgraph isomorphism.

Definition 3 (Consistency for one arc). *Let G_1 and G_2 be two digraphs. We say that a labeling $\mathcal{I}:V(G_1) \rightarrow 2^{V(G_2)}$ is consistent with an arc $(x, y) \in A(G_1)$, iff $\mathcal{I}(x) \rightsquigarrow \mathcal{I}(y)$.*

In the example of Figure 1, a vertex is designated by a letter and a number: the letter is the label of the vertex and the number is only an identification number. In this example the labeling $\mathcal{I}:\mathcal{I}(a_0)=\{a_4, a_{10}\}$ and $\mathcal{I}(b_1)=\{b_5, b_9\}$, is consistent with the edge (a_0, b_1) since $\mathcal{I}(a_0) \rightsquigarrow \mathcal{I}(b_1)$.

Definition 4 (AC-projection $\rightarrow$). *Let G_1 and G_2 be two digraphs. A labeling $\mathcal{I}$ from G_1 into G_2 is an AC-projection iff $\mathcal{I}$ is consistent with all the arcs $e \in A(G_1)$. We note it $G_1 \rightarrow G_2$*

The name "AC-projection" comes from the classical AC (arc consistency) used in [12].

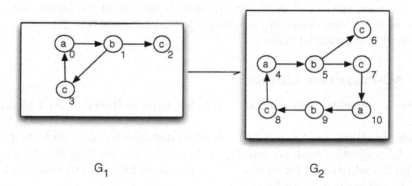

G_1 G_2

Fig. 1. AC-Projection: Example

Consider the labelling $\mathcal{I}(a_0):\{a_4, a_{10}\}$, $\mathcal{I}(b_1):\{b_5, b_9\}$, $\mathcal{I}(c_2):\{c_6, c_7, c_8\}$,$\mathcal{I}(c_3):\{c_7, c_8\}$. We verify $\mathcal{I}(a_0) \rightsquigarrow \mathcal{I}(b_1)$, $\mathcal{I}(b_1) \rightsquigarrow \mathcal{I}(c_2)$, $\mathcal{I}(b_1) \rightsquigarrow \mathcal{I}(c_3)$, $\mathcal{I}(c_3) \rightsquigarrow \mathcal{I}(a_0)$. Then $\mathcal{I}$ is an AC-projection from G_1 into G_2 since $\mathcal{I}$ is a labelling consistent with all arcs of G_1.

2.2 AC-Projection Properties

We have defined a new mapping relation between graphs. In this paragraph we study the properties of this relation (complexity, interpretation).

We recall the classical homomorphism definition for digraphs.

Definition 5 (homomorphism $\mapsto$). *A homomorphism of a digraph G_1 to a digraph G_2 is a mapping of the vertex sets $f{:}V(G_1) \to V(G_2)$ which preserves the arcs and labels, i.e such that $(x,y) \in A(G_1) \Rightarrow (f(x),f(y)) \in A(G_2)$ and $\forall x$, $l(x)=l(f(x))$.*
Notation: $G_1 \mapsto G_2$

Note that for digraph $(f(x),f(y)) \in A(G_2)$ implies that $f(x) \neq f(y)$, since each edge of $A(G_2)$ consists of two distinct elements.

We have the following proposition which links the AC-projection to the Homomorphism.

Proposition 1. *For two digraphs G_1 and G_2, if $G_1 \mapsto G_2$ then $G_1 \to G_2$.*

Proof. See [3]

This proposition is the foundation of many CSP resolution methods. These methods are based on the classical arc consistency algorithm AC1 used in CSP, which has been improved (AC2 ... AC5), the actual minimal complexity is: $O(ed^2)$ where e is the number of arcs and d the size of the largest domain [12].

In our case, the size of the largest domain is the size of the largest subset of nodes with the same label. So an AC-projection between two digraphs can be computed in polynomial time.

2.3   AC-Projection Algorithm

We give a simple AC-projection algorithm for digraphs (based on AC1 algorithm [12]).

This algorithm *Arc-Consistency* takes two digraphs G_1, G_2 and tests if there is an AC-projection from G_1 into G_2. It begins by the creation of a first rough labeling $\mathcal{I}$ and reduces, for each vertex x, the given lists $\mathcal{I}(x)$ to consistent lists using the procedure *ReviseArc*.

The consistency check fails if some $\mathcal{I}(x)$ becomes empty; otherwise the consistency check succeeds and the algorithm gives the labeling $\mathcal{I}$ which is an AC-projection $G_1 \to G_2$.

Procedure: ReviseArc
Data: An arc $(x,y) \in V(G_1)$
Data: A labeling $\mathcal{I}$ from G_1 into G_2
Data: A digraph G_2
Result: A new labeling $\mathcal{I}'$ from G_1 into G_2
$\mathcal{I}':= \mathcal{I}$;
$\mathcal{I}'(x):= \mathcal{I}(x)$ - $\{x' \in V(G_2) \mid \not\exists y' \in \mathcal{I}(y)$ with $(x',y') \in A(G_2)\}$;
$\mathcal{I}'(y):=\mathcal{I}(y)$ - $\{y' \in V(G_2) \mid \not\exists x' \in \mathcal{I}(x)$ with $(x',y') \in A(G_2)\}$;
return $\mathcal{I}'$

Procedure: Arc-Consistency

Data: Two digraphs G_1 and G_2

Result: An AC-projection $\mathcal{I}$ from G_1 into G_2 if there is one else an empty set $\emptyset$

// Initialisation

for $x \in V(G_1)$ **do**

| $\mathcal{I}(x) = \{y \in V(G_2) \mid l(x) = l(y))\}$;

end

$S := A(G_1)$;

while $S \neq \emptyset$ **do**

    Choose an arc (x,y) from S; // In general the first element of S

    $\mathcal{I}':=\text{ReviseArc}((x,y),\mathcal{I},G_2)$;

    //If for one vertex x $\in$ V(G_1) we have $\mathcal{I}'$(x)= $\emptyset$ then there is no arc consistency

    if $(\mathcal{I}'(x) = \emptyset)$ *or* $(\mathcal{I}'(y) = \emptyset)$ **then**

    | **return** $\emptyset$;

    end

    // $\mathcal{I}'$ is consistent now with the arc (x, y); but it can be non-consistent with some other previously tested arcs so we have to verify and change (if necessary), the consistency of all these arcs.

    if $\mathcal{I}(x) \neq \mathcal{I}'(x)$ **then**

    | $S := S \bigcup \{(x',y') \in V(G_1) \mid x' = x \text{ or } y' = x\}$;

    end

    if $\mathcal{I}(y) \neq \mathcal{I}'(y)$ **then**

    | $S := S \bigcup \{(x',y') \in V(G_1) \mid x' = y \text{ or } y' = y\}$;

    end

    Remove (x,y) from S;

    $\mathcal{I}:=\mathcal{I}'$;

end

return $\mathcal{I}$;

The *Arc-Consistency* algorithm has a polynomial time complexity [3,12] and gives, if there is one, an AC-projection $\mathcal{I}$ from G_1 into G_2 verifying: for all AC-projection $\mathcal{I}'$ from G_1 into G_2, we have $\forall$ x $\in$ V(G_1), $\mathcal{I}'$(x) $\subseteq$ $\mathcal{I}$(x) [3].

3   AC-Projection and Machine Learning

In [10], we have studied the construction of a concept lattice, where the extension part is a subset of the set of example but where the intension part is described by a digraph. In the context of machine learning, the automatic bottom up construction of such a hierarchy can be viewed as an unsupervised conceptual classification method. In this paper the generalization partial order was based on homomorphism relation between digraph. To deal with the homomorphism complexity, we proposed a class of digraph with a polynomial homomorphism operation. This limit the generality of the description language.

In that paper we propose to put the bias on the projection operator. Since the complexity of the AC-projection is polynomial, our idea is to use the AC-projection algorithm instead of the homomorphism projection.In doing so, we need a structural interpretation of the results. In the case of the subgraph isomorphism relation between two graphs, there is no interpretation problem, because it is an "inclusion" relationship. For the homomorphism relation the interpretation is less natural since two vertices can get the same image. The structural interpretation of the AC-projection seems unnatural. For example, see Figure 1 and seek for the substructures which are in G_1 and G_2.

In fact, in the paper [3], the author gives the following proposition:

Proposition 2. *Let G_1 and G_2 be any labelled digraphs with $G_1 \rightarrow G_2$. If an directed labeled tree T satisfies $T \mapsto G_1$ then $T \mapsto G_2$.(recall $\mapsto$ is the homomorphism relation)*

A limited interpretation of the proposition 2 is: every subtree of G_1 has an homomorphic image in G_2. So all the covering trees of the digraph G_1 of the Figure 1, are homomorphic with G_2.

3.1   AC-Projection and Generalization

The basic building blocks of concept learning is the notion of example and description language. In our framework, each example, of the set of example, is described by one digraph. So we have a set of digraphs $\mathcal{E}$. We want to find a set of labeled digraphs which have an AC-projection with a subset of the labelled

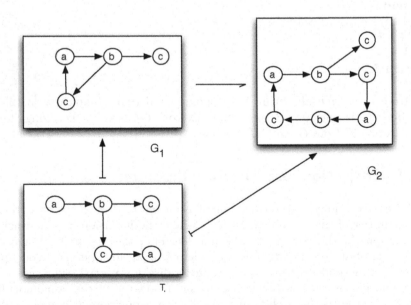

Fig. 2. AC-projection and interpretation

digraphs in $\mathcal{E}$. It is a classical unsupervised classification problem. A generalization algorithm uses a generalization operator: from two graphs we search for the
more specific graph which generalizes two graphs (least general generalization
[14]). Our generalization order will use the AC-projection relation.

First, we have to specify the generalisation relation between digraphs.

Definition 6 (Generalisation relation)
For two digraphs G_1,G_2, we consider that G_1 is more general than G_2 iff $G_1 \rightharpoonup G_2$.

This relation is only a pre-order because the antisymmetry property is not fulfilled.

The same problem occurs in Inductive Logic Programming. To get rid of this
problem, Plotkin [14] defined equivalence classes of clauses, and showed that
there is a unique representative of each clause, which he named 'the reduced
clause'. For this purpose, we define the following equivalence relation between
two digraphs.

Definition 7 (AC-equivalence graphs)
Two digraphs G_1 and G_2 are AC-equivalent, denoted by $G_1 \leftrightharpoons G_2$, iff both $G_1 \rightharpoonup G_2$ and $G_2 \rightharpoonup G_1$.

For example in the Figure 1 we have $G_1 \rightharpoonup G_2$ but we also have $G_2 \rightharpoonup G_1$ with
the labelling $\mathcal{I}$: $\mathcal{I}(a_4) = \{a_0\}, \mathcal{I}(a_{10}) = \{a_0\}, \mathcal{I}(b_5) = \{b_1\}, \mathcal{I}(b_9) = \{b_1\}, \mathcal{I}(c_7) = \{c_3\}, \mathcal{I}(c_8) = \{c_3\}, \mathcal{I}(c_6) = \{c_2, c_3\}$.

Using this equivalence relation, we can define equivalence classes of digraph.

3.2   AC-Projection and Reduction

We have an equivalence relation between graphs using the AC-projection. In this
paragraph we study the properties of this operation and search for a reduced
element in an equivalence class of graphs. For this purpose, we define two reduction operators. Using these operators we construct an AC-equivalent digraph by
removing (first operator) or merging (second operator) vertices.

Definition 8 (AC-redundant vertex)
*For a digraph G, for a vertex $x \in V(G)$, if $G \leftrightharpoons G\text{-}x$ then x is an AC-redundant
vertex. (With $G\text{-}x = G'$ s.t $V(G')=V(G) - \{x\}$ and $A(G')=A(G)\text{-}\{(y,z) \mid y=x$
or $z=x\}$).*

In the Figure 3 the node 1 labelled "c" is AC-redundant.

Definition 9 (AC-equivalent vertices)
*For a digraph G, we say that $x_1,x_2 \in V(G)$ are AC-equivalent iff for the AC-
projection $\mathcal{I}$: $G \rightharpoonup G$, $\mathcal{I}(x_1)=\mathcal{I}(x_2)$.*

In the Figure 3 the nodes with same label are, in this case, AC-equivalent.
These two definitions give a reduction operator.

340   M. Liquiere

Procedure:$\mathcal{R}$

Data: a labelled digraph G
Result: a labelled digraph G' //with G' $\leftrightharpoons$ G
G':=G;
next := true;
while *next* **do**
    next:=false;
    if *(there is a AC-redundant vertex $x \in V(G')$)* **then**
        G':=G'-x;
        next:=true;
    end
    if *there is a set of AC-equivalent vertices $E=\{x_1,... x_n\} \in V(G')$ with*
    $|E| > 1$ **then**
        // Merge of the AC-equivalent vertices of E
        add a vertice x to V(G') with P(x)= $\bigcup$ P(x_i) and S(x)=$\bigcup$ S(x_i);
        // remove all $x_i \in E$ from G'
        G':=G'-E ;
        next:=true;
    end
end
return G';

This $\mathcal{R}$ operation is polynomial because the AC-projection is polynomial.

Proposition 3 ($\mathcal{R}$ equivalence)
$\mathcal{R}(G) \leftrightharpoons G$

Proof.
1) If we remove an AC-redundant vertex $x \in V(G)$, by definition G $\leftrightharpoons$ G-x.
2) if we merge a set of AC-equivalent vertices $E=\{x_1,... x_n\}$ with $|E| > 1$ in a vertex x, we obtain a new graph G'. We have to prove that G $\leftrightharpoons$ G'
We have a AC-projection $\mathcal{I}$ from G into G
a) G $\rightarrow$ G'
We construct a labeling $\mathcal{I}'$ from G into G' with
for $x_i \in E$, $\mathcal{I}'(x_i)$=x
for $y_j \notin E$, if $\mathcal{I}(y_j) \bigcap E \neq \emptyset$ then
$\mathcal{I}'(y_j)= (\mathcal{I}(y_j) - E) \bigcup x$
else
$\mathcal{I}'(y_j)=\mathcal{I}(y_j)$
We know that $\forall x_i \in E$, and $\forall y_j \in S(x),\mathcal{I}(x_i) \sim \succ \mathcal{I}(y_j)$. Since S(x)=$\bigcup$ S(x_i) | $x_i \in E$ we have $\mathcal{I}'(x_i)=\{x\} \sim \succ \mathcal{I}'(y_j)$ and reciprocally for the predecessors.
So $\mathcal{I}'(y_j)$ is an AC-projection.
b) G' $\rightarrow$ G
We construct a labeling $\mathcal{I}''$ from G' into G with
for $x \in V(G')$, $\mathcal{I}''(x)$=E
for $y \in V(G')$ and $y \neq x$, $\mathcal{I}''(y)=\mathcal{I}(y)$

We know that for all $x_i \in E$ and $y_j \in S(x_i)$, $\mathcal{I}(x_i) \leadsto \succ \mathcal{I}(y_j)$ (and reciprocally for the predecessors).

Since $\mathcal{I}''(x) = \mathcal{I}(x)$ and $\mathcal{I}''(y) = \mathcal{I}(y)$, we have $\mathcal{I}''(x_i) \leadsto \succ \mathcal{I}''(y_j)$ then $\mathcal{I}''$ is an AC-projection.

The Figure:3 shows the application of the $\mathcal{R}$ reduction operator on a digraph.

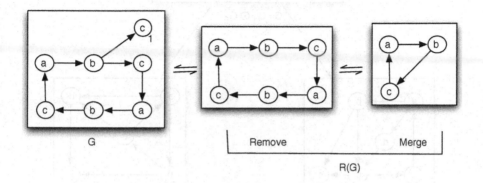

Fig. 3. $\mathcal{R}(\text{G})$

3.3 AC-Projection and Generalization Operator

There are some pairs, (representation languages and generality relations), which have a least general generalization operator . For logic formula and θ-subsumption, this operator is the classical lgg (or rlgg) introduced by plotkin [14]. For graph and homomorphism this operator is the graph product [10]. In mathematics this kind of operator is defined as a product operator [15].

Definition 10 (Product operator). *A binary operator $\bullet$ is a product operator iff for a pre order (or a partial order)$\geqslant$ between element E_i*

- $E_1 \bullet E_2 \geqslant E_1$
- $E_1 \bullet E_2 \geqslant E_2$
- *if $E \geqslant E_1$ and $E \geqslant E_2$ then $E \geqslant E_1 \bullet E_2$*

For digraph we have the following product operator ($\otimes$) for the homomorphism pre-order [16,3,10].

Definition 11 (Product operator $\otimes$ for digraphs and homomorphism)
For two digraphs G_1 and G_2 We construct $G = G_1 \otimes G_2$ with

- $L(G) = L(G_1) \bigcap L(G_2)$
- $V(G) \subseteq V(G_1) \times V(G_2) = \{x \mid x = (x_1, x_2) \text{ with } l(x) = l(x_1) = l(x_2)\}$
- $A(G) = \{(x,x') \mid x = (x_1, x_2), x' = (x'_1, x'_2) \text{ and } (x_1, x'_1) \in V(G_1), (x_2, x'_2) \in V(G_2)\} \subseteq A(G_1) \times A(G_2)$

For the AC-projection we have also a generalization operator.

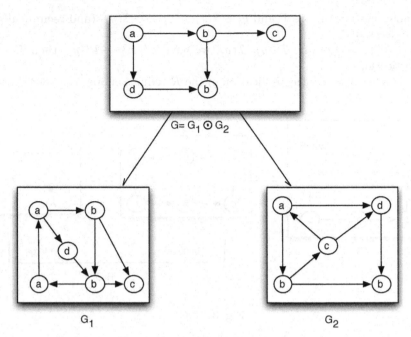

Fig. 4. Product ⊙ of two digraphs

Proposition 4 (Product operator ⊙ for digraphs and AC-projection)
For two digraphs G_1, G_2. The binary operator $G_1 \odot G_2 = \mathcal{R}(G_1 \otimes G_2)$ is a product operator.
So for $G = G_1 \odot G_2$ we have:
1) $G \rightarrow G_1$, $G \rightarrow G_2$
2) for a digraph G' if $H \rightarrow G_1$ and $G' \rightarrow G_2 \Rightarrow G' \rightarrow G$

Proof
1) For two digraphs G_1, G_2, and $G = G_1 \otimes G_2$ we have $G \mapsto G_1$ and $G \mapsto G_2$ (by property of the digraph product operation [10]). And we know:

- $\mathcal{R}(G) \leftrightharpoons G$ (proposition 5).
- if $G \mapsto G_1 \Rightarrow G \rightarrow G_1$ (proposition 1)

2) We know that $G' \rightarrow G_1$ and $G' \rightarrow G_2$. So there is two labelings $\mathcal{I}_1$ and $\mathcal{I}_2$ with for each $x \in V(G')$, $\mathcal{I}_1(x) = \{x_1^1, ..., x_1^n\} \subseteq V(G_1)$ and $\mathcal{I}_2(x) = \{x_2^1, ..., x_2^m\} \subseteq V(G_2)$.

In $G_1 \otimes G_2$ we have all the couples $(x_1^1, x_2^1), (x_1^1, x_2^2), ...(x_1^n, x_2^m)$ by construction. We define the following labelling $\mathcal{I}: G' \rightarrow G_1 \otimes G_2$ with $\mathcal{I}(x) = \mathcal{I}_1(x) \times \mathcal{I}_2(x)$. If $(x,y) \in V(G')$ we have to prove that $\mathcal{I}(x) \sim \succ \mathcal{I}(x)$.

For each $(x_1^i, x_2^j) \in \mathcal{I}(x)$ there is $(y_1^a, y_2^b) \in \mathcal{I}(x)$ with $(x_1^i, y_1^a) \in V(G_1)$ and $(x_2^j, y_2^b) \in V(G_2)$. Because $x_1^i \in \mathcal{I}_1(x)$ there is, at least, one $y_1^a \in \mathcal{I}_1(y)$ with $(x_1^i, y_1^a) \in V(G_1)$. Since $x_2^j \in \mathcal{I}_2(x)$ there is, at least, one $y_2^b \in \mathcal{I}_2(y)$ with $(x_2^j, y_2^b) \in V(G_2)$. Since, the graph product $G_1 \otimes G_2$ builds all the couple (x_1^i, x_2^j) and

(y_1^a, y_2^b) (with same labels). $((x_1^i, x_2^j), (y_1^a, y_2^b)) \in$ V(G) iff $(x_1^i, y_1^a) \in$V(G_1) and $(x_2^j, y_2^b) \in$V(G_2). So for $(x_1^i, x_2^j) \in \mathcal{I}(x)$ there is, by definition of $\mathcal{I}$, $(y_1^a, y_2^b) \in \mathcal{I}(y)$ then $\mathcal{I}(x) \sim \succ \mathcal{I}(y)$. $\mathcal{I}$ is an AC-projection from G' $\rightharpoonup G_1 \otimes G_2$. Since $\mathcal{R}(G) \leftrightharpoons$ G (by construction) then $\odot$ is a product operator.

In Figure:4 G is the product of G_1 and G_2. It represents all the different[1] subtrees common at G_1 and G_2.

4   Concept Lattice and AC-Projection

We have a generalization operator and a pre-order between digraphs. With this knowledge, we can define the notion of concept [13].

Definition 12 (concept, $\vee$, $\geq$)
For a set of examples $\mathcal{E}$, each example $e \in \mathcal{E}$ is described by a digraph $d(e) \in D$ (description space).
  For a digraph G, we note $\alpha(G) = \{e_i \in \mathcal{E} \mid G \rightharpoonup d(e_i)\}$.
  For $E_1 \subseteq \mathcal{E}$, we note $\beta(E_1) = \odot_{e \in H} d()$.
  A concept is a couple (E_1, G_1) with $E_1 \subseteq E$, G_1 a digraph with $\alpha(G_1) = E_1$ and $\beta(E_1) \leftrightharpoons G_1$.
  For two concepts (E_1, G_1), (E_2, G_2):
  $(E_1, G_1) \vee (E_2, G_2) = (\alpha(G), G = G_1 \odot G_2) (E_1, G_1) \geq (E_2, G_2)$ iff $G_1 \rightharpoonup G_2$

Proposition 5 (AC concept lattice)
For a set of examples $\mathcal{E}$, each example $e \in \mathcal{E}$ is described by a digraph $d(e) \in D$. The correspondance α, β defines a Galois connection between $2^{\mathcal{E}}$ and D.

Proof. see [10]

This proposition gives the structure of the search space (a concept semilattice). The size of the concept lattice is limited by the minimum of SD and SP where SD is the size of the description space and SP the size of the partition space. The size SD is very large for relational description but SP is limited by 2^n where n is the number of examples.

If we use our method on the set of examples of [10] we obtain the following join-semilattice Figure:5. In this concept semilattice, each node represent a concept with an extension part: a subset of the set of examples and an intension part :a digraph. The partial order between the elements of the lattice is based on AC-projection, then the digraph, intension part of a concept, can be interpreted as a compact description of a very large (potentially infinite) set of trees. But, thanks to AC-projection, we don't have to explore all the elements of this set as in classical tree mining method [8]. Using this example, we obtain a lattice which is isomorphic with the one given by Graal [10] but it is not always the case. This comes from the fact that, for this set of graphs, the set of included paths is enough to obtain this lattice.

---

[1] For the homomorphism relation.

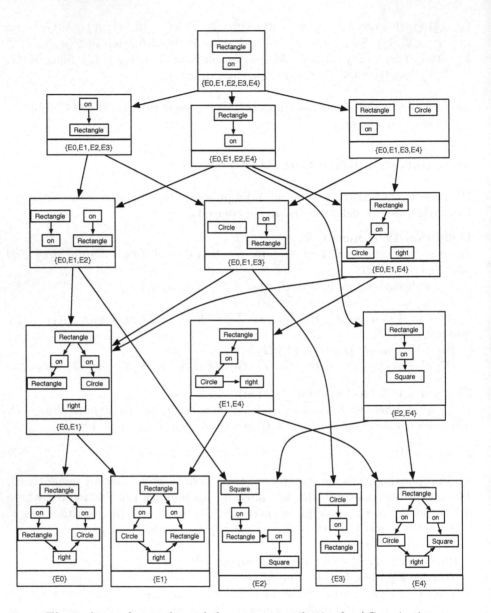

Fig. 5. A set of examples and the concept semilattice for AC-projection

5   Conclusion

This study has attempted to merge ideas from different communities: Graph,
ILP, CSP. In these communities, the same problem is examined from different
points of view: graph and homomorphism, logic and θ-subsumption, and constraint satisfaction problems and resolution.

Merging these knowledges we have obtained:

- the definition of a new generality relation between graphs with polynomial complexity.
- the definition of a least general generalization operator.

These results can be used for the construction of a concept lattice where the intention part of a concept is described by a digraph. The advantages of this approach compare with the approach [10] are:

- We use general digraph for the description of the examples. In the paper [10] we have a polynomial complexity only for a specific class of graphs.
- All the operations are polynomial (equivalence, reduction, product and concept order).
- We find graphs which express a large set of trees in a compact form.
- The size of the concept lattice is smaller.

But, since the AC projection is less precise, the classifications obtained are also less precise. However, we have a generalization operator, between two digraphs, which gives a digraph which represents all the homomorphic trees belonging at this two digraphs. So, a large part of their common structure is caught.

Acknowledgements. The author is grateful to C. Bessiere and F. Koriche for their constructive remarks.

References

1. Sowa, J.F.: Conceptual structures: information processing in mind and machine. Addison-Wesley Longman Publishing Co. Inc. Boston, MA, USA (1984)
2. Baget, J.F., Mugnier, M.L.: Extension of simple conceptual graphs: the complexity of rules and constraints. Journal of Artificial Intelligence Research 16, 425–465 (2002)
3. Hell, P., Nesetril, J.: Graphs and homomorphism. In: Oxford Lecture Series in Mathematics and its Application, vol. 28, Oxford University Press, Oxford (2004)
4. Mugnier, M., Chein, M.: Polynomial algorithms for projection and matching. In: ICCS. LNCS (LNAI), vol. 754, pp. 239–251. Springer, Heidelberg (1992)
5. Croitoru, M., Compatangelo, E.: A combinatorial approach to conceptual graph projection checking. In: 24th Int'l Conf. of the British Computer Society's Specialist Group on Art'l Intell, pp. 239–251. Springer, Heidelberg (2004)
6. Pfeiffer, H.: A comparison of different conceptual structures projection algorithms, Submitted to ICCS'07 (2007)
7. Maloberti, J., Sebag, M.: Fast theta-subsumption with constraint satisfaction algorithms. Machine Learning 55, 137–174 (2004)
8. Zaki, M.: Efficiently mining frequent trees in a forest. In: 8th Intl. Conf. knowledge discovery and data mining, pp. 71–80 (2002)
9. Cook, D.J., Holder, L.B.: Substructure discovery using minimum description length and background knowledge. Journal of Artificial Intelligence Research 1, 231–255 (1994)

10. Liquiere, M., Sallantin, J.: Structural machine learning with galois lattice and graphs. In: Shavlik, J.W. (ed.) ICML, pp. 305–313. Morgan Kaufmann, San Francisco (1998)
11. Sebag, M., Rouveirol, C.: Resource-bounded relational reasoning: induction and deduction through stochastic matching. Machine Learning 38, 41–62 (2000)
12. Bessiere, C.: Constraint propagation. In: Rossi, F., van Beek, P., Walsh, T. (eds.) Handbook of Constraint Programming, Elsevier, Amsterdam (2006)
13. Ganter, B., Wille, R.: Formal Concept Analysis: Mathematical Foundations. Springer-Verlag, New York, Inc. Secaucus, NJ, USA, Translator-C. Franzke (1997)
14. Plotkin, G.: A note on inductive generalization. Machine Intelligence 5, 153–163 (1970)
15. Crole, R.: Categories for Types. Cambridge Mathematical Textbooks. Cambridge University Press, Cambridge (1993)
16. Weichsel, P.M.: The kronecker product of graphs. Proc.Am.Math.Soc. 13, 47–52 (1962)

Mining Frequent Closed Unordered Trees Through Natural Representations*

José L. Balcázar, Albert Bifet, and Antoni Lozano

Universitat Politècnica de Catalunya,
{balqui,abifet,antoni}@lsi.upc.edu

Abstract. Many knowledge representation mechanisms consist of link-based structures; they may be studied formally by means of unordered trees. Here we consider the case where labels on the nodes are nonexistent or unreliable, and propose data mining processes focusing on just the link structure. We propose a representation of ordered trees, describe a combinatorial characterization and some properties, and use them to propose an efficient algorithm for mining frequent closed subtrees from a set of input trees. Then we focus on unordered trees, and show that intrinsic characterizations of our representation provide for a way of avoiding the repeated exploration of unordered trees, and then we give an efficient algorithm for mining frequent closed unordered trees.

1 Introduction

Trees, in a number of variants, are basically connected acyclic undirected graphs, with some additional structural notions like a distinguished vertex (root) or labelings on the vertices. They are frequently a great compromise between graphs, which offer richer expressivity, and strings, which offer very efficient algorithmics. From AI to Compilers, through XML dialects, trees are now ubiquitous in Informatics.

One form of data analysis contemplates the search of frequent (or the so-called "closed") substructures in a dataset of structures. In the case of trees, there are two broad kinds of subtrees considered in the literature: subtrees which are just induced subgraphs, called *induced subtrees*, and subtrees where contraction of edges is allowed, called *embedded subtrees*. In these contexts, the process of "mining" usually refers, nowadays, to a process of identifying which common substructures appear particularly often, or particularly correlated with other substructures, with the purpose of inferring new information implicit in a (large) dataset. In our case, the dataset would consist of a large set (more precisely, bag) of trees; algorithms for mining embedded labeled frequent trees include TreeMiner [22], which finds all embedded ordered subtrees that appear with a

---

* Partially supported by the 6th Framework Program of EU through the integrated project DELIS (#001907), by the EU PASCAL Network of Excellence, IST-2002-506778, by the MEC TIN2005-08832-C03-03 (MOISES-BAR), MCYT TIN2004-07925-C03-02 (TRANGRAM), and CICYT TIN2004-04343 (iDEAS) projects.

U. Priss, S. Polovina, and R. Hill (Eds.): ICCS 2007, LNAI 4604, pp. 347–359, 2007.

frequency above a certain threshold, and similarly SLEUTH [23], for unordered trees. Algorithms for mining induced labeled frequent trees include FreqT [2], which mines ordered trees, and uFreqT [15], uNot [3], HybridTreeMiner [8], and PathJoin [18] for unordered trees. FreeTreeMiner [11] mines induced unordered free trees (i.e. there is no distinct root). A comprehensive introduction to the algorithms on unlabeled trees can be found in [17] and a survey of works on frequent subtree mining can be found in [9].

Closure-based mining refers to mining closed substructures, in a sense akin to the closure systems of Formal Concept Analysis; although the formal connections are not always explicit. For trees, a closed subtree is one that, if extended in any manner, leads to reducing the set of data trees where it appears as a subtree; and similarly for graphs. Frequent closed trees (or sets, or graphs) give the same information about the dataset as the set of all frequent trees (or sets, or graphs) in less space. Yan and Han [19,20] proposed two algorithms for mining frequent and closed graphs. The first one is called gSpan (graph-based Substructure pattern mining) and discovers frequent graph substructures without candidate generation. gSpan builds a new lexicographic order among graphs, and maps each graph to a unique minimum DFS code as its canonical label. Based on this lexicographic order, gSpan adopts the depth-first search strategy to mine frequent connected subgraphs. The second one is called CloseGraph and discovers closed graph patterns. CloseGraph is based on gSpan, and is based on the development of two pruning methods: equivalent occurrence and early termination. The early termination method is similar to the early termination by equivalence of projected databases method in CloSpan [21], an algorithm of the same authors for mining closed sequential patterns in large datasets. However, in graphs there are some cases where early termination may fail and miss some patterns. By detecting and eliminating these cases, CloseGraph guarantees the completeness and soundness of the closed graph patterns discovered.

Chin et al. proposed CMTreeMiner [10], the first algorithm to discover all closed and maximal frequent labeled induced subtrees without first discovering all frequent subtrees. CMTreeMiner shares many features with CloseGraph, and uses two pruning techniques: the *left-blanket* and *right-blanket* pruning. The *blanket* of a tree is defined as the set of immediate supertrees that are frequent, where an *immediate supertree* of a tree t is a supertree of t that has one more vertex than t. The *left-blanket* of a tree t is the blanket where the vertex added is not in the right-most path of t (the path from the root to the rightmost vertex of t). The *right-blanket* of a tree t is the blanket where the vertex added is in the right-most path of t. Their method is as follows: it computes, for each candidate tree, the set of trees that are occurrence-matched with its blanket's trees. If this set is not empty, they apply two pruning techniques using the left-blanket and right-blanket. If it is empty, then they check if the set of trees that are transaction-matched but not occurrence matched with its blanket's trees is also empty. If this is the case, there is no supertree with the same support and then the tree is closed.

Arimura and Uno [1] considered closed mining in attribute trees, which is a subclass of labeled ordered trees and can also be regarded as a fragment of description logic with functional roles only. These attribute trees are defined using a relaxed tree inclusion. Termier et al. [16] considered the frequent closed tree discovery problem for a class of trees with the same constraint as attribute trees.

We propose a representation of ordered trees in Section 2.1 and describe a combinatorial characterization and some properties. In Section 3 we present the new algorithm for mining ordered frequent trees. In Section 4 we extend our method to unordered trees and show that intrinsic characterizations of our representation provide for a way of avoiding the repeated exploration of unordered trees. We propose in Section 5 a new efficient algorithm for mining rooted induced closed subtrees taking advantage of some combinatorial properties of our new representation of trees.

These mining processes can be used for a variety of tasks. Let us describe some possibilities. Consider web search engines. Already the high polysemy of many terms makes sometimes difficult to find information through them; for instance, a researcher of soil science may have a very literal interpretation in mind when running a web search for "rolling stones", but it is unlikely that the results are very satisfactory; or a computer scientist interested in parallel models of computation has a different expectation from that of parents-to-be when a search for "prams" is launched. A way for distributed, adaptive search engines to proceed may be to distinguish navigation on unsuccessful search results, where the user follows a highly branching, shallow exploration, from successful results, which give rise to deeper, little-branching navigation subtrees. Indeed, through an artificially generated bimodal dataset, consisting of shallow trees (modeling unsuccessful searches, with average fanout 8, depth 3) and deeper, slender trees (modeling successful searches, fanout 2, depth 8), the closures found by our algorithms distinguish quite clearly about 2/3 of the shallow trees; they fail to distinguish small shallow trees from small deep trees, which is not surprising since when the size is small, trees are both slender and shallow. However, a large majority of the shallow trees got clearly separated from the others.

As another example, consider the KDD Cup 2000 data [14]. This dataset is a web log file of a real internet shopping mall (gazelle.com). This dataset, of size 1.2GB, contains 216 attributes. We used the attribute 'Session ID' to associate to each user session a unique tree. The trees record the sequence of web pages that have been visited in a user session. Each node tree represents a content, assortment and product path. Trees are not built using the structure of the web site, instead they are built following the user streaming. Each time a user visits a page, if he has not visited it before, we take this page as a new deeper node, otherwise, we backtrack to the node this page corresponds to, if it is the last node visited on a concrete depth. The resulting dataset consists of 225, 558 trees. On them, an improved variation of our algorithms was considerably faster than the only alternative algorithm now available for the task, CMTreeMiner. Further analysis and improvements will be reported detailedly in [6].

2 Preliminaries

Our *trees* will be finite rooted trees with nodes of unbounded arity, and we will consider two kinds of trees: *ordered trees*, in which the children of any node form a sequence of siblings, and *unordered trees*, in which they form a set of siblings. A *bottom-up subtree* of a tree t is any connected subgraph rooted at some node v of t which contains all the descendants of v in t and no more. The *depth* of a node is the length of the path from the root to that node (the root has depth 0). A bottom-up subtree of a tree t is at depth d if its root is at depth d in t.

In order to compare link-based structures, we will also be interested in a notion of subtree where the root is preserved. A tree t' is a *top-down subtree* (or simply a *subtree*) of a tree t if t' is a connected subgraph of t which contains the root of t.

Given a finite dataset D of unlabeled rooted trees, we say that one of them s_i *supports* a tree t if t is a subtree of s_i. The number of indices i whose s_i in the dataset D supports t is called the *support* of the tree t. A subtree t is called *frequent* if its support is greater than or equal to a given threshold min_sup specified. The frequent subtree mining problem is to find all frequent subtrees in a given dataset. Any subtree of a frequent tree is also frequent and any supertree of an nonfrequent tree is also nonfrequent.

We define a frequent tree t to be *closed* if none of its proper supertrees has the same support as it has. Generally, there are much fewer closed sets than frequent ones. In fact, we can obtain all frequent subtrees with their support from the set of closed frequent subtrees with their supports. So, the set of closed frequent subtrees maintains the same information as the set of all frequent subtrees.

2.1 Natural Representations

Sequences of natural numbers will play a technical role in our development in order to represent both ordered and unordered trees.

Definition 1. *A natural sequence is a finite sequence of natural numbers, starting and ending by 0, and where each difference of consecutive numbers is either +1 or -1.*

In order to describe our representation of ordered trees, we will consider the following operations of general sequences of natural numbers (not necessarily natural sequences).

Definition 2. *Given two sequences of natural numbers x, y, we represent by*

- *$x \cdot y$ the sequence obtained as concatenation of x and y*
- *$x_{i:j}$ the subsequence of numbers starting at position i up to position j in x (where positions range from 1 to $|x|$, the length of x)*
- *$x + i$ $(x - i)$ the sequence obtained adding (subtracting) i to each component of x*

We represent by x^+ the sequence $x + 1$ and consider that $+$ has precedence over concatenation.

For example, if $x = (0) \cdot (0, 1, 0)^+ \cdot (0)$, then $x = (0, 1, 2, 1, 0)$, and $x_{3:5} = (2, 1, 0)$. Now, we can represent trees by means of natural sequences.

Definition 3. *We define a function $\langle \cdot \rangle$ from the set of ordered trees to the set of natural sequences as follows. Let t be an ordered tree. If t is a single node, then $\langle t \rangle = (0)$. Otherwise, if t is composed of the trees $t_1, \ldots, t_k$ joined to a common root r (where the ordering $t_1, \ldots, t_k$ is the same of the children of r), then*

$$\langle t \rangle = (0) \cdot \langle t_1 \rangle^+ \cdot (0) \cdot \langle t_2 \rangle^+ \cdot \ldots \cdot (0) \cdot \langle t_k \rangle^+ \cdot (0).$$

We say that $\langle t \rangle$ is the natural representation of t.

Note that, first of all, the above function is well defined: the recursive definition of $\langle t \rangle$ ensures that it is a natural sequence. It is also easy to see that it is a bijection between the ordered trees and the natural sequences and that $\langle t \rangle$ basically corresponds to a pre-post-order traversal of t where each number of the sequence represents the depth of the current node in the traversal. As an example, the natural representation of the tree in the Figure 1 is the

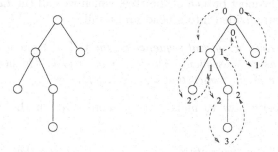

Fig. 1. A tree on the left and its pre-post-order traversal on the right

natural sequence $(0, 1, 2, 1, 2, 3, 2, 1, 0, 1, 0)$. Note that, for example, the subsequence $(1, 2, 1, 2, 3, 2, 1)$ corresponds to the bottom-up subtree rooted at the left son of the root. We can state this fact in general:

Proposition 1. *Let $x = \langle t \rangle$, where t is an ordered tree. Then, t has a bottom-up subtree r at depth $d > 0$ if and only if $(d-1) \cdot (\langle r \rangle + d) \cdot (d-1)$ is a subsequence of x.*

Proof. We prove it by induction on d. If $d = 1$, then since

$$(d-1) \cdot (\langle r \rangle + d) \cdot (d-1) = (0) \cdot \langle r \rangle^+ \cdot (0)$$

the property holds by the recursive definition of natural representation. For the induction step, let $d > 1$. To show one direction, suppose that r is a bottom-up subtree of t at depth d. Then, r must be a bottom-up subtree of one of the bottom-up subtrees corresponding to the children of the root of t. Let t' be the bottom-up subtree at depth 1 that contains r. Since r is at depth $d-1$ in t', the induction hypothesis states that $(d-2) \cdot (\langle r \rangle + d - 1) \cdot (d-2)$ is a subsequence of $\langle t' \rangle$. But $(0) \cdot \langle t' \rangle^+ \cdot (0)$ is also, by definition, a subsequence of x. Combining both facts, we get that $(d-1) \cdot (\langle r \rangle + d) \cdot (d-1)$ is a subsequence of x, as desired. The argument also works in the contrary direction, and we get the equivalence.   □

3   Mining Frequent Subtrees in the Ordered Case

Our approach here is similar to gSpan: we represent the potential subtrees to be checked for frequent on the dataset in such a way that extending them by one single node corresponds to a clear and simple operation on the representation. The completeness of the procedure is mathematically proved, that is, we argue that all trees can be obtained in this way with our notion of extension, defined below. This allows us to avoid extending trees that are found to be already nonfrequent.

We show now that our representation allows us to traverse the whole subtree space by an operation of extension by a single node, in a simple way. The possible places where the new node can be added in our natural representation belong to the right-most branch of the tree, which we call the *tail* —our natural representations are divided into a head and a tail—.

Definition 4. *Given a natural sequence x, the* tail *of x, denoted by* tail(x), *is the longest decreasing subsequence which is a suffix of x. The* head *of x, denoted by* head(x), *is the prefix y such that $x = y \cdot$ tail(x).*

Now we can specify how to make one or more steps in the generation of (a representation of) a tree.

Definition 5. *Given two natural sequences x, y, we say that x yields y in one step (in symbols, $x \vdash^1 y$) if y is obtained from x by replacing some element e in* tail(x) *by the subsequence $(e, e+1, e)$.*

Definition 6. *A succession of natural sequences $x_1, x_2, \ldots, x_n$ such that $x_1 \vdash^1 x_2 \vdash^1 \cdots \vdash^1 x_n$ is called a* generating path *from x_1 to x_n. Given two natural sequences x, y, if there is a generating path from x to y, we write $x \vdash^* y$ and we say that x yields y.*

Our terminology with trees will be the following: if t and t' are two trees and $\langle t \rangle$ yields $\langle t' \rangle$, then we say that t' *is extended from t (in one step* if $\langle t \rangle$ yields $\langle t' \rangle$ in one step). All the trees extended from t are called the *extensions of t*. Now we can state the main result that validates the algorithm above.

Theorem 1. *For every natural sequence x, there is a unique generating path from (0) to x.*

Proof. The existence of a generating path for a natural sequence x is straightforward by induction on $n = (|x| + 1)/2$. Note that all natural sequences are of odd length. If $n = 1$, then $|x| = 1$, and the only possibility is that $x = (0)$;

therefore, there is a generating path (of length 0). Now suppose that $n > 1$, and let $k = |\mathsf{tail}(x)|$. Note that $k \geq 2$. By eliminating the first two elements of $\mathsf{tail}(x)$ in x, we get the natural sequence

$$x' = \mathsf{head}(x) \cdot \mathsf{tail}(x)_{3:k}$$

where $\mathsf{tail}(x)_{3:k}$ is the empty sequence if $|\mathsf{tail}(x)| = 2$. Note that the last number e in $\mathsf{head}(x)$ belongs to the tail of x', so we can replace it by $(e, e+1, e)$, obtaining x. Therefore $x' \vdash^1 x$. Since $(|x'| + 1)/2 = n - 1$, by induction hypothesis there is a generating path from (0) to x' and, therefore, to x.

To show uniqueness, suppose that y, with $|y| = n$, is a natural representation obtained by two different generating paths from (0). Let x be the natural representation in both paths from which the two different paths split. Then, there are two natural sequences x' and x'' such that $x' \neq x''$ and

$$x \vdash^1 x' \vdash^* y \quad \text{and} \quad x \vdash^1 x'' \vdash^* y.$$

Since x yields x' and x'' in one step, $|x'| = |x''|$ and $\mathsf{tail}(x') \neq \mathsf{tail}(x'')$ (different replacements in $\mathsf{tail}(x)$ produce different new tails), the lengths of the tails $\mathsf{tail}(x')$ and $\mathsf{tail}(x'')$ must be different. W.l.o.g., suppose that $|\mathsf{tail}(x')| < |\mathsf{tail}(x'')| = k$. Now we claim that

$$x'_{n-k+1} < x''_{n-k+1}, \tag{1}$$

the reason being that x''_{n-k+1} belongs to the tail of x'' (contrarily to x'_{n-k+1}), and so it has the maximum possible value at position $n - k + 1$. Consider the following properties, which are straightforward and we state without proof.

Property 1. If $u \vdash^* v$ for two natural sequences u and v, then $\mathsf{head}(u)$ is a prefix of v.

Property 2. If $u \vdash^* v$ for two natural sequences u and v, then $u_i \leq v_i$ for every $i \leq |u|$.

Now, we have the following. The element x'_{n-k+1} of x' does not belong to $\mathsf{tail}(x')$ (because its length is less than k). Therefore, by Property 1 and the fact that $x' \vdash^* y$ we have that $y_{n-k+1} = x'_{n-k+1}$. On the other hand, since $x'' \vdash^* y$, we have by Property 2 that $x''_{n-k+1} \leq y_{n-k+1}$. Finally, by inequality 1, we get the contradiction

$$y_{n-k+1} = x'_{n-k+1} < x''_{n-k+1} \leq y_{n-k+1}. \qquad \square$$

For this section we could directly use gSpan, since our structures can be handled by that algorithm. However, our goal is the improved algorithm described in the next section, to be applied when the ordering in the subtrees is irrelevant for the application.

Indeed, we have designed our representation in such a way that it will allow us to check only canonical representatives for the unordered case, thus saving the computation of support for all (except one) of the ordered variations of the same unordered tree. Figures 2 and 3 show the gSpan-based algorithm, which is as follows: beginning with a tree of a single node, it calls recursively the FREQUENT_ORDERED_SUBTREE_MINING algorithm doing one-step extensions and checking that they are still frequent.

FREQUENT_ORDERED_MINING(D, min_sup)

    Input: A tree dataset D, and min_sup.
    Output: The frequent tree set T.

  1   $t \leftarrow$ one node tree
  2   $T \leftarrow \emptyset$
  3   $T \leftarrow$ FREQUENT_ORDERED_SUBTREE_MINING(t, D, min_sup, T)
  4   **return** T

Fig. 2. The Frequent Ordered Mining algorithm

FREQUENT_ORDERED_SUBTREE_MINING(t, D, min_sup, T)

    Input: A tree t, a tree dataset D, and min_sup.
    Output: The frequent tree set T.

  1 insert t into T
  2 **for** every t' that can be extended from t in one step
  3   **do if** support(t') $\geq min\_sup$
  4   **then** $T \leftarrow$ FREQUENT_ORDERED_SUBTREE_MINING(t', D, min_sup, T)
  5 **return** T

Fig. 3. The Frequent Ordered Subtree Mining algorithm

4   Mining Frequent Subtrees in the Unordered Case

The main result of this section is a precise mathematical characterization of the natural representations that correspond to canonical variants of unordered trees.

In unordered trees, the children of a given node form sets of siblings instead of sequences of siblings. Therefore, ordered trees that only differ in permutations of the ordering of siblings are to be considered the same unordered tree. We select one of them to act as canonical representative of all the ordered trees corresponding to the same unordered tree: by convention, this canonical representative has larger trees always to the left of smaller ones. More precisely,

Definition 7. *Let t be an unordered tree, and let $t_1, \ldots, t_n$ be all the ordered trees obtained from t by ordering in all possible ways all the sets of siblings of t. The canonical representative of t is the ordered tree t_0 whose natural representation is maximal (according to lexicographic ordering) among the natural representations of the trees t_i, that is, such that*

$$\langle t_0 \rangle = \max\{\langle t_i \rangle \mid 1 \leq i \leq n\}.$$

We can use, actually, the same algorithm as in the previous section to mine unordered trees; however, much work is unnecessarily spent in checking repeatedly ordered trees that correspond to the same unordered tree as one already checked. A naive solution is to compare each tree to be checked with the ones already checked, but in fact this is an inefficient process, since all ways of mapping siblings among them must be tested.

A far superior solution would be obtained if we could count frequency only for canonical representatives. We explain next one way to do this: the use of natural representations allows us to decide whether a given (natural representation of a) tree is canonical, by using an intrinsic characterization which is not stated in terms of the representations of all other ordered trees; but only in terms of the natural representation itself.

For the characterization, it is convenient to use the notion of splitting triplet:

Definition 8. *Given a natural sequence* $(x_1, \dots, x_n)$, *a splitting triplet* (i, j, k) *represents three different positions* $i < j < k$ *of the sequence having the same values* $(x_i = x_j = x_k)$ *and with all intermediate values strictly higher.*

The notion of splitting triplet is the natural-representation analogue of "consecutive" bottom-up subtrees.

Lemma 1. *Let* t *be an ordered tree and* $x = \langle t \rangle$. *Then,* x *contains a splitting triplet* (i, j, k) *if and only if there is a node in* t *at depth* x_i *which is the parent of two bottom-up subtrees* t_1, t_2 *such that* $\langle t_1 \rangle = x_{i:j} - x_i$ *and* $\langle t_2 \rangle = x_{j:k} - x_i$, *and* t_1 *appears immediately before* t_2.

Proof. Suppose that x contains the splitting triplet (i, j, k), and let $d = x_i + 1$. Consider the subsequence $x_{i:j}$, which can be expressed as

$$(x_i) \cdot x_{i+1:j-1} \cdot (x_j) = (d - 1) \cdot y \cdot (d - 1)$$

where all the values in y are strictly higher than $d-1$ (because of the conditions of splitting triplets). Since y is a subsequence of x, $y-d$ must be a natural sequence which, by Theorem 1, has a generating path and, therefore, corresponds to a tree t_1. So, $y = \langle t_1 \rangle + d$ and we get that

$$x' = (d - 1) \cdot (\langle t_1 \rangle + d) \cdot (d - 1)$$

is a subsequence of x. Similarly, it can be argued that

$$x'' = (d - 1) \cdot (\langle t_2 \rangle + d) \cdot (d - 1)$$

is a subsequence of x, for some tree t_2. By Proposition 1, both x' and x'' correspond to two bottom-up subtrees of t at depth d. Since they share the position j (as the end of x' and beginning of x''), the roots of t_1 and t_2 must be siblings, and t_1 must appear immediately before t_2.

The other direction can be shown by a similar argument: the existence of the two subtrees t_1 and t_2 as described in the statement implies, by Proposition 1, that their natural representations must appear in x (modulo an increase of d in their values) as is said above. ☐

Theorem 2. *A natural sequence x corresponds to a canonical representative if and only if, for every splitting triplet (i, j, k), $x_{i:j} \geq x_{j:k}$ in lexicographic order.*

Proof. If x corresponds to a canonical representative, it is clear that the condition about the splitting triplets must hold. Otherwise, the existence of a triplet (i, j, k) in x where $x_{i:j} < x_{j:k}$, would imply, by Lemma 1, the possibility of a better ordering of the two corresponding bottom-up subtrees, resulting in a tree t' with $\langle t' \rangle > x$, which is a contradiction.

To show the other direction, suppose that x does not correspond to a canonical representative. Therefore, a reordering of two bottom-up subtrees whose roots are consecutive siblings must give a larger natural representation. Again by Lemma 1, this corresponds to the existence of a triplet (i, j, k) in x which does not satisfy the above condition, that is, such that $x_{i:j} < x_{j:k}$.   □

We must point out here that a further, later refinement of the notion of natural representation described here, however, has given us improved performance for the mining algorithms. Hence, the major contribution of this work lies in these algorithms, and not in the present section, and we suggest the interested reader to consult [6] (soon to be available at the authors web pages) for the most attractive results on how to check for unordered tree representatives in natural representations.

Figure 4 shows the gSpan-based algorithm for unordered trees (for which we use a similar calling procedure as for the ordered case shown in Figure 2). The main difference with the algorithm for the case of ordered trees is that FREQUENT_UNORDERED_SUBTREE_ MINING checks at the beginning that the input tree is its canonical representative using the method implicit in Theorem 2.

FREQUENT_UNORDERED_SUBTREE_MINING(t, D, min_sup, T)

    Input: A tree t, a tree dataset D, and min_sup.
    Output: The frequent tree set T.

```
1   if not CANONICAL_REPRESENTATIVE(t)
2       then return T
3   insert t into T
4   C ← ∅
5   for every t' that can be extended from t in one step
6       do if support(t') ≥ min_sup
7           then insert t' into C
8   for each t' in C
9       do T ← FREQUENT_UNORDERED_SUBTREE_MINING(t', D, min_sup, T)
10  return T
```

Fig. 4. The Frequent Unordered Subtree Mining algorithm

5 Closure-Based Mining

In [5] we aim at clarifying the properties of closed trees, providing a more detailed justification of the term "closed" through a closure operator obtained from a Galois connection, along the lines of [12], [7], [13], or [4] for unstructured or otherwise structured datasets. Also, we designed two algorithms for finding the intersection of two subtrees: the first one in a recursive way, and the second one using dynamic programming.

In this section, we propose a new algorithm to mine unordered frequent closed trees. Figure 5 illustrates the framework.

Figure 6 shows the pseudocode of CLOSED_UNORDERED_SUBTREE_ MINING. It is similar to FREQUENT_UNORDERED_SUBTREE_MINING, adding a checking of closure in lines 9-10.

CLOSED_UNORDERED_MINING(D, min_sup)
    Input: A tree dataset D, and min_sup.
    Output: The closed tree set T.

  1 $t \leftarrow$ one node tree
  2 $T \leftarrow \emptyset$
  3 $T \leftarrow$ CLOSED_UNORDERED_SUBTREE_MINING(t, D, min_sup, T)
  4 **return** T

Fig. 5. The Closed Unordered Mining algorithm

CLOSED_UNORDERED_SUBTREE_MINING(t, D, min_sup, T)
    Input: A tree t, a tree dataset D, and min_sup.
    Output: The closed frequent tree set T.

  1 **if not** CANONICAL_REPRESENTATIVE(t)
  2   **then return** T
  3 $C \leftarrow \emptyset$
  4 **for** every t' that can be extended from t in one step
  5   **do if** support(t') $\geq min\_sup$
  6   **then** insert t' into C
  7   **do if** support(t') $=$ support(t)
  8   **then** t is not closed
  9 **if** t is closed
10   **then** insert t into T
11 **for each** t' in C
12   **do** $T \leftarrow$ CLOSED_UNORDERED_SUBTREE_MINING(t', D, min_sup, T)
13 **return** T

Fig. 6. The Closed Unordered Subtree Mining algorithm

References

1. Arimura, H., Uno, T.: An output-polyunomial time algorithm for mining frequent closed attribute trees. In: Kramer, S., Pfahringer, B. (eds.) ILP 2005. LNCS (LNAI), vol. 3625, pp. 1–19. Springer, Heidelberg (2005)
2. Asai, T., Abe, K., Kawasoe, S., Arimura, H., Sakamoto, H., Arikawa, S.: Efficient substructure discovery from large semi-structured data. In: SDM (2002)
3. Asai, T., Arimura, H., Uno, T., Nakano, S.-I.: Discovering frequent substructures in large unordered trees. Discovery Science, 47–61 (2003)
4. Baixeries, J., Balcázar, J.L.: Discrete deterministic data mining as knowledge compilation. In: Workshop on Discrete Math. and Data Mining at SIAM DM Conference (2003)
5. Balcázar, J.L., Bifet, A., Lozano, A.: Intersection algorithms and a closure operator on unordered trees. In: MLG 2006, 4th International Workshop on Mining and Learning with Graphs (2006)
6. Balcázar, J.L., Bifet, A., Lozano, A.: Mining frequent closed rooted trees, 2007 (submitted)
7. Balcázar, J.L., Garriga, G.C.: On Horn axiomatizations for sequential data. In: Eiter, T., Libkin, L. (eds.) ICDT 2005. LNCS, vol. 3363, pp. 215–229. Springer, Heidelberg (2004)
8. Chi, Y., Yang, Y., Muntz, R.R.: HybridTreeMiner: An efficient algorithm for mining frequent rooted trees and free trees using canonical forms. In: Chi, Y., Yang, Y., Muntz, R.R. (eds.) SSDBM '04: Proceedings of the 16th International Conference on Scientific and Statistical Database Management (SSDBM'04), Washington, DC, USA, 2004, p. 11. IEEE Computer Society Press, Los Alamitos (2004)
9. Chi, Y., Muntz, R., Nijssen, S., Kok, J.: Frequent subtree mining – an overview. Fundamenta Informaticae XXI, 1001–1038 (2001)
10. Chi, Y., Xia, Y., Yang, Y., Muntz, R.: Mining closed and maximal frequent subtrees from databases of labeled rooted trees. Fundamenta Informaticae XXI, 1001–1038 (2001)
11. Chi, Y., Yang, Y., Muntz, R.R.: Indexing and mining free trees. In: ICDM '03: Proceedings of the Third IEEE International Conference on Data Mining, Washington, DC, USA, 2003, p. 509. IEEE Computer Society, Los Alamitos (2003)
12. Ganter, B., Wille, R.: Formal Concept Analysis. Springer, Heidelberg (1999)
13. Garriga, G.C.: Formal methods for mining structured objects. PhD Thesis (2006)
14. Kohavi, R., Brodley, C., Frasca, B., Mason, L., Zheng, Z.: KDD-Cup 2000 organizers report: Peeling the onion. SIGKDD Explorations 2(2), 86–98 (2000)
15. Nijssen, S., Kok, J.N.: Efficient discovery of frequent unordered trees. In: First International Workshop on Mining Graphs, Trees and Sequences, pp. 55–64 (2003)
16. Termier, A., Rousset, M.-C., Sebag, M.: DRYADE: a new approach for discovering closed frequent trees in heterogeneous tree databases. In: Perner, P. (ed.) ICDM 2004. LNCS (LNAI), vol. 3275, pp. 543–546. Springer, Heidelberg (2004)
17. Valiente, G.: Algorithms on Trees and Graphs. Springer, Heidelberg (2002)
18. Xiao, Y., Yao, J.-F., Li, Z., Dunham, M.H.: Efficient data mining for maximal frequent subtrees. In: ICDM '03: Proceedings of the Third IEEE International Conference on Data Mining, Washington, DC, USA, 2003, p. 379. IEEE Computer Society, Los Alamitos (2003)
19. Yan, X., Han, J.: gSpan: Graph-based substructure pattern mining. In: ICDM '02: Proceedings of the 2002 IEEE International Conference on Data Mining (ICDM'02), Washington, DC, USA, 2002, p. 721. IEEE Computer Society, Los Alamitos (2002)

20. Yan, X., Han, J.: CloseGraph: mining closed frequent graph patterns. In: KDD '03: Proceedings of the ninth ACM SIGKDD international conference on Knowledge discovery and data mining, pp. 286–295. ACM Press, New York, NY, USA (2003)
21. Yan, X., Han, J., Afshar, R.: CloSpan: Mining closed sequential patterns in large databases. In: SDM (2003)
22. Zaki, M.J.: Efficiently mining frequent trees in a forest. In: 8th ACM SIGKDD International Conference on Knowledge Discovery and Data Mining (2002)
23. Zaki, M.J.: Efficiently mining frequent embedded unordered trees. Fundam. Inform. 66(1-2), 33–52 (2005)

Devolved Ontology for Smart Applications

Iain Duncan Stalker, Nikolay Mehandjiev, and Martin Carpenter

School of Informatics,
University of Manchester,
PO Box 88, Manchester,
M60 1QD, UK
iain.stalker@manchester.ac.uk

Abstract. Many smart applications allow enterprises to communicate effectively with and through interconnected computing resources, however, successful communication presupposes a shared understanding; a so-called *semantic alignment*. Devolved ontology was developed to promote semantic alignment in agile partnerships. We consider the approach to be promising for *any* environment where multiple contexts interface and co-locate, including, for example, the Pragmatic Web, virtual organisations and indeed smart applications. We motivate and introduce devolved ontology and show how to use this to foster semantic alignment.

1 Introduction

While there is yet no widely agreed definition of a "smart application", there are identifiable traits possessed by recognised instances. For example, many smart applications allow enterprises to communicate effectively with and through interconnected computing resources, thus promoting interactions and transactions which might otherwise be missed. However, successful communication presupposes a shared understanding; a so-called *semantic alignment*. With the advent of Semantic Web and related technologies, the path to semantic alignment is typically sought through ontology. However, in open environments, such as for many smart applications, ontology does not necessarily provide a simple route to semantic alignment. For example, there may be a number of candidate ontologies for a particular domain or there might be the need for communication among parties from multiple domains, each with its own ontology.

Our interest in semantic alignment derives from our work using agent-based systems to support agile partnerships, where partners combine their respective strengths opportunistically to improve competitiveness; and typically form with a target project in mind, i.e., are goal-oriented. We are particularly interested in using ontologies to represent knowledge and information in dynamic, evolving domains in which discourse makes use of concepts from multiple (application) contexts. We find strong, inviting parallels between the motivation for our work and the communication demands made by many smart applications.

Agile partnerships are dynamic, open networks of entities which assemble opportunistically to fulfill a particular purpose; examples include Virtual Enterprises, Supply Chain Networks and eMarketplaces. Enabling software technologies for such partnerships must both capture the distribution of intelligence or expertise and facilitate meaningful communication. Multiagent systems offer much to foster the open nature of agile

U. Priss, S. Polovina, and R. Hill (Eds.): ICCS 2007, LNAI 4604, pp. 360–373, 2007.
© Springer-Verlag Berlin Heidelberg 2007

partnerships. For example, *ad hoc* interaction with new arrivals is supported through *agent communication languages* and *interaction protocols* [5,25]. Nevertheless, there are limitations: communication in multiagent systems presupposes a common ontology, which is typically fixed in both content and semantics. Yet, the nature of agile partnerships suggests neither a fixed ontology nor a unique semantics is appropriate. In agile partnerships ontologies do indeed co-evolve with their communities of use. We have developed an approach which supports this and which we see as a potentially useful contribution to the development of knowledge architectures intended to foster semantic alignment, especially in environments where multiple contexts interface and co-locate, including, for example, the Pragmatic Web [19], knowledge management in virtual organisations [20] and indeed smart applications.

2   Preliminaries

In this section we present the formal apparatus used to implement the particular devolved ontology model we introduce in Section 5. Specifically, we briefly introduce Formal Concept Analysis, Partially Shared Views and we recall some relevant aspects of the Theory of Utility. We assume a familiarity with some aspects of order theory (we recommend [4]) and with multiagent systems [25,5], though this is more intuitive.

2.1   Formal Concept Analysis

Formal Concept Analysis (FCA) [6] is a powerful, elegant method of analysis which identifies (conceptual) structures within data sets. The qualifier *formal* emphasises that these are mathematical notions, which do not necessarily capture the everyday use of the terms. in particular, it distinguishes the vocabulary of FCA from namesakes in philosophy; or cognitive science. In the sequel we dispense with the qualifier for convenience.

Definition 1. (Context and Concept) *A* context *is a triple* (G,M,I) *where G and M are sets and* $I \subseteq G \times M$. *G is the set of* objects, *M is the set of* attributes *and I is an* incidence relation. *We write gIm for* $(g,m) \in I$.

Let $A \subseteq G$ *and* $B \subseteq M$. *Define* $A^{\rhd} = \{m \in M \mid gIm, \forall g \in A\}$, *then* $A^{\rhd}$ *is the set of attributes shared by all objects in the set A. Similarly define* $B^{\lhd} = \{g \in G \mid gIm, \forall m \in B\}$, *then* $B^{\lhd}$ *is the set of all objects possessing the attributes in the set B. These maps are called* derivation operators. *A* concept *of the context* (G,M,I) *is a pair* (A,B), *such that* $A^{\rhd} = B$ *and* $A = B^{\lhd}$. *The* extent *of the concept* (A,B) *is A and the* intent *is B.*

Definition 2. (Concept Lattice) *Denote the set of all concepts of a context* $\mathcal{B}(G,M,I)$, *or simply* $\mathcal{B}$ *where the context is clear. Define a partial order,* $\leq$, *on* $\mathcal{B}$ *as follows:* $(A_1,B_1) \leq (A_2,B_2) \Leftrightarrow A_1 \subseteq A_2 \Leftrightarrow B_1 \supseteq B_2$. *Then* $(\mathcal{B}, \leq)$ *is called the* associated complete lattice of concepts, *or simply* concept lattice, *of the context* (G,M,I).

We illustrate the basics of FCA through a simple example. Table 1 illustrates a simple context for set of natural and artificial bodies of water, inspired by [14]; categorising these according to attributes of origin, size and the nature of the motion of the main body of the water. Consider {Lake, Mere}$^{\rhd}$ = {natural, standing, large, medium} and

Table 1. A Simple Context for Bodies of Water; inspired by [14]. A cross in a cell indicates that the attribute in the column is often applied to the object in the row. For this example, we do not require a discrete choice for size, thus a river can be either a large- or medium-sized body of water. Similarly, we recognise the possibility of bodies of water which arise both naturally or are created, e.g. Lake.

	Origin		Motion		Size		
	Natural	Artificial	Flowing	Standing	Large	Medium	Small
Brook	X		X				X
Stream	X		X				X
River	X		X		X	X	
Mere	X			X	X	X	
Pond	X	X		X		X	X
Lake	X	X		X	X	X	
Ditch		X		X			X
Canal		X	X		X	X	
Reservoir		X		X	X	X	

{natural, standing, large, medium}$^{\triangleleft}$ = {Lake, Mere}. Thus ({Lake, Mere}, {natural, standing, large, medium}) is a concept of the simple context of Table 1. Similarly, ({River, Lake, Mere, Reservoir, Canal}, {medium, large}) is a concept of the simple context of Table 1. Moreover, since ({Lake, Mere}, {natural, standing, large, medium}) $\leq$ ({River, Lake, Mere, Reservoir, Canal }, {medium, large}) the former is a subconcept of the latter.

We can provide pictorial representation of the concepts of our context and their inter-relations using a Hasse diagram [4]; see Fig. 1[1]. The concept lattice for a given context provides a direct manner in which to identify whether a relationship exists between two given concepts; and further, clarifies the nature of this relationship. For example, the concept lattice for a given context allows us to identify the immediate subconcept (respectively, superconcept) of any two concepts of a given context.

2.2 Partially Shared Views

Partially Shared Views (PSV) is a scheme to facilitate communication among disparate groups [10]. It arose in the context of template-based office communication systems. Central to the scheme is a number of semistructured templates for different types of objects. The term *type* is used to denote a particular class of objects and this notion corresponds directly to the notion of *concept* in an ontology. Five cases of communication are presented, ranging from no common language to a coincident common language, where *language* is used in a restricted sense to denote a set of (object) types. Of particular interest to our application is the fourth case (*Internal Common Language*, [10] p.13). Here a common language is included in each of the group languages. A type hierarchy

---

[1] The node presentations provide useful information concerning filters and ideals [6] furnished by the tool used to produce this figure, *Concept Explorer* (http://sourceforge.net/projects/conexp). This information is additional to our current purposes, thus we do not discuss here.

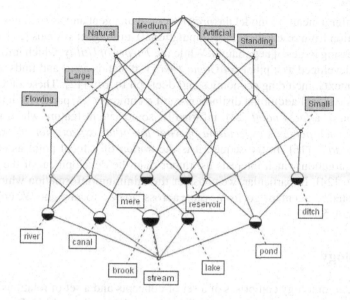

Fig. 1. A Concept Lattice for Bodies of Water from Table 1. The concept lattice is read in the following way: objects accumulate from the bottom upwards; and attributes accumulate from the top downwards. For example, the concept at the node marked "Large" includes "Medium" and "Large" as attributes; and "river", "lake", "mere", "reservoir" and "canal" as objects; the concept at the node marked "Flowing" has this as its sole attribute with "river", "brook", "stream" and "canal" as objects.

identifies a partially ordered set of types: typically, ordered according to attributes or properties. The idea is to find a common superconcept to support communication between agents using different specialisations of the common ontology. Suppose *Agent A* uses ontology *Application Ontology A* to construct a message, message, which contains for convenience a single concept, and sends this to *Agent B*. Further suppose that *Agent B* uses *Application Ontology B* and that each application ontology extends a common ontology, but is otherwise different. Notwithstanding available translation rules, there are two cases to consider. If the concept in message *is* contained in the common ontology, then *Agent B* understands the communication. Otherwise, the concept in message *is not* contained in the common ontology, and *Agent B* will not understand the communication. In this case, the concept is mapped to an appropriate superconcept in the common ontology: this used in its place; *Agent B* understands the revised communication, which we note is strictly an approximation of the initial message. Common interapplication terms means that the mapping to a superconcept need not always go to the common ontology.

2.3 Theory of Utility

Formally, a *Utility Function* (for an agent $\mathscr{A}$), $u_{\mathscr{A}}$, associates with each possibility from a set of outcomes or states, $\omega \in \Omega$, a measure—a real number $u_{\mathscr{A}}(\omega) \in \mathbb{R}$—to reflect the "enjoyment" which would be derived from each state. That is, $u_{\mathscr{A}} : \Omega \to \mathbb{R}$. Utility functions offer a means to model the *preferences* of an agent and as such have attracted much attention from researchers in (computational) multiagent systems [25], especially when addressing issues of negotiation. While the *Theory of Utility*, which informs utility functions, developed as a pillar for *Game Theory*, it stands apart and finds application in other contexts, including economics and decision theory [12]. These other contexts often provide useful additional distinctions and techniques. Of particular interest to us is the notion of a *focal point* [24]: informally, some salient feature which provides a focus *"for each person's expectation of what the other expects him to expect to be expected to do"* [16]. In the sequel, it is the *notion* of a focal point as captured in the above quotation which interests us, rather than the development of the *Theory of Focal Points* [24]. In particular, we consider the minimum information which must be communicated by agent to "get its message across" to be a focal point. We enlarge upon this in Section 5.

3   Ontology

Informally, an ontology comprises of a set of concepts and a set of relations which describe and constrain how the concepts refer, interrelate and combine. Recent interest in ontologies has led to a number of definitions of the term "ontology", see e.g. [13] or [7], but we prefer that offered by Guarino: an explicit, partial account of a conceptualisation, where a conceptualisation identifies "a set of informal rules that constrain the structure of a piece of reality, which an agent uses in order to isolate and organize relevant objects and relevant relations" [8].

The value—in terms of reusability and portability—of a conceptualisation and *a fortiori* an ontology derives in part from its dependence on a given viewpoint. Informally, a viewpoint signifies the position taken by some agent when considering some "piece of reality" or domain of interest; and accommodates, *inter alia*, any perceptual, societal, environmental, linguistic, technological and cognitive constraints which appertain, including the intended use of that knowledge. An ontology deriving from a shared conceptualisation is likely to be more generic, perhaps more widely applicable and thus more valuable. We consider this prime motivation for a negotiated formalisation, especially for a domain of discourse.

In Computer Science, ontologies are typically used for one of two purposes: to formalise a domain of interest; or to support communication through a controlled, unambiguous vocabulary. While it is possible and often instructive to view the second as a special case of the first—in that we formalise a domain of discourse—their respective, underlying motives are fundamentally different.

1. *Formalising a Domain.* This is an exercise in (knowledge) engineering. We build an abstract model or construct a theory which ideally gives a precise and accurate account of the salient aspects of a domain of interest; which can be substantiated by practice or experiment. Thus, *objectivity*, i.e., independence of the account from the observer, is of primary importance. Typically, defining a substantive concept within a given domain involves agreement at two levels: we must identify what objects exist in our (shared) conceptualisation; and how these objects are characterised.

Implicit in our theory is an ontological commitment: by describing some phenomenon through the use of denoting symbols, we are committed to the existence of certain entities and relations among these. This echoes perhaps the most familiar theory of ontological commitment; that of Quine, which claims in essence that one is committed to an entity if one refers to it directly or indirectly; cf. [15].

2. *Supporting Communication.* Supporting communication is an exercise in pragmatics. *Pragmatics* is a subfield of linguistics which investigates the nature of communication in concrete situations. In particular, it distinguishes two intents within a given communicative act—usually verbal, but these apply in a wider sense—namely [11,18]: *informative intent* or the (interpretive or referential) meaning of the sentence; and *communicative intent* or the intended meaning of the speaker. Of especial interest in supporting communication are the so-called *deictic* aspects, which, in a general sense, confirm that valid interpretation demands knowledge of the context in which the communication occurs. This suggests that we must assume the viewpoint of the agent responsible for a given communicative act to receive the communicative intent for the specific, concrete situation; and conversely, that, to ensure that the communicative intent is conveyed, a communicating agent should not presuppose that its viewpoint prevails in a domain of discourse.

The nature of ontological commitment in supporting communication differs markedly from that arising when formalising a domain: fewer concepts and relations are necessary; and importantly less structure is required.

In our opinion, the failure to maintain this dichotomy is one, significant cause of the delay in delivering on the promise of ontologies for communication; and frustrates much of the interaction between those active in the two different aspects. This is particularly evident when, as a first step to communication among partners from different domains, ontological alignment is sought in a manner which is tantamount to formalising the domain of discourse. There is a perceived need to agree on precise concept definitions and much is made of the merging of ontologies to achieve this. Accordingly, independently of method, agreement is sought at two levels: the identification of what objects exist in the (shared) conceptualisation; and how these objects are structurally defined. Yet, for a given domain of discourse, we—as individuals acting upon the world—are capable of entertaining simultaneously a number of conceptualisations which may be inconsistent, even contradictory or at different levels of granularity. We choose the most appropriate to the task at hand: we select according to context. As such, it is not convenient nor desirable to fix a unique characterisation of the domain of discourse. Indeed, such a choice often proves to be an impediment. Thus, in a practical sense maintaining the dichotomy means that we treat communication as a *de facto* exchange of a minimal sets of *essential* tokens of information; and we do not impose our ontology onto the communication.

4   Our Approach: A Devolved Ontology Model

Informally, a *Devolved Ontology Model* comprises of a core ontology and a number of extensions of this into peripheral and interapplication domain ontologies. It is a structure to facilitate ontological and semantic alignment among communicating entities.

The core ontology provides a common ground for understanding among partners and is central to the partnership. The concepts included within this are agreed through negotiation of all partners. As such, the responsibility for the evolution and maintenance of the core is shared by the partners. Each peripheral ontology represents an extension of the core ontology into an application domain. The responsibility for the evolution and maintenance of each peripheral ontology *devolves* upon the appropriate partner or partners. This includes the responsibility for extending the core into the particular context and ensuring that the peripheral ontology remains consistent with the core. Since two partners may share a number of concepts which are not part of the core, we recognise the existence of interapplication domains and ontologies. The responsibility for the initial extension of the core into the interapplication ontology devolves upon two agents jointly; for further extension into each application ontology devolves onto the appropriate single agent.

Crucially, devolving responsibility upon the appropriate partner (respectively partners) includes leaving the choice of appropriate syntactic structure to it (respectively them). Therefore, the first step in creating a formal devolved ontology model is the removal of syntactic aspects: structures are initially flattened. We propose that a given concept has a number of tokens, e.g. a set of attributes, associated with it. The tokens used to represent the concept (at a particular instant) are selected according to context, projecting away from those which are redundant to leave only an essential subset. We refer to the full set of tokens as the global (domain) concept: this may include inconsistencies. In the special case where the tokens are the same for each participant, we call this a common (domain) concept. To compensate for the removal of syntactic structure, it is imperative that we find some "natural" structure and allow this to emerge. In the development of the model in Section 5, we use *Formal Concept Analysis* (FCA) [6], *Closure Operators* (see, for example, [4]), and *Lattice Theory* [2] to capture these ideas; and to provide sufficient rigour for systematic treatment. The selection of tokens as needed and the appeal to a "natural" structure allow our ontologies to be minimal and *self-constructing*.

5   Concept Negotiation

We assume that an agent is reluctant to alter its knowledge base unless there is some (positive) payoff. Moreover, once motivated to revise its knowledge base, it will seek to minimise the extent of any change. Accordingly, in any concept negotiation we have a natural focal point, cf. [24], for each agent: namely, those *essential details* which must be conveyed to ensure that the transaction is *appropriately informative*. For the sender, this represents a "lower bound" for the concept under negotiation: any acceptable alternatives reside "between" this and the original concept. For the receiver, the closer he gets to this lower bound, the better, since he makes the minimum necessary change to obtain the essential information. Typically, he will not know what this is, thus it is in his interests at each stage to strip away attributes.

5.1   Using FCA and PSV to Create a Devolved Ontology Model

In PSV a *view* is defined to be *"a set of object types and their relations. A view V_2 is subtype of V_1 if some of the message types in V_2 are specializations of ("children* *of") the message types in V_1"* (p.16) [10]. FCA provides use with an appropriate formalism through which to realise PSV. Suppose that the structure of our domain (i.e. its ontology) is comprised of two (main) application ontologies, A and B, which share and thus generalise a, perhaps notional, common ontology. Generalisation identifies a subontology relationship and so the common ontology is a proper subset of each of the application ontologies. In fact, the common ontology is a (common) subcontext of each application ontology [6]. Further, a (notional) global ontology, which includes all concepts in the domain, subsumes each application ontology and similary provides a supercontext of each application ontology.

We return to Table 1; also see Fig. 2. As our application ontology A, we categorise the bodies of water according to the attributes nature and size with the additional information that we explicitly identify whether the main body of water flows. As our application ontology B, we categorise the bodies of water according to attributes of nature, "standing" and "large". Thus, our common ontology emerges as the categorisation of the bodies of water according to the attributes of origin and "large". In each case, we enlarge the common ontology by (order-) embedding it into the particular augmented ontology. As such, the common ontology is a sublattice of each augmented ontology. Fig. 1 illustrates our (notional) global ontology and includes all of our concept lattices.

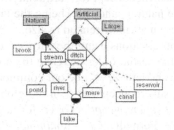

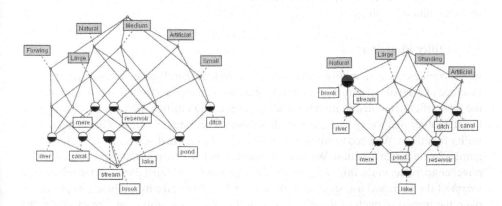

Fig. 2. Common ontology (top), application ontology A (left) and application ontology B (right). The common concept lattice is a factor lattice of each application domain lattice cf. [6]. Moreover, the application domain lattices are factor lattices of the global domain lattice of Fig. 1 and provide an *atlas decomposition* of this [6].

Each of the common ontology and the augmented ontologies is a sublattice of the notional global ontology.

Borrowing the term *view* from [10] we consider the above ontologies as (defining) views of the bodies of water. *Common View*, V_C: categorisation of bodies of water according to the attributes of origin and "large". *Augmented View A*, V_A: an augmentation of the common view to include a consideration of all attributes of size and also the identification of flowing bodies of water. *Augmented View B*, V_B: an augmentation of the common view to include identification of standing bodies of water. *Global View*, V_G: an all-encompassing view which we associate with the domain as a whole: essentially, a superset of V_A, V_B and V_C.

Consider, *Agent A* sends a message, message $= (\ldots \{\text{medium, artificial}\} \ldots)$, which *Agent B* does not understand, as the concept does not exist in its view, V_B. Thus, if appropriate, we replace it with closest superconcept in the common view, V_C: from Fig. 2 we see is {artificial}. This represents an approximation of the original message and may or may not contain sufficient information. In the case that some notion of size is required, the superconcept from the common ontology provides a starting point. Moreover, we find that in application ontology A "medium" never arises as the sole attribute for size: it always occurs with either "large" or "small". Thus, it may be possible to use {artificial, large} as a replacement concept for {artificial}: which is seen directly from the common ontology as the natural candidate, being the closest common subconcept of {artificial} which includes an attribute related to size.

We can formalise these observations using the partial order. Consider the notion of closest superconcept using the partial order: let C be the concept of interest, $(\mathscr{B}_G, \leq)$ be our global concept lattice and let $(\mathscr{B}_C, \leq)$ be our common concept lattice. Let $C_G^u \subseteq \mathscr{B}_G$ denote the set of all of superconcepts of C in the global concept lattice $\mathscr{B}_G$ (the *upward closure*). The *closest superconcept* is C_T (T for target), where $C_T \in C_G^u \cap \mathscr{B}_C$ and $C_T \leq C'$, $\forall C' \in C_G^u \cap \mathscr{B}_C$. Formally, C_T is a *minimal element* of $C_G^u \cap \mathscr{B}_C$. This is not necessarily unique, but we omit discussion of this here. Essentially, the mapping to the appropriate superconcept is a projection away from those attributes which are not in the common ontology.

5.2   Utility Functions

FCA provides a way to realise aspects of PSV and together these give rise to a particular instance of a devolved ontology model. While this is an elegant model, it provides merely the *what* of concept negotiation for a set of interacting entities, leaving us to determine through other methods *when* and *why* these should seek to negotiate. This is the role of utility functions. We need to equip our agents with these. We discuss this with simple examples in the current section. We assume that the decision to negotiate when faced with a novel concept depends, *inter alia*, on the *importance* of the third party(ies) involved; the *worth* of the (current) transaction; and the *cost-benefit* of admitting the concept. Moreover, the import of each of these depends on the stage of negotiation. For example, the cost-benefit of admitting the concept is unknown in the initial stages and has little impact on the decision to proceed with negotiation. When admitting the concept, the cost-benefit is a dominant factor. *Both* the receiver and sender can choose whether or not to enter into a negotiation over a novel concept. Thus, each could be equipped with a utility function.

Table 2. Criteria for Cost-Benefit. Figure in cell ij indicates the relationship between criteria i and j as follows: 1 - indifferent; 3 - i is slightly more important; 5 - i is more important; 7 - i is significantly more important; 9 - i dominates. 2,4,6, and 8 are intermediates. ji is the reciprocal of ij.

	Frequency f_r	Frequency f_t	Concepts N_A
Frequency (relative) (f_r)	1	5	3
Frequency (time) (f_t)	1/5	1	1/7
Auxiliary Concepts (N_A)	1/3	7	1

The decision to admit a novel concept belongs to the ontology agent associated with the ontology to which the novel concept would be admitted.

We take a simple approach. For each of *importance, worth* and *cost-benefit*: we identify a number of criteria; we allow the user to compare and rank these and we normalise the user rankings to provide a set of weights, $w_i \in [0,1]$, with $\sum_{i=1}^{n} w_i = 1$, where n is the number of criteria. For each utility function, we allow the user to set a threshold value, $u \in [0,1]$, which must be exceeded (for the utility measure to be worthwhile). Thus, we derive utility functions of the form: $U(U_i, U_w, U_c) = w_1 f_1(U_i) + w_2 f_2(U_w) + w_3 f_3(U_c)$, where U_i, U_w and U_c denote the utility (sub)functions for *importance, worth* and *cost-benefit*, respectively; the $w_i, i = 1,2,3$ denote weights; and $f_i, i = 1,2,3$ are functions (of the appropriate arguments) which return a value in [0,1]. Each of the utility (sub)functions takes a form similar to the total utility functions.

As a simple example, consider *cost-benefit*. Suppose we identify and rank the criteria as in Table 2. Normalising and then averaging[2] the entries of Table 2 leads to a set of average values which we use to construct a cost-benefit utility (sub)function to reflect our preferences: $U_c(f_r, f_t, N_A) = 0.59 c_1(f_r) + 0.08 c_2(f_t) + 0.33 c_3(N_A)$, where, for simplicity, we might choose simple threshold functions for c_1, c_2 and c_3. Weights and utility functions for *importance* and *worth* are derived analogously.

5.3 Negotiation Protocols

Having presented the *what*, the *why* and the *when*, it remains to show the *how*. Negotiation protocols provide this. For our purposes, a *protocol* is (simply) a prescribed sequence of message exchanges, i.e., a "conversation" template. The manner in which an agent responds when faced with a potential case for (concept) negotiation is informed by the nature of the relationship with the third party(ies) involved. This includes considerations of trust, vested interests, the degree of acquaintance, and so forth. The intangible nature of these often proves an impediment to the construction of satisfactory models.[3] We consider this information beyond the more immediate, objective measures

---

[2] *Normalisation.* Let $e_1, \ldots, e_n$ denote the entries in a given column. The normalised entries are $\tilde{e}_1, \ldots, \tilde{e}_n$, where $\tilde{e}_j = e_j / \sum_{i=1}^{n} e_i, j = 1, \ldots, n$.

Averaging. Let $\tilde{e}_{k1}, \ldots, \tilde{e}_{km}$ denote the normalised values in row k. The average for row k is $\bar{e}_k = \sum_{i=1}^{m} \tilde{e}_{ki} / m$.

[3] Naturally, the same argument can be levelled at notions of importance and worth presented above, but we feel that a greater degree of objectivity obtains for these.

captured in (our) utility functions and thus provide a choice of protocol through which to negotiate. For example, if one trusts implicitly the third party, then one might comfortably seek his opinion of the usefulness of a concept in future communications, secure in the knowledge that a fair response is obtained, cf. *Protocol B,* below The choice of protocol can be derived from an agent's list of acquaintances, cf. [5], or from the values in the utility functions when deciding whether to negotiate, cf. Subsection 5.2, or a combination of these. We assume the utility functions discussed in Subsection 5.2. For simplicity, we also assume a single third party. We present an example protocol motivated by "future usefulness", which for convenience we call *Protocol B.*

5.4   Protocol B: "Future Usefulness"

Protocol B is driven by "future usefulness" as indicated by sender. As such a certain degree of trust is vested in the sender. The protocol begins once the two agents, *Sender* and *Receiver* (say), have agreed to negotiate over the novel concept in the message.

The protocol, summarised in Fig. 3, takes place between the Ontology Agent of the *Receiver* and the *Sender* of the message containing an unknown concept. If the *Sender* has an Ontology Agent on its platform, then this may (also) participate in the negotiation. The primary aim of the negotiation is to decide on the best way to treat the unknown concept, and it will produce one of the following three outcomes:

1. If the novel concept will be used by *Sender* in future transactions, then it is *internalised* by (the Ontology Agent of) the *Receiver.*
2. If negotiation with the *Sender* is frustrated, for instance, the sender does not know the negotiation protocol, then the concept will be temporarily asserted, if necessary supporting constructs are available.
3. If the novel concept is to be used on an infrequent basis, then the negotiation will attempt to find an appropriate superconcept acceptable to both parties. Details of this aspect of the protocol are omitted. If agreement can not be reached, then we would revert to a temporary assertion.

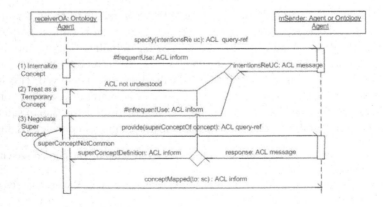

Fig. 3. Sequence Diagram for Protocol B

6 Discussion and Concluding Remarks

We have introduced the notion of an evolvable devolved ontology, a formal model which we developed initially to address relations among ontological structures which arise in agile partnerships. We have shown how to use this to promote *ad hoc* semantic interoperability and thereby support communication in open environments where parnerships are opportunistic. We have chosen to present only the theoretical aspects of our approach as we believe that it is the synthesis of ideas, i.e. the framework, which will be of particular interest to smart applications rather than a particular implementation. In our actual implementation, we have used agents, see, for example, [20], but the techniques lend themselves to implementation in other instances of intelligent software, or can even be approached manually. The principles are general and we consider the approach to be promising for *any* environment where multiple contexts interface and co-locate, including, for example, the Pragmatic Web [19], knowledge management in virtual organisations [20] and indeed smart applications. Naturally, there are limitations and we briefly mention a couple here. The FCA approach to ontology structuring presupposes that we can agree upon a shared set of tokens (e.g. attributes, properties) through which to describe the domain objects of interest; and that these *mean* the same thing to participants. The negotiation mechanisms assume that the agents share a common communication language and in particular a common content language (cf. www.fipa.org).

The approach we have presented is in fact a particular instance of a more general model. FCA can be related to the *information system* of Dana Scott [17]. The set of attributes M of a (formal) context (G, M, I) constitutes a set of *information tokens* and gives rise to a *derived information system* [26]. The set of attributes constitutes a set of descriptors which can be applied to a set of objects of interest, thus creating a formal context. Thus, descriptions of the objects using the information tokens induce a closure on the derived information system. This is simply another way in which to view the creation of a concept lattice. From this perspective, we can discern three ways in which new knowledge and information becomes available in dynamic contexts:

1. *New information* arises from the introduction and application of a new information token, i.e. a new descriptor or attribute, to the existing objects;
2. *New knowledge* arises from the introduction of a new object which is described using a novel combination of the information tokens available; and
3. *New information and new knowledge* arises when the two happen simultaneously.

Summarising a (general) devolved ontology model of from this perspective, we have for a partnership $\mathscr{P} = \{\mathscr{M}_1, \ldots, \mathscr{M}_n\}$: an evolving partnership ontology $\mathscr{O}(\bigvee \mathscr{P})$, which is a closure system on the derived information system obtained from all tokens used throughout the partnership domains; a number of peripheral and interapplication ontologies, $\mathscr{O}(\mathscr{I})$ and $\mathscr{O}(\mathscr{M})$ where $\mathscr{I} \subset \mathscr{P}$ and $\mathscr{M} \in \mathscr{P}$, each being a closure on the partnership ontology; and an evolving core ontology $\mathscr{O}(\bigwedge \mathscr{P})$ which reflects the common view on the partnership ontology and is embedded into each ontology: each ontology extends it. The common view offers a *semantics approximation* of each of the ontologies and as such provides the basis for any negotiation over concepts. All of these views on the partnership ontology are connected in a formal way: through an *adjunction*, of which the Galois connection is a particular example. The evolutions arise through the introduction (and also removal) of new information tokens and new objects: these arise when partners introduce new descriptors or items or both.

Another important point is that while we dispense with the need to agree upon structures for the domain of discourse—appealing instead to a "natural" structure—the original structures remain in the actual application domains. These are only temporarily "forgotten" while a semantic alignment is sought; this is achieved rigorously and in a "reversible" manner (also) through adjunctions. We defer discussion to future presentations.

There many related works to which we could make reference. This is to be expected in such an active area of research; for example, the applicability of Formal Concept Analysis to ontology construction and management,is recognised by many, e.g., [22] and [3]. In the interests of space and brevity, we mention four specific works of particular relevance. First, *Partially Shared Views* [10], which we have applied and discussed above. Second, *Exploiting Partially Shared Ontologies for Multi-Agent Communication* [21], which exhibits some strong parallels in organising ontologies, for example, the author (independently) proposes an approach analogous to PSV. Third, the ontology merging techniques developed in *FCA-Merge* [23], offer a way in which to synthesise a global ontology (cf. Fig. 1) from application domain ontologies (cf. Fig. 2), in a manner analogous to *gluing* [6]. Fourth, the *Ontology Negotiation Protocol (ONP)* of Bailin and Truszkowski, e.g. [1], addresses issues arising when "agents communicate in solving tasks when they encounter each other on the web": we observe a number of parallels. A state-of-the-art survey in the use of various formal mechanisms for ontologies can be found in [9].

References

1. Bailin, S.C., Truszkowski, W.: Ontology negotiation as a basis for opportunistic cooperation between intelligent information agents. In: Klusch, M., Zambonelli, F. (eds.) CIA 2001. LNCS (LNAI), vol. 2182, Springer, Heidelberg (2001)
2. Birkhoff, G.: Lattice Theory, 3rd edn. AMS Colloquium Publication, Providence, RI (1967)
3. Cimiano, P., Hotho, A., Stumme, G., Tane, J.: Conceptual knowledge processing with formal concept analysis and ontologies. In: ICFA 2004, pp. 189–207 (2004)
4. Davey, B.A., Priestley, H.A.: An Introduction to Lattices and Order, 2nd edn. Cambridge University Press, Cambridge (2002)
5. Ferber, J.: Multi-Agent Systems. An Introduction to Distributed Artificial Intelligence. Addison-Wesley, Reading (1999)
6. Ganter, B., Wille, R.: Formal Concept Analysis. Mathematical Foundations. Springer, Heidelberg (1999)
7. Goméz-Pérez, A., Fernández-López, M., Cocho, O.: Ontological Engineering. In: Advanced Information and Knowledge Processing, Springer, Heidelberg (2004)
8. Guarino, N., Giaretta, P.: Ontologies and knowledge bases: towards and terminological clarification. In: Mars, N. (ed.) Towards Very Large Knowledge Bases: Knowledge Building and Knowledge Sharing, IOS Press, Amsterdam (1995)
9. Kalfoglou, Y., Schorlemmer, M.: Ontology mapping: the state of the art. The Knowledge Engineering Review 18, 1–31 (2003)

10. Lee, J., Malone, T.W.: Partially shared views: A scheme for communicating among groups that use different type hierarchies. ACM Transactions on Information Systems 8(1) (1990)
11. Leech, G.: Principles of Pragmatics. Longman, London (1983)
12. Luce, R.D., Raiffa, H.: Games and Decisions. Introduction and Critical Survey. John Wiley & Sons, New York (1957) (Dover Reprint 1989)
13. Noy, N., Klein, M.: Ontology evolution: Not the same as schema evolution. Knowledge and Information Systems 6, 428–440 (2004)
14. Obitko, M., Snášel, V., Smid, J.: Ontology design with formal concept analysis. In: Snášel, V., Bělohlávek, R. (eds.) CLA 2004, Technical University of Ostrava, pp. 111–119 (2004)
15. Quine, W.V.O.: On what there is. Review of Metaphysics 2, 21–38 (1948)
16. Schelling, T.C.: The Strategy of Conflict. Harvard University Press, Cambridge, Mass. (1960)
17. Scott, D S: Domains for denotational semantics. In: Nielsen, M., Schmidt, E.M. (eds.) Automata, Languages, and Programming. LNCS, vol. 140, pp. 577–613. Springer, Heidelberg (1982)
18. Sperber, D., Wilson, D.: Relevance: Communication and Cognition. Blackwell, Oxford (1986)
19. Stalker, I.D., Mehandjiev, N.D.: A devolved ontology model for the pragmatic web. In: Proceedings of the First International Pragmatic Web Conference (PragWeb) 2006, Stuttgart (September 2006)
20. Stalker, I.D., Mehandjiev, N.D., Carpenter, M.R.J., Gledson, A.: Dynamic knowledge management in open multiagent environments. In: Proceedings of AMKM 2005, Workshop of AAMAS 2005 (July 2005)
21. Stuckenschmidt, H.: Exploiting partially shared ontologies for multi agent communication (2002)
22. Stumme, G.: Using ontologies and formal concept analysis for organizing business knowledge (2001)
23. Stumme, G., Maedche, A.: FCA-MERGE: Bottom-up merging of ontologies. In: IJCAI, pp. 225–234 (2001)
24. Sugden, R.: A theory of focal points. The Economic Journal 105(430), 533–550 (1995)
25. Wooldridge, M.: An Introduction to Multiagent Systems. John Wiley & Sons Ltd, Chichester (2002)
26. Zhang, G.Q.: Chu spaces, concept lattices and domains. In: 19th Conference on the Mathematical Foundations of Programming Semantics. Electronic Notes in Theoretical Computer Science, vol. 83, Montreal, Canada (March 2003)

Historical and Conceptual Foundation of Diagrammatical Ontology

Peter Øhrstrøm, Sara L. Uckelman, and Henrik Schärfe

Department of Communication and Psychology
Aalborg University
Institute for Logic, Language, and Computation
Universiteit van Amsterdam
Department of Communication and Psychology
Aalborg University

Abstract. During the Renaissance there was a growing interest for the use of diagrams within conceptual studies. This paper investigates the historical and philosophical foundation of this renewed use of diagrams in ontology as well as the modern relevance of this foundation. We discuss the historical and philosophical background for Jacob Lorhard's invention of the word 'ontology' as well as the scientific status of ontology in the 16th and 17th century. We also consider the use of Ramean style diagrams and diagrammatic ontology in general. A modern implementation of Lorhard's ontology is discussed and this classical ontology is compared to some modern ontologies.

Keywords: Ontology, diagrammatical reasoning, conceptual structures.

It is commonplace in modern computer science to present ontologies in terms of diagrams. In this way the ontologies are supposed to be more readable than they would be if presented as sets of logical formulae. In addition, the use of diagrams has been supposed to facilitate and support conceptual reasoning. According to Peirce, the use of diagrams in logic can be compared with the use of experiments in chemistry. Just as experimentation in chemistry can be described as "the putting of questions to Nature", the conceptual experiments upon diagrams may be understood as "questions put to the Nature of the relations concerned" (CP: 4.530). This should not be misunderstood. Logic is not psychology. Peirce made it very clear that logic is not "the science of how we do think", but it determines "how we ought to think" (CP: 2.52). In this way, logic is not descriptive, but, according to Peirce, it should be seen as a normative science. In fact, he considered diagrammatical reasoning as "the only really fertile reasoning", from which not only logic but every science could benefit (CP: 4.571).

However, logicians have had similar views for centuries, although the points may not have been stated so elegantly as Peirce did. In particular, diagrammatical representation has been regarded as useful within the study of ontology. An early example of this is the often cited 'Tree of Porphyry'. Whether Porphyry actually did use diagrams, we cannot say for certain, but the literature on this particular structure points in general to a rendering by Peter of Spain from the 13th century. Diagrams were used in medieval discussion of conceptual structures, but the emphasis on the

U. Priss, S. Polovina, and R. Hill (Eds.): ICCS 2007, LNAI 4604, pp. 374–386, 2007.

importance of diagrammatical reasoning within conceptual studies became much stronger during the Renaissance. In this paper we intend to discuss the historical and conceptual foundation of this renewed use of diagrams in ontology. We intend to show that scientists working with the development of ontologies may benefit from reflections on this historical and philosophical foundation of their enterprise. In section 1, we discuss the historical and philosophical background for Jacob Lorhard's invention of the word 'ontology'. In section 2, we consider the scientific status of ontology in the 16th and 17th century. In section 3, we shall focus on the use of Ramean style diagrams in science in general and in ontology in particular. In section 4 we discuss selected elements of Lorhard's diagrammatic ontology. In section 5, we discuss how Lorhard's ontology can be implemented in a modern context using the Amine platform, and compare Lorhard's ontology with some modern ontologies. Finally, we discuss the modern relevance of the beliefs incorporated in the ontology of the 16th and 17th century.

1 The Invention of the Word 'Ontology'

The word 'ontologia' is not an original Greek word, i.e., it was never used in ancient philosophy. As we have argued in [Øhrstrøm, Andersen, Schärfe 2005] the word was constructed in the beginning of the 17th century by Jacob Lorhard (1561-1609), who, probably mainly for pedagogical reasons, wanted to present metaphysics, i.e., the conceptual structure of the world, in a diagrammatical manner. In a sense, Lorhard used 'ontology' as a synonym for 'metaphysic'. But by introducing the new word he probably also wanted to indicate that the field was being renewed.

Jacob Lorhard was born in 1561 in Münsingen in South Germany. We do not know much about his life. But it appears that the 10 years younger Johannes Kepler met him at Tübingen University, where Kepler is known to have studied in the period 1587-91. At that time Lorhard was probably a young teacher. Kepler listed Lorhard as one of the persons whom he regarded as hostile to him, and he added: "Lorhard never communicated with me. I admired him, but he never knew this, nor did anyone else". [Koestler: 235-6]

Lorhard was (like Kepler) a Protestant, and he was involved in various religious studies and discussions. In fact, the new way of treating and presenting conceptual structures signaled by the introduction of the word 'ontology' can easily been seen in the context of the general openness that characterized academic life within the Protestant circles in the late 16th century. This general and scientific openness was clearly essential for many of the important contributions to the new approach to science which was being developed during the same period, with Kepler as one its most important representatives. Clearly, this new approach to science could easily be related to discussions regarding worldview in general, and thereby also to metaphysics and ontology.

Lorhard was deeply interested in metaphysics, understood as the study of the conceptual structure of the world. In 1597 he published his *Liber de adeptione*, in which he wrote:

Metaphysica, quae res omnes communiter considerat, quatenus sunt οντα, quatenus summa genera & principia, nullis sensibilibus hypothesibus subnixa. [1597: 75] Metaphysica, which considers all things in general, as far as they are existing and as far as they are of the highest genera and principles without being supported by hypotheses based on the senses. (Our translation.)

Lorhard came to the Protestant city St. Gallen in 1602, where he worked as a teacher and a preacher. The year after, in 1603, he became 'Rektor des Gymnasiums' in the protestant city of St. Gallen. He was accused of alchemy and also a heretical view on baptism. He was, however, able to defend himself rather convincingly, and his statements of belief were in general accepted by the church of St. Gallen. (See [Hofmeier et al. 1999: 28 ff.] and [Bätscher 1964: 171 ff.]) In 1606 he published his *Ogdoas scholastica*, a volume consisting of eight books dealing with Latin and Greek grammar, logic, rhetoric, astronomy, ethics, physics, and metaphysics (or ontology), respectively.

Although Lorhard only used his new word a few times in the book, he did present his new term in a very prominent manner letting "ontologia" appear in the frontispiece of *Ogdoas scholastica*. This was probably the very first use ever of the term 'ontology' in a book. The title of the book is stated as "Metaphysices seu ontologiæ" indicating that 'ontologia' is to be used synonymously with 'metaphysica'.

As suggested by Marco Lamanny [2006], it is very likely that Lorhard's book on ontology in *Ogdoas scholastica* is in fact mainly based on Clemens Timpler's *Metaphysicae Systema methodicum* [1604], which was published in Steinfurt. Lamanny [2006] has convincingly demonstrated that all the essential philosophical terms in the book also appear in Timpler's book with the same mutual relations. However, it is evident that Jacob Lorhard in composing his version of the metaphysical system made two very important contributions to the understanding and presentation of the field:

1) He introduced the new word "ontology", which has been important since then in philosophical discourse and much more recently also in computer science.

2) He presented his material (in fact, all eight books of *Ogdoas scholastica*) in diagrammatical manner representing the conceptual structure in terms of graphical relations.

As we shall see in section 3, Lorhard did his work under the influence of the works of Peter Ramus. It should be emphasized that Lorhard in transforming Timpler's metaphysical ideas into Ramean style diagrams did in fact make original contributions relevant for the understanding and presentation of the conceptual framework of reality.

In 1607, i.e., the year after the publication of *Ogdoas scholastica*, Lorhard received a calling from Landgraf Moritz von Hessen to become professor of theology in Marburg. At that time Rudolph Göckel (1547-1628) was also professor in Marburg in logic, ethics, and mathematics. Göckel apparently also paid great attention to Timpler's work. In fact, he had written a preface of Timpler's book [Timpler 1604]. It seems to be a likely assumption that Lorhard and Göckel met one or several times during 1607, and that they shared some of their findings with each other. In this way the sources suggest that Göckel during 1607 may have learned about Lorhard's new term 'ontologia' not only from reading *Ogdoas scholastica* but also from personal conversations with Lorhard. For some reason, however, Lorhard's stay in Marburg became very short and after less than a year he returned to his former position in St. Gallen. Lorhard died on 19 May, 1609. Later, in 1613, Lorhard's book was printed in a second and revised edition under the title *Theatrum philosophicum*. In this new edition the word 'ontologia' had disappeared from the front cover, whereas it has been maintained inside the book. In 1613, however, the term is also found in Rudolph Göckel's *Lexicon philosophicum*. Here the word 'ontologia' is only mentioned briefly as follows: "ontologia, philosophia de ente seu Transcendentibus" (i.e., "ontology, the philosophy of being or the transcedentals"). It is very likely that Göckel included this term in his own writings due to inspiration from Lorhard.

2 The Scientific Status of Ontology

Lorhard introduced metaphysics (or ontology) using the Greek term επιστημη for which we in [2005: 429] suggested the translation 'knowledge'. However, as argued by Claus Asbjørn Andersen [personal communication], it appears from the context that Lorhard must have used επιστημη as corresponding to the Latin *scientia*. Taking this into account, Lorhard's definition of 'ontology' becomes "the science of the intelligible as intelligible insofar as it is intelligible by man by means of the natural light of reason without any concept of matter" [1606: Book 8, p.1]. This science is obviously not just any 'knowledge' among many other branches of human knowledge. Being "the science of the intelligible" it is clearly logically and systematically prior to other discipline of the human intellect, i.e., a first philosophy.

As mentioned above, ontology according to Lorhard is about what can be understood by man "by means of the natural light of reason without any concept of matter", and as emphasized in his *Liber de adeptione*, it should not rely on assumptions based on the senses primarily. This means that in working with the ontology we should not involve any concept of 'matter'. As convincingly argued by Claus Asbjørn Andersen [2004: 96 ff.], Göckel's presentation of ontology includes an even stronger emphasis of the importance of abstraction from the material. In this way ontology may be characterized as the study of what can be understood by the human intellect organized in a system reflecting the order of the conceptual understanding in a proper manner.

It is an important guiding principle in Göckel's ontology that the fundamental terms in the structure are organized in pairs of concepts. The same is clearly the case in Lorhard's ontology. His system is presented in terms of dichotomies whenever possible, i.e., he probably wanted to divide any complex class of concepts into two subclasses characterized by contradictory terms.

Lorhard's approach to ontology was probably very much inspired by the Peter Ramus (1515-72), who had strongly criticized Aristotelian scholasticism, and who had suggested that the liberal arts should be organised and presented in a new manner. Ramus emphasized the importance of mathematics in the contexts of knowledge in general, but he also insisted on a practical and operational approach to mathematics. As emphasized by R. Hooykaas [1987] Ramus was interested in how the making of instruments could support the application of mathematics in the study of reality. This interest was probably based on the belief in a mathematical structure of the physical and conceptual universe. This view when taken together with the practical approach mathematics turned out to be essential for the rise of modern natural science.

In 1562 Ramus converted to Calvinism, and he was murdered in Paris in the St. Bartholomew's Massacre on August 26, 1572. The fact that he was considered to be a Protestant martyr made many intellectual Protestants interested in his ideas. In fact, his religious and scientific ideas became very influential in the Protestant world during the 16th and 17th century.

Lorhard (like Ramus) accepted the idea that we may understand reality (or at least important aspects of reality) by means of the natural light of reason, i.e., we have as rational beings access to necessary truth in mathematics and in reality in general. Ontology is the science of the structure of the conceivable truth about the material and immaterial world. In this way, ontology may be seen as included in natural theology according to which man as a rational being may understand essential aspects of the world without having to base his understanding on any special revelation. If seen in this way, ontology must be something universal, in principle accessible to every rational human being. In addition, ontology does not depend on anything physical, although as a science it is certainly very important, since it forms the background for our interaction with the world. Given this kind of practical importance, it was obvious to Lorhard that ontology should be one of the sciences taught to young people early in the education.

3 The Diagrammatical Approach to Ontology

As noted above, Lorhard's approach to ontology and in particular his use of diagrams, was probably very much inspired by Peter Ramus (1515-72), who had argued that scientific knowledge at least, for pedagogical reasons, should be simplified using diagrams organised in dichotomies.

Walter J. Ong [1959: 436 ff.] has pointed out that there seems to be an interesting relation between invention of printing and the impact of the development associated with Ramus' ideas. Shortly after the invention of printing the use of tables of dichotomies or bracketed outlines of subjects became very famous. As in Lorhard's books the subjects were often organised as long series of dichotomies presented in terms of brackets. This way of organising and presenting subjects can also be found in manuscripts written before that time, but they seem to have been relatively rare before the invention of printing. It is very likely that the new technology of printing facilitated the spread of what was considered to be a very impressive and powerful way of presenting a subject matter. According to Ong [1959: 437] there was a kind of "addiction to such outlines" during the 16th and 17th century. The ideology behind this tendency seems to have been that the diagram in a very effective manner, can make the conceptual relations clear to us, and that the very conceivability of a term may fundamentally depend on its relations to other terms or concepts, i.e., that "words are made intelligible by being diagrammatically related to one another" [Ong 1959: 437].

Ramus himself often used diagrams based on dichotomies. As argued by Stephen Triche and Douglas McKnight [2004], his main purpose for representing knowledge in terms of diagrams was pedagogical. In fact, he argued that following his ideas and pedagogical logic the various studies of the liberal arts could be united in one course. Triche and McKnight state:

Ramus's primary intellectual accomplishment was the refinement of the art of dialectic by transforming dialectical reasoning into a single method of pedagogical logic for organizing and demonstrating all knowledge. In addition, his invention of method completes humanism's transformation of medieval scholasticism's courses of study in the liberal arts into a recursive singular course of studies called curriculum. [2004: 40]

According to Ramus this kind of new order in the higher studies should be established using the laws of logic (dialectic). Given that logic operates with two truth-values, true and false (corresponding to yes/no), this can easily lead to the idea of dichotomies. In this way, he believed, that every subject can be represented in terms of a diagram of dichotomised concepts. Also, the order in which the concepts appear in the diagram is not arbitrary. According to Ramus there is a natural order of the concepts, which should used in the construction of the diagram. This order should be taken into account when teaching the subject in question. In his own words:

> Through the light of artistic method, everything is more clearly taught and much more easily understood, since universal, general matters come first with subsidiary parts following, and all things arrainged by that wonderful, linking organization of antecedents and consequents (Quoted from [Triche & McKnight 2004, p.46]).

It is obvious that this view may lead to a high degree of standardisation in teaching, since it follows from the Ramean view that there is only one optimal way of organising the subject in question, and since every teacher should take this order of concepts into account.

The Ramean use of dichotomies has often been discussed e.g. in confessing his own "leaning to the number Three in philosophy", Peirce noted that other numbers have had their champions, and he gives as an example that "Two was extolled by Peter Ramus" [CP: 1.355]. It is in fact quite obvious that Ramus believed that every subject can be presented in terms of his dichotomistic diagrams. As pointed out by Bruce MacLennan "the Ramean Tree (or Ramean Epitome) proceeds by logical dichotomy from the most general term of any subject matter. In effect the Ramean Tree is an abstract geometrical diagram of the (supposed) essential structure of reality" [MacLennan 2006: 96].

4 Elements of Lorhard's Ontology

Jacobus Lorhard presented his ontology in terms of connected Ramean style diagrams written in an elaborated manner. This means that he wanted to use the principle of dichotomy as far as possible. Fig. 1 is a translation of the first page of his ontology, and the chapter continues with 58 pages of similar structures. The capital letters (A, B, C, EE, RRR) refer to continuations on subsequent pages in a way that almost resembles modern day hypertext.

In the presentation of his ontology, Lorhard uses the Ramean style bracket as his basic representational tool. However, he uses these brackets in three distinct ways. Most commonly, the brackets are a tool for dividing complex terms into two or more disjunctive subsets represented by contrasting terms. For example, infinity is either absolute or restricted (§I), necessity is either absolute or hypothetical (§L), goodness is either apparent or true (§O), and so on.

The second way that these brackets are used is in introducing explanatory notes. This usage occurs only in the very top levels of the tree, and instead of the brackets dividing a complex term into two subsets, one branch of the bracket gives a further gloss on how a term should be understood and the other then introduces how the term may be further divided. For example, before dividing 'the intelligibles' into 'nothing' and 'something', there is a note (λόγος) defining what intelligbles are. (See Fig. 1).

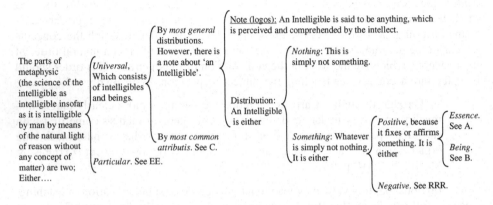

Fig. 1. First page of Lorhard's ontology

He uses the brackets in a third way not to divide one complex term into two more specific terms, but to gather two sub-terms back together before dividing them as a group. For example, when Lorhard is discussing time, he first divides it into the subgroups of momentary time and successive time. However, members of both of these classes are either real or imaginary, and he indicates this by having opposite-facing brackets collect the categories of successive time and momentary time together before dividing the entire group into that which is real and that which is imaginary (§D). See Fig. 2.

Wherever possible, Lorhard divides terms into two, exclusive and exhaustive, sub classes. However, there are cases where this is not possible, such as when he divides respective or relative goodness into the three categories of 'honor', 'utility', and 'jocundity' (§P). In these cases, it is no longer immediately clear that the chosen categories do in fact exhaustively represent the space. Certainly it is not obvious to a 21st century person that these three types of respective goodness are the only three types, or even that they are mutually exclusive (which Lorhard appears to think they are).

One thing which is clear is that Lorhard, in making this tree, is not attempting to give *definitions* of classes, but rather *divisions* (or, as he sometimes says *distributions*) of classes. This is easily seen, for example in Fig. 1, when he divides intelligibles into the two classes 'something' and 'nothing', as he describes 'nothing' as that which isn't something, and 'something' is glossed as that which isn't nothing; or when a 'principle' is glossed as that on which a principiate depends, and a 'principiate' is glossed as that which depends on a principle (§§VV, vv). If these glosses are taken as

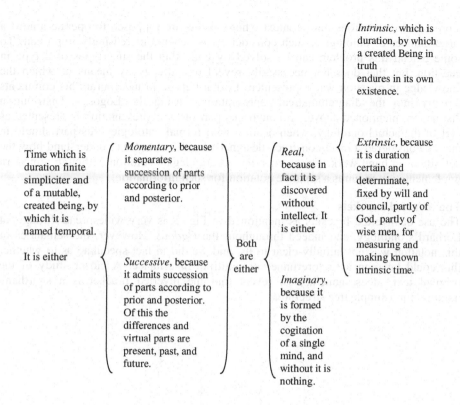

Time which is duration finite simpliciter and of a mutable, created being, by which it is named temporal.

It is either

{
Momentary, because it separates succession of parts according to prior and posterior.

Successive, because it admits succession of parts according to prior and posterior. Of this the differences and virtual parts are present, past, and future.
}

Both are either

{
Real, because in fact it is discovered without intellect. It is either

Imaginary, because it is formed by the cogitation of a single mind, and without it is nothing.
}

{
Intrinsic, which is duration, by which a created Being in truth endures in its own existence.

Extrinsic, because it is duration certain and determinate, fixed by will and council, partly of God, partly of wise men, for measuring and making known intrinsic time.
}

Fig. 2. Fragment concerning time

definitions of the terms, then circularity results. One must know in advance the meanings of the terms before one can proceed to classifying and codifying the relationships between the classes.

5 A Modern Implementation of Lorhard's Ontology

As part of this investigation, Lorhard's ontology was translated into English, and also into a present-day notation. The problems related to translating the ontology from Renaissance Latin to English is discussed in the annotated translation [Lorhard 2007]. Here, we shall report some of the most interesting aspects of turning this 400 year old system of thought into a modern ontology. Lorhard's text was represented using the Amine platform and resulted in a formal ontology, understood here as a hierarchy of types. We are assuming that the Ramean brackets correspond to a subtype relation, that is: for the most parts. Certain aspects of the notation will be discussed below.

The use of meta-constructs

In modern ontologies it can be very difficult to see how distinctions are made, and types are derived from these distinctions. In particular, it is often difficult to see clearly what the author(s) of a given ontology was aiming at through their distinctions, which again makes it difficult to decipher the intention behind the represented distinctions. In a context where agents are supposed to operate amidst a large number of ontologies, such considerations become increasingly important. To some extend this problem can be solved by collecting the supertypes of a type in question, but that does not necessarily reveal the strategy by means of which the knowledge in question was represented. Lorhard chose to incorporate his comments directly into the diagrammatical representation, using the Logos – Distribution distinction, mentioned above. Although this part of the representation is presented as part of the actual ontology, when dealing with formal ontologies this part should in fact be considered a meta-construct, designed to aid the reader to understand how the definitions at hand work. This seems like a very elegant solution, although a modern implementation requires a separate notation for such information.

The inverted brackets
The use of inverted brackets as mentioned in Fig. 2. is very widespread throughout Lorhard's ontology, and indeed throughout the *Ogdoas*. However, the semantics of this notation was not initially clear to us, and we did in fact speculate as to whether this could be seen as a forerunner of multiple inheritance. A closer study of the original texts does, nonetheless, reveal that the inverted bracket is a shorthand notation for a simple tree structure.

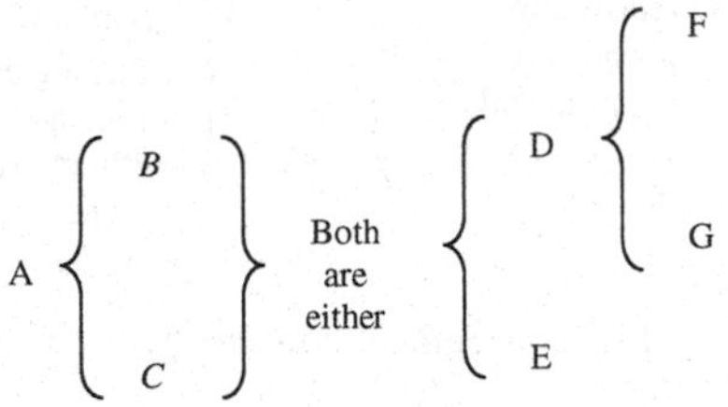

Fig. 3. The inverted brackets

So in fact the structure shown here can be unfolded as seen below in the left, which again corresponds to the more modern graphical of a hierarchy shown below to the right.

It is striking that the Renaissance texts all show the hierarchies written from left to right, whereas in modern representations it is usual to draw hierarchies in a vertical

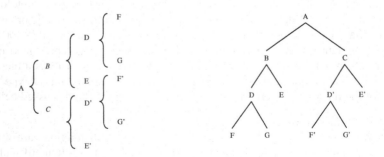

Fig. 4. Rendering of the inverted brackets

manner. Quite probably, this manner of representation is inherited from Ramus, and possibly also became conventionalized through printing practises of that time. The shift in style of representation from horizontal to vertical is however interesting because it reflects our conceptualization of the models at hand – a condition that is also reflected in our use of language, e.g. *sub*-types. However, the history of such preferences in representation style must be left for enquiry elsewhere.

In terms of translating this ontology into a contemporary system, the shorthand notation requires a separate naming of the types that are part of the structure to be duplicated.

The top ontology

The layout of the original text does not offer a single overview of the top structure of the ontology. The elaborate system of references guides the reader through the pages from section to section. Each section is organized as if one was traversing a tree. The top structure can therefore be extracted and reproduced as in fig. 5.

It is worth noticing that Lorhard's ontology does not begin with a distinction between *physical* and *abstract*, as many other ontologies do, but rather the first top distinction is between *universals* and *particulars*. Universal is then divided into a class of the *general intelligible* and a class defined by *common attributes*. The *particular* is divided into *substantial* (on its own) and *accidental* (through something else). It turns out that these distinctions are rather typical in Lorhard's thinking, and it hints at a guiding principle for the construction of many subsequent divisions, as will be described next.

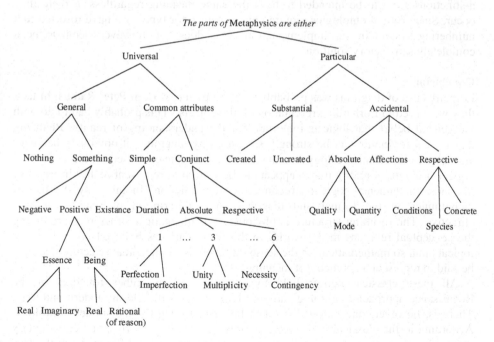

Fig. 5. The top distinctions

Iterations

Throughout the ontology there is an extensive use of repeated terms. As discussed at the end of section 4, the reader of the diagrams must follow the path from the earlier distinctions on order to grasp the meaning of mentioned of recurring terms, such as *real* and *imaginary*. See fig. 2 and 5. Such contextual readings are obviously not practical in computational environments, but do point to a guiding principle of Lorhard's thinking.

In terms of Knowledge Representation (KR), some of these recurring distinctions can be said to belong to a KR meta-language, employing such terms as: *generic* and *specific* (which occur 7 times) *universal* and *singular, immanent* and *transcendent etc.* Other distinctions are of a more striking, and, in our opinion, also more revealing nature since they seem to reflect Lorhard's metaphysical beliefs and thus give rise to a more detailed understanding of his world view. For example, the distinction between *created* and *uncreated* occurs 3 times, and the distinction between *real* and *imaginary* occurs 6 times, all in the *universal* section of the ontology. In the *particular* section of the ontology, *real* is more often opposed to *rational*, that is: things that exist in their own right versus things that exist through some intelligence.

The matters of the ontological status of the real and the imaginary certainly deserve further investigation in another context. Here, we shall confine ourselves to suggest that the extensive use of iterations indicates a principle for handling complex knowledge representations, namely that a few select distinctions are applied frequently rather that once and for all in a top distinction. This constitutes a problem when the ontology is translated into a contemporary KR system in that the repeated distinctions seem to be intended to have the same meaning regardless of there they occur. Since formal ontologies require unique names for types, we have resorted to a numbering system in our implementation. This does not, however, seem to be a completely satisfactory solution.

Conclusion

Lorhard's use of diagrams was probably inspired by the work of Peter Ramus. In fact, they were used in Lorhard's presentation of all subjects. This probably had to do with the general belief that logic is important for the understanding of reality. Realising that logical reasoning can be strongly supported by diagrams, it obviously becomes attractive to represent ontology in a diagrammatical manner. Based on the belief in a logical structure of reality it also appears to be natural to represent reality in terms of the most fundamental logical structure, the contradiction. In this way the use of dichotomies in the formal and diagrammatical description of reality becomes attractive. The resulting structure is obviously a mathematical structure representing the conceptual relations in the world. In this way reality is believed not only to be logical but also mathematical, in the sense that there is a conceptual structure that may be said to represent a geometry of meaning.

All these classical beliefs held by Lorhard and the other founders of the Renaissance approach to ontology are to a large extent still held in modern ontology. There is, however, one major difference between Lorhard's and modern ontology. According to the classical belief there is only one ontology corresponding to reality and truth. There may of course be other suggested structures different from the true ontology, but they will simply be false descriptions of reality. According to the classical view there will be no room for the alternative ontologies fit for different purposes. In other words, whereas an ontological structure in a modern context may be seen as a model or a tool fit for certain purposes and unfit for others, an ontological structure will classically be much more than a tool. It will be an attempted description of reality, which is true or false.

Acknowledgements

We would like to express our gratitude to Jørgen Albretsen, who pointed us to the information on Kepler, to Claus Asbjørn Andersen, Albertus Magnus Institut, Bonn for discussions of issues pertaining to our previous work. The authors also acknowledge the stimulus and support of the 'European project on delimiting the research concept and the research activities (EURECA)'sponsored by the European Commission, DG-Research, as part of the Science and Society research programme — 6th Framework.

References

Andersen, C.A.: Philosophia de ente seu transcendentibus. Die Wissenschaft vom Seienden als solchem und von den Transzendentalien in der spätscholastischen Metaphysik. Eine Untersuchung im Ausgang von Franciscus Suárez. Konferensspeciale. Københavns Universitet (2004)

Bätscher, T.W.: Kirchen- und Schulgeschichte der Stadt St. Gallen, Erster Band, pp. 1550–1630. Tschudy-Verlag, St. Gallen (1964)

Göckel (Goclenius), R.: Lexicon philosophicum, quo tanquam clave philosophicae fores aperiuntur. Francofurti (1613), Reprographic reproduction, Georg Olms Verlag, Hildesheim (1964)

Hofmeier, T., Gantenbein, U.L., Gamper, R., Ziegler, E., Bachmann, M.: Alchemie in St. Gallen. Sabon-Verlag, St. Gallen (1999)

Hooykaas, R.: The Rise of Modern Science: When and Why? British Journal for the History of Science 453–473 (1987)

Koestler, A.: The Sleepwalkers: A History of Man's Changing Vision of the Universe. Arkana, New York (1989) http://www.formalontology.it/essays/correspondences_timpler-lorhard.pdf

Lorhard, J.: Liber de adeptione veri necessarii, seu apodictici. Tubingae (1597)

Lorhard, J.: Ogdoas scholastica. Sangalli (1606)

Lorhard, J.: Theatrum philosophicum, Basilia (1613)

Lorhard, J.: Diagraph of Metaphysic or Ontology. Trans. by Sara L. Uckelman. IMPACT - an electronic journal on formalisation in media, text and language. (forthcoming 2007), http://www.impact.aau.dk/articles.html

MacLennan, B.: From Pythagoras to the Digital Computer: The Intellectual Roots of Symbolic Artificial Intelligence (2006) (forthcoming)

Ong, W.J.: From Allegory to Diagram in the Renaissance Mind: A Study in the Significance of the Allegorical Tableau. The Journal of Aesthetics and Art Criticism 17(4), 423–440 (1959)

Peirce, C.S.: Collected Papers of Charles Sanders Peirce, (CP). In: Hartshorne, C., Weiss, P., Burke, A. (eds.), vol. I-VIII, Harvard University Press (1931-1958)

Timpler, C.: Metaphysicae Systema methodicum. Steinfurt (1604)

Triche, S., McKnight, D.: The quest for method: the legacy of Peter Ramus. History of Education 33(1), 39–54 (2004)

Øhrstrøm, P., Andersen, J., Schärfe, H.: What Has Happened to Ontology. In: Dau, F., Mugnier, M.-L., Stumme, G. (eds.) ICCS 2005. LNCS (LNAI), vol. 3596, pp. 425–438. Springer, Heidelberg (2005)

Learning Common Outcomes of Communicative Actions Represented by Labeled Graphs

Boris A. Galitsky[1], Boris Kovalerchuk[2], and Sergei O. Kuznetsov[3]

[1] LogLogic Inc. 3061B Zanker Rd San Jose CA 95134
bgalitsky@loglogic.com
[2] Dept. of Computer Science, Central Washington University,
Ellensburg, WA, 98926, USA
borisk@cwu.edu
[3] Higher School of Economics, Moscow, Russia
skuznetsov@yandex.ru

Abstract. We build a generic methodology based on learning and reasoning to detect specific attitudes of human agents and patterns of their interactions. Human attitudes are determined in terms of communicative actions of agents; models of machine learning are used when it is rather hard to identify attitudes in a rule-based form directly. We employ scenario knowledge representation and learning techniques in such problems as predicting an outcome of international conflicts, assessment of an attitude of a security clearance candidate, mining emails for suspicious emotional profiles, mining wireless location data for suspicious behavior, and classification of textual customer complaints. A preliminary performance estimate evaluation is conducted in the above domains. Successful use of the proposed methodology in rather distinct domains shows its adequacy for mining human attitude-related data in a wide range of applications.

1   Introduction: Reasoning with Conflict Scenarios

Scenarios of interaction between agents are an important subject of study in AI. An extensive body of literature addresses the problem of logical simulation of behavior of autonomous agents and assistants, taking into account their beliefs, desires and intentions [5,15]. A substantial advancement has been achieved in building the scenarios of multiagent interaction, given properties of agent including their attitudes. Recent work in agent communications has been in argumentation [3], in dialog games [1,2], in formal models of dialog [9], in conversation policies [11], in social semantics [12] and in collaborative learning [4]. In terms of temporal conceptual semantic system [14] interaction between agents can be considered as a life track of a temporal system consisting of agents.

However, means of automated comparative analysis for interaction scenarios for *human* agents are still lacking. The comparative analysis of interaction scenarios between human agents for automated decision making, decision support and recommendations is needed in many applications. In this paper we build a representation machinery and continue our development of a machine learning technique [7,10] towards operating with a *wide range of scenarios* which include a sequence of

U. Priss, S. Polovina, and R. Hill (Eds.): ICCS 2007, LNAI 4604, pp. 387–400, 2007.
© Springer-Verlag Berlin Heidelberg 2007

communicative actions. We also propose a framework for classifying scenarios of inter-human conflicts and prediction of their outcomes. Formalized inter-human conflict is a special case of formal scenario where the agents have inconsistent and dynamic goals; a negotiation procedure is required to achieve a compromise. In this paper we explore a series of domains of various natures with respect to how the structure of conflict resolution and negotiation can be visually represented and automatically learned within a unified framework. We follow along the line of our previous studies demonstrating that it is possible to judge about consistency of these scenarios based on the extracted communicative actions [7].

The paper is organized as follows. The introduction of the domain of conflict scenarios is followed by a formal treatment of communicative actions, defining a conflict scenario as a graph, and machine learning of such graphs. We then present our domains and give respective examples of a variety of graphs consisting from communicative actions. The paper is concluded with comparative analysis of graph learning results in these domains.

2   Formalizing Conflict Scenarios for Learning

In this section we present our model of multiagent scenarios oriented to the use in a machine learning setting. Here we develop a knowledge representation methodology based on approximation of a natural language description of a conflict (Galitsky 2003). Further details are available online in the full version of the paper [8].

To form a data structure for machine learning, we approximate an inter-human interaction scenario as a sequence of communicative actions, ordered in time, with a causal relation between certain communicative actions (more precisely, the *subjects* of these actions). Scenarios are simplified to allow for effective matching by means of graphs: only communicative actions remain as a most important component to reflect the dialogue structure and express similarities between scenarios. Each vertex corresponds to a communicative action, which is performed by either *proponent*, or *opponent*. An arc (oriented edge) denotes a sequence of two actions.

In our model mental actions have two parameters: *agent name* and *subject* (information transmitted, a cause addressed, a reason explained, an object described, etc.). Representing scenarios as graphs, we take into account both parameters. Arc types bear information whether the subject stays the same. Thick arcs link vertices that correspond to communicative actions with the same subject; thin arcs link vertices that correspond to communicative actions with different subject. The curve arcs denote a causal link between the arguments of mental actions, e.g., [*ask*]- *the service is not as advertised* ⇒ [*disagree*]- *failures in the service contract* (and, therefore, *the service is not as advertised*). Let us consider an example of a scenario and its graph (Figure 1). Further examples are available in the extended version of this paper [8].

One of the most important tasks in assisting negotiations and resolving inter-human conflicts is the *validity* assessment. A scenario (in particular, a complaint) is *valid* if it is plausible, internally consistent, and also consistent with available domain-specific knowledge. In case of inter-human conflicts or negotiations, such domain-specific knowledge is frequently unavailable. In this study we demonstrate that a

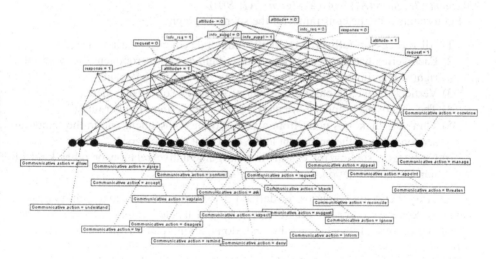

- **I asked** why the service was not as advertised
- They **explained** that I did not understand the advertised features properly
- I **disagreed** and **confirmed** the particular failures in a service contract
- They **agreed** with my points and **suggested** compensation
- I **accepted** it and **requested** to send it to my home address together with explanations on how it happened.
- They **promised** to send it to me.
- In a month time I **reminded** them to mail it to me
- After two months I **asked** what happened with my compensation...

Fig. 1. A sample complaint scenario and its graph representation. * stands for an arbitrary action, ? – the action to be predicted.

scenario can be assigned to a class *valid* or *invalid* based on communicative actions *only* with the accuracy sufficient for deployment in decision-support systems. To provide a framework for learning communicative actions, we need to select their attributes (Figure 2).

Fig. 2. The concept lattice for communicative actions

Based on the speech act theory, we selected the attributes of communicative actions to provide an adequate coverage of their meanings (further details are in [8]).

3   Defining Scenarios as Graphs

Each scenario includes multiple interaction *steps*, each consisting of mental actions with the alternating first attribute {*request – respond - additional request or other follow up*}. A step comprises one or more consequent actions with the same subject. Within a step, vertices for mental actions with common argument are linked with *thick* arcs.

For example, *suggest* from scenario V2 (Figure 3) is linked by a thin arc to mental action *ignore*, whose argument is not logically linked to the argument of *suggest* (the subject of suggestion). The first step of V2 includes *ignore-deny-ignore-threaten*; these mental actions have the same subject (it is not specified in the graph of conflict scenario). The vertices of these mental actions with the same argument are linked by the *thick* arcs. For example, it could be **ignored** *refund because of a wrong mailing address,* **deny** *the reason that the refund has been ignored [because of a wrong mailing address],* **ignore** *the denial [...concerning a wrong mailing address], and* **threatening** *for that ignorant behavior [...concerning a wrong mailing address].* We have *wrong mailing address* as the common subject *S* of mental actions *ignore-deny-ignore-threaten* which we approximate as

ignore(A1, S) & deny(A2,S) & ignore(A1,S) & threaten(A2, S), keeping in mind the scenario graph . In such approximation we write *deny(A2, S)* for the fact that *A2 denied the reason that the refund has been ignored because of S.* Indeed, *ignore(A1, S) & deny(A2,S) & ignore(A1,S) & threaten(A2, S).* Without a scenario graph, the best representation of the above in our language would be

ignore(A1, S) & deny(A2, ignore(A1, S)) & ignore(A1, deny(A2, ignore(A1, S))) & threaten(A2, ignore(A1, deny(A2, ignore(A1, S)))).

Let us enumerate the constraints for the scenario graph:

1) All vertices are fully ordered by the temporal sequence (earlier-later);
2) Each vertex has a special label relating it either to the proponent (drawn on the right side in Figure 3) or to the opponent (drawn on the left side);
3) Vertices denote actions either of the proponent or of the opponent;
4) The arcs of the graph are oriented from earlier vertices to later ones;
5) Thin and thick arcs point from a vertex to the subsequent one in the temporal sequence (from the proponent to the opponent or vice versa);
6) Curly arcs, staying for causal links, argumentative relation or other kind of non-temporal relation, can jump over several vertices.

Similarity between scenarios is defined by means of maximal common sub-scenarios. Since we describe scenarios by means of labeled graphs, first we consider formal definitions of labeled graphs and domination relation on them (see, e.g., [6,10]).

Given ordered set *G* of graphs *(V,E)* with vertex- and edge-labels from the sets $(\mathcal{L}_V, \doteq)$ and $(\mathcal{L}_\varepsilon, \doteq)$. A labeled graph Γ from *G* is a quadruple of the form *((V,l),(E,b))*, where *V* is a set of vertices, *E* is a set of edges, *l: V → $\mathcal{L}_V$* is a function assigning labels to vertices, and *b: E → $\mathcal{L}_\varepsilon$* is a function assigning labels to edges. We do not distinguish isomorphic graphs with identical labelings.

The order is defined as follows: For two graphs $\Gamma_1 := ((V_1,l_1),(E_1,b_1))$ and $\Gamma_2 := ((V_2,l_2),(E_2,b_2))$ from *G* we say that Γ_1 **dominates** Γ_2 or $\Gamma_2 \le \Gamma_1$ (or Γ_2 is a **subgraph** of Γ_1) if there exists a one-to-one mapping $\varphi: V_2 \to V_1$ such that it

- respects edges: $(v,w) \in E_2 \Rightarrow (\varphi(v), \varphi(w)) \in E_1$,
- fits under labels: $l_2(v) \doteq l_1(\varphi(v))$, $(v,w) \in E_2 \Rightarrow b_2(v,w) \doteq b_1(\varphi(v), \varphi(w))$.

Note that this definition allows generalization ("weakening") of labels of matched vertices when passing from the "larger" graph G_1 to "smaller" graph G_2.

Now, generalization Z of a pair of scenario graphs X and Y (or their similarity), denoted by $X \sqcap Y = Z$, is the set of all inclusion-maximal (in terms of relation $\doteq$) common subgraphs of X and Y, each of them satisfying the following additional conditions:

- To be matched, two vertices from graphs X and Y must denote mental actions of the same agent;
- Each common subgraph from Z contains at least one thick arc.

This definition is easily extended to finding generalizations of several graphs (e.g., see [6, 10]). We denote $X \equiv Y$ if $X \sqcap Y = \{X\}$.

4 Nearest-Neighbor Classification

The following conditions hold when a scenario graph U is assigned to a class (we consider positive classification, i.e., to valid complaints, the classification to invalid complaints is made similarly):

1) U is similar to (has a nonempty common scenario subgraph of) a positive example R^+. It is possible that the same graph has also a nonempty common scenario subgraph with a negative example R^-. This means that the graph is similar to both positive and negative examples.

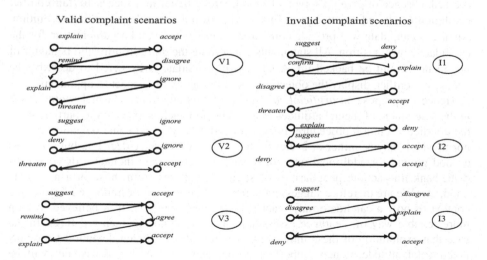

Fig. 3. A scenario with unassigned complaint status and the procedure of relating this scenario to a class

2) For any negative example R^-, if U is similar to R^- (i.e., $U \sqcap R^- \neq \varnothing$) then $U \sqcap R^- \sqsubseteq U \sqcap R^+$. This condition introduces the measure of similarity and says that to be assigned to a class, the similarity between the unknown graph U and the closest (in terms of $\sqsubseteq$) scenario from the positive class should be higher than the similarity between U and each negative example (i.e., representative of the class of invalid complaints).

5   Supporting Scenarios of Security Clearance Assessment

It is well known that assessment of mental attitude of a security clearance candidate is an important feature which is worth developing and automation. Obviously, there is some correlation between basic parameters of candidates such as bad habits, problematic career history, distrustful relationships, education failures etc. and skills/capabilities required to obtain a security clearance. However, there is no direct link between these parameters and a mental attitude of candidates; and the role of the latter is crucial to clearance-related decisions. Therefore, assessment of mental attitude, which is independent of the history of candidate career and personal life, is desirable for the purpose of the clearance-related decision.

In accordance to psychological studies, **inter-personal conflicts** may serve as an adequate means to assess such personal qualities of individuals as their mental attitudes. In the course of conflict, an individual with proper mental attitude is expected to demonstrate a stable and clear desire to resolve the conflict, cooperation with opponents and other involved parties when/if their actions are intended to assist conflict resolution, treating involved parties honestly and with respect. A candidate will show a stable emotional profile in the course of conflict resolution: absence of being depressed, absence of give up – type of mood, strong belief in a successful/fair conflict resolution result and an attempt to find her/his own active role in conflict resolution. Also, this candidate will provide consistent, concise, and valid argumentation for the candidate's own position. All statements concerning the untruthful/invalid behavior of opponents should be backed up. A successful candidate is expected to describe the history of conflict, display the objectivity, and fairness with respect to opponents.

Hence we propose *an artificial conflict resolution environment* which would assist in the assessment of mental attitudes of a candidate which is expected to participate in the conflict resolution procedure. For each candidate, we find some deviation from a norm, which may be minor or irrelevant to a security clearance decision, but serves as a good ground for additional questions. Such deviation may include a driving accident, bank transaction, peculiarities of spending patterns from those in a neighborhood, etc. We are therefore suggesting using likely irrelevant or minor red flags in the context of how a candidate may react to associated conflicts. It is believed to be a more reliable way of clearance assessment then just ignoring such red flags. In the case that exploration of these minor red flags reveals significant deviation from normal mental attitude, a new important component for security assessment will be available.

We outline a possible framework and scenario for the assessment.

A candidate *submits* (requests consideration of) an application for security clearance.

In response the candidate *receives* the following:

"Thank you for applying. We regret to *inform* you that in the course of consideration we have discovered certain circumstances which may negatively impact the decision with regard to the security clearance award. Our concern goes back to your years in college/military service/probation period in a company/performance in a company... which we believe may compromise your eligibility for the security clearance. If you believe we obtain this information in error or believe that it is irrelevant to the decision with regard to the clearance, please *contact* Mr #1 who is your caseworker."

Then the candidate *contacts* the mentioned caseworker with explanations. In *response*, the candidate gets the following letter:

"Thank you for your attempts to clarify the situation and your explanation that the evidence available has been obtained by us in error or irrelevant. However, in accordance to the other case worker, Mr #2 the facts provided by yourself do not fully exclude the possibility that what we have found is not plausible at all. I would encourage yourself to contact Mr #2 and clear this out. In case of positive decision with Mr #2 we will proceed with your case.

The candidate is then expected to contact Mr #2 with *request* for further details about his ambiguous circumstances. Having *received* no definitive response from Mr #2 (being *ignored*), the candidate sends a message to Mr #1 requesting a response from him or Mr #2. Mr #1 comes back to the candidate *claiming* that another piece of evidence has been found that compromises candidate's eligibility to the security clearance.

The candidate is expected to *respond* to Mr #1 with *explanation* and argumentation against the second piece of evidence. Meanwhile, Mr #2 *responds* to the candidate confirming that the candidate's explanation defeating the first piece of evidence has been *accepted* and the respective application unacceptability claim has been *dismissed*. Also, Mr #2 states that he believes that the second piece of evidence could be *dismissed* in his opinion as well, but there is a *disagreement* with Mr #1 who still believes that the second piece of evidence is valid. Mr #2 then *encourages* the candidate to address a number of points regarding the second piece of evidence. Then the candidate is nevertheless *expected* to communicate the raised issues with both agents which would lead to the successful *dismissal* of the second piece of evidence as well. Finally, the candidate is *requested* to describe the conflict and resolution strategy.

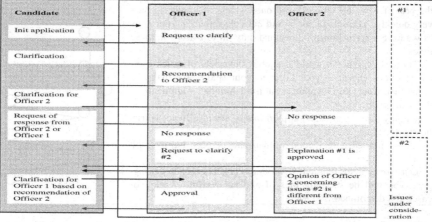

Fig. 4. Scenario representation for security clearance assessment

The interaction between the candidate and officers is shown at Figure 4.

6   Revealing Suspicious Emotional Profiles of Agents

In this section we introduce the idea of building emotional profile of an email message to characterize the emotional possible distress of the author. Emotional profile is a way to combine meanings of individual words in sentences and then to merge expressions for emotions in these sentences for deriving a high-level characteristic of emotional load of a textual message. It turns out that explicit expressions for emotions are amplified by the words which are not explicit indications of emotions but characterize interaction between involved agents (their communicative actions, Searle; 1969).

We call *Emotional profile* a formal representation of a sequence of emotional states through a textual discourse. *Intensity* of linguistic expressions for emotions has been the subject of extensive psychological studies (see references in [8]); we base our categorization of emotions and qualitative expression for emotion intensity in these studies. We apply computational treatment to our observations in the domain of customer complaints [7] that emotions are *amplified* by communicative actions. For example the expression *I was upset because of him* is considered to express a weaker intensity of emotion than the expression *He ignored my request and I got upset* with

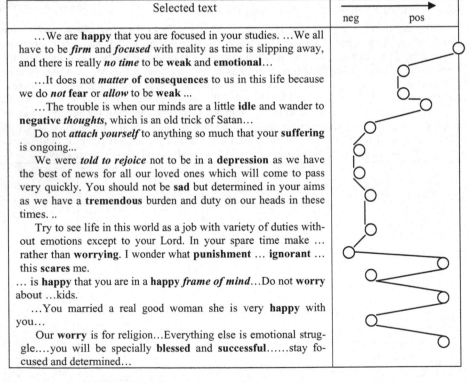

Selected text	neg pos
...We are **happy** that you are focused in your studies. ...We all have to be *firm* and *focused* with reality as time is slipping away, and there is really *no time* to be **weak** and **emotional**...	
...It does not *matter* **of consequences** to us in this life because we do *not* **fear** or *allow* to be **weak** ...	
...The trouble is when our minds are a little **idle** and wander to **negative** *thoughts*, which is an old trick of Satan...	
Do not *attach yourself* to anything so much that your **suffering** is ongoing...	
We were *told to rejoice* not to be in a **depression** as we have the best of news for all our loved ones which will come to pass very quickly. You should not be **sad** but determined in your aims as we have a **tremendous** burden and duty on our heads in these times. ..	
Try to see life in this world as a job with variety of duties without emotions except to your Lord. In your spare time make ... rather than **worrying**. I wonder what **punishment** ... **ignorant** ... this **scares** me.	
... is **happy** that you are in a **happy** *frame of mind*...Do not **worry** about ...kids.	
...You married a real good woman she is very **happy** with you...	
Our **worry** is for religion...Everything else is emotional struggle....you will be specially **blessed** and **successful**......stay focused and determined...	

Fig. 5. Example of an email message where a detection of emotional distress could prevent a would-be terrorist attack. On the right: emotion intensity profile, negative to positive from left to right.

communicative actions *request-upset*. In our formal representation of the latter case the communicative action *ignore* is substituted into the emotion *upset* as the second parameter: *upset(i, ignore(he, request(i,_)))*. Emotional profile of a textual scenario includes one or more expressions in predicates for emotions, communicative actions and mental states for each sentence from this scenario mentioning emotional state. Moreover, we compute the intensity of emotion for each such sentence.

To access the emotion level of the whole scenario, we track the evolution of the intensity of emotions. If it goes up and then goes down, one may conclude that a conflict occurred, and then has been resolved. A monotonous increase of emotion intensity would happen in case of an unresolved conflict (dispute). Conversely, a decrease in intensity means that involved parties are coming to an agreement. An oscillating intensity profile indicates more complex pattern of activity, and in most cases it reveals a strong emotional distress.

As an example, we present a fragment of correspondence between a would-be British suicide bomber and his relatives, who have been charged in connection to failing to notify authorities of a potential terrorist attack (Fig.5). We show expressions for emotions in bold, and associated expressions for communicative actions or mental states in bold italic. As the reader observes, emotional profile in this email is very peculiar. Primarily, there are very strong oscillations of the emotional intensity. These oscillations are medium at the beginning of message, stay negative at the middle portion of it and become very volatile towards the end of the message.

7 Revealing Suspicious Behavior of Cell Phone Users

In this section we introduce the idea of using telecommunication data for detection possible suspicious behavior of cell phone users. Providing telecommunication services is heavily dependent on the accurate determination of the handset locations to promptly switch from one service station to another. Telecommunication servers accumulate huge amount of data that includes the recording of locations of handsets at certain time intervals. Also, the phone numbers of both callers and call addressees are recorded. Crimes might be prevented and networks of criminals groups with peculiar inter-connections identified if it were possible to discover sets of unusual patterns of coordinated movement for groups of cell phones.

The raw data for our analysis includes the series of absolute locations (detected with certain accuracy at certain time intervals) for wireless subscribers (agents) and the selected locations where these agents are making a call or a receiving a call. We assume that conversation recordings are unavailable due to privacy of conversations, expensive recordings and unreliable speech recognition techniques. Having obtained the location data vs time, it is possible to extract the patterns of movement on a rule-based basis. The set of movement patterns we use is *turn right/lef, U-turn, keep going, stop*. Detecting movement patterns, we distinguish ones which were deliberately selected, and ones where a vehicle just follow a road. In our further considerations the default movement patterns for is *turn right/le* and *U-turn* will be deliberate. We use a labeled graph representation of a sequence of movements and phone calls as abstract communicative actions. If movements and phone calls are coordinated, the sequence of calls and movements is important to hypothesize on possible intentions of the agents in involved vehicles. Discovering correspondence between the movement patterns and communications, we attempt to determine what is the direction of information transmission between agents, and how is this information linked to what has been observed in connection.

The purpose of the analysis is to understand whether surveillance (as a partial case of a suspicious behavior) is taking place, and to discover the roles of involved agents. Obviously, the earlier it is possible to detect a suspicious behavior of parties, the sooner an interception is possible to assure security. Initially we do not know which agent is leading which is reporting, and we hypothesis about this assignment in the course of determining whether activity of these agents is normal or suspicious.

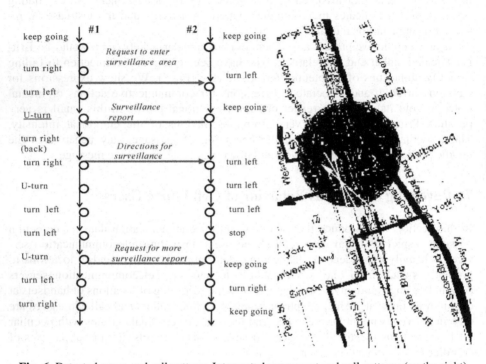

Fig. 6. Detected move and call patterns Interpreted movement and call patterns (on the right)

We show the trajectory of movements for two agents (Fig.6 on the right) and the constructed communication scenario graph (on the left). In this example, we suspect that there is an area surveillance by two agents. One can see that the leading agent is #1 and reporting is #2. The agent #1 investigates a number of approaches to the high security area (depicted by a circle) and leads the agent # 2 through this area (#1 is on the left at the map and in the graph, and #2 is on the right).

8 Evaluation of Representation Adequateness

To demonstrate that the proposed representation language of labeled graphs is adequate to represent scenarios of interactions between human agents in various domains, we performed the evaluation of coding to graph / decoding from graph and evaluate distortion of communicative action-related information. We conducted the evaluation with respect to the criteria on how the suggested model based on communicative actions can represent real-world scenarios including complaints, conflict between communities of agents, emotional interactions, induced conflict interactions while security clearance assessment, and wireless interaction under possible suspicious behavior.

We start the evaluation from textual complaints which were downloaded from the public website PlanetFeedback.com in 2005. For the purpose of this evaluation, each complaint was manually coded as a sequence of communicative actions, being assigned with a particular status. We formed the dataset for three banks, each of which consisted of 20 complaints. The usability and adequacy of our formalism was evaluated on the basis of a team of individuals divided into three classes: complainants, company representatives and judges.

Complainants had a task to read a textual complaint and draw a graph so that another team member (a company representative) could comprehend it (and briefly sketch the plot as a text). A third team member (judge) then compared the original complaint and the one written by the company representative as perceived from the form. The result of this comparison was the judgment on whether the scenario structure has been dramatically distorted in respect to the validity of a given complaint. It must be noted that less than 15% of complaints were hard to capture by means of communicative actions. We also observed that about a third of complaints lost important details and could not be adequately restored (although they might still be properly related to a class). Nevertheless, one can see that the proposed representation mechanism is adequate for representing so complex and ambiguous structures as textual complaints in most cases.

Note that in our approach the role of defeat relationships and causal links between the subjects of communicative actions is to represent common features of scenarios, and not to determine the validity of claims being communicated. Communicative actions of one scenario are matched against those of another scenario, and attack relationships between arguments are matched against those of another scenario, irrespectively of the validity of these arguments.

Conducting the evaluation of adequateness in other domains, we split the members of evaluation team into reporters, assessors and judges. Reporters represented scenarios as graphs, and assessors decoded the perceived structure of communicative actions back into text. Finally, the judges compared the original description (be it text or other media in the case of wireless interaction) with the respective originals.

For the banks, one can track deviation of one dataset versus another, which is 10-15% of the third set versus the first two sets. This is due to the lower variability of scenarios, which makes it easier to represent and reconstruct it (classification accuracy is comparable). Recognition for banking complaints is almost as accurate as coding via graph (representation), but not the reconstruction of the structure of interactions between complainants and their opponents.

Coding emotional profiles via graphs similar to Fig.5 was not as expressive as in the case of complaints, and classification accuracy is closer to the scenario reconstruction than to the scenario representation accuracy. Indeed, the proposed language via communicative actions captures peculiarity of emotional profiles in a lesser degree than the structure of complaint scenarios. We were unable to evaluate the security

Table 1. Evaluation of the adequacy of complaint representation language

Domain/dataset	Number of scenarios	% of scenarios which were successfully represented as graphs by experts	% of scenarios which were (at least partially) reconstructed from the	% of scenarios which were properly represented and reconstructed	% of properly related to a class (being adequately represented), 2 classes
Compalints-Bank 1 (Galitsky 2006)	20	85	75	65	72
Complaints-Bank 2	20	80	75	60	75
Complaints-Bank 3	20	95	85	75	78
Conflict between communities of agents (presented in [8])	2	50	50	50	No eval
Domain Sect. 5	12	75	67	58	60
Domain Sect. 6	4	No eval	No eval	No eval	No eval
Domain Sect. 7	38	84	74	55	61
Average	18.7	78.2	71	60.5	69.2

assessment scenarios in real world; however we obtained sufficient data to track the accuracy for wireless interactions. In terms of representation it is as good as complaint scenarios, but the reconstruction (which is the most important operation) accuracy is lower than for complaints, and the accuracy of classification lies in between representation and reconstruction. In such domain as Wireless interaction and Emotional interaction there is much higher loss of information then in the other domains, however proper classification (with providing background on *why* a given scenario is related to a class) gives a little bit better results. For complaints, where the representation and classification machinery was tuned, the accuracy is naturally higher than in the other domains we started to tackle recently, and the available dataset is rather limited.

Hence for an average number of almost 19 scenarios per dataset, almost 80% can be somehow represented via labeled graphs, about 70% reconstructed from graph without major loss of the conflict structure, and 60% both correct representation and reconstruction. The classification accuracy of relating to one out of two classes is close to the reconstruction accuracy. Note that the setting of the Nearest Neighbor classification is different from random classification which gives 50% for two classes.

9 Conclusions

We explored the role of communicative actions in representing various kinds of conflicts in multiagent systems and discovered that proper formalization of communicative actions are essential to judge on conflicts. A machine learning approach to relate a formalized conflict scenario to a class is proposed, which takes into account structures of communicative actions represented via labeled graphs. It has been developed and evaluated in the domain of customer complaints in our previous studies, and then used in other domains of inter-human conflicts of distinct natures. The representation language is that of labeled directed acyclic graphs with generalization operator on them. For machine learning, the scenarios are represented as a sequence of communicative actions attached to agents; these actions are grouped by subjects. Causal and argumentation defeat relationships between the subjects of communicative actions are coded in the graph and used by machine learning as well.

The structure of graphs, as well as the number and structure of classes depend on a domain, but the criteria of sequences of communicative actions have been shown useful to express commonalities between scenarios. Hence domain-independent communicative actions' representation via labeled graphs, once developed, can be reused from one conflict domain to another. At the same time, having the common representation language, scenarios from one domain are dissimilar to the ones from another domain, so only the knowledge about communicative actions structure is common between these domains. In each domain, graph structures are different, so we cannot export experience from domain to domain.

Based on speech act theory, we designed a set of attributes for communicative actions and showed how the procedure of relating a complaint to a class can be implemented as Nearest Neighbor learning machinery. The approach to learn scenarios of inter-human interactions (encoded as sequences of communicative actions) is believed to be original on one hand and universal on the other hand. We believe that rather few computational approach has been applied to such problem as understanding customer complaints, and the other domains where mining for communicative actions is useful, have not been tackled computationally either.

We believe that suggested approach is appropriate for deployment in decision support settings in the respective domains. One needs to integrate scenario encoding into graphs, classification and predication, and visualization [13] components to assist human experts in making decisions in the explored domains.

References

1. Baker, M.J.: A Model for Negotiation in Teaching-Learning Dialogues. Journal of Artificial Intelligence in Education 5(2), 199–254 (1994)
2. Boella, G., Hulstijn, V., van der Torre, L.: Persuasion strategies in dialogue. In: Grasso, F., Reed, C. (eds.) Proc. of the ECAI workshop on Computational Models of Natural Argument (CMNA'04), Valencia (2004)
3. Chesnevar, C., Maguitman, A.G., Simari, G.R.: Argument-based critics and recommenders: A qualitative Perspective on user support system. DKE 59-2 293-319 (2006)
4. Dillenbourg, P.: Collaborative Learning, Cognitive and computational approaches. Pergamon Press, Oxford (1999)
5. Ferguson, G.: AAAI SS on Intentions in Intelligent Systems Stanford Univ. (2007)
6. Ganter, B., Kuznetsov, S.O.: Pattern Structures and Their Projections. In: Delugach, H.S., Stumme, G. (eds.) ICCS 2001. LNCS (LNAI), vol. 2120, pp. 129–142. Springer, Heidelberg (2001)

7. Galitsky, B.: Reasoning about mental attitudes of complaining customers. In: Knowledge-Based Systems, vol. 19(7), pp. 592–615. Elsevier, Amsterdam (2006)
8. Galitsky, B., Kovalerchuk, B., Kuznetsov, S.O.: Extended version of the current paper (2007), http://www.knowledge-trail.com/complaint
9. Johnson, M.W., McBurney, P., Parsons, S.: A mathematical model of dialog. Electronic Notes in Theoretical Computer Science, 2005 (in the press)
10. Kuznetsov, S.O.: Learning of Simple Conceptual Graphs from Positive and Negative Examples. In: Żytkow, J.M., Rauch, J. (eds.) Principles of Data Mining and Knowledge Discovery. LNCS (LNAI), vol. 1704, pp. 384–392. Springer, Heidelberg (1999)
11. Nodine, M., Unruh, A.: Constructing Robust Conversation Policies in Dynamic Agent Communities. In: Dignum, F., Creaves, M. (eds.) Issues in Agent Comm. Springer, Heidelberg (2000)
12. Singh, M.P.: A Social Semantics for Agent Communication Languages. In: Dignum, F., Greaves, M. (eds.) Issues in Agent Communication, Springer, Heidelberg (2000)
13. Kovalerchuk, B., Schwing, J. (eds.): Visual and Spatial Analysis: Advances in Data Mining, Reasoning, and Problem Solving. Springer, Heidelberg (2005)
14. Wolff, K.E.: Basic Notions in Temporal Conceptual Semantic Systems. In: Kuznetsov, S.O., Schmidt, S. (eds.) ICFCA 2007. LNCS (LNAI), vol. 4390, pp. 60–72. Springer, Heidelberg (2007)
15. Yorke-Smith, N.: AAAI Symp Series Interaction Challenges for Intelligent Assistants. Stanford Univ. (2007)

Belief Flow in Assertion Networks

Sujata Ghosh[1,2,*], Benedikt Löwe[2,3,4], and Erik Scorelle[2]

[1] Department of Mathematics
Visva Bharati
Santiniketan, West Bengal, India
[2] Institute for Logic, Language and Computation
Universiteit van Amsterdam
Plantage Muidergracht 24, 1018 TV Amsterdam, The Netherlands
{sujata,bloewe,escorell}@science.uva.nl
[3] Mathematisches Institut
Rheinische Friedrich-Wilhelms-Universität Bonn
Beringstraße 1, 53115 Bonn, Germany
[4] Department Mathematik
Universität Hamburg
Bundesstrasse 55, 20146 Hamburg, Germany

Abstract. We define an abstract model of belief propagation on a graph based on the methodology of the revision theory of truth together with the *Assertion Network Toolkit*, a graphical interface designed to test our semantics.

1 Introduction

Formulas as labelled graphs

Formulas of ordinary propositional logic can be seen as trees, with the propositional variables as terminal nodes. For example,

reads as $\neg(p \,\&\, \neg(q \,\&\, r)) \equiv \neg p \vee (q \,\&\, r) \equiv p \rightarrow (q \,\&\, r)$.

In order to compute the truth value of a molecular formula of propositional logic, we take an assignment of truth values for the propositional variables (the terminal nodes of the tree) and let the truth value develop backwards along the edges of the tree by the usual rules of propositional logic.[1] In that sense, we can visualize truth as flowing from the leaves to the root of the tree.

---

[*] The first author acknowledges the travel support of the ILLC in Amsterdam for a research visit in June 2006, and a **Rubicon** grant of the NWO (680-50-0504) for her visit in the academic year 2006/07.
[1] Exact definitions will follow in the proof of Theorem 2.

U. Priss, S. Polovina, and R. Hill (Eds.): ICCS 2007, LNAI 4604, pp. 401–414, 2007.
© Springer-Verlag Berlin Heidelberg 2007

Cyclic graphs

In a logic that allows self-reference, we are not dealing with trees anymore, but with arbitrary, possibly cyclic graphs. The simple flow of truth values from the leaves to the roots may now become a complicated and potentially never-ending pattern. Consider the Liar sentence which corresponds to the graph

While this creates serious problems for sets of sentences with self-reference, the revision theory of truth developed by Herzberger, Gupta, and Belnap [He82a, GuBe93] and the Gaifman Pointer Semantics [Ga88, Ga92] have extended the idea of a set of sentences as a graph to the case involving circularity. These theories consider infinite, possibly (if the graph is infinite) transfinite sequences of revision along the edges of the graph, and then singles out those patterns that are stable as the semantics of the set of sentences. The two characteristic components of revision theory are *backward propagation of truth along the edges* ("revision") and *identification of the stable values*. This was discussed in the survey talk *Revision Forever!* at ICCS 2006 [Lö06].[2]

Even though the revision theory of truth has a wide range of applications (*cf.* [Lö06, § 6]), its basic concepts are determined by the fact that it is a theory of truth: the nodes in the graph represent statements and we want to investigate the circular nature of truth in situations with self-reference. A typical example for this would be the *Nested Liars*:

> A: *"Everything that B says is false!"*
> B: *"Everything that A says is false!"*

A simple analysis tells us that exactly one of A and B speaks the truth. Without additional information, we cannot determine which one is the liar, but we can rule out certain truth-value patterns. While the nested liars are not paradoxical in the strict sense, as they admit consistent valuations, they still pose a problem: the nested liars are completely symmetric, but none of the consistent valuations is. If you think of the nested liars as an abstract, disembodied system of sentences, then they *should* have a symmetric valuation, but they don't!

But in real life, we are not dealing with abstract, disembodied systems of sentences. Behind the letters A and B, there are agents in a communication situation, and depending on who they are and how much we trust their judgement, the situation might turn out not to be symmetric after all. Even on Smullyan's logic island, if one of them is a Knight and the other one is a Knave, we might have independent information that leads us to believe which one is the Knight.

Thinking of nodes in the graph as agents rather than sentences leads from an analysis of truth to an analysis of a belief network. Consider the following example: Suppose a reasoning agent is sitting in an office without windows. Next to him is his colleague, also located in the office without windows; the agent is simultaneously talking on the phone to his friend who is sitting in a street café.

---

[2] For more details, *cf.* [Kü+05].

Friend: Everything your colleague says is false; the sun is shining!
Colleague: Everything your friend says is false; it is raining!

This situation can be described by the following graph, interpreting an arrow labelled + as "everything uttered is true", an arrow labelled − as "everything uttered is false", and S as "the sun is shining":

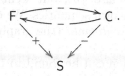

As in the *Nested Liars*, there are two consistent truth value assignments, but the context makes sure that one of them is intuitively preferred, as the agent's friend has first hand experience of the weather in the street café. Based on this preference, we experience *beliefs* flowing from the leaves through the graph: the contextually based stronger belief in the positive arrow F → S leads us to disbelieve the negative arrows C → S (since it is in conflict with F → S) and C → F (since we believe one of the statements that F makes, the utterance "Everything your friend says is false" must be disbelieved), and then in the next step to believe the arrow F → C.

This example shows that even in the context of belief and agents, the underlying idea of pointer semantics with its backward propagation of truth values is still fruitful.

However, as soon as we interpret the values assigned to nodes and arrows as *belief values* rather than truth values, a new phenomenon occurs that was not present in the case of the theory of truth: *forward propagation of belief along the edges*: If a trusted source A states "φ is false", but the reasoning agent believes in φ, then his belief in φ should influence his trust in A, but also the agent's trust in A should influence his belief in φ.

Aim of the paper. Related work

Our aim in this paper is to provide a formal model that handles both backward and forward influence of values in the graph while keeping the spirit of revision semantics. Based on this model, we define a belief semantics via stability and check that this captures our intuitions of belief dynamics.

Of course, our model is neither the first attempt to provide a formal background for reasoning about beliefs in a multi-agent setting nor the first one to consider the evolution of numbers on a graph network.

There are many approaches to reasoning about beliefs (and knowledge) with many agents. As an example, consider dynamic epistemic logic [Ge99] or announcement logics [Pl89]. This theory uses graphs in the spirit of modal logic in order to represent states of the world. The dynamics results in changing of the graph (as new belief states are created by actions or deleted by generated knowledge). This is very different from our approach. Another example, closer to traditional belief revision, is the work of Dragoni and Giorgini on multi-source belief revision [DrGi01] which shares some features with our set-up.

Also the idea of attaching numbers representing beliefs to nodes of a graph is not at all new: there is a large body of literature on *Bayesian Belief Nets* (see, e.g., [Pe88, Wi05]) with a lot of interesting technical results. The set-up (assigning probabilities to the nodes and using the Bayesian rule for propagation) is close to our approach (see also §5), but seems to be lacking the notion of stability that is crucial to our own semantics. We do find the notion of stability in the research area of *graph automata* (the graph version of cellular automata as described in [ToKuMu02]).

A very intriguing connection can be found to the area called *social network analysis* which ranges from empirical social science to computer science [WeBe88, BrEr05].

What is novel about our approach is the combination of number propagation (as known in neural networks or cellular automata) and the semantics derived from stability (as used in the revision theory of truth).

Outline of the paper

In §2, we provide the abstract formal background called **assertion network semantics** that will then have to be instantiated by concrete functions and operations. The aim is then to find concrete functions that make assertion network semantics recover our natural intuitions. This is necessarily an empirical study, and we shall see an extensive example in §3. Since there are so many potential concretizations of our assertion network semantics, we decided to support our work with a software tool called *Assertion Network Toolkit* which is described in §4. In our conclusion in §5, we list the further projects coming out of the work described in this paper.

2   The Most Abstract Model

In our paper, a **(directed) labelled graph** G is a triple $G = \langle V, E, \ell \rangle$, where V is a set of nodes, $E \subseteq V \times V$ is the set of edges, and $\ell : E \to \{+, -\}$ is a labelling function. We write

$$\text{In}(v) := \{e \in E \,;\, \exists w(e = \langle w, v \rangle)\}, \text{ and}$$
$$\text{Out}(v) := \{e \in E \,;\, \exists w(e = \langle v, w \rangle)\},$$

and let $\text{indeg}(v) := |\text{In}(v)|$ and $\text{outdeg}(v) := |\text{Out}(v)|$ denote the indegree and outdegree of v, respectively. We denote the set of terminal nodes by $T := \{v \in V \,;\, \text{outdeg}(v) = 0\}$.

In our graphs, the terminal nodes stand for facts (like "the sun is shining"). The non-terminal nodes correspond to agents making statements, either about a fact or about an agent. Labelled edges are these statements, where an edge labelled $+$ corresponds to a positive statement ("is true" or "Everything he or she says is true") and an edge labelled $-$ corresponds to a negative statement ("is false" or "Everything he or she says is false"). This intended semantics is captured in the following definition: a function $I : V \rightarrow \{0,1\}$ is called an **interpretation (on a labelled graph)**. Let $* : \{+,-\} \times \{0,1\} \rightarrow \{0,1\}$ be defined by $+ * 0 = - * 1 = 0$ and $+ * 1 = - * 0 = 1$. We say that an interpretation **respects** a labelled graph G if for all nonterminal vertices v, we have $I(v) = \bigwedge_{w \in \text{Out}(v)} \ell(v,w) * I(w)$. Obviously, every function $I : T \rightarrow \{0,1\}$ has at most one extension to an interpretation that respects G.

Our labelled graphs are similar and yet different from Bolander's dependency graphs and the corresponding pointer semantics [Bo03, Chapter 5]: nonequivalent sets of clauses can have the same dependency graph whereas the labels of the edges make the semantics of the labelled graph unequivocal.[3]

The following theorem connects labelled graphs with the above semantics to pointer semantics.

Theorem 1. *Labelled graphs interpret pointer semantics in the following sense: for every set of clauses Σ with propositional variables $\{p_0, ..., p_n\}$ there is a labelled graph G with vertices $\{v_0, ..., v_n\}$ such that an interpretation (in the sense of footnote 3) $I : \mathbb{N} \rightarrow \{0,1\}$ respects Σ if and only if the interpretation (on the labelled graph) $I^* : V \rightarrow \{0,1\}$ defined by $I^*(v_i) := I(i)$ respects G.*

Proof. Fix a set of clauses $\Sigma = \{i{:}E_i \, ; \, 0 \leq i \leq n\}$ and write all expressions E_i in disjunctive normal form, *i.e.*, $E_i = \bigvee_k \bigwedge_j \ell_{ijk}$ where ℓ_{ijk} is either a propositional variable or a negation of a propositional variable.

To start, let us notice that our intended semantics for the labelled graphs only deals with conjunction, not disjunction ("everything he says is true", "everything he says is false"). Therefore, we have to define disjunction via de Morgan's laws, where $\varphi \vee \psi$ is represented by

We construct a labelled graph to express Σ. We start with the set of vertices: for each i, we take a vertex v_i representing p_i, then for each conjunctive clause $\bigwedge_j \ell_{ijk}$ occurring in one of the expressions, we take a vertex w_{ik}. In order to deal with the disjunctions, we add vertices x_i^0 and x_i^1 for every expression E_i.

---

[3] Here, we refer to the pointer semantics of a propositional language as described in [Lö06, § 2]: Every propositional variable is an **expression**; $\bot$ and $\top$ are expressions; if E and F are expressions, then $\neg E$, $E \wedge F$, and $E \vee F$ are expressions. If E is an expression and n is a natural number, then $n{:}E$ is a **clause**.

We say that an **interpretation** is a function $I{:}\mathbb{N} \rightarrow \{0,1\}$ assigning truth values to propositional letters. Obviously, an interpretation extends naturally to all expressions. Now, if $n{:}E$ is a clause and I is an interpretation, we say that I **respects** $n{:}E$ if $I(n) = I(E)$, and I respects a set of clauses Σ if it respects every element of Σ.

Now the edges are drawn as follows: If for some j, we have $\ell_{ijk} = p_{i*}$, we draw an edge $w_{ik} - + \succ v_{i*}$, if for some j, we have $\ell_{ijk} = \neg p_{i*}$, we draw an edge $w_{ik} - - \succ v_{i*}$, and for all i and k, we draw edges $x_i^0 - - \succ w_{ik}$, $x_i^1 - - \succ x_i^0$, and $v_i - + \succ x_i^1$. Obviously, this general construction produces a labelled graph with the desired property.[4]   q.e.d.

As mentioned in the introduction, we shall now combine the idea of (stable) revision semantics with forward propagation of belief along edges. Fix some metric space $\langle \Lambda, d \rangle$ of **values**. We call a function $H : E \cup V \to \Lambda$ a **hypothesis**. This can be interpreted as a state of beliefs (about the communication situation) of the reasoning agent.

For an edge $e = \langle v, w \rangle$, let $M_e := \text{In}(v) \cup \{v, e, w\} \cup \text{Out}(w)$ and $n_e := \text{Card}(M_e)$. For a vertex v, let $M_v := \text{In}(v) \cup \{v\} \cup \text{Out}(v)$ and $n_v := \text{Card}(M_v)$. If $M_x = \langle x_1, ..., x_m \rangle$, we write $H(M_x) := \langle H(x_1), ..., H(x_m) \rangle$. For every $x \in E \cup V$, we fix an evaluation function $\Psi_x : \Lambda^{n_x} \to \Lambda$ which we can now use to define a **revision sequence** in the spirit of revision theory:

Fix an initial hypothesis H. We define the sequence $\langle H_i ; i \in \omega \rangle$ at some $x \in E \cup V$ by simultaneous recursion as follows:

$$H_0(x) := H(x)$$
$$H_{i+1}(x) := \Psi_x(H(M_x)).$$

Inspired by the stability concept of revision theory, we can now define a partial stability semantics for our labelled graph. Suppose you have an initial hypothesis H, some $x \in E \cup V$ and some $\lambda \in \Lambda$ such that $\lim_{i \to \infty} H_i(x) = \lambda$, then we say

---

[4] In order to illustrate the procedure, let us give an example: Consider the set of clauses $\{0:(\neg p_0 \wedge p_1) \vee (\neg p_1 \wedge p_2), 1:\neg p_2 \vee (p_0 \wedge p_1), 2:\neg p_2\}$. By the rules given above, this transforms into the following labelled graph:

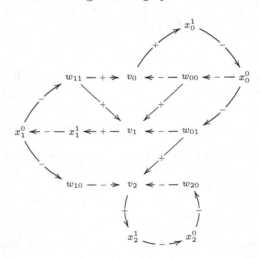

that **the value of** x **is stably** λ **in situation** H. We call this the **assertion network semantics** A_H defined by

$$A_H(x) := \begin{cases} \lambda & \text{if the value of } x \text{ is stably } \lambda \text{ in situation } H, \text{ and} \\ \text{undefined} & \text{if } \langle H_i(x) \, ; \, i \in \omega \rangle \text{ diverges.} \end{cases}$$

Theorem 2. *The stable truth predicate of revision semantics is a special case of assertion network semantics, i.e., for every set of clauses Σ there is a labelled graph G and there are evaluation functions such that A_H coincides with the (partial) stable truth predicate on Σ.*

Proof. We shall give evaluation functions Ψ_x such that the assertion network semantics recover the (partial) stable truth predicate defined by

$$S_H(p_i) := \begin{cases} \text{true} & \text{if } p_i \text{ is stably true in the sequence starting with } H, \\ \text{false} & \text{if } p_i \text{ is stably false in the sequence starting with } H, \text{ and} \\ \text{undefined} & \text{otherwise.} \end{cases}$$

Here, we are using Theorem 1 by dealing with revision semantics in labelled graphs instead of dependency graphs.[5]

As values, we choose $\Lambda := \{\text{true}, \text{false}\}$ with the discrete metric. In the classical revision semantics, edges don't play a rôle, so we shall now identify the values of an edge $e = \langle v, w \rangle$ and w.

For a given hypothesis H, we define the following variant H^* on the edges of the labelled graph that incorporates the values of the labels: $H^*(e) := H(e)$ if $\ell(e) = +$ and $H^*(e) := \neg H(e)$ if $\ell(e) = -$. Now, for each $e = \langle v, w \rangle \in E$, we let

$$\Psi_e(H(M_e)) := \begin{cases} \text{true if for all } e^* \in \text{Out}(w), \text{ we have } H^*(e^*) = \text{true, and} \\ \text{false otherwise.} \end{cases}$$

Clearly, with the functions Ψ_e, the assertion network semantics gives back the stable semantics as given above. $\quad$ q.e.d.

3   Concretization

The goal of this enterprise was to develop a semantics to deal with beliefs in assertion situations with reference according to our intuitions. Now, the abstract assertion network semantics depends on the proper choice of the evaluation functions Ψ_x, and without fixing these functions, it is not concrete enough to allow checking the resulting semantics against our intuitions.

The search for a concretization of the abstract model that conforms to our intuitions is an empirical question, and we shall return to this in § 4.

For the time being, let us now concentrate on one rather natural example. We choose $\Lambda := [-1, 1]$ where we interpret 1 as "the agent firmly believes",

---

[5] The partial semantics S_H corresponds to the lightface version of the stable truth predicate in [Lö06, p. 27] as opposed to the boldface version integrating over all starting hypotheses. *Cf.* [Kü+05].

−1 as "the agent firmly disbelieves", and 0 as "the agent has no evidence either way". We'll focus on the belief change in the edges (representing the statements) and the terminal nodes, and consider the nonterminal nodes as auxiliary. In the following, since we want to keep our values between −1 and 1, we shall be using the function $\mathsf{R}(r) := \min(\{\max(\{-1, r\}), 1\})$, and if e is an edge, we let

$$s_e := \begin{cases} 1 & \text{if } \ell(e) = +, \text{ and} \\ -1 & \text{if } \ell(e) = -. \end{cases}$$

1. If $x \in V \backslash T$, and $\{e_0, ..., e_{n_x}\} = \mathrm{Out}(x)$, then we let

$$\Psi_x(\lambda_1, ..., \lambda_{n_x}) := \frac{\sum_{0 \leq i \leq n_x} \lambda_i}{n_x}.$$

2. For a terminal node t, the value of Ψ_t depends on the value of t and the values of the incoming edges. Let $\{e_0, ..., e_{n_t}\} = \mathrm{In}(t)$, then

$$\Psi_t(\lambda, \lambda_0, ..., \lambda_{n_t}) := \mathsf{R}\left(\lambda + \frac{\sum_{1 \leq i \leq n_t} s_i \lambda_i}{n_t}\right),$$

where λ is the value of t and λ_i is the value of e_i.

3. Finally, for an edge $e = \langle v, w \rangle$, we let the value depend on the value λ of e, the value $\widehat{\lambda}$ of w and the values $\lambda_0, ..., \lambda_{n_e}$ of $\mathrm{In}(v)$.

$$\Psi_e(\lambda, \widehat{\lambda}, \lambda_0, ..., \lambda_{n_e}) := \begin{cases} \frac{1}{2} \cdot \left(\mathsf{R}\left(\lambda + \frac{\sum_{1 \leq i \leq n_e} s_i \lambda_i}{n_t}\right) + \widehat{\lambda}\right) & \text{if } \ell(e) = +, \\ \frac{1}{2} \cdot \left(\mathsf{R}\left(\lambda + \frac{\sum_{1 \leq i \leq n_e} s_i \lambda_i}{n_t}\right) - \widehat{\lambda}\right) & \text{if } \ell(e) = -. \end{cases}$$

Note that this choice of evaluation functions is not unique, but rather natural, including all of the relevant information in the computation of Ψ_x without discrimination, except that the original value of a node gets more influence than the incoming information.

To see this choice of evaluation functions at work, let us consider the following communication situation:

Professors Jones, Miller and Smith are colleagues in a computer science department. Jones and Miller dislike each other without reservation and are very liberal in telling everyone else that "everything that the other one says is false". Smith just returned from a trip abroad and needs to find out about two committee meetings on Monday morning. He sends out e-mails to his colleagues and to the department secretary. He asks all three of them about the meeting of the faculty, and Jones and the secretary about the meeting of the library committee (of which Miller is not a member).

Jones replies: "We have the faculty meeting at 10am and the library committee meeting at 11am; by the way, don't believe anything that Miller says, as he is always wrong."

Miller replies: "The faculty meeting was cancelled; by the way don't believe anything that Jones says, as he is always wrong."

The secretary replies: "The faculty meeting is at 10 am and the library committee meeting is at 11 am. But I am sure that Professor Miller told you already as he is always such an accurate person and quick in answering e-mails: everything Miller says is correct."

This situation is described by the following diagram:

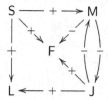

Given the described situation, we assign an initial hypothesis to this graph as follows: the agent has no evidence for any of the statements except that he believes that the secretary (being well-informed about administrative matters) is correct about the two meetings, so we assign the value 0 to everything except for the edges $S \rightarrow L$ and $S \rightarrow F$ where we assign $+0.5$. When we run the revision, all values converge (up to two decimal digits precision) after at most 16 steps of revision, giving the following stability pattern:

S	M	J	F	L	$S \rightarrow L$	$S \rightarrow F$	$J \rightarrow L$	$J \rightarrow F$	$M \rightarrow F$	$J \rightarrow M$	$M \rightarrow J$	$S \rightarrow M$
0	0	0	0	0	+0.5	+0.5	0	0	0	0	0	0
↓	↓	↓	↓	↓	↓	↓	↓	↓	↓	↓	↓	↓
+0.7	−1.0	+1.0	+1.0	+1.0	+1.0	+1.0	+1.0	+1.0	−1.0	+1.0	−1.0	−1.0

This corresponds perfectly to our intuitions: if we give belief primacy to the two factual announcements of the secretary, then Jones is speaking the truth twice and Miller is uttering a falsehood. To get an even better feeling of how the system behaves, we shall now gradually change the initial hypothesis by increasing all values that are in favour of Miller (i.e., the values of $M \rightarrow F$, $M \rightarrow J$, and $S \rightarrow M$. Let us define the hypothesis H^λ by

$$H^\lambda(x) := \begin{cases} +0.5 & \text{if } x \in \{S \rightarrow F, S \rightarrow L\}, \\ \lambda & \text{if } x \in \{M \rightarrow F, M \rightarrow J, S \rightarrow M\}, \text{ and} \\ 0 & \text{otherwise.} \end{cases}$$

The special case H^0 is the case discussed above. We shall now slowly increase λ and observe the stability behaviour of the system. For $\lambda = +0.1$, all limits stay the same, even though the rate of convergence is a bit slower in some cases. But for $\lambda = +0.2$, we can observe the first change, as the value of F now converges to -1.0 (instead of $+1.0$). When λ is $+0.5$, the values of both F and L converge to -1.0.

Independent Evidence

Our assertion networks allow us to formalize (abstract) independent evidence. For any given labelled graph $\langle V, E, \ell \rangle$ with initial hypothesis H, we define **independent evidence** for $v \in V$ **of strength** n to be represented by a labelled graph $\langle V^*, E^*, \ell^* \rangle$ with $V^* = V \cup \{x_0, ..., x_{n-1}\}$, $E^* := E \cup \{\langle x_i, v \rangle \,;\, i < n\}$, and ℓ^* extending ℓ with $\ell^*(\langle x_i, v \rangle) = +$, and an initial hypothesis H^* extending

H with $H^*(x_i) = H^*(\langle x_i, v \rangle) = +1.0.$[6] In the following diagram, independent evidence of strength 1 has been added to Jones (J):

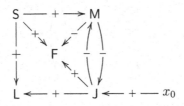

Adding independent evidence has the expected effect. For instance, if we add independent evidence of strength 1 to J as in the above picture, then the effect of increasing λ in the initial hypothesis H^λ is weakened. In order to get a stable value -1.0 for F, you have to go to $H^{+0.3}$ (instead of $H^{+0.2}$ without independent evidence), and in order to get stable values -1.0 for both terminal nodes, you have to move to $H^{+0.6}$ (instead of $H^{+0.5}$ in the case without independent evidence).

4   The Assertion Network Toolkit

As mentioned in § 3, the search for the right functions Ψ_x in order to capture our intuitions is a largely empirical endeavour. In order to test candidate functions, we have to go through a large number of examples, in particular more complicated examples that require a sufficiently large number of steps to converge. The example discussed in § 3 has five vertices and eight edges. Already in this case, doing all computations by hand is cumbersome, for larger examples, it is virtually impossible.

As a consequence, we decided to implement a piece of graphical interface software that allows us to play around with the functions and values. This will be used in the future to investigate the advantages and weaknesses of assertion network semantics.

The **Assertion Network Toolkit** is written solely in C/C++ and built on the Boost Graphing Library (BGL), GTK, GTKmm, and the Cairo graphics

---

[6] The proper modelling of independent evidence requires forward propagation of belief as the following example shows:

Consider the graph $A - + \twoheadrightarrow B$ with $H(B) = -1$. If we use a revision theoretic model that only allows backward propagation (*e.g.*, the one given in the proof of Theorem 2), then the outcome would be disbelief in A, $A \to B$ and B. In a semantics with only backward propagation (*i.e.*, $H_{i+1}(B)$ depends only on $H_i(B)$), we could not determine the difference between this scenario and all of the following scenarios:

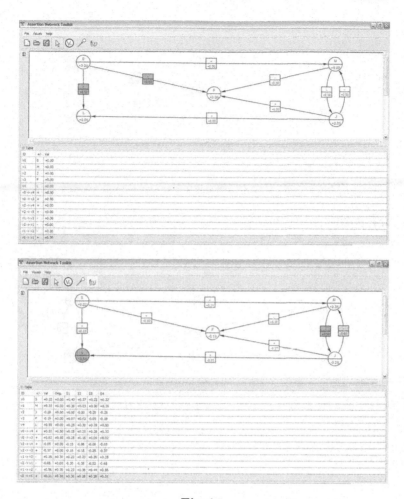

Fig. 1.

library. Boost property maps contain all the data of a graph object while the
drawing area and table widgets serve as a view for the values stored in the maps.
Stepping through a function locks the map and updates the view according to
the new values. These properties and functions are currently predefined, but due
to the flexibility of BGL and the independence of the graph from the views, the
next version of the Assertion Network Toolkit will allow users to create graphs
with custom properties and define the behavior of a function. The software is
platform independent and will remain so as it is developed more.

In the Assertion Network Toolkit, the set-up of a labelled graph is user-friendly
and simple. It allows to set the evaluation functions and the initial hypothesis, and
then runs the revision according to the functions. In the following example, we look
at the graph discussed as an example above with the initial hypothesis $H^{+0.3}$.

In the first picture of Figure 1, we can see the finished graph with the initial
values according to $H^{+0.3}$. The edges $S \rightarrow L$ and $S \rightarrow F$ are marked blue (darker

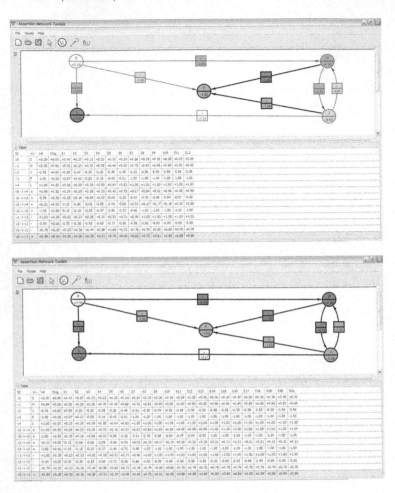

Fig. 2.

shade of grey in black-and-white print) as they have a positive value higher than a threshold value set by the user.

We start the revision procedure by clicking the "step" button. Immediately, the two edges marked blue fall below the threshold value and lose their blue marking. After four steps (the next picture), we can now see that the edge M → J and the node L are above the threshold of +0.5 and are therefore marked in blue. Similarly, the edge J → M is below the lower threshold value and is consequently marked in red (lighter shade of grey in black-and-white print).

After twelve steps (the first picture of Figure 2), we see that a number of nodes and edges has now reached the threshold values and has been marked red or blue. In addition, we now see that the values of F and L have been constant for the last five iterations. Again, this is a definition of convergence that can be set by the user. In this particular example, we defined no change for five iterations as convergence which is marked by setting the respective vertex or edge in bolded lines.

After twenty-one iterations (the final picture), all vertices and edges have reached that status and are marked in bolded lines.

5 Conclusion

In § 2, we have presented an abstract framework for a revision semantics with forward flow of belief. The abstract framework has to be made concrete by the choice of evaluation functions Ψ_x. Whether a given choice of function is appropriate, i.e., conforms with our intuitions about human belief formation, is an empirical question. This empirical question can and should be tested with many examples; for these tests, we provided the *Assertion Network Toolkit* in § 4. In § 3, we have provided one rather natural specification of the abstract framework and have discussed how it behaves for our given example.

The most natural task for the future is the **experimental** project of testing various natural choices of transition functions (including the one given in § 3) in many examples to investigate their behaviour. These experiments will also include a closer investigation of the effects of independent evidence. On the more mathematical side, there is the **logical** task of investigating the logic of our semantics: after fixing transition functions and a notion of *validity* (e.g., $H(v) > 0.75$ or $H(v) = +1.0$), what are the validities exhibited by our semantics? The broadest project is the **comparative** project of connecting our approach to the other models involving graphs and belief (e.g., belief nets). Can one of the models be obtained as a special case of the other (as in Theorem 2)?

References

[Bo03]   Bolander, T.: Logical Theories for Agent Introspection, PhD thesis, Technical University of Denmark (2003)

[BrEr05]   Brandes, U., Erlebach, T. (eds.): Network Analysis. LNCS, vol. 3418. Springer, Heidelberg (2005)

[DrGi01]   Dragoni, A.F., Giorgini, P.: Revising Beliefs received from Multiple Sources. In: Williams, M.-A., Rott, H. (eds.) Frontiers in Belief Revision, pp. 429–442. Kluwer Academic Publishers, Dordrecht (2001) [Applied Logic Series 22]

[Ga88]   Gaifman, H.: Operational Pointer Semantics: Solution to Self-referential Puzzles I. In: Vardi, M. (ed.) Proceedings of the 2nd Conference on Theoretical Aspects of Reasoning about Knowledge, Pacific Grove, CA, March 1988, pp. 43–59. Morgan Kaufmann, San Francisco (1988)

[Ga92]   Gaifman, H.: Pointers to Truth. Journal of Philosophy 89, 223–261 (1992)

[Ge99]   Gerbrandy, J.: Dynamic epistemic logic. In: Moss, L.S., Ginzburg, J., de Rijke, M. (eds.) Proceedings of the 2nd Conference on Information-theoretic Approaches to Logic, Language and Computation (ITALLC) held at Regent's College, London, July 1996, vol. 2, pp. 67–84. CSLI Publications, Stanford, CA (1999) [CSLI Lecture Notes 96]

[GuBe93]   Gupta, A., Belnap, N.: The Revision Theory of Truth, CambridgeUniversity Press, Cambridge, MA (1993)

[He82a]   Herzberger, H.G.: Naive Semantics and the Liar Paradox. Journal of Philosophy 79, 479–497 (1982)

[He82b]   Herzberger, H.G.: Notes on Naive Semantics. Journal of Philosophical Logic 11, 61–102 (1982)

[Kü+05]   Kühnberger, K.-U., Löwe, B., Möllerfeld, M., Welch, P.: Comparing inductive and circular definitions: parameters, complexities and games. Studia Logica 81, 79–98 (2005)

[Lö06]   Löwe, B.: Revision Forever! In: Schärfe, H., Hitzler, P., Øhrstrøm, P. (eds.) ICCS 2006. LNCS (LNAI), vol. 4068, pp. 22–36. Springer, Heidelberg (2006)

[Pe88]   Pearl, J.: Probabilistic reasoning in intelligent systems: networks of plausible inference Morgan Kaufmann, San Francisco (1988) [The Morgan Kaufmann Series in Representation and Reasoning]

[Pl89]   Plaza, J.A.: Logics of Public Communications. In: Emrich, M.L., Pfeifer, M.S., Hadzikadic, M., Ras, Z.W. (eds.) Proceedings of the Fourth International Symposium on Methodologies for Intelligent Systems: Poster Session Program, Oak Ridge National Laboratory, pp. 201–216 (1989) [ORNL/DSRD-24]

[Sc92]   Scott, J.P.: Social Network Analysis. Sage, Thousand Oaks (1992)

[ToKuMu02]   Tomita, K., Kurokawa, H., Murata, S.: Graph automata: natural expression of self-reproduction. Physica D: Nonlinear Phenomena 171, 197–210 (2002)

[WeBe88]   Wellman, B., Berkowitz, S.D.: Social Structures: A network approach. Cambridge University Press, Cambridge (1988)

[Wi05]   Williamson, J.: Bayesian nets and causality, Philosophical and computational foundations. Oxford University Press, Oxford (2005)

Conceptual Fingerprints: Lexical Decomposition by Means of Frames – a Neuro-cognitive Model

Wiebke Petersen and Markus Werning

Heinrich-Heine University Düsseldorf
Universitätsstr. 1, 40225 Düsseldorf, Germany
{petersen,werning}@phil.uni-duesseldorf.de

Abstract. Frames, i.e., recursive attribute-value structures, are a general format for the decomposition of lexical concepts. Attributes assign unique values to objects and thus describe functional relations. Concepts can be classified into four groups: sortal, individual, relational and functional concepts. The classification is reflected by different grammatical roles of the corresponding nouns. The paper aims at a cognitively adequate decomposition, particularly, of sortal concepts by means of frames. Using typed feature structures, an explicit formalism for the characterization of cognitive frames is developed. The frame model can be extended to account for typicality effects. Applying the paradigm of object-related neural synchronization, furthermore, a biologically motivated model for the cortical implementation of frames is developed. Cortically distributed synchronization patterns may be regarded as the fingerprints of concepts.

1 Introduction

If one does not want to assume lexical atomism – the view that the possession of any concept expressible by a simple word is completely independent of the possession of any other concept – the question arises in which particular way the possession of some lexical concepts depends on the possession of other concepts. An explicit answer to that question should ideally be (i) in accordance with linguistic data, (ii) formally explicit, (iii) cognitively plausible, and (iv) neurobiologically realistic.

In this paper we will outline a theory of lexical decomposition that attempts to fulfil the four desiderata. Driven by linguistic considerations on the grammatical role of nouns, we will begin with a classification of lexical concepts into four groups. For our account of lexical decomposition, we will use Barsalou's (1992) cognitive frame theory as a point of departure. We will show how frames can be rendered by labeled graphs and how this graphical structure is transformed into a formally explicit typed-feature structure. Concentrating on frames for concepts which linguistically are expressed by nouns, our project aligns with well-established graph-based knowledge representation formalisms that focus on *situations* as in frame semantics (Fillmore, 1982) and *propositions* as with conceptual graph theory (Sowa, 1984). Our formalism is guided by Guarino's (1992) considerations on the ontological status of attributes in frames. To match our

U. Priss, S. Polovina, and R. Hill (Eds.): ICCS 2007, LNAI 4604, pp. 415–428, 2007.
© Springer-Verlag Berlin Heidelberg 2007

approach with psychological results on categorization (J. D. Smith & Minda, 2000, for review), we digress from a classical Aristotelian interpretation of lexical decomposition, which proposes a definitional relation between concepts and their constituents. Instead, our theory will accommodate cognitive typicality effects regarding concept satisfaction. In contrast to decompositional approaches in prototype theory (E. E. Smith & Medin, 1981), which render concepts by flat feature lists, our frame-theoretic approach allows for a much deeper hierarchical structure. The last part builds on neurobiological evidence that in earlier work has already been proposed to support semantic structure (Werning, 2005). Using oscillatory neural networks as a model, we will show how frames might be implemented in the cortex.

2   Nouns and the Classification of Concepts

Concepts can be distinguished with respect to both arity and referential uniqueness (Löbner, 1985). Sortal and individual concepts are of arity 1 and thus typically have no possessor argument. Sortal concepts (e.g., 'apple') denote classical categories and fail to have unique referents. Individual concepts (e.g., 'Mary'), in contrast, have unique referents. Concepts with arity greater than 1 comprise all relational concepts including functional concepts. It is characteristic for relational concepts (e.g., 'brother') that their referents are given by a relation to a possessor ('brother of Tom'), while unique reference is not generally warranted. Functional concepts (e.g., 'mother') form a special case of uniquely referring relational concepts. They establish a right-unique mapping from possessors to referents (Fig. 1).

	non-unique reference	unique reference
arity:1	**SC**: sortal concepts: *person, house, verb, wood*	**IC**: individual concepts: *Mary, pope, sun*
arity:>1	**RC**: (proper) relational concepts: *brother, argument, entrance*	**FC**: functional concepts: *mother, meaning, distance, spouse*

Fig. 1. The classification of concepts

The classification of concepts typically corresponds to specific grammatical properties of the expressing noun itself or its context. In English, nouns expressing concepts without unique reference (SCs and RCs) are typically used without definite article. Nouns expressing concepts of higher arity (RCs and FCs) are typically used in possessive constructions, where the possessor is specified by a genitive (*the cat's pow*) or prepositional phrase (*the pow of the cat*).

There is considerable variation in the expression of definiteness and possession across languages. Languages that lack definite articles often employ other strategies to indicate definiteness. In Russian, e.g., word order can be used to signal that a noun refers unambiguously. Here, the preverbal position of a noun phrase hints at a definite interpretation. Hence a question such as *What is on the table?* is likely to be answered by

(1) *Na stol'-e   l'ež-it   knig-a.*
on table-PREP lie-3SG.PRES book-NOM.SG
'There is a book on the table'

In contrast, *Where is the book?* is likely to be answered by

(2) *Knig-a   l'ež-it   na stol'-e.*
book-NOM.SG lie-3SG.PRES on table-PREP
'The book is on the table'

To express possession, a manifold of strategies is used as well. Hungarian, e.g., displays morphological agreement of the possessum with the possessor (quoted from Ortmann, 2006):

(3) a. *a   te   kalop-od*   b. *a   Péter   kalop-ja*
DF PRON.2SG hat-P'OR.2SG   DF PRON.1SG hat-P'OR.3SG
'your hat'   'Peter's hat'

A suffix (here, *-od* and *-ja*) is attached to the possessed noun, thus specifying agreement with the possessor with respect to the features number and person.

Languages with alienability splits such as the Hokan language Eastern Pomo distinguish overtly whether the concept of the possessed object is conceptualized as being of arity equal or greater one (quoted from Ortmann, 2006):

(4) a. *wí-bayle*   b. *wáx   ša?ri*
1SG-husband   PRON.1SG.GEN BASKET
'my husband'   'my basket'

If the noun is conceptualized as being relational ('husband'), it will enter the inalienable possessive construction: the possessor is simply realized by a prefix attached to the possessed noun. In case of alienable possession by contrast the noun ('basket') is not conceptualized as being relational. Possession cannot be expressed on the word level, but rather on the phrase level, by means of a free pronoun.

The type of a given concept may be shifted according to context: The noun *father*, which, in its normal use, has unique reference and arity 2 and thus expresses a functional concept, can be used in contexts like *Fathers don't like cooking* or *The fathers of the constitution were wise*, in a way expressing a sortal or a relational concept, respectively. In some languages (e.g., Yucatec Maya) those type shifts are even overtly realized (Ortmann, 2006).

3   Non-relational Frames

Following Minsky (1975) and Barsalou (1992), frames as recursive attribute-value structures are a general format to account for mental concepts. Guided by the

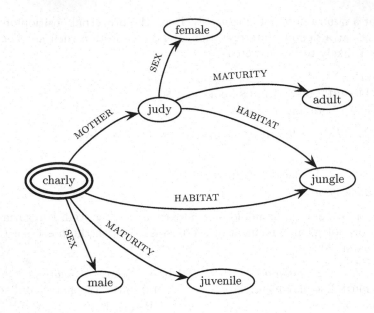

Fig. 2. A knowledge representing frame drawn as a labeled graph

above mentioned demands that our concept-decomposition framework should be
formally explicit and cognitively adequate, we aim at keeping our frame model
as simple and rigid as possible. We do not want to introduce any elements in our
model language solely due to technical or computational reasons or for the sake
of generality and expressibility. In section 4 and section 5, rather, we will point
to cognitive and neuro-biological evidence for our model language. As our aim is
to decompose *concepts* we rest our frame model on the restrictive theory of typed
feature structures (Carpenter, 1992) and not on the much wider framework of
conceptual graphs (Sowa, 1984). We are aware, though, that all our frames can be
easily transformed into conceptual graphs (but not *vice versa*). In accordance
with Barsalou (1992) we, for our frame model, assume that attributes assign
unique values to objects and thus describe functional relations. The values can
be structured frames themselves. Attributes in frames are therefore functional
concepts and embody the concept type on which the categorization is based.

We model non-relational frames as connected directed acyclic rooted graphs
with labeled nodes (value types) and arcs (attributes).[1] Fig. 2 shows the graph
of an example frame representing knowledge about a young male gorilla Charly
living with his mother Judy in the same jungle. The double-encircled node
'charly' points out that the graph represents a frame about Charly. The outgoing
arcs of the 'charly'-node stand for the attributes of Charly and point to their

---

[1] The definition of frames as directed rooted graphs enables us to adopt the theory
of typed feature structures, which is well-established in computational linguistics.
Our definitions follow Carpenter (1992) as closely as possible, except for definition
5, which digresses in one point fundamentally.

values. Hence, the sex of Charly is male and the maturity of Charly is juvenile. The value of the attribute 'mother' is a complex frame itself, describing that Judy, the mother of Charly, is female and adult. The fact that Judy and Charly live in the same jungle is indicated by the single 'jungle'-node to which the two 'habitat'-arcs from 'charly' and 'judy' point.

Definition 1. *Given a set* TYPE *of types and a finite set* ATTR *of attributes. A non-relational frame is a tuple* $F = (Q, \bar{q}, \theta, \Phi)$ *where:*

- *Q is a finite set of nodes,*
- *$\bar{q} \in Q$ is the root node,*
- *$\theta : Q \to$ TYPE is the total node typing function,*
- *$\Phi :$ ATTR $\times Q \to Q$ is the partial transition function.*

Furthermore, for each $q \in Q$ there be a finite sequence of attributes $A_1 \ldots A_n \in$ ATTR with $\Phi(A_n, \ldots, \Phi(A_2, \Phi(A_1, \bar{q})) \ldots) = q$, i.e., q and $\bar{q}$ are connected by a finite path; and for no $q \in Q$ there be a finite sequence of attributes $A_1 \ldots A_n \in$ ATTR* with $\Phi(A_n, \ldots, \Phi(A_2, \Phi(A_1, q)) \ldots) = q$, i.e., the graph is acyclic.*

The root node of a non-relational frame is its referring node. If $\theta(\bar{q}) = t$, we say that the frame is of type t. A node with no outgoing arcs is called an *end node* of the frame. To be able to speak of the paths of a frame, we need the following definition:

Definition 2. *Given a set* TYPE *of types, a finite set* ATTR *of attributes, and a non-relational frame $F = (Q, \bar{q}, \theta, \Phi)$. A sequence of attributes $A_1 \ldots A_n \in$ ATTR* is a path of F if $\Phi(A_n, \ldots, \Phi(A_2, \Phi(A_1, \bar{q})) \ldots)$ is defined. The set of all paths of a frame F is denoted by Π_F. A path $\pi \in \Pi_F$ is said to be* maximal *in F if $\pi A \notin \Pi_F$ for all attributes A.* MAXPATH$_F$ *denotes the set of maximal paths in F. The node typing function θ can be extended to the* path typing function $\Theta : \Pi_F \to$ TYPE *in a natural way:*

$$\Theta(A_1 \ldots A_n) = \theta(\Phi(A_n, \ldots, \Phi(A_2, \Phi(A_1, \bar{q})) \ldots)).$$

Since the information represented by a frame does not depend on the concrete set from which the nodes are drawn, we can abstract away from this set and focus on how the nodes are connected by labeled arcs. Fig. 3 shows the frame of Fig. 2 represented as an recursive attribute-value-matrix (AVM). The AVMs are constructed as follows: Frames are enclosed in square brackets with an index denoting the type of the root node. Each first-level attribute is stated in the brackets followed by a colon and followed by the value of the attribute. The values are either complex frames themselves or unstructured (i.e., not specified by further attributes). In the case of an unstructured value, we write 'ATTRIBUTE:type' instead of 'ATTRIBUTE:[ ]$_{type}$'. The symbol ① indicates that the path [MOTHER:HABITAT:] starting from the root node points to the same node as the path [HABITAT:], i.e., the two paths share the same value.

The types are ordered in a type hierarchy, which induces a subsumption order on frames. Cognitively the types correspond to categories and the type hierarchy to an IS-A-hierarchy.

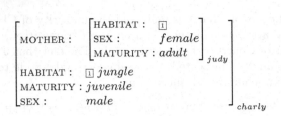

Fig. 3. AVM-abstraction of the frame-graph of Fig. 2

Definition 3. *A type hierarchy* (TYPE, $\sqsupseteq$) *is a partial ordered set which forms a join semilattice, i.e., for any two types there exists a least upper bound.*

A type t_1 is a subtype of a type t_2 if $t_1 \sqsupseteq t_2$. A type t is said to be minimal *if it has no subtypes. The set of minimal types is denoted by* MINTYPE.

Definition 4. *Given a type hierarchy* (TYPE, $\sqsupseteq$) *and a finite set* ATTR *of attributes. A frame $F = (Q, \bar{q}, \theta, \Phi)$ subsumes a frame $F' = (Q', \bar{q}', \theta', \Phi')$, notated as $F \sqsubseteq F'$, iff there exists a total function $h : Q \to Q'$ such that:*

- $h(\bar{q}) = \bar{q}'$,
- *for each $q \in Q$: $\theta(q) \sqsubseteq \theta'(h(q))$,*
- *if $q \in Q$, $a \in$ ATTR, and if $\Phi(a, q)$ is defined, then $h(\Phi(a, q)) = \Phi'(a, h(q))$.*

The following example shows an unspecified 'ape'-frame subsuming an unspecified 'gorilla'-frame, which subsumes the fully specified 'charly'-frame (see the type hierarchy in Fig. 4):

$$\begin{bmatrix} \text{HABITAT} : habitat \\ \text{SEX} : \quad sex \end{bmatrix}_{ape} \sqsubseteq \begin{bmatrix} \text{HABITAT} : jungle \\ \text{SEX} : \quad sex \end{bmatrix}_{gorilla} \sqsubseteq \begin{bmatrix} \text{HABITAT} : jungle \\ \text{SEX} : \quad male \end{bmatrix}_{charly}.$$

As Guarino (1992) points out, frame-based knowledge engineering systems as well as feature-structure-based linguistic formalisms normally force a radical choice between attributes and types. As a consequence, generic frames like

$$\begin{bmatrix} \text{MATURITY} : maturity \\ \text{HABITAT} : \quad jungle \end{bmatrix}_{gorilla}$$

occur frequently in which the unspecified value 'maturity' is assigned to the attribute 'MATURITY'. The parallel naming of the attribute 'MATURITY' and the type 'maturity' pretends a systematic relationship between the attribute and the type which is not intended by the formalism.

A second problem addressed in Guarino (1992) concerns the question which binary relations should be expressed by attributes. If one allows attributes to be unrestricted arbitrary binary relations, this leads to frames like the following one, which was first discussed in Woods (1975):

$$\begin{bmatrix} \text{HEIGHT} : 6feet \\ \text{HIT} : \quad mary \end{bmatrix}_{john}.$$

Although 'HEIGHT' and 'HIT' can be represented by binary predicates, the ontological status of the link they establish between 'john' and '6 feet' and between 'john' and 'mary' respectively differs fundamentally.

Our main thesis on frames is that non-relational frames decompose non-relational concepts into functional concepts. But our definition of non-relational frames only uses attributes for the decomposition. Hence, the question arises how attributes and functional concepts are connected. All sample attributes we have used so far (MOTHER, SEX, ...) correspond to functional concepts. Guarino (1992) distinguishes between the *denotational* and the *relational* interpretation of a relational concept. This distinction can be used to explain how functional concepts can act as concepts and as attributes: Let there be a universe $\mathcal{U}$ and a set of functional concepts $\mathcal{F}$. A functional concept (like any concept) denotes a set of entities:

$$\delta : \mathcal{F} \to 2^{\mathcal{U}}$$

(e.g., $\delta(\text{mother}) = \{m \mid m \text{ is the mother of someone}\}$).

A functional concept also has a relational interpretation:

$$\varrho : \mathcal{F} \to 2^{\mathcal{U} \times \mathcal{U}}$$

(e.g., $\varrho(\text{mother}) = \{(p, m) \mid m \text{ is the mother of } p\}$).

The denotational and the relational interpretation of a functional attribute have to respect the following *consistency postulate*: Any value of a relationally interpreted functional concept is also an instance of the denotation of that concept. (If $(p, m) \in \varrho(\text{mother})$, then $m \in \delta(\text{mother})$). Furthermore, the relational interpretation of a functional concept f is a function, i.e., if $(a, b), (a, c) \in \varrho(f)$, then $b = c$.

These considerations allow us, to clarify the ontological status of attributes in frames: Attributes in frames are relationally interpreted functional concepts! Hence, attributes are not frames themselves and are therefore unstructured. Frames of non-relational concepts decompose into relationally interpreted functional concepts.

In order to restrict the class of admissible frames, the plain type hierarchy can be enriched by an appropriateness specification. It regulates which attributes are appropriate for frames of a special type and restricts the values of the appropriate attributes.[2] Our definition of type signatures consequently dismisses the artificial distinction between attributes and types in contrast to the standard definition (Carpenter, 1992): the attribute set is merely a subset of the type set. Hence, attributes occur in two different roles: as names of binary functional relations between types and as types themselves.

---

[2] Type signatures can be automatically induced from sets of untyped non-relational frames, i.e. frames in which only the maximal paths are typed. With FCAType an implemented system for such inductions is available, which uses formal concept analysis (Kilbury, Petersen, & Rumpf, 2006; Petersen, 2006, 2007).

Definition 5. *Given a type hierarchy* (TYPE, $\sqsupseteq$) *and a set of attributes* ATTR $\subseteq$ TYPE. *An* appropriateness specification *on* (TYPE, $\sqsupseteq$) *is a partial function* Approp : ATTR $\times$ TYPE $\to$ TYPE *such that for each* $a \in$ ATTR *the following holds:*

- *attribute introduction: There is a type* Intro(a) $\in$ TYPE *with:*
  - Approp(a, Intro(a)) = a *and*
  - *for every* $t \in$ TYPE: *if* Approp(a, t) *is defined, then* Intro(a) $\sqsubseteq t$.
- *specification closure: If* Approp(a, s) *is defined and* $s \sqsubseteq t$, *then* Approp(a, t) *is defined and* Approp(a, s) $\sqsubseteq$ Approp(a, t).
- *attribute consistency: If* Approp(a, s) = t, *then* $a \sqsubseteq t$.

A type signature is a tuple (TYPE, $\sqsupseteq$, ATTR, Approp), *where* (TYPE, $\sqsupseteq$) *is a type hierarchy,* ATTR $\subseteq$ TYPE *is a set of attributes, and* Approp : ATTR $\times$ TYPE $\to$ TYPE *is an appropriateness specification. A type* t *is said to be* atomic *if* Approp(a, t) *is undefined for any* $a \in$ ATTR.

The first two conditions on an appropriateness specification are standard in the theory of type signatures (Carpenter, 1992), except that we tighten up the attribute introduction condition by claiming that the introductory type of an attribute a carries the appropriateness condition '$a : a$'. By the attribute consistency condition we ensure that Guarino's consistency postulate holds and that Barsalou's view on frames, attributes, and values is modeled appropriately:

> At their core, frames contain attribute-value sets. Attributes are concepts that represent aspects of a category's members, and values are subordinate concepts of attributes, (Barsalou, 1992).

Hence, the possible values of an attribute are subconcepts of the denotationally interpreted functional concept. This is reflected in the type signature by the condition that the possible values of an attribute are restricted to subtypes of the type corresponding to the attribute.

A small example type signature is given in Fig. 4. The appropriateness specification is split-up into single appropriateness conditions:[3] The expression 'SEX:sex' at type 'ape' means that the attribute 'SEX' is appropriate for frames of type 'ape' and its value is restricted to 'sex', hence, Approp(SEX, ape) = sex. The attribute conditions are passed on downwards. Hence, the type 'gorilla' inherits the appropriateness condition 'SEX:sex' from its upper neighbor 'ape'. It also inherits the appropriateness condition 'HABITAT:habitat', but tightens it up to 'HABITAT:jungle', which is permissible by the specification closure condition. The definition of the type signature makes sure that the permissible values of an attribute are subtypes of the attribute type. Hence, the possible values of SEX, i.e., 'female' and 'male', are subtypes of the type 'sex'. Notice that the subtypes of an attribute type are not generally attribute types themselves.

---

[3] To improve readability we mark the two roles of attributes in our frame notation: attributes used as types are written in small letters and attributes used as attributes in capitals.

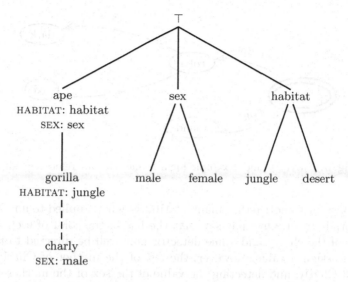

Fig. 4. Example type signature

A frame whose end nodes are all labeled by atomic minimal types is said to be a *fully-specified frame*. We call a non-relational frame *well-typed* with respect to a type signature if all attributes of the frame are licensed by the type signature and if additionally the attribute values are consistent with the appropriateness specification. The definition of the appropriateness specification guarantees that every arc in a well-typed frame points to a node which is typed by a subtype of the type corresponding to the attribute labeling the arc. The decomposition of concepts into frames requires that the frame in question be well-typed.

4   Frames and Typicality

One of the main virtue of frames is that they allow the decomposition of sortal and individual concepts by means of functional concepts. This decomposition now enables us to explain how a subject may subsume a perceived or otherwise given object under a sortal or individual concept. The degree, between 0 and 1, to which an object of the universe $\mathcal{U}$ instantiates a certain type is given by the function:

$$d : \text{TYPE} \times \mathcal{U} \to [0, 1].$$

In every frame the root node corresponds to the decomposed concept ('charly', 'cherry'). The set of maximal paths MAXPATH $\subseteq \Pi$ is well-defined for every frame. In a fully specified frame, end nodes, e.g, 'red', are atomic minimal types and are identified by maximal paths, i.e., [COLOR:HUE:] beginning at the root node, i.e., 'cherry' (Fig. 5).

It is natural to assume that the cognitive subject is endowed with a detector system that for all atomic minimal types renders the degree d to which it is instantiated by a given object. These might be hue detectors, sex detectors etc.

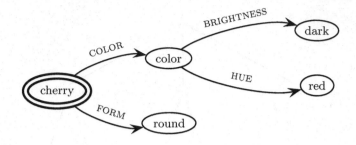

Fig. 5. Cherry frame. Example for a frame without reference shift.

It is important to notice that many attributes when applied to an object shift the referential object. One may say that the hue of the color of a cherry is still a property of the cherry and a hue detector may well be directed to the cherry in order to assign a value. However, the sex of the mother of Charly is not a property of Charly, and detecting the value of the sex of the mother of Charly, requires a potential sex detector to be directed to the mother. It is hence useful to introduce a reference-shifting function

$$\sigma : \mathcal{U} \times \varPi \rightarrow \mathcal{U}$$

that maps every object of the universe relative to the path in question onto the same or another object of the universe.

In the classical bi-valued case, the values of d are restricted to 0 and 1. Here, for a fully specified frame, where $\Theta(\text{MAXPATH}) \subseteq \text{MINTYPE}$, we can conclude that an object x is to be subsumed under the decomposed concept C if and only if all the types of the end nodes are properly instantiated:

$$d(C, x) = \min_{m \in \text{MAXPATH}} d(\Theta(m), \sigma(x, m)).$$

A cognitively more realistic picture, however, is attained if we specify how typical a certain minimal type is for instances of the concept. Red may, e.g., be more typical than green as the hue of the color of cherries. Nevertheless the hue of the color of some cherries still is green. We can achieve this by considering alternative types for each maximal path. For each maximal path m we then have a set $\text{ALT}(m)$ containing the minimal type $\Theta(m)$ and all its alternative types with regard to the path. I.e., provided that $m \in \text{ATTR}^*$ is a maximal path of a frame with $m = A_1 \ldots A_n$, $\text{ALT}(m)$ is the set of all atomic minimal subtypes of the type A_n. Our definition of the type signature guarantees that $\text{ALT}(m)$ covers all possible maximally specified values of A_n. For each of the types $t \in \text{ALT}(m)$ we can then specify a typicality value relative to the maximal path m of the fully specified frame for a decomposed concept C. The typicality value $\tau(C, m, t)$ tells how typical the type t is for the object $\sigma(x, m)$ given that x instantiates C. With these conventions we can apply previous results of Werning and Maye (2005, 2007) and, on the basis of the detector outputs, estimate to which degree an object x instantiates the decomposed concept C:

$$d(C, x) \geq \min_{m \in \text{MaxPath}} \max_{t \in \text{ALT}(m)} \tau(C, m, t) \, d(t, \sigma(x, m)).$$

The proof of the theorem, firstly, requires that the types of $\text{ALT}(m)$, for each maximal path m, are a quasi partitioning of the referentially appropriately shifted universe $\mathcal{U}$. I.e., for all $x \in \mathcal{U}$

$$\sum_{t \in \text{ALT}(m)} d(t, \sigma(x, m)) = 1.$$

In the equation the $\geq$-direction reflects exhaustivity and the $\leq$-direction reflects exclusivity. A cognitively appropriate type signature guarantees that each subtype of an attribute type is a reasonable value of the attribute. Furthermore, the attribute consistency condition on appropriateness warrants that all values of an attribute are subtypes of the attribute type. However, it does not follow that the minimal subtypes of an attribute type exhaust the values of the attribute ('red' could have two subtypes 'light red' and 'dark red' such that something red could be neither light red nor dark red). However, such a situation is excluded if we only consider type signatures automatically induced from untyped object frames by the system of Petersen (2007). Exclusivity is warranted by the fact that all types in $\text{ALT}(m)$ are minimal. A case where 'light red', 'dark red', and 'red' all occur as types in $\text{ALT}(m)$ is thus excluded.

Secondly, we have to presuppose that the extension of the decomposed concept is completely determined by the extensions of the minimal types at the end nodes of the fully specified frame. This is to say that the extensions of the types at intermediate nodes do not independently bear on the extension of the decomposed concept. However, it is not to say that the extension of the decomposed concept does not depend on the extensions of types at intermediate nodes. For, the extensions of those in turn depend on the extensions at the end nodes. The condition of complete determination again holds trivially for type signatures induced from sets of untyped frames as described by Petersen (2007).

5 Neuro-cognitive Interpretation

For many attributes (HUE, BRIGHTNESS, ORIENTATION, DIRECTION, SIZE, etc.) involved in the course of visual processing – we call them qualitative attributes – one can anatomically identify so-called neuronal feature maps (Hubel & Wiesel, 1968). These are structures of neurons that exhibit a certain topological organization. With regard to one attribute or feature dimension one finds a pinwheel-like structure for each receptive field (i.e., a specific region of the stimulus). This structure is called a hypercolumn. For each receptive field and each such attribute (e.g., HUE) we find a hypercolumn such that neurons for the entire spectrum of subtypes ('red', 'green', etc.) of that attribute fan out around a pin-wheel center. Neurons of a hypercolumn with a tuning for one and the same feature or subtype (e.g., 'red') form a so-called column. We may assume that such neurons function as detectors and thus evaluate atomic minimal types for a given stimulus object.

More than 30 cortical areas forming feature maps are experimentally known to be involved in the visual processing of the monkey (Felleman & van Essen, 1991). These findings justify the hypothesis that in the cortex there may be neural correlates of attributes and their subtypes.

The fact that subtypes of different attributes may be instantiated by the same stimulus object, but are processed in distinct regions of cortex poses the problem of how this information is integrated in an object-specific way. How can it be that the horizontality and the redness of a red horizontal bar are represented in distinct regions of cortex, but still are part of the representation of one and the same object? This is the binding problem in neuroscience (Treisman, 1996).

A prominent and experimentally well supported solution postulates oscillatory neuronal synchronization as a mechanism for binding (von der Malsburg, 1981; Gray, König, Engel, & Singer, 1989): Clusters of neurons that are indicative for different properties sometimes show synchronous oscillatory activity, but only when the properties indicated are instantiated by the same object in the perceptual field; otherwise they are firing asynchronously. Synchronous oscillation, thus, might be regarded to fulfill the task of binding together various property representations in order to form the representation of an object as having these properties (for a review see Singer, 1999).

Using oscillatory networks (Schillen & König, 1994; Maye & Werning, 2004) as models, the structure of object-related neural synchronization could be interpreted (Werning, 2005) as providing a conceptual structure expressible in a first-order predicate language. To show this, an eigenmode analysis of the network dynamics is computed. Per eigenmode, oscillation functions play the role of object representations or concepts. Clusters of feature sensitive neurons play the role of property representations or predicate concepts. Werning (2003) extends this approach from an ontology of objects to an ontology of events. Werning and Maye (2006) discuss ambiguous and illusionary representations. The following theorem (Werning & Maye, 2007) nicely links the results of this paper to previous results on the neural implementation of conceptual structure. The degree to which the object x is represented as instantiating the atomic type t by a network eigenmode is given by the equation:

$$d(t, x) = \max\{\Delta(\alpha(\mathrm{x}), f_j) | \mathbf{f} = \beta(t) c \mathbf{v}\}.$$

Here $\alpha(\mathrm{x})$ is the oscillation function representing the object x, and $\beta(t)$ is a matrix identifying the neural clusters which function as detectors for the type t. $\mathbf{v}$ and c are the results of the eigenmode analysis and account for the spatial, respectively, temporal variation of the network activity in that eigenmode. Δ is defined as the normalized inner product of two square-integrable time-dependent functions in a given temporal interval and measures the degree of synchrony between an object-related oscillation and the actual oscillatory activity in a neural cluster. $d(t, x)$ approaches 1 if the oscillation function $\alpha(\mathrm{x})$, which represents the object x, is highly synchronous with some component oscillatory activity f_j of f − i.e., the vector containing the eigenmode-relative temporal evolution of the type-related cluster of detector neurons.

If we conjoin the estimation of $d(C, x)$ in terms of type-specific detector outputs $d(t, x)$ with the identification of the latter with particular oscillatory network activity, we may conclude with the following hypothesis: Provided that a concept is completely decomposable into a fully specified frame with detectors for each type of a maximal path, the degree to which the cortex represents an object as an instance of the concept can be estimated by a general pattern of synchronizing neural activity distributed over various feature-selective neural clusters that correspond to the atomic types of the frame. This pattern may be called the cortical fingerprint of the concept.

References

Barsalou, L.W.: Frames, concepts, and conceptual fields. In: Lehrer, A., Kittay, E.F. (eds.) Frames, fields, and contrasts, pp. 21–74. Erlbaum, Hillsdale, NJ (1992)

Carpenter, B.: The logic of typed feature structures. Cambridge University Press, Cambridge (1992)

Fillmore, C.J.: Frame semantics. In: Linguistics in the morning calm, pp. 111–137. Hanshin Publishing Co. Seoul (1982)

Gray, C., König, P., Engel, A.K., Singer, W.: Oscilliatory responses in cat visual cortex exhibit inter-columnar synchronization which reflects global stimulus properties. Nature 338, 334–337 (1989)

Gray, C., König, P., Engel, A.K., Singer, W.: Oscilliatory responses in cat visual cortex exhibit inter-columnar synchronization which reflects global stimulus properties. Nature 338, 334–337 (1989)

Guarino, N.: Concepts, attributes and arbitrary relations: some linguistic and ontological criteria for structuring knowledge bases. Data Knowl. Eng. 8(3), 249–261 (1992)

Hubel, D.H., Wiesel, T.N.: Receptive fields and functional architecture of monkey striate cortex. Journal of Physiology 195, 215–243 (1968)

Kilbury, J., Petersen, W., Rumpf, C.: Inheritance-based models of the lexicon. In: Wunderlich, D. (ed.) Advances in the theory of the lexicon, pp. 429–477. Mouton de Gruyter, Berlin (2006)

Löbner, S.: Definites. Journal of Semantics 4(4), 279–326 (1985)

Maye, A., Werning, M.: Temporal binding of non-uniform objects. Neurocomputing 58–60, 941–948 (2004)

Minsky, M.: A framework for representing knowledge. In: Winston, P.H. (ed.) The psychology of computer vision, McGraw-Hill, New York (1975)

Ortmann, A.: (In) Alienabilitätssplits und die Typologie der adnominalen Possession (Manuscript, University of Düsseldorf) (2006)

Petersen, W.: FCAType − a system for type signature induction. In: Yahia, S.B., Nguifo, E.M. (eds.) 4th international conference on concept lattices and their applications (2006)

Petersen, W.: Inducing type signatures from decomposition lattices. International Journal of Foundations of Computer Science, 2007 (submitted)

Schillen, T.B., König, P.: Binding by temporal structure in multiple feature domains of an oscillatory neuronal network. Biological Cybernetics 70, 397–405 (1994)

Singer, W.: Neuronal synchrony: A versatile code for the definition of relations? Neuron 24, 49–65 (1999)

Smith, E.E., Medin, D.L.: Categories and concepts. Harvard University Press, Cambridge, MA (1981)

Smith, J.D., Minda, J.: Thirty categorization results in search of a model. Journal of Experimental Psychology: Learning, Memory and Cognition 26(1), 3–27 (2000)

Sowa, J.: Conceptual structures. Addison-Wesley, Reading MA (1984)

Treisman, A.: The binding problem. Current Opinion in Neurobiology 6, 171–178 (1996)

von der Malsburg, C.: The correlation theory of brain function (Internal Report Nos. 81-2). MPI for Biophysical Chemistry, Göttingen (1981)

Werning, M.: Ventral vs. dorsal pathway: the source of the semantic object/event and the syntactic noun/verb distinction. Behavioral and Brain Sciences 26(3), 299–300 (2003)

Werning, M.: The temporal dimension of thought: Cortical foundations of predicative representation. Synthese 146(1/2), 203–224 (2005)

Werning, M., Maye, A.: Frames, coherency chains and hierarchical binding: The cortical implementation of complex concepts. In: Bara, B.G., Barsalou, L., Bucciarelli, M. (eds.) Proceedings of the twenty-seventh annual Conference of the Cognitive Science Society, pp. 2347–2352. Erlbaum, Mahwah, NJ (2005)

Werning, M., Maye, A.: The neural basis of the object concept in ambiguous and illusionary perception. In: Sun, R., Miyake, N. (eds.) Proceedings of the Twenty-Eighth Annual Conference of the Cognitive Science Society, pp. 876–881. Erlbaum, London (2006)

Werning, M., Maye, A.: The cortical implementation of complex attribute and substance concepts: Synchrony, frames, and hierarchical binding. Chaos and Complexity Letters 2(2) 2007 (in press)

Woods, W.A.: What's in a link: Foundations for semantic networks. In: Bobrow, D.G., Collins, A.M. (eds.) Representation and understanding: Studies in cognitive science, Academic Press, New York (1975)

Constants and Functions in Peirce's Existential Graphs

Frithjof Dau

University of Wollongong, Australia
dau@uow.edu.au

Abstract. The system of Peirce's existential graphs is a diagrammatic version of first order logic. To be more precise: As Peirce wanted to develop a logic of *relatives* (i.e., relations), existential graphs correspond to first order logic with relations and identity, but without constants or functions. In contemporary elaborations of first order logic, constants and functions are usually employed. In this paper, it is described how the syntax, semantics and calculus for Peirce's existential graphs has to be extended in order to encompass constants and functions as well.

1 Motivation and Introduction

It is well-known that Peirce (1839-1914) extensively investigated a *logic of relations* (which he called 'relatives'). Much of the third volume of the collected papers [HB35] is dedicated to this topic (see for example "Description of a Notation for the Logic of Relatives [...]" (3.45–3.149, 1870) "On the Algebra of Logic" (3.154–3.251, 1880), "Brief Description of the Algebra of Relatives" (3.306–3.322, 1882), and "the Logic of Relatives" (3.456–3.552, 1897)). As Burch writes, in Peirce's thinking 'reasoning is primarily, most elementary, reasoning about *relations*' ([Bur91], p. 2, emphasis by Burch).

Starting in 1896, Peirce invented a diagrammatic form of formal logic, namely his system of existential graphs [Zem64, Rob73, Shi02, PS00, Dau06b]. The Beta part of this system corresponds to first order logic (FO) [Zem64, Dau06b]. To be more precise: As Peirce investigated a logic of relations, the Beta part of existential graphs is equivalent to FO with relations and identity, but without constants or functions. In contrast to that, contemporary symbolic formalizations of FO are intended to represent statements about constants, relations, and functions. This paper shows how the the syntax, semantics, and the calculus of existential graphs has to be extended in order to cover constants and and functions as well.

This paper is part of the author's research on Sowa's conceptual graphs and Peirce's existential graphs [Dau02, Dau03, Dau06d, Dau06a, Dau06c, Dau06b]. It aims to provide a sufficiently formal elaboration of the paper's goal. For this reason, a formal elaboration of existential graphs, including their syntax, semantics, and calculus, would be needed. Due to space limitations, this is not possible. To resolve this problem, only those definitions and theorems of [Dau03, Dau06b] which are needed to keep this paper almost self-contained will be given.

U. Priss, S. Polovina, and R. Hill (Eds.): ICCS 2007, LNAI 4604, pp. 429–442, 2007.
© Springer-Verlag Berlin Heidelberg 2007

In contrast to concept graph with cuts (CGwCs)[1] or formulas of FO, existential graphs are not per se discrete structures. To formalize them, [Dau06b] takes a two-step approach. First, discrete structures, so-called EXISTENTIAL GRAPH INSTANCES (EGIs), are introduced. An EGI can be best understood as one (of many) possible discrete formalizations of a given existential graph. Then all different EGIs which formalize the same (naive) existential graph are aggregated in a class, and each of these classes is called a FORMAL EXISTENTIAL GRAPH. For further details, see [Dau06b]. Due to space limitation, the scrutiny in this paper is not carried out on formal existential graphs, but on EGIs instead.

Sec. 2 provides a short overview of the definitions and theorems of [Dau03, Dau06b] which are needed in this paper for defining the syntax and semantics of EGIs. The main task is to extend the calculus. In Sec. 3, the general methodology for extending the calculus is provided. Then new rules for constants and function names are given in Sec. 4, and their soundness and completeness is proven. In Sec. 5, a short example for a formal proof within the extended system of EGIs is provided. Finally, Sec. 6 discusses the results of the paper.

2   Syntax and Semantics

We start with the underlying structure for EGIs and CGwCs, namely *relational graphs with cuts*, and a quasiorder $\leq$ on all elements of such graphs.

Definition 1 (Relational Graphs with Cuts). *A* RELATIONAL GRAPH WITH CUTS *is a structure* $(V, E, \nu, \top, Cut, area)$*, where*

- V*,* E *and* Cut *are pairwise disjoint, finite sets whose elements are called* VERTICES EDGES *and* CUTS*, respectively,*
- $\nu : E \to \bigcup_{k \in \mathbb{N}_0} V^k$ *is a mapping,*
- $\top$ *is a single element with* $\top \notin V \cup E \cup Cut$*, the* SHEET OF ASSERTION*, and*
- *area* $: Cut \cup \{\top\} \to \mathfrak{P}(V \cup E \cup Cut)$ *is a mapping with a)* $c_1 \neq c_2 \Rightarrow area(c_1) \cap area(c_2) = \emptyset$ *, b)* $V \cup E \cup Cut = \bigcup_{d \in Cut \cup \{\top\}} area(d)$*, and c)* $c \notin area^n(c)$ *for each* $c \in Cut \cup \{\top\}$ *and* $n \in \mathbb{N}$ *(with* $area^0(c) := \{c\}$ *and* $area^{n+1}(c) := \bigcup \{area(d) \mid d \in area^n(c)\}$*).*

For an edge $e \in E$ *with* $\nu(e) = (v_1, \ldots, v_k)$ *we set* $|e| := k$*. The vertices, edges and cuts will be called the* ELEMENTS *of the graph. The elements of* $Cut \cup \{\top\}$ *are called* CONTEXTS*. Finally, as for every* $x \in V \cup E \cup Cut$ *we have exactly one context* $c \in Cut \cup \{\top\}$ *with* $x \in area(c)$*, we can write* $c = area^{-1}(x)$ *for every* $x \in area(c)$*, or even more simple and suggestive:* $c = ctx(x)$*.*

---

[1] CGwCs are a formal elaboration of simple conceptual graphs [Sow84, Sow92, Sow00, CM92, CM95], where the cuts of Peirce's existential graphs are added to allow for negation of subgraphs.

Definition 2 (Ordering on the Contexts, Enclosing Relation). *Let* $\mathfrak{G} :=$ $(V, E, \nu, \top, Cut, area)$ *be a relational graph with cuts. We define a mapping* $\beta :$ $V \cup E \cup Cut \cup \{\top\} \to Cut \cup \{\top\}$ *by* $\beta(x) := x$ *for* $x \in Cut \cup \{\top\}$, *and* $\beta(x) :=$ $ctx(x)$ *for* $x \in V \cup E$. *Next we set* $x \leq y :\Longleftrightarrow \exists n \in \mathbb{N}_0.\beta(x) \in area^n(\beta(y))$ *We define* $x < y :\Longleftrightarrow x \leq y \wedge y \not\leq x$ *and* $x \lesssim y :\Longleftrightarrow x \leq y \wedge y \neq x$. *For a context* $c \in Cut \cup \{\top\}$, *we set furthermore* $\leq[c] := \{x \in V \cup E \cup Cut \cup \{\top\} \mid x \leq c\}$ *and* $\lesssim[c] := \{x \in V \cup E \cup Cut \cup \{\top\} \mid x \lesssim c\}$. *Each element* x *of* $\bigcup_{n \in \mathbb{N}} area^n(c)$ *is said to be* ENCLOSED BY c, *and vice versa:* c *is said to* ENCLOSE x. *For each element of* $area(c)$, *we moreover say that it is* DIRECTLY ENCLOSED BY c.

The relation $\leq$ is indeed a quasiorder. Moreover, on the contexts, it is a tree. The proof for the following lemma can be found in [Dau03] and [Dau06b].

Lemma 1 ($\leq$ Induces a Tree on the Contexts). *For a relational graph with cuts* $\mathfrak{G} := (V, E, \nu, \top, Cut, area)$, $\leq$ *is a quasiorder. Furthermore,* $\leq \big|_{Cut \cup \{\top\}}$ *is an order on* $Cut \cup \{\top\}$ *which is a tree with* $\top$ *as greatest element.*

When defining the semantics, vertices which are deeper nested than some edge they are incident with cannot be evaluated. So this case has to be ruled out. For this reason, the next definition is needed.

Definition 3 (Dominating Nodes). *If* $ctx(e) \leq ctx(v)$ *($\Leftrightarrow e \leq v$) for every* $e \in E$ *and* $v \in V_e$, *then* $\mathfrak{G}$ *is said to have* DOMINATING NODES.

Next, we will define EGIs to be relational graphs with cuts, where the edges are additionally labelled with names. If EGIs are used to formalize existential graphs, we would only need relation names. For the purpose of this paper, we will introduce an alphabet with names for constants, functions and relations.

Definition 4 (Alphabet with Constants, Functions and Relations). *An* ALPHABET *is a structure* $(\mathcal{C}, \mathcal{F}, \mathcal{R}, ar)$ *of* CONSTANT NAMES, FUNCTION NAMES *and* RELATION NAMES, *resp., together with an arity-function* $ar : \mathcal{F} \cup \mathcal{R} \to \mathbb{N}$ *which assigns to each function name and relation name its arity. To ease the notation, we set* $ar(C) = 1$ *for each* $C \in \mathcal{C}$. *We assume that the sets* $\mathcal{C}, \mathcal{F}, \mathcal{R}$ *are pairwise disjoint. The elements of* $\mathcal{C} \cup \mathcal{F} \cup \mathcal{R}$ *are the* NAMES *of the alphabet. Let* $\doteq \in \mathcal{R}_2$ *be a special name which is called* IDENTITY.

Later on, we will interpret an n-ary function F to be an n-ary relation which satisfies a specific property, namely: For each n objects $o_1, \ldots, o_{n-1}$ exists exactly one object o_n with $F(o_1, o_2, \ldots, o_{n-1}, o_n)$. So, functions can be understood as special relations. Please note that we adopt the arity of relations for functions. That is, an n-ary function assigns a value to $n-1$ arguments. This understanding of the arity of a function is not the common one, but it will ease the forthcoming notations. Analogously, even an constant o can be understood as a special relation, namely the relation $\{(o)\}$. That is: constants correspond to unary relations which contain exactly one element (or to functions with zero arguments).

Now we are prepared to define existential graph instances (EGIs).

Definition 5 (Existential Graph Instance over $(\mathcal{C}, \mathcal{F}, \mathcal{R}, ar)$). *An* EXISTENTIAL GRAPH INSTANCE (EGI) OVER AN ALPHABET $\mathcal{A} = (\mathcal{C}, \mathcal{F}, \mathcal{R}, ar)$ *is*

a structure $\mathfrak{G} := (V, E, \nu, \top, Cut, area, \kappa)$ where $(V, E, \nu, \top, Cut, area)$ is a relational graph with cuts and dom. nodes, and $\kappa : E \to \mathcal{C} \cup \mathcal{F} \cup \mathcal{R}$ is a mapping
such that $|e| = ar(\kappa(e))$ for each $e \in E$. The elements of E with $\kappa(e) \doteq$ are
called IDENTITY-EDGES. The system of all EGIs over $\mathcal{A}$ will be denoted by $\mathcal{EGI}^{\mathcal{A}}$.

As said in the introduction, existential graphs are not per se discrete structures.
The major problem in formalizing existential graphs is caused by lines of identities and networks of lines of identities (i.e., *ligatures*). Peirce understood a line of
identity to be composed of bold dots, which can be interpreted to denote existentially quantified objects. These dots overlap, and the overlapping is interpreted
that the objects denoted by the dots are identical. This understanding of the
'inner structure' of a line of identity gives rise to the discrete EGIs, where dots
are formalized by the vertices, and overlapping of dots is formalized by edges
labelled with $\doteq$. But depending on how many dots we assign to a line of identity,
different EGIs can formalize a given existential graph. Note that an existentially
quantified object is syntactically formalized in CGs by a concept box $\boxed{\top : *}$.
Due to this obsertavion, EGIs can in turn understood to be those CGwCs where
only concept boxes of the form $\boxed{\top : *}$ appear.

Below, the proposition 'there is a cat which is not cute or which is not on a
mat' is depicted in several ways. First, an existential graph is provided. Next,
two possible EGI-formalizations of this graph are given. As just mentioned, they
only differ in the number of dots assigned to the lines of identity. For this reason,
in the formalization of [Dau06b], these two EGIs are members of the class which
formalize the given existential graph. The first EGI has in fact the minimal
number of vertices, the second EGI contains redundant vertices (the calculus for
EGIs is much easier to formalize if redundant vertices are allowed, for this reason, EGIs with redundant vertices are considered as well). In the diagrammatic
representation of EGIs, the vertices, as usual in graph theory, are drawn as bold
dots. Note that identity-edges are drawn as simple lines connecting the respective
bold dots. Finally, for the first EGI, the corresponding CGwC is depicted.

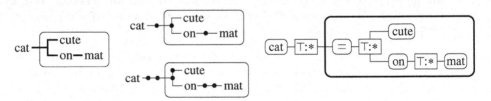

Next we define isomorphisms and partial isomorphisms between EGIs. The formal definition of an isomorphism is canonical. The rules of the calculus (like
the rules of Peirce, i.e. erasure, insertion, double cut, iteration and deiteration,
or the new rules presented in this paper for constants and functions) modify a
graph within a given context. For this reason, we furthermore have a notion of
two EGIs being isomorphic except a context.

Definition 6 ((Partial) Isomorphism). *For $i = 1, 2$, let two EGIs $\mathfrak{G}_i :=$
$(V_i, E_i, \nu_i, \top_i, Cut_i, area_i, \kappa_i)$ be given.*

An ISOMORPHISM $f = f_V \,\dot\cup\, f_E \,\dot\cup\, f_{Cut}$ *is composed of three bijective mappings* $f_V : V_1 \rightarrow V_2$, $f_E : E_1 \rightarrow E_2$ *and* $f_{Cut} : Cut_1 \cup \{\top_1\} \rightarrow Cut_2 \cup \{\top_2\}$ *which satisfy* $f_E(v_1, \ldots, v_n) = (f_V(v_1), \ldots, f_V(v_n))$ *for each* $e = (v_1, \ldots, v_n) \in E_1$, $f[area_1(c)] = area_2(f(c))$ *for each* $c \in Cut_1 \cup \{\top_1\}$ *(with* $f[area_1(c)] = \{f(k) \mid k \in area_1(c)\}$*), and* $\kappa_1(e) = \kappa_2(f_E(e))$ *for all* $e \in E_1$.

Let furthermore two contexts $c_i \in Cut_i \cup \{\top_i\}$ $i = 1, 2$, *be given. For each* i, *let* $V_i' := \{v \in V_i \mid v \not\le c_i\}$, $E_i' := \{e \in E_i \mid e \not\le c_i\}$, *and* $Cut_i' := \{d \in Cut_i \cup \{\top_i\} \mid d \not< c_i\}$. *Let* $\mathfrak{G}_i'$ *be the restriction of* $\mathfrak{G}_i$ *to these sets, i.e., for* $area_i' := area_i|_{Cut_i'}$ *and* $\kappa_i' := \kappa_i|_{E_i'}$, *let* $\mathfrak{G}_i' := (V_i', E_i', \nu|_{E_i'}, \top_i, Cut_i', area_i', \kappa_i')$. *If* $f = f_{V_1'} \,\dot\cup\, f_{E_1'} \,\dot\cup\, f_{Cut_1'}$ *is an isomorphism between* $\mathfrak{G}_1'$ *and* $\mathfrak{G}_2'$ *with* $f_{Cut}(c_1) = c_2$, *then* f *is called* (PARTIAL) ISOMORPHISM FROM $\mathfrak{G}_1$ TO $\mathfrak{G}_2$ EXCEPT FOR c_1 AND c_2.

In this definition, for the restrictions $area_i'$ and κ_i', we of course agree that the ranges of these functions are restricted to $V_i' \cup E_i' \cup Cut_i'$ as well. Moreover, note that this definition relies on the graph to have dominating nodes (otherwise it might happen that the structures $\mathfrak{G}_i'$ are no well-defined EGIs).

After defining the syntax for EGIs, we now turn to the semantics. First the models are defined in the usual manner known from formal logic.

Definition 7 (Relational Structures over $(\mathcal{C}, \mathcal{F}, \mathcal{R}, ar)$). *A* RELATIONAL STRUCTURE OVER AN ALPHABET $\mathcal{A} = (\mathcal{C}, \mathcal{F}, \mathcal{R}, ar)$ *is a pair* $\mathcal{M} := (U, I)$ *consisting of a nonempty* UNIVERSE U *and a function* $I := I_\mathcal{C} \cup I_\mathcal{F} \cup I_\mathcal{R}$ *with*

1. $I_\mathcal{C} : \mathcal{C} \rightarrow U$,
2. $I_\mathcal{F} : \mathcal{F} \rightarrow \bigcup_{k \in \mathbb{N}} \mathfrak{P}(U^k)$ *is a mapping where for each* $F \in \mathcal{F}$ *with* $ar(F) = k$, $I(F) \in U^k$ *is (total) function* $I(F) : U^{k-1} \rightarrow U$, *and*
3. $I_\mathcal{R} : \mathcal{R} \rightarrow \bigcup_{k \in \mathbb{N}} \mathfrak{P}(U^k)$ *is a mapping where for each* $R \in \mathcal{F}$ *with* $ar(R) = k$, *we have* $I(R) \in U^k$. *The name* '$\doteq$' *is mapped to the identity relation on* U.

When an EGI is evaluated in a relational structure (U, I), we have to assign objects of our universe of discourse U to its vertices. This is done by valuations.

Definition 8 (Valuations). *Let an EGI* $\mathfrak{G} := (V, E, \nu, \top, Cut, area, \kappa)$ *be given and let* (U, I) *be a relational structure over* $\mathcal{A}$. *Each mapping* $ref : V' \rightarrow U$ *with* $V' \subseteq V$ *is called a* PARTIAL VALUATION OF $\mathfrak{G}$. *If* $V' = V$, *then* ref *is called* (TOTAL) VALUATION OF $\mathfrak{G}$. *Let* $c \in Cut \cup \{\top\}$. *If* $V' \supseteq \{v \in V \mid v > c\}$ *and* $V' \cap \{v \in V \mid v \le c\} = \emptyset$, *then* ref *is called* PARTIAL VALUATION FOR c. *If* $V' \supseteq \{v \in V \mid v \ge c\}$ *and* $V' \cap \{v \in V \mid v < c\} = \emptyset$, *then* ref *is called* EXTENDED PARTIAL VALUATION FOR c.

The semantics for EGIs is based on Peirce's endoporeutic method. He read and evaluated existential graphs from the outside, hence starting with the sheet of assertion, and proceeded inwardly. During this evaluation, he assigned successively values to the lines of identity. This idea is adopted in the next definition.

Definition 9 (Endoporeutic Evaluation of Graphs). *Let an EGI $\mathfrak{G}$:=*
$(V, E, \nu, \top, Cut, area, \kappa)$ be given and let (U, I) be a relational structure over $\mathcal{A}$. In-
ductively over the tree $Cut \cup \{\top\}$, we define $(U, I) \models \mathfrak{G}[c, ref]$ for each context
$c \in Cut \cup \{\top\}$ and every partial valuation $ref : V' \subseteq V \to U$ for c:
$(U, I) \models \mathfrak{G}[c, ref] :\Longleftrightarrow$

> *ref can be extended to an partial valuation $\overline{ref} : V' \cup (V \cap area(c)) \to U$*
> *(i.e., $\overline{ref}$ is an extended partial valuation for c with $\overline{ref}(v) = ref(v)$ for*
> *all $v \in V'$), such that the following conditions hold:*
>
> - *$\overline{ref}(e) \in I(\kappa(e))$ for each $e \in E \cap area(c)$ (edge condition))*
> - *$(U, I) \not\models \mathfrak{G}[d, \overline{ref}]$ for each $d \in Cut \cap area(c)$ (cut condition and*
>   *iteration over $Cut \cup \{\top\}$))*

For $(U, I) \models \mathfrak{G}[\top, \emptyset]$ we write $(U, I) \models \mathfrak{G}$. If $\mathfrak{H}$ is a set of EGIs and if $\mathfrak{G}$ is an
EGI such that $(U, I) \models \mathfrak{G}$ for each model (U, I) that satisfies $(U, I) \models \mathfrak{G}'$ for
each $\mathfrak{G}' \in \mathfrak{H}$, we write $\mathfrak{H} \models \mathfrak{G}$.

Finally, we assume that we have a sound and complete calculus for EGIs where
only relation names occur (i.e., over alphabets $(\emptyset, \emptyset, \mathcal{R}, ar)$). Moreover, we assume that this calculus is based on Peirce's rules for existential graphs (erasure,
insertion, double cut, iteration and deiteration). As EGIs can be understood to
be CGwCs over alphabets without names for constants or types, we can adopt
the CGwCs-calculus of [Dau03] for this purpose. A similar calculus, directly for
EGIs, is provided in [Dau06b]. Both calculi contain Peirce's rules[2] and have additional rules which are needed to handle identity edges. Due to space limitations,
no calculus is given here.

The rules of the common calculi for FO (Hilbert-style calculi, natural deduction, sequent calculi) allow only modifications of formulae at their top-level.
In contrast to that, the rules of Peirce allow modifications of a graph inside
arbitrarily deep contexts. Due to this, Peirce's rules are much more powerful, and their soundness proofs can turn out to be rather complex. For this
reason, both in [Dau03] and [Dau06b], two lemmata are provided which ease
the soundness proofs. The lemma which is needed in this paper is given
below.

Theorem 1 (Main Thm. for Soundness, Equivalence Version). *Let EGIs*
$\mathfrak{G} := (V, E, \nu, \top, Cut, area, \kappa)$, $\mathfrak{G}' := (V', E', \nu', \top', Cut', area', \kappa')$ be given and
let f be an isomorphism between $\mathfrak{G}$ and $\mathfrak{G}'$ except for $c \in Cut$ and $c' \in Cut'$.
Set $Cut_c := \{d \in Cut \cup \{\top\} \mid d \not< c\}$. Let $\mathcal{M}$ be a relational structure and let
$P(d)$ be the following property for contexts $d \in Cut_c$: Every partial valuation
ref for d satisfies $\mathcal{M} \models \mathfrak{G}[d, ref] \Longleftrightarrow \mathcal{M} \models \mathfrak{G}'[f(d), f(ref)]$. Then, if P holds
for c, then P holds for each $d \in Cut_c$. Particularly, If P holds for c, we have
$\mathcal{M} \models \mathfrak{G} \Longleftrightarrow \mathcal{M} \models \mathfrak{G}'$.

---

[2] The formal iteration rule in [Dau06b] is more powerful than the formal iteration rule
in [Dau03] and, as it is discussed in [Dau06b], resembles better Peirce's notion of the
iteration rule for existential graphs.

3   General Logical Background

When considering constant names and function names instead of relation names only, we have new entailments between graphs. For example, if C is a constant name, the empty sheet of assertion (semantically) entails the graph $\bullet\!\!-\!\!C$. Thus it must be possible to derive this graph from the empty sheet of assertion (which would not be possible if C was an 1-ary relation name). The new entailments must be reflected by the calculus, thus the calculus has to be extended in order to capture the specific properties of constants and functions. There are basically two approaches: Firstly, we can add axioms, secondly, we can add new rules to the calculus. Besides the empty sheet of assertion, Peirce's calculus for existential graphs has no axioms. To preserve this property, we will adopt the second approach. This section describes the methodology how this shall be done.

As already mentioned, constant names and function names can be understood as relation names which are mapped to relations with specific properties. If we have an alphabet $\mathcal{A}' = (\mathcal{C}, \mathcal{F}, \mathcal{R}, ar)$ with constants and function names, we can then consider the alphabet $\mathcal{A} := (\emptyset, \emptyset, \mathcal{C} \,\dot\cup\, \mathcal{F} \,\dot\cup\, \mathcal{R}, ar)$, where each name is now understood as relation name. In this understanding, each EGI over $\mathcal{A}'$ is an EGI over $\mathcal{A}$ as well. Moreover, if $\mathcal{M}' := (U, I')$ with $I' := I'_{\mathcal{C}} \cup I'_{\mathcal{F}} \cup I'_{\mathcal{R}}$ is relational structure over the alphabet $\mathcal{A}'$, then $\mathcal{M} := (U, I)$ with $I(F) := I'_{\mathcal{F}}(F)$ for each $F \in \mathcal{F}$, $I(R) := I'_{\mathcal{R}}(R)$ for each $R \in \mathcal{R}$, and $I(C) := \{(I'_{\mathcal{C}}(C)\}$ for each $C \in \mathcal{C}$ is the corresponding model over the alphabet $\mathcal{A}$. We implicitly identify $\mathcal{M}$ and $\mathcal{M}'$. Due to this convention, each model over $\mathcal{A}'$ is an model over $\mathcal{A}$ as well. But the models for $\mathcal{A}'$ form a subclass of the models for $\mathcal{A}$. That is, if we denote the models for $\mathcal{A}'$ with $\mathfrak{M}_2$ and the models for $\mathcal{A}$ with $\mathfrak{M}_1$, we have $\mathfrak{M}_2 \subsetneq \mathfrak{M}_1$.

Thus we have to deal with two classes of models, which yield two entailment relations. If $\mathfrak{H}$ is a set of EGIs and if $\mathfrak{G}$ is an EGI such that $\mathcal{M} \models \mathfrak{G}$ for each relational structure $\mathcal{M} \in \mathfrak{M}_i$ with $\mathcal{M} \models \mathfrak{G}'$ for each $\mathfrak{G}' \in \mathfrak{H}$, we write $\mathfrak{H} \models_i \mathfrak{G}$.

In Sec. 2, we assumed to have a sound and complete calculus for EGIs where only relation names occur; that is, for EGIs which are evaluated in $\mathfrak{M}_1$. In the following, this calculus shall be denoted by $\vdash_1$. The soundness and completeness of $\vdash_1$ can be now stated as follows: If $\mathfrak{H} \cup \{\mathfrak{G}\}$ is a set of EGIs over $\mathcal{A}$, we have

$$\mathfrak{H} \vdash_1 \mathfrak{G} \quad \Longleftrightarrow \quad \mathfrak{H} \models_1 \mathfrak{G} \tag{1}$$

We seek a calculus $\vdash_2$ which extends $\vdash_1$ (that is, $\vdash_2$ has new rules, which will be denoted by $\vdash_2 \,\supseteq\, \vdash_1$) and which is sound and complete with respect to $\mathfrak{M}_2$.

The calculus $\vdash_1$, and hence $\vdash_2$ as well, encompasses the 5 basic-rules of Peirce. Thus for both calculi, the deduction theorem (see Lemma 6.5 of [Dau03] or Lemma. 8.7 of [Dau06b]) holds, i.e., for $i = 1, 2$, we have

$$\mathfrak{G}_a \vdash_i \mathfrak{G}_b \quad \Longleftrightarrow \quad \vdash_i \left(\mathfrak{G}_a \overline{\left(\mathfrak{G}_b \right)} \right) \tag{2}$$

We will extend $\vdash_1$ to $\vdash_2$ as follows: First of all, the new rules in $\vdash_2$ have to be sound. Then for a set of graphs $\mathfrak{H}$ and an EGI $\mathfrak{G}$ we have

$$\mathfrak{H} \vdash_2 \mathfrak{G} \quad \Longrightarrow \quad \mathfrak{H} \models_2 \mathfrak{G} \tag{3}$$

On the other hand, let us assume that for each $\mathcal{M} \in \mathfrak{M}_1 \backslash \mathfrak{M}_2$, there exists a graph $\mathfrak{G}_\mathcal{M}$ with

$$\vdash_2 \mathfrak{G}_\mathcal{M} \quad \text{and} \quad \mathcal{M} \not\models \mathfrak{G}_\mathcal{M} \tag{4}$$

If the last two assumptions (3) and (4) hold, we obtain that $\vdash_2$ is an adequate calculus, as the following theorem shows.

Theorem 2 (Completeness of $\vdash_2$). *A set* $\mathfrak{H} \cup \{\mathfrak{G}\}$ *of EGIs over* $\mathcal{A}$ *satisfies*

$$\mathfrak{H} \models_2 \mathfrak{G} \quad \Longrightarrow \quad \mathfrak{H} \vdash_2 \mathfrak{G}$$

Proof: Let $\mathfrak{H}_2 := \{\mathfrak{G}_\mathcal{M} \mid \mathcal{M} \in \mathfrak{M}_1 \backslash \mathfrak{M}_2\}$. From (3) we conclude: $\models_2 \mathfrak{G}_\mathcal{M}$ for all $\mathfrak{G}_\mathcal{M} \in \mathfrak{H}_2$. Now (4) yields:

$$\mathfrak{M}_2 = \{\mathcal{M} \in \mathfrak{M}_1 \mid \mathcal{M} \models \mathfrak{G} \text{ for all } \mathfrak{G} \in \mathfrak{H}_2\} \tag{5}$$

Now let $\mathfrak{H} \cup \{\mathfrak{G}\}$ be an arbitrary set of graphs. We get:

$\mathfrak{H} \models_2 \mathfrak{G} \overset{\text{Def}}{\Longleftrightarrow}$ f.a. $\mathcal{M} \in \mathfrak{M}_2$: if $\mathcal{M} \models \mathfrak{G}'$ for all $\mathfrak{G}' \in \mathfrak{H}$, then $\mathcal{M} \models \mathfrak{G}$

$\overset{(5)}{\Longleftrightarrow}$ f.a. $\mathcal{M} \in \mathfrak{M}_1$: if $\mathcal{M} \models \mathfrak{G}'$ for all $\mathfrak{G} \in \mathfrak{H}_2 \cup \mathfrak{H}$, then $\mathcal{M} \models \mathfrak{G}$

$\Longleftrightarrow \mathfrak{H} \cup \mathfrak{H}_2 \models_1 \mathfrak{G}$

$\overset{(1)}{\Longleftrightarrow} \mathfrak{H} \cup \mathfrak{H}_2 \vdash_1 \mathfrak{G}$

$\Longleftrightarrow$ there are $\mathfrak{G}_1, \ldots, \mathfrak{G}_n \in \mathfrak{H}$ and $\mathfrak{G}'_1, \ldots, \mathfrak{G}'_m \in \mathfrak{H}_2$ with

$\mathfrak{G}_1 \quad \mathfrak{G}_2 \quad \ldots \quad \mathfrak{G}_n \quad \mathfrak{G}'_1 \quad \mathfrak{G}'_2 \quad \ldots \quad \mathfrak{G}'_m \quad \vdash_1 \mathfrak{G}$

$\overset{(2)}{\Longleftrightarrow}$ there are $\mathfrak{G}_1, \ldots, \mathfrak{G}_n \in \mathfrak{H}$ and $\mathfrak{G}'_1, \ldots, \mathfrak{G}'_m \in \mathfrak{H}_2$ with

$\vdash_1 \left(\mathfrak{G}_1 \, \mathfrak{G}_2 \ldots \, \mathfrak{G}_n \, \mathfrak{G}'_1 \, \mathfrak{G}'_2 \ldots \, \mathfrak{G}'_m \, \boxed{\mathfrak{G}_b} \right)$

$\overset{\vdash_2 \supseteq \vdash_1 , (4)}{\Longrightarrow}$ there are $\mathfrak{G}_1, \ldots, \mathfrak{G}_n \in \mathfrak{H}$ and $\mathfrak{G}'_1, \ldots, \mathfrak{G}'_m \in \mathfrak{H}_2$ with

$\vdash_2 \mathfrak{G}'_1 \ldots \, \mathfrak{G}'_m \left(\mathfrak{G}_1 \, \mathfrak{G}_2 \ldots \, \mathfrak{G}_n \, \mathfrak{G}'_1 \, \mathfrak{G}'_2 \ldots \, \mathfrak{G}'_m \, \boxed{\mathfrak{G}_b} \right)$

$\overset{\text{deit.}}{\Longleftrightarrow}$ there are $\mathfrak{G}_1, \ldots, \mathfrak{G}_n \in \mathfrak{H}$ and $\mathfrak{G}'_1, \ldots, \mathfrak{G}'_m \in \mathfrak{H}_2$ with

$\vdash_2 \mathfrak{G}'_1 \ldots \, \mathfrak{G}'_m \left(\mathfrak{G}_1 \, \mathfrak{G}_2 \ldots \, \mathfrak{G}_n \, \boxed{\mathfrak{G}_b} \right)$

$\overset{\text{era.}}{\Longrightarrow}$ there are $\mathfrak{G}_1, \ldots, \mathfrak{G}_n \in \mathfrak{H}$ with $\vdash_2 \left(\mathfrak{G}_1 \ldots \, \mathfrak{G}_n \, \boxed{\mathfrak{G}_b} \right)$

$\overset{(2)}{\Longleftrightarrow}$ there are $\mathfrak{G}_1, \ldots, \mathfrak{G}_n \in \mathfrak{H}$ with $\mathfrak{G}_1, \ldots, \mathfrak{G}_n \vdash_2 \mathfrak{G}$

$\overset{\text{Def.}}{\Longrightarrow} \mathfrak{H} \vdash_2 \mathfrak{G}$ $\qquad\qquad\qquad \square$

4   Extending the Calculus

In this section, the calculus is extended in order to capture the specific properties of constants and functions. We start the scrutiny with functions.

The following EGI holds in a model (U, I) exactly if F is interpreted as an n-ary (total) function $I(F) : U^{n-1} \to U$:

$$\mathfrak{G}_F := $$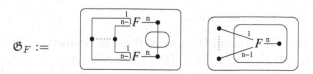

More precisely: The left subgraph is satisfied if F is interpreted as partial function (that is, to objects $o_1, \ldots, o_{n-1}$ exist at most one o_n with $I(F)(o_1, \ldots, o_n)$), the right subgraph is satisfied if for objects $o_1, \ldots, o_{n-1}$ exist at least one o_n with $I(F)(o_1, \ldots, o_n)$. In other words: The left subgraph guarantees the uniqueness, the right subgraph the existence of function values.

According to the last subsection, we have to find rules which are sound and which enable us to derive each graph $\mathfrak{G}_F$ with $F \in \mathcal{F}$. They are given below.

Definition 10 (New Rules for Function Names). *Let $F \in \mathcal{F}$ be an n-ary function name. Then all rules of the calculus, where F is treated like a relation name, may be applied. Moreover, the following additional transformations may be performed:*

- **Functional Property Rule (uniqueness of values)** *Let e, f be n-ary edges with $\nu(e) = (v_1, \ldots, v_{n-1}, v_e)$, $\nu(f) = (v_1, \ldots, v_{n-1}, v_f)$, $ctx(e) = ctx(v_e)$, $ctx(f) = ctx(v_f)$, and $\kappa(e) = \kappa(f) = F$. Let c be a context with $c \leq ctx(e)$ and $c \leq ctx(f)$. Then arbitrary identity-links id with $\nu(id) = (v_e, v_f)$ may be inserted into c or erased from c.*
- **Total Function Rule (existence of values)** *Let $v_1, \ldots, v_{n-1}$ be vertices, let c be a context with $c \leq ctx(v_1), \ldots, ctx(v_{n-1})$. Then we can add a vertex v_n and an edge e to c with $\nu(e) = (v_1, \ldots, v_n)$ and $\kappa(e) = F$. Vice versa, if v_n and e are a vertex and an edge in c with $\nu(e) = (v_1, \ldots, v_n)$ and $\kappa(e) = F$ such that v_n is not incident with any other edge, e and v_n may be erased.*

We have to show that these rules are sound are complete. We start with the soundness of the rules.

Lemma 2 (The Total Function Rule is Sound). *If $\mathfrak{G}$ and $\mathfrak{G}'$ are two EGIs over $\mathcal{A} := (\mathcal{C}, \mathcal{F}, \mathcal{R}, ar)$, $\mathcal{M} := (U, I)$ is a relational structure with $\mathcal{M} \models \mathfrak{G}$ and $\mathfrak{G}'$ is derived from $\mathfrak{G}$ with the total function rule, then $\mathcal{M} \models \mathfrak{G}'$.*

Proof: Let $\mathfrak{G}'$ be obtained from $\mathfrak{G}$ by adding a vertex v_n and an edge e to c according to the total function rule. We want to apply Lemma 1 to c, so let ref be a valuation for the context c.

Let us first assume that we have $\mathcal{M} \models \mathfrak{G}[c, ref]$, i.e., there is an extension $\overline{ref}$ of ref to $V \cap area(c)$ with $\mathcal{M} \models \mathfrak{G}[c, \overline{ref}]$. Let $o := I(F)(ref(v_1), \ldots, ref(v_n))$. Then $\overline{ref}' := \overline{ref} \cup \{(v_n, o)\}$ is a extended partial valuation for c in $\mathfrak{G}'$ which satisfies $\mathcal{M} \models \mathfrak{G}[c, \overline{ref}']$, as the additional edge condition for e in the context c of $\mathfrak{G}'$ holds due to the definition of $\overline{ref}'$. Particularly, we obtain $\mathcal{M} \models \mathfrak{G}'[c, ref]$.

Now let $\mathcal{M} \models \mathfrak{G}'[c, ref]$, i.e., there is an extension $\overline{ref}'$ of ref to $V \cap area(c)$ with $\mathcal{M} \models \mathfrak{G}'[c, \overline{ref}']$. For $\overline{ref} := \overline{ref}' \backslash \{(v_n, \overline{ref}'(v_n))\}$ we have $\mathcal{M} \models \mathfrak{G}[c, \overline{ref}]$, thus $\mathcal{M} \models \mathfrak{G}[c, ref]$.

Now Lemma 1 yields the lemma. □

Lemma 3 (The Functional Property Rule is Sound). *If $\mathfrak{G}$ and $\mathfrak{G}'$ are two EGIs over $\mathcal{A} := (\mathcal{C}, \mathcal{F}, \mathcal{R}, ar)$, $\mathcal{M} := (U, I)$ is a relational structure with $\mathcal{M} \models \mathfrak{G}$ and $\mathfrak{G}'$ is derived from $\mathfrak{G}$ with the functional property rule, then $\mathcal{M} \models \mathfrak{G}'$.*

Proof: Let $\mathfrak{G}'$ be obtained from $\mathfrak{G}'$ by inserting an identity-link id with $\nu(id) = (v_e, v_f)$ into c. We set $c_e := ctx(e)$ and $c_f := ctx(f)$. The EGIs $\mathfrak{G}$ and $\mathfrak{G}'$ are isomorphic except for the context c. First note that the contexts c_e and c_f must be comparable. W.l.o.g. we assume $c_e \geq c_f \geq c$.

We first consider the case $c_e = c_f = c$. We want to apply Lemma 1 to c, so let ref_c be a partial valuation for c. In $\mathfrak{G}'$ in the context c, we have added the edge id, thus for c, there is one more edge condition to check. So it suffices to prove

$$(U, I) \models \mathfrak{G}[c, ref_c] \quad \Longrightarrow \quad (U, I) \models \mathfrak{G}'[c, ref_c] \tag{6}$$

Let $(U, I) \models \mathfrak{G}[c, ref_c]$. That is, there is an extension $\overline{ref_c}$ of ref_c to $V \cap area(c)$ with $\mathfrak{G} \models \mathfrak{G}[c, \overline{ref_c}]$, i.e., $\overline{ref_c}$ satisfies all edge- and cut-conditions in c. Particularly, it satisfies the edge-conditions for e and f, that is:

$$(\overline{ref_c}(v_1), \dots \overline{ref}(v_{n-1}), \overline{ref_c}(v_e)) \in I(\kappa(e)) \quad \text{and}$$
$$(\overline{ref_c}(v_1), \dots \overline{ref}(v_{n-1}), \overline{ref_c}(v_f)) \in I(\kappa(f))$$

i.e., $\overline{ref_c}(v_e) = I(F)(\overline{ref_c}(v_1), \dots \overline{ref_c}(v_{n-1})) = \overline{ref_c}(v_f)$. So the additional edge condition for id in $\mathfrak{G}'$ is satisfied by $\overline{ref_c}$. We obtain $\mathfrak{G}' \models \mathfrak{G}[c, \overline{ref_c}]$, hence $\mathfrak{G}' \models \mathfrak{G}[c, ref_c]$, thus Eqn. (6) holds. Now Lemma 1 yields $\mathcal{M} \models \mathfrak{G} \Longleftrightarrow \mathcal{M} \models \mathfrak{G}'$.

Next we consider the case $c_e = c_f > c$. We want to apply Lemma 1 to c_e, so let ref_{c_e} be a partial valuation for c_e. To apply Lemma 1, it it suffices to prove

$$\mathfrak{G} \models \mathfrak{G}[c_e, \overline{ref_{c_e}}] \quad \Longleftrightarrow \quad \mathfrak{G}' \models \mathfrak{G}[c_e, \overline{ref_{c_e}}] \tag{7}$$

for each extension $\overline{ref_{c_e}}$ of ref_{c_e} to $area(c_e) \cap V$. So let $\overline{ref_{c_e}}$ be such an extension, If $\overline{ref_{c_e}}$ does not satisfy the edge-conditions for e and f, we have $\mathfrak{G} \not\models \mathfrak{G}[c, \overline{ref_{c_e}}]$ and $\mathfrak{G}' \not\models \mathfrak{G}[c, \overline{ref_{c_e}}]$, thus Eqn. (7) holds. So let $\overline{ref_{c_e}}$ satisfy the edge-conditions for e and f. Analogously to the case $c_e = c_f = c$ we obtain $\overline{ref_{c_e}}(v_e) = \overline{ref_{c_e}}(v_f)$. Moreover, for each extension ref_c of $\overline{ref_{c_e}}$ to a partial valuation of c, we obtain $\mathfrak{G} \models \mathfrak{G}[c, ref_c] \Longleftrightarrow \mathfrak{G}' \models \mathfrak{G}[c, ref_c]$. This can be seen analogously to the case $c_e = c_f = c$, as $\mathfrak{G}$ and $\mathfrak{G}'$ differ only by adding the edge edge id in c, but for each extension of ref_c to $area(c) \cap V$, the edge-condition for id is due to $\overline{ref_{c_e}}(v_e) = \overline{ref_{c_e}}(v_f)$ fulfilled. Now it can easily be shown by induction that for each context d with $c_e > d \geq c$ and each extension ref_d of $\overline{ref_{c_e}}$ to $area(d) \cap V$, we have $\mathfrak{G} \models \mathfrak{G}[d, ref_d] \Longleftrightarrow \mathfrak{G}' \models \mathfrak{G}[d, ref_d]$. This yields $\mathfrak{G} \models \mathfrak{G}[c_e, \overline{ref_{c_e}}] \Longleftrightarrow \mathfrak{G}' \models \mathfrak{G}[c_e, \overline{ref_{c_e}}]$, i.e., Eqn. (7) holds again.

Next we consider the case $c_e > c_f > c$. The basic idea of the proof is analogous to the last cases, but we have two nested inductions. Again we want to apply

Lemma 1 to c_e, so let ref_e be a partial valuation for c_e. Again we show that Eqn. (7) holds for each extension $\overline{ref_e}$ of ref_e to $area(c_e) \cap V$. Similarly to the last case, we assume that $\overline{ref_e}$ satisfies the edge-condition for e. It is sufficient to show that

$$\mathfrak{G} \models \mathfrak{G}[c_f, ref_f] \iff \mathfrak{G}' \models \mathfrak{G}[c_f, ref_f] \tag{8}$$

holds for each each extension ref_f of $\overline{ref_e}$ to $area(c_f) \cap V$: Then similarly to the last case, an inductive argument yields that for each context d with $c_e > d \geq c_f$ and each extension ref_d of $\overline{ref_{c_e}}$ to $area(d) \cap V$, we have $\mathfrak{G} \models \mathfrak{G}[d, ref_d] \iff \mathfrak{G}' \models \mathfrak{G}[d, ref_d]$. This yields $\mathfrak{G} \models \mathfrak{G}[c_e, \overline{ref_e}] \iff \mathfrak{G}' \models \mathfrak{G}[c_e, \overline{ref_e}]$. That is, Eqn. (7) holds.

It remains to show that Eqn. (8) holds. Let us consider an extension ref_f of $\overline{ref_e}$ to $area(c_f) \cap V$. To prove Eqn. (8), it is sufficient to show that

$$\mathfrak{G} \models \mathfrak{G}[c_f, \overline{ref_f}] \iff \mathfrak{G}' \models \mathfrak{G}[c_f, \overline{ref_f}] \tag{9}$$

holds for each extension $\overline{ref_f}$ of ref_f to $area(c_f) \cap V$. Now we can perform the same inductive argument as in the last case. If $\overline{ref_f}$ does not satisfy the edge-condition for f, we are done. Otherwise we have $\overline{ref_f}(v_e) = \overline{ref_f}(v_f)$. For each extension ref_c of $\overline{ref_f}$ to $area(c) \cap V$, we obtain $\mathfrak{G} \models \mathfrak{G}[c, ref_c] \iff \mathfrak{G}' \models \mathfrak{G}[c, ref_c]$. Now from the usual inductive argument we obtain that for each context d with $c_f > d \geq c$ and each extension ref_d of $\overline{ref_f}$ to $area(d) \cap V$, we have $\mathfrak{G} \models \mathfrak{G}[d, ref_d] \iff \mathfrak{G}' \models \mathfrak{G}[d, ref_d]$. From this we conclude that Eqn. (9), thus Eqn. (8), holds. This finishes the proof for the case $c_e > c_f > c$.

Finally, the cases $c_e > c_f = c$ and $c_f > c_e = c$ can be handled analogously. $\square$

Next, the new rules for constants are introduced. As constants correspond to that functions f with zero arguments, a distinction between constants and function names is, strictly speaking, not necessary. So the rules for constant names correspond to rules for 1-ary functions (i.e. functions f with $dom(f) = \emptyset$).

Definition 11 (New Rules for Constant Names). *Let $C \in \mathcal{C}$ be a constant name. Then all rules of the calculus, where F is treated like a relation name, may be applied. Moreover, the following additional transformations may be performed:*

- **Constant Identity Rule** *Let e, f be two unary edges with $\nu(e) = (v_e)$, $\nu(f) = (v_f)$, $ctx(v_e) = ctx(e)$, $ctx(v_f) = ctx(f)$, and $\kappa(e) = \kappa(f) = C$. Let c be a context with $c \leq ctx(e)$ and $c \leq ctx(f)$. Then arbitrary identity-links id with $\nu(id) = (v_e, v_f)$ may be inserted into c or erased from c.*
- **Existence of Constants Rule** *In each context c, we may add a fresh vertex v and an fresh unary edge e with $\nu(e) = (v)$ and $\kappa(e) = C$. Vice versa, if v and e are a vertex and an edge in c with $\nu(e) = (v)$ and $\kappa(e) = F$ such that v is not incident with any other edge, e and v may be erased from c.*
  That is: Devices $\bullet\!\!-\; C$ may be inserted into or erased from c.

It remains to prove the completeness of the extended calculus.

Theorem 3 (Extended Calculus is Complete). *Each set $\mathfrak{H} \cup \{\mathfrak{G}\}$ of EGIs over $\mathcal{A} := (\mathcal{C}, \mathcal{F}, \mathcal{R}, ar)$ satisfies $\mathfrak{H} \models \mathfrak{G} \Rightarrow \mathfrak{H} \vdash \mathfrak{G}$.*

Proof: Due to the remark before Def. 11 and Thm. 2, it is sufficient to show that for each $F \in \mathcal{F}$, the graph $\mathfrak{G}_F$ can be derived with the new rules. The functional property rule (fp) enables us to derive the left subgraph of $\mathfrak{G}_F$ as follows:

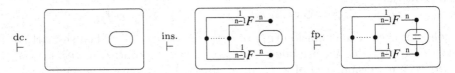

The right subgraph of $\mathfrak{G}_F$ can be derived with the total function rule (tf):

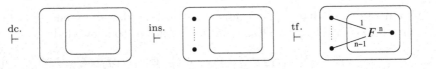

5   An Example for a Proof with Constants and Functions

In this section, an example for a formal proof with EGIs is provided. We prove a trivial fact in group theory, namely the uniqueness of neutral elements. Assume that e_1 and e_2 are neutral elements, i.e. we have $\forall x : x \cdot e_1 = e_1 = e_1 \cdot x$ and $\forall x : x \cdot e_2 = e_2 = e_2 \cdot x$. From this we can conclude $e_1 = e_2$.

In the following, a formal proof with EGIs for this fact is provided. We assume that e_1, e_2 are employed as constant names and $\cdot$ as function name.

We start with the assumption that e_1, e_2 are neutral elements, i.e.

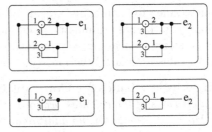

Erasure yields:

First, we insert e_1 and e_2 (i.e., edges which are labeled with e_1 and e_2) as follows:

The edges are iterated:

Now we can remove the identity edges with the constant identity rule.

The next graph is derived with the existence of constants rule.

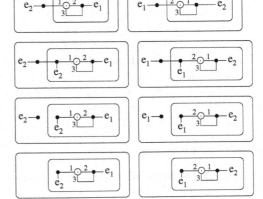

Next, we remove the double cuts and rearrange the graph.

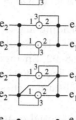

We can insert identity edges with the constant identity rule.

The functional property rule now allows to add another identity edge.

The erasure rule finally yields: $e_2 \bullet\!\!-\!\!-\!\!-\!\!\bullet e_1$ □

6 Discussion and Outlook

We have shown how existential graphs have to be modified to cover constants and functions as well. Though the approach of this paper is somewhat generic, the set of the new rules depends on the syntactical implementation of constants and functions. In CGwCs, constant names are assigned to the vertices instead of the edges. Although the expressivity of the system remains the same, we have new syntactical possibilities to express a given statement. For this reason, further rules in the calculus are needed. A discussion on this can be found in [Dau06b].

Existential graphs should not be understood as a diagrammatic version of the specific form of FO where only relations are used. As this paper shows, they can be tailored to formalize other kinds of logic as well. Another example is Description Logics. In [DE06], the syntax and semantics of a fragment of existential graphs is provided which corresponds to the Description Logic $\mathcal{ALCI}$. A calulus for this system is provided in a paper which has recently be submitted to the conference on visual languages and human centric computing. Similar to this paper, this calculus is based on Peirce's original calculus, augmented with additional rules. Together with the general, formal elaboration of existential graphs in [Dau06b], these results show that the system of exististential graph conforms the needs of different forms of contemporary formal logic.

References

[Bur91] Burch, R.W.: A Peircean Reduction Thesis: The Foundation of Topological Logic. Texas Tech. University Press, Texas, Lubbock (1991)

[CM92] Chein, M., Mugnier, M.-L.: Conceptual graphs: Fundamental notions. Revue d'Intelligence Artificiell 6, 365–406 (1992)

[CM95] Chein, M., Mugnier, M.-L.: Conceptual graphs are also graphs. Technical report, LIRMM, Université Montpellier II, Rapport de Recherche 95003 (1995)

[Dau02] Dau, F.: An embedding of existential graphs into concept graphs with negations. In: Priss, U., Corbett, D.R., Angelova, G. (eds.) ICCS 2002. LNCS (LNAI), vol. 2393, pp. 15–19. Springer, Heidelberg (2002)

[Dau03]   Dau, F.: The Logic System of Concept Graphs with Negations and its Relationship to Predicate Logic. In: Dau, F. (ed.) The Logic System of Concept Graphs with Negation. LNCS (LNAI), vol. 2892, Springer, Heidelberg
(2003)

[Dau06a]   Dau, F.: Fixing shin's reading algorithm for peirce's existential graphs. In:
Barker-Plummer, D., Cox, R., Swoboda, N. (eds.) Diagrams 2006. LNCS
(LNAI), vol. 4045, pp. 88–92. Springer, Heidelberg (2006)

[Dau06b]   Dau, F.: Mathematical logic with diagrams, based on the existential
graphs of peirce. Habilitation thesis (to be published), Available at:
http://www.dr-dau.net

[Dau06c]   Dau, F.: The role of existential graphs in peirce's philosophy. In: Schärfe,
H., Hitzler, P., Øhrstrøm, P. (eds.) ICCS 2006. LNCS (LNAI), vol. 4068,
pp. 28–41. Springer, Heidelberg (2006)

[Dau06d]   Dau, F.: Some notes on proofs with alpha graphs. In: Schärfe, H., Hitzler,
P., Øhrstrøm, P. (eds.) ICCS 2006. LNCS (LNAI), vol. 4068, pp. 172–188.
Springer, Heidelberg (2006)

[DE06]   Dau, F., Eklund, P.: Towards a diagrammatic reasoning system for description logics. Submitted to the Journal of Visual Languages and Computing,
Elsevier (2006), Available at http://www.kvocentral.org

[HB35]   Hartshorne, W., Burks (eds.): Collected Papers of Charles Sanders Peirce,
Harvard University Press, Cambridge, Massachusetts (1931-1935)

[PS00]   Peirce, C.S., Sowa, J.F.: Existential Graphs: MS 514 by Charles Sanders
Peirce with commentary by John Sowa, 1908 (2000), Available at:
http://www.jfsowa.com/peirce/ms514.htm

[Rob73]   Roberts, D.D.: The Existential Graphs of Charles S. Peirce. Mouton, The
Hague, Paris (1973)

[Shi02]   Shin, S.-J.: The Iconic Logic of Peirce's Graphs. Bradford Book, Massachusetts (2002)

[Sow84]   Sowa, J.F.: Conceptual structures: information processing in mind and machine. Addison-Wesley, Reading, Mass. (1984)

[Sow92]   Sowa, J.F.: Conceptual graphs summary. In: Nagle, T.E., Nagle, J.A., Ger-
holz, L.L., Eklund, P.W. (eds.) Conceptual Structures: current research and
practice, pp. 3–51. Ellis Horwood (1992)

[Sow00]   Sowa, J.F.: Knowledge Representation: Logical, Philosophical, and Computational Foundations, Brooks Cole, Pacific Grove, CA (2000)

[Zem64]   Zeman, J.J.: The Graphical Logic of C. S. Peirce. PhD thesis, University of
Chicago, (1964), Available at: http://www.clas.ufl.edu/users/jzeman/

Revelator Game of Inquiry:
A Peircean Challenge for Conceptual Structures in Application and Evolution

Mary Keeler

Center for Advanced Research Technology in the Arts and Humanities
University of Washington, Seattle
mkeeler@u.washington.edu

Abstract. In unpublished manuscripts from Peirce's last decade, he emphasizes his dialogic and interactive view of logic-as-semeotic, exemplified by the Existential Graphs. Recently published research of these manuscripts solidly supports the project of creating a game for instituting his pragmatic methodology to demonstrate his full semeotic logic. Revelator is my conception of that game, to pursue Peirce's ideas for *improving the economy of inquiry*. Revelator's design somewhat resembles many well-known games, such as bridge, chess, crossword puzzles, and even poker, but its core purpose is to reveal complex relations among the conditional propositions, by which players represent their conjectures as plays in the game. The game design invites the application and evolution of Conceptual Structures technology to aggregate, integrate, and display the *complex logical behavior* of these propositions. Plays are treated as rule-defined agents that can adapt in complex conceptual environments to form multi-agents, promoting the emergence of collaboratively formulated and selected models of possible knowledge (or robust hypotheses). Peirce's full vision of a dynamic logic continues to challenge Conceptual Structures to become an engine of inquiry.

1 Introduction

Successful inquiry is a complex phenomenon, an experience that requires imagination in conjecturing, in devising ways of gathering and checking the evidence as exhaustively as possible, and in avoiding potential sources of error. Good inquirers are careful, skillful and persistent in collecting relevant evidence and discovering new evidence, with the intellectual honesty to avoid temptation to discount unfavorable evidence that threatens to undermine their conjectures. They need both rigorous reasoning, to predict the consequences of their conjectures, and good judgment, for assessing the significance of the evidence and guarding against the tendency of wishful thinking. Because of these demands, even scientific inquiry progresses in a "ragged and uneven way," as Susan Haack describes it, and yet still finds "new truths, better instruments, better vocabulary, etc., and ways to build on them; so that over the centuries the sciences have built a great edifice of well-warranted claims and theories (even though, to be

U. Priss, S. Polovina, and R. Hill (Eds.): ICCS 2007, LNAI 4604, pp. 443–459, 2007.
© Springer-Verlag Berlin Heidelberg 2007

sure, the trash-heap of discarded concepts and theories is larger by far)" [Haack (2003: 338; see 340-41].

Particularly in 21st century science, explains C. Dyke, the necessary increase of multi-disciplinary inquiry requires that "everyone is fully, self-consciously, aware of the 'rules and regulations' governing serious contribution, both in the home discipline and in the more extended one" [Dyke (1988): 3-4]. He invokes Warren Weaver's manifesto delineating the evolution of inquiry in science leading to its current imperative.

> [T]he science of the enlightenment taught us how to deal with organized simplicity. Nineteenth century science (Boltzmann, etc.) taught us to deal with disorganized complexity. The challenge for twentieth century science is to learn how to deal with organized complexity (without, I would add, pretending that it is simply conjunctive simplicity). ... Not only are the phenomena to be studied complex, but scientific practice itself is a phenomenon of organized complexity. The complexity of investigation must be studied along with the complexities investigated. The old positivist philosophy of science was a canon of simplicity, providing no room for a clear understanding of complexity. Insofar as working scientists (and social scientists) continue to understand their own activity in a positivist way (as many do), they will not find the space to meet Weaver's challenge. [Dyke: 5]

Although Dyke concurs that inquiry needs, in place of the positivist explanatory framework of simplicity and linearity, a new framework for dealing with organized complexity, he insists that it be capable of accommodating "the firm results obtained by the sciences of organized simplicity and disorganized complexity." These stand as foundations upon which to build a new integrated approach, he says, if we pursue strategies that: "(a) are consistent with and legitimated by our earlier successful practice; (b) take full advantage of the resources and the models we have at our disposal; (c) do not foreclose any legitimate options that we might want reopened at a later stage; and (d) do not leave us with a tangled mess of hypotheses incapable of being integrated or even compared" [11].

Ahti-Veikko Pietarinen finds plentiful evidence in Peirce's late manuscripts that his work on logic, semeotic [his preferred spelling (see *CP* 8.377, 1909)], and Existential Graphs (EGs) was pointed toward what in the 20th century became game theory and model theory, on the way to a new framework for scientific inquiry. Pietarinen maintains: "Understanding of Peirce's logic is only just evolving. This is mainly due to unavailability of published material from his last and very prolific epoch. ... I believe that the connections between, say, the emergence of existential assumptions in quantification, the reduction thesis concerning relational notions, the dialogical approaches to semantics, the tenet of constructivism, and the theory of modalities are all destined to find solid logical home in Peirce's overall semeiotic programme." [Pietarinen: 181]. He concludes, "Much more is to be expected from applying and eventually injecting Peirce's ideas into the modern theories of games and rational behaviour than is currently realised" [462]. Revelator is my attempt to apply Peirce's ideas in a game context toward creating a framework for 21st-century inquiry that can integrate successes of the previous frameworks [see Keeler (2000, 2003-2006); *CP* 7.328-335] (1873); *CP* 1.372 MS 909, 1887-88; MS 298 (c. 1905); MS 318 (c. 1907)].

2   Revelator: Game of Inquiry

Revelator is conceived as a game for improving Inquiry. We conduct inquiry when we confront a puzzling situation and attempt to resolve the puzzle by constructing hypotheses. Hypotheses formulate our conjectures about what we anticipate might solve the puzzling situation. Hypotheses are "educated guesses" that may become firmer when they improve our anticipation: when certain conditions are fulfilled as we guessed, to produce the consequence we expect. A hypothesis formulates an answer to the pragmatic question: Under what (specified) circumstances would my belief (about something) be true?

In playing Revelator, players make their conjectures explicit by formulating them in terms of conditional propositions that attempt to answer puzzling questions. For a very simple example, if you saw some unfamiliar animals, you might conjecture that they are birds and make the claim: "Those animals are birds, because they can fly." Any conjecture may serve as an explicit hypothesis, if it is formulated in a conditional proposition whose antecedent specifies a course of action to be performed and whose consequent describes certain consequences to be expected. The primary rule of this game of inquiry is that players must use the conditional form to relate claims and reasons supporting them in the explicit form of hypotheses; for example: "If I observe those animals flying, then they are birds." Notice that the claim and reason above can be easily reformulated into this "if ... then" form.

The Revelator Game of Inquiry is to be played among a group of inquirers, who confront a puzzling situation and want to construct hypotheses collaboratively. Although not played collaboratively, the US television game "Jeopardy" may begin to suggest the format of this game. Several Jeopardy contestants compete by formulating questions in response to answers displayed in a matrix of answers categorized under topics. Instead of Jeopardy's arbitrary format requiring questions from contestants to match answers posed in topical order, Revelator requires players to use the format of a conditional proposition, composed of a claim and a reason for that claim, as a legal play in the game. Plays in this logical form can serve as inferences to be related to and articulated with other players' claims and reasons contributed as conditional propositions in the progress of the game.

Through inquiry, we gain knowledge of which conjectures are justified to be considered candidate hypotheses. In expressing a conjecture as a claim, we assert a real possibility of an event we can imagine, which would be realized under certain describable conditions. Descriptions of these conditions are reasons that might justify our claims as contributions to the process of inquiry, by referring to evidence that can be checked to support the claims [see Keeler (2004); (2005)].

Peirce explains that although any claim that pretends to disclose a new fact without basing it on new evidence cannot possibly be correct; that observation cannot serve alone. "[F]or if it did the only active part which we should have to play in this method of inquiry would be simply the willingness to observe, and there would be no distinction of a wrong method and a right method of investigation." There must also be "*an elaborative process of thought by which the ideas given by observation produce others in the mind*" [my emphasis]. Furthermore, observations widely vary and are never exactly repeated or reproduced. Not only can no one make another's observations, or reproduce them; but no one can make at one time those observations which that same person makes at another time. "They belong to the particular situation of the observer, and the particular instant of time. ... Since, therefore, the likeness of these thoughts consists entirely in the result of comparison, and comparison is not observation, it follows that observations are not alike except so far as there is a possibility of some mental process besides observation" [*CP* 7.329-33 (c. 1873)].

Especially in collaborative inquiry, interpretations of evidence and the inferences relating them can quickly become complex. How can participants efficiently construct the evidentially soundest and inferentially most fruitful hypotheses from the countless possible conjectures asserted by all? How can participants "put their heads together" in collaboration, combining their individual "best guesses," to construct hypotheses that incorporate possibly all their conjectures? What logical augmentation tools might facilitate that aggregation and integration process? And what self-corrective habits in their interactions might participants cultivate by engaging in such a process? Revelator's purpose is to address these questions in the spirit of Peirce's "economy of research" [see *CP* 1.122 (1896), 7.158-61 (1902), 7.83 (1902), 7.219 (1901)]. The ultimate research question will be: Can Conceptual Structures technology augment the process of aggregating and integrating inferences to reveal collaborative hypotheses, to improve the process of inquiry?

3   The Economics of Inquiry

According to Peirce's theory of inquiry, while deduction can discover the hidden complexities of our concepts and induction is "the sole court of last resort in every case," only by abductive reasoning can inquirers originate a proposition. He maintains that although even careless abduction will eventually suggest a true hypothesis, "The whole service of logic to science ... is of the nature of an economy. ... it follows that the rules of scientific abduction ought to be based exclusively upon the economy of research" [*CP* 3.363 (1885); 4.581 (1906); 7.220 (1901); Keeler (2006); see Note 1; also see Tursman]. Instructed by Peirce's ideas, a pragmatic game for improving the economy of collaborative inquiry would induce players to find the most promising initial and strongest unifying claims, consider new and provocative evidence, foster the requisite technical skills (including those for effective expression). Broadly, its purpose would be to promote awareness of the patience, time, and persistence needed to add inference to inference for steady advancement and to encourage players to remain on the lookout for techniques to cope with these factors more effectively. Overall, the game must help investigators routinely self-correct — that is, to form habits that minimize error. Finding and reducing errors (both in interpreting evidence and in constructing inferences) is crucial in constructing good hypotheses for the economy of inquiry [see Weiner: 178; and *CP* 1.120 (1896), see Note 2].

Inquiry has many possible sources of error that cannot be completely captured by any proper logical model for finding the truth. The work of investigation is difficult and inquirers are fallible, sometimes because of prejudices or entrenched and unexamined commitments to poorly warranted background beliefs, such as in stereotypical thinking. Inquiry, unlike advocacy, is an attempt to discover the truth of some question, whatever that truth may be — but without expecting omniscience to reveal the complete truth. In advocacy, people negotiate about whose perspective should prevail. Advocates attempt to make a case for some opinion, by selecting and emphasizing whatever evidence favors that opinion and ignoring or playing down any that does not. Inquirers must do their best to discover some truth about the puzzles that concern them, regardless of whether that truth advances any personal interests. They must seek out and assess the worth of relevant evidence by a process in which they understand that their claims are fallible, revisable, and seldom impartial, because the social context of their work can affect even what questions are considered worthy of investigation and what solutions occur to them [see Haack (2003): 338-41; *CP* 1.43-49 (1896)].

When we as inquirers make our beliefs explicit in the conditional form of claims and reasons, we consciously distinguish between the possibility that something is in fact true and *how we think we know it is true*. The result of inquiry, accordingly, is not that our belief becomes true, but that we gain knowledge of how justified the belief is. In other words, inquiry requires us to distinguish between *identification* and *classification*: no two things are in fact identical, but we may be justified in classifying our representations of them as being related in some way that can be explicitly expressed. A particular hypothesis can be judged correct, then, to the extent that we have perceived and effectively represented a correspondence between a description of some consequence of our classification and the identified actual occurrence of that expected consequence. Our knowledge is built of these justified relations among our *representations of* described and classified *experiences* of what we call "facts" [see Keeler (2006): 319-20].

Whatever relations we claim to be among things we observe, and call "facts," are conditionally dependent on how we perceive and conceive them. Because hypotheses explicitly express this conditional dependency, when you as an inquirer assert a hypothesis you become responsible for its claims, as though you had placed a wager on it [see *CP* 5.543 (1903)]. If in the process of inquiry these claims are found to be correct, you win the wager; but just as with plays in the game of poker, the significance of those claims together with other justified claims, as knowledge, must wait to be *revealed* in the evolution of further inquiry. Since none of us is omniscient, the more other inquirers engage in contributing and evaluating claims and reasons, and building justified relations among them, the greater our chances of constructing strong hypotheses and reliable knowledge based on more experience. At the same time, the complexity of conceived claims and reasons represented in collaborative inquiry makes economy an even greater challenge.

4   Inquiry's Intricate Forms of Inference

Peirce explains three qualities, "Caution, Breadth, and Incomplexity," as the economic considerations in the intricate evaluation among hypotheses.

> In respect to caution, the game of twenty questions is instructive. ... The secret of the business lies in the caution which breaks a hypothesis up into its smallest logical components, and only risks one of them at a time. What a world of futile controversy and of confused experimentation might have been saved if this principle had guided investigations into the theory of light! Correlative to the quality of caution is that of breadth. For when we break the hypothesis into elementary parts, we may, and should, inquire how far the same explanation accounts for the same phenomenon when it appears in other subjects. [*CP* 7.220-21 (1901)]

He further explains how an incomplex and even rough hypothesis can be more robust and do what a more elaborate one would fail to do [see *CP* 7.222 (1901)]. And he often identifies incomplexity with the dialogic purpose of his EGs in "the central problem of logic, [which is] to say whether one given thought is truly, i.e., is adapted to be, a development of a given other or not" [*CP* 4.9 (1906)].

To avoid advocacy, inquiry should proceed only from claims that can be subjected to careful scrutiny of their reasons (as evidence), and inquirers should rely on a "multitude and variety" of many claims and reasons that can be *conceptually articulated*, rather than the apparent conclusiveness of any one claim. As Peirce explains, reasoning in inquiry should not form a "chain of inferences" (which is no stronger than its weakest link) but rather a *cable*, "whose fibers may be ever so slender, provided they are sufficiently numerous and intimately connected" [*CP* 5.3 (1902)]. The minutest details formulated as claims and reasons can collectively turn out to be crucial contributions in constructing strong arguments. Although this process of inquiry cannot be fully automated, technology can perform functions of representation, bookkeeping, and logical articulation that are tedious and error-prone for humans, to clarify and reveal hidden conceptual complexities.

To grasp or understand a concept is to have practical mastery of inferences in a network involving that concept—and *evolving* its application. Fully grasping complex inferential networks of conditional relations is a significant challenge for inquirers, especially in collaborative inquiry. Asserting a responsible claim requires understanding at least some of its consequences, and realizing what other claims it relates to and what other evidence relates to it. In a game of inquiry, players' develop research strategies in making plays that can justify other plays, can be justified by still other plays, and that close off or precludes still further plays. Players in the game need logical augmentation to help them develop the practical mastery of inferential articulation for this conceptual content [see Keeler (2004); and (2005) for a scenario of players].

5 Play of the Game

Scoring the plays in Revelator involves keeping track of each player's properly contributed conjectures (each asserted conditional proposition increases a player's score by one point). Strategy in the game involves learning to evaluate all contributions. A player must be able to keep track "upstream," to find what other claims may have implied or justified any claim in question, and also "downstream," to keep track of what else any claim in question implies or justifies as consequences. Overall, players must keep track of the interactions of claims and reasons, especially those that are inferentially or interpretationally incompatible, indicating that more investigation is needed.

Unlike many "normal-form" games identified by game theory, in which each player chooses a strategy once and for all, Revelator is an "extensive-form game," in which new strategies are developed *as more general claims and reasons*, calculated to incorporate or select other players' claims and reasons. Pietarinen explains that in the traditional theory of games (formulated in von Neumann & Morgenstern (1944), strategic interaction is static and that "the truly dynamic theory of games is still under intense development" [448]. He points out in Peirce's terms, "strategies are instructions that evaluate actions, and hence are species of thirdness. They indicate what the actions of a player or an agent ought to be in an inventive manner. In their capacity of providing functions that evaluate individual choices, they also provide a route by which one might hope to be able to understand how intelligence emerges, namely through the constant evaluation of individual action, and with the aid of the associated notions of learning and recognition of new concepts as implied by these actions" [442].

Since conditional propositions are the counters that increase a player's score, each counter must be linked to its player's identification, and appear in a collection of that player's conjectures in the game. Strategically, any conjecture is a player's agent, and should provide motivation for what else is or might be claimed. The play of the game reveals the possible "strategies" of "conjecture-agents" (that is, the logical consequences of their combined implications) among all "agents" (or plays) in the game. A form of "controlled English" can be used to accomplish the translation of the "if ... then" form of plays into formal logical expressions (See example: <www.ifi.unizh.ch/attempto>). In the operation of a real game of inquiry, relations among the plays would become complex. In the earlier example, "If those animals can fly, they might be birds," the reason "those animals can fly" would be articulated with other reasons related to claims that an animal is a bird, and also to any other claims that are justified with the capability to fly expressed in a reason.

In imposing constraints on the linguistic form of plays, Revelator is somewhat like the game of bridge, as Dyke analyzes that challenge: "to accept the lean vocabulary with its rigid constraints, and to shape and manage it so that it gains the capacity to do its limited job elegantly and precisely" [Dyke: 80]. Dyke compares playing bridge to a laboratory experiment in which experts carry out a dialogic, goal directed, and limited but intellectually complex activity [see Dyke: 74]. His concept of *information space* conceptualizes the constraints and limitations on the legitimate discourse of the bridge auction [Dyke: 83]. The matrix used to represent the calls in a bridge auction can be used to trace the path through the information space leading to a final contract, which makes the game seem like the perfect place to evaluate the rationality of paths [see Dyke: 89]. However, he explains, as in many cases of evolutionary ecology or the genetic code:

> [m]any possible pathways are adequate for particular hands, and particular pathways are adequate for many possible hands. (Brooks and Wiley [1986] remark that evolution is not the survival of the fittest, but the survival of the adequate.) No management of the limited information space is possible which univocally matches bidding matrices with hands. There are, however, ways of grouping hands and matrices to provide criteria for *reasonable* matchings. Were this not so, bidding skillfully and choosing a bidding system would be impossible. Determinism is absent here, so skill finds an essential role. [Dyke: 90]

Skilled inquirers evaluate each conjecture for: what it implies, what other conjectures are consistent with it, what others are inconsistent, and how it stands up to the evidence (that is, what consequences should follow from its truth, to what degree it is confirmed by any consequences that do follow, how it is false if the consequences do not follow). Whether we are investigative journalists, detectives, historians, house inspectors, dog breeders, theater set designers, or just making our way through life, we use such skill more or less explicitly. Formal inquiry is conducted to improve the skill of ordinary everyday inquiry, by overcoming our sensory and cognitive limitations and our fragility of commitment to finding out. Science has been remarkably successful because of the steady evolution in its enhancements of imagination aids, of sensory and reasoning capabilities, and of evidence-sharing and intellectual honesty, which are intricately related in the operation of its inquiry [see Haack (2003): 341]. Revelator is intended to reveal these multi-dimensional complexities.

6   Complexities of Inquiry in Operation

Haack uses the analogy of a crossword puzzle to represent the nonlinear character of inquiry, its "weaving of interconnected threads" making mutual support among conjectures possible, without vicious circularity [see Haack 1993, 2003]. Determining progress in a game of inquiry is more like determining the reasonableness of entries in a crossword puzzle with their pervasive mutual support, than like judging the soundness of an essentially one-directional mathematical proof.

Crossword clues are analogous to inquirers' reasons for believing based on experiential evidence, and any already filled-in entries are analogous to claims already established with some certainty. Although the clues don't depend on the entries, the entries are somewhat interdependent. Relations among clues and entries are also analogous to the *asymmetries* between experiential evidence and asserted claims that must be based upon that evidence. Confidence in the correctness of any entry in a crossword puzzle depends on: how much support the clue gives that entry, along with support from any intersecting entries that have already been filled in; confidence that those intersecting entries are correct, independent of the entry in question; and the extent to which intersecting entries have been filled in. Justifying an entry or a play in the game then must be partly *causal* (requiring evidential verifiability) and partly *evaluative* (requiring logical validity), and the crossword analogy illustrates how the "explanatory integration" of these two parts depends on how favorable, how secure, and how comprehensive any supporting evidence is [see Haack (1993): 81-82].

An especially successful play in the game of inquiry then would be like completing a long central entry in a crossword, making other entries significantly easier to fill-in: a substantial contribution to the *explanatory integration* of "a web of conjectures." At the same time, such a play also scores well with *experiential anchoring*: a conjecture is more justified the better it is anchored in experience and supported by other conjectures that are integrated components of an explanatory story and also anchored in experience. Such a "breakthrough" may even make further breakthroughs feasible, consolidating or generalizing over many dependent conjectures. Conversely, discovering a wrong crossword entry resembles what might be called a "breakdown" in the game of inquiry, when a key claim turns out to be confirmed invalid by all players.

Figuring out how reasonable our confidence in some crossword entry is, comes down to not only how well some entry is supported by others, but how well it is supported by its clue. Analogously, appraisal of how justified a particular conjecture is depends on both how justified are other conjectures that it depends on (how dependently supported it is), but also on how justified the reasons are for that conjecture (how independently secure it is). Justification for conjectures cannot be proclaimed categorically, but must be ascertained in degrees.

Furthermore, both degree of support and degree of independent security are not sufficient to determine the degree of justification. Eventually the appraisal reaches a point where the issue is not how well some conjecture is supported by others, but how well it is supported by experiential evidence. Devastating evidence, such as demonstrating that an initial, foundational conjecture was based on an illusory observation, can "wipe-out" an entire construct of conjectures. The comprehensiveness of the evidence for (or against) a conjecture must also be taken into account in determining its justification. This would include failures to take relevant evidence into account (including to look closely enough, to check from different angle, etc.). So a conjecture is more justified, the more supported and the more independently secure it is, and also the more comprehensively relevant evidence is taken into account. Distinguishing the error- and ignorance-related aspects of our fallibility, through explicit inquiry, reveals that they are pervasively interdependent and complex [see Haack 2003].

The crossword and other game analogies only begin to show the intricacies and complexities of formal inquiry. Without that formality and responsible conduct, our everyday careless inquiry often becomes what Haack [2003] calls "pseudo-inquiry." These are really forms of advocacy that are ubiquitous in academe, politics, and elsewhere, they include "sham reasoning" (when we make a case for the truth of beliefs to which we are already steadfastly committed) and "fake reasoning" (when we make a case for the truth of beliefs to which we are indifferent but believe will benefit us). Sham and fake reasoning show how inquiry can be perverted to give beliefs support and security, without comprehensive evidence for their justification. Such pursuits reduce knowledge to a sort of "map" for the "virtual territory" of limited purposes and advocate that *representation* as all there is to knowledge. Inquiry then becomes the sort of "game" in which we "mistake the map for the territory." In genuine inquiry, we understand the role of such "map-making" as the construction of coherent accounts or models to carry out exploratory, conceptual investigations.

7   Inquiry as a Complex Adaptive System

Players in the Revelator game would construct these model representations, by which to "prune, filter, and select" among all the contributed claims and reasons, toward formulation of collaboratively constructed hypotheses (or robust models). Its game format would serve as an effective method for inquiry in several ways that resemble the skills-building features of familiar games. First, the game would formalize the strategic process of inquiry, explicitly and sportively. Second, it would encourage collaborators to engage in the conceptual discipline of formulating model hypotheses. Third, it would induce responsible conduct among players by requiring explicit reasons for their claims, and to encourage competition within a stable pattern of cooperation [see Axelrod (1984)]. Revelator leaves the burden of constructing and checking intricate logical relations among contributed claims and reasons to automated conceptual processing, which would keep score and track individual contributions, identifying each with its originating player, to create an automatic credit system that promotes fair competition among inquiring players.

Since inquiry's purpose is to construct hypotheses that are reliable enough to serve as stable strategies in the evolution of further inquiry, within Revelator's game context competing claims and reasons could behave as players' agents in complex adaptive systems (*cas*) [see Holland 1995, 1998]. The building blocks for evolving the stable strategies in *cas* are interacting agents, described in terms of rules (expressed as "if ... then" statements). In *cas*, any agent must adapt to other adaptive agents as part of its adaptation to an environment, just as a player's contributed conjecture (expressed in "if ... then" form) must adapt to others contributed in the game. Agents adapt their behavior by changing their rules as experience accumulates; in the same way, hypotheses must change claims and reasons as evaluations and evidence accumulate.

Analogous to the children's game of building blocks, the game of inquiry has propositional "building blocks," with logical constraints rather than physical ones. These conditional-proposition agents (as "if ... then" rules) establish the "dimensions," in place of the dimensions of physical blocks. Geometrical and gravitational (forceful) constraints are replaced with inferential and evidential (factual). These conditionally-related building blocks must "behave" as complex systems adapting to a conceptual "environment," in which fallibility would serve as gravity does in physical systems, within the "dynamics" of conjectures. Players could explore future possibilities and continually bring the state of the model up to date as new claims are contributed, to improve the *faithfulness* of the model they construct. Revelator is *explicitly* a game of inquiry, so players remain aware that: "uncertainty lies in the model's *interpretation*, the mapping between the model and the world" [Holland (1998): 44-48].

At the beginning of Peirce's last decade, in a series of lectures at Harvard, he struggled to explain thought (or Thirdness) as an active factor in the real world, against the common assumption that the inviolable laws of dynamics determine all motion, and explain whatever happens in material universe, leaving no room for the influence of thought. He stressed that the laws of dynamics are different from such laws as gravitation and elasticity, and may even be precisely like logical principles: "They only say how bodies will move after you have said what the forces are. They permit any forces, and therefore any motions." Finally, he asked how anyone can be certain that we have sufficient knowledge of these laws to be reasonably confident that they are so absolutely eternal and immutable that they escape the "great law of evolution"?

> Each hereditary character is a law, but it is subject to development and to decay. Each habit of an individual is a law; but these laws are modified so easily by the operation of self-control, that it is one of the most patent of facts that ideals and thought generally have a very great influence on human conduct. That truth and justice are great powers in the world is no figure of speech, but a plain fact to which theories must accommodate themselves. [*CP* 1.348 (1903)]

In Pietarinen's view, these *easily modified* habits are evolutionary strategies that include: "rules, responses, guides, customs, dispositions, cognitive conceptions, generalisations, and institutions that have influenced [conduct] through evolutionary time." *Interpretation* is the evolutionary strategy by which Peirce "attempted to illustrate the emergence of experience as dialogical action between the inner and the outer, or the potential and the actual" [442, 191]. Without this *evaluative* function, complex adaptive systems cannot bridge the gap between rule-governed habits and truly inventive habits (between Secondness and Thirdness, in Peirce's terms). John Holland concludes: "we will not truly understand complex adaptive systems until we understand the emergent phenomena that attend them" [(1998): 242].

8   Holland's Explanatory Framework

Researchers in collaborative inquiry often jointly uncover possibilities unsuspected by any one participant, as do players in a game. And like regular players of a game, investigators begin to recognize certain kinds of patterns that become "building blocks" for longer-term, subtle strategies (something like "forks," "pins," and "discovered attacks" in chess). Holland identifies this "getting more out than you put in" as a ubiquitous "emergent" feature in the world around us: in rules of thumb for farming, ant colonies, networks of neurons, the immune system, the Internet, and in our understanding of the physical world, which has emerged from a small corpus of equations originated by Newton and Maxwell. Holland's work investigates the enigma of this feature: "how can the interactions of agents produce an aggregate entity that is more flexible and adaptive than its component agents" [(1998): 215, 248]?

Holland began in the 1970s to develop his "framework for understanding many important facets of learning in organisms and machines, ranging in complexity from conditioning in rats to scientific discovery." He collaborated with philosophers, cognitive scientists, and AI researchers in the attempt to integrate the ideas of several disciplines and construct a systematic approach to the study of induction: "all inferential processes that expand knowledge in the face of uncertainty." The basis for his framework was derived from his earlier *classifier systems* [1978]. "Classifier systems are a kind of rule-based system with general mechanisms for processing rules in parallel, for adaptive generation of new rules, and for testing the effectiveness of existing rules. These mechanisms make possible performance and learning without the 'brittleness' characteristic of most expert systems in AI." The resulting "pragmatic framework" denied the sufficiency of purely syntactic accounts of equivalence between inferences, and insisted that "sensible inferential rules take into account the kinds of things being reasoned about" [(1986): 1-6].

His more recent, simulation work demonstrates that a small number of rules or laws can generate systems of surprising complexity—but not just of random patterns. These "emergent systems" have recognizable features, a dynamic flux of patterns, and perpetual novelty. Emergent phenomena are recognizable and recurring, or *regular*, although not easily recognized or explained. If the origin of these regularities and their relations to one another can be understood, Holland thinks we might hope to comprehend emergent phenomena in complex systems. "The crucial step is to extract the regularities from incidental and irrelevant details" [(1998): 4]. Knowing that it took centuries of study to recognize the patterns of play in the game of chess, we should not expect to find the patterns of emergent systems simply by discovering underpinning laws of dynamics. Holland reminds us, however, that mathematical descriptions in a modeling process can help in discerning patterns and that a well-conceived model makes possible prediction and planning, to *reveal* new possibilities. Games and maps are historical antecedents of modeling, and computers make possible even more complex and dynamic models [see (1998): 28-52].

Holland's framework for the study of emergence from complexity specifies mechanisms and procedures for combining them. His use of "mechanism" extends beyond overtly mechanical to mean something like an elementary particle in physics for mediating interactions. Mechanisms provide a precise way of describing the elements (the agents, rules, and interactions) for defining complex systems, a common way (across disciplines) of describing the diverse rule-governed systems that exhibit emergence. In particular, mechanisms for recombination of elementary "building blocks" play a critical role. These interacting component mechanisms, called "constraint generating procedures" (*cgp*'s), have no central control, which increases the flexibility of their interactions, which then rapidly increases the possibilities for emergence [see Ibid (1998): 125-26].

Holland identifies the mechanisms and interactions necessary for advanced modeling of emergence in his model system, Echo, where complex multiagents can evolve from a single free agents, and then into specific aggregates of multiagents from single seed multiagents. Models can employ rules to allow a range of control (as in flight simulators), by which players can see and manipulate the mechanisms and interactions underlying the models, and use their intuition to explore plausible regimes. In simulators, models can reveal what amounts to the crossword "breakthroughs" and "wipeouts" that could, as Holland describes, "appear and reappear under a wide variety of assumptions," without committing players to real consequences [(1998): 141, 243].

Sometimes, in scientific inquiry, it is possible to follow the classic "hypothesize, test, and revise" pattern but, as Holland argues, real innovation requires more than incremental revision. In his framework, there are two major steps: "(a) discovery of relevant building blocks, and (b) construction of coherent, relevant combinations of those building blocks." He speculates that the selection mechanisms in this creative process "are akin to those of evolutionary selection, simply running on a much faster time-scale." He even conjectures that there could be a "game" with the rigor of a *cgp* that would permit insightful combinations of symbols as building blocks for creating models—as well as metaphors [(1998): 217, and see 202].

9   Tentative Conclusions and Future Challenges

This prologue to more careful examination of Holland's models of emergence indicates that the design goal of Revelator should be to enhance the *creative* process of inquiry (or abduction), even though this emergent phenomenon is still in "a shroud of conjecture," as Holland puts it. Players *create* rules in a game of Revelator, with each responsible and legal play. These *agent-rules* are the building blocks from which players must select and construct *generators* as "winning combinations," multiagents with dynamic (logical) trajectories [(1998): 129]. In a normal game, such as checkers, what counts as winning is pre-established in the pre-set game environment (checker board with checkers). In Revelator, as in any inquiry, the players *create* their game environment by the rules they contribute, and winning involves strategically selecting and combining those agent-rules to formulate multiagents that reveal adaptive, higher-order behavior hidden in the complexity of their conceptual environment. Another way of saying this is: players contribute and attempt to aggregate and integrate their selected rules (or agent-conjectures) as the mechanisms that might generate a model (or multiagent-hypothesis).

The selective exploration of different possible combinations is quite like finding the strategies in playing any other game. Like good play in checkers, sophisticated actions in complex adaptive systems depend on crediting *anticipation* and *stage setting* (or *pragmatic* actions) [see Holland (1998): 54]. In selecting rules (or conjectures) that combine as mechanisms to specify a model (or hypothesis), how could players (with limited capacity for tracing out complexities) manage to identify generators of higher-level organization, the "levers" that make "breakthroughs" possible (remembering the crossword analogy)? Under Holland's framework, the process would start from a complex pattern of related conjectures from which players may have no idea what might emerge. In their selection process, induction must "mediate the transition between the patterns of interest and the rules that attempt to model those patterns." Knowing what details to ignore is *not* a matter of derivation or deduction; it *is* a matter of experience and discipline, as in any artistic or creative endeavor. When this process goes well, the resulting description reveals repeated elements and symmetries that suggest rules or mechanisms [see (1998): 230].

Rather than viewing rules as a set of facts about the agent's environment, which must be kept consistent with one another by consistency checking, Holland views rules as hypotheses that undergo testing and confirmation. "On this view, the object is to provide contradictions rather than to avoid them ... [and] rules amount to alternative, competing hypotheses. When one hypothesis fails, competing rules are waiting in the wings to be tried" [(1998): 53]. His technique for resolving the competition is experience-based (closely related to the concept of building confirmation statistically): a rule's winning ability depends on its usefulness in the past. Each rule is assigned credit strength that over time comes to reflect the rule's usefulness to the system, changing the system's performance as it gains experience (for adaptation, by credit assignment). An agent-rule's value is then based on its interactions rather than on some predetermined fitness function [see (1998): 97]. The goal is the improvement of relations among rules, not some pre-determined optimality [see (1998): 216]. "What actions and interactions between these individual agents produced an organized aggregate that persisted? What were the adaptive mechanisms that favored the emergence of this aggregate?" [(1995): 97]. Furthermore, "Only persistent patterns will have directly traceable influence on future configurations in generated systems. The rules of the system, of course, assure causal relations among all configurations that occur, but the persistent patterns are the only ones that lend themselves to a consistent observable ontogeny" [(1998): 225].

Holland's pragmatic approach encourages Pietarinen's hope that Peirce's final efforts might eventually be rewarded in a general framework for his rudimentary forms of strategic interaction, the EGs. Pietarinen concludes that while the CGs system of knowledge representation is "foundationally rich," it fails to be genuinely dynamic and interactive: "Instead, CGs throw light on what goes on in the one-sided case of a single bearer of a sentence, or in the monologic comprehension of discourse" [104]. He stresses that we will not realize the value of such graphical systems until we can make their "dynamic and dialogical character revealed in the apparatus of extensive games" [171]. If we are to understand how Peirce's EGs are a method that can "break to pieces all the really serious barriers ... to the logical analysis of thought," and really accomplish the rendering of the operation of thinking as "moving pictures of thought," we must first appreciate that "thinking always proceeds in the form of a dialogue ... essentially composed of signs, as its matter, in the sense in which a game of chess has the chessmen for its matter" [*CP* 4.6 (1898); and see Sowa (2005): 61-67].

Can CGs, together with (the more interactive) Formal Concept Analysis (FCA) evolve to meet this application challenge [see Sowa (2000) and Gerhing (2006)]? Taking the physical building-block analogy further, could we eventually have "GIS" and "GPS" technology for virtual exploration of the conduct of inquiry in a "semeotic game terrain?" Such virtual terrain with "global scope" could provide for the continuity of inquiry, as Peirce foresaw it: "there is no real reason why there must be a limit to the size of our hypotheses ... to maintain a single proposition tentatively should be no easier than to maintain a consistent set" [in Feibleman: 334; *CP* 6.277 (c. 1893)]. Rather than becoming merely "tools" in "the researcher's digital toolkit" [see especially Shum, et al., in Kirschner: 186], can Conceptual Structures technology become an *engine* for Revelator as a pragmatic methodological framework for continuing to improve its applications in their evolution [see Keeler (2006)]?

10   Notes

[1] An abductive argument has a relation of similarity between the facts stated in the premises and the facts stated in the conclusion, without compelling one to accept the truth of the conclusion when the premisses are true. Peirce goes on to say that the facts in the premises of an abductive argument constitute an icon of the facts in the conclusion, asserted positively and admitted with suitable inclination. It is in this sense that abduction starts a new idea; in Peirce's words, it is "originary."

Deduction is, in Peirce's words, "an argument representing facts in the Premiss, such that when we come to represent them in a Diagram we find ourselves compelled to represent the fact stated in the Conclusion." The notion of index arises here, in that "the Conclusion is drawn in acknowledgment that the facts stated in the Premiss constitutes an Index of the fact which it is thus compelled to acknowledge." It is in this sense that deduction is demonstrative reasoning, "obsistent" and "compulsive" in Peirce's terms.

Induction is an argument starting from a hypothesis that is a result of abduction, interspersed with results of possible experiments deduced from hypotheses and selected independently of any epistemic access to its truth value. Peirce called them "virtual predictions." The hypothesis is concluded "in the measure in which those predictions are verified, this conclusion, however, being held subject to probable modification to suit future experiments." The relation between the facts stated in the premises and the facts stated in the conclusion of inductive arguments is symbolic, as "the significance of the facts stated in the premises depends upon their predictive character, which they could not have had if the conclusion had not been hypothetically entertained." In Peirce's terminology, inductive arguments are "transuasive" in the assurance of the amplification of positive knowledge [*CP* 2.96; Pietarinen: 26-27].

[2] The best hypothesis, in the sense of the one most recommending itself to the inquirer, is the one which can be the most readily refuted if it is false. This far outweighs the trifling merit of being likely. For after all, what is a likely hypothesis? It is one which falls in with our preconceived ideas. But these may be wrong. Their errors are just what the scientific man is out gunning for more particularly. But if a hypothesis can quickly and easily be cleared away so as to go toward leaving the field free for the main struggle, this is an immense advantage. [*CP* 1.120]

References

General Note: For all *CP* references, Collected Papers of Charles Sanders Peirce, 8 vols., ed. Arthur W. Burks, Charles Hartshorne, and Paul Weiss (Cambridge: Harvard University Press, 1931-58).

Axelrod, R.: The Evolution of Cooperation. Basic Books, New York (1984)
Dyke, C.: Evolutionary Dynamics of Complex Systems: A Study in Biosocial Complexity. Oxford University Press, Oxford (1988)
Feibleman, J.K.: On the Future of Some of Peirces Ideas. In: Wiener, P.P., Young, F.H. (eds.) Studies in the Philosophy of Charles S. Peirce. Harvard University Press, Cambridge, MA (1952)
Gehring, P., Wille, R.: Semantology: Basic Methods for Knowledge Representation. In: Schärfe, H., Hitzler, P., Øhrstrøm, P. (eds.) ICCS 2006. LNCS (LNAI), vol. 4068, pp. 215–228. Springer, Heidelberg (2006)
Haack, S.: Evidence and Inquiry: Towards Reconstruction in Epistemology. Blackwell, Oxford (1993)
Haack, S.: Defending Science: within Reason. Prometheus Books (2003)
Holland, J.H., Holyoak, K.J., Nisbett, R.E., Thagard, P.R.: Induction: Processes of Inference, Learning, and Discovery. MIT Press, Cambridge (1986)
Holland, J.H., Holyoak, K.J., Nisbett, R.E., Thagard, P.R.: Adaptation in Natural and Artificial Systems. The MIT Press, Cambridge (1992)
Holland, J.H., Holyoak, K.J., Nisbett, R.E., Thagard, P.R.: Emergence: from Chaos to Order. Basic Books, New York (1998)
Holland, J.H., Holyoak, K.J., Nisbett, R.E., Thagard, P.R.: Hidden Order: How Adaptation Builds Complexity. Basic Books, New York (1995)
Hovy, E.: Methodology for the Reliable Construction of Ontological Knowledge. In: Dau, F., Mugnier, M.-L., Stumme, G. (eds.) ICCS 2005. LNCS (LNAI), vol. 3596, pp. 91–106. Springer, Heidelberg (2005)
Keeler, M.: Pragmatically Yours. In: Amin, A., Pudil, P., Ferri, F.J., Iñesta, J.M. (eds.) SPR 2000 and SSPR 2000. LNCS, vol. 1876, pp. 82–99. Springer, Heidelberg (2000)
Keeler, M.: Hegel in a Strange Costume: Reconsidering Normative Science in Conceptual Structures Research. In: Ganter, B., de Moor, A., Lex, W. (eds.) ICCS 2003. LNCS, vol. 2746, pp. 37–53. Springer, Heidelberg (2003)
Keeler, M.: Using Brandoms Framework to Do Peirces Normative Science. In: Wolff, K.E., Pfeiffer, H.D., Delugach, H.S. (eds.) ICCS 2004. LNCS (LNAI), vol. 3127, Springer, Heidelberg (2004)

Keeler, M., Pfeiffer, H.D.: Games of Inquiry for Collaborative Concept Structuring. In: Dau, F., Mugnier, M.-L., Stumme, G. (eds.) ICCS 2005. LNCS (LNAI), vol. 3596, pp. 396–410. Springer, Heidelberg (2005)

Keeler, M., Pfeiffer, H.D: Building a Pragmatic Methodology for KR Tool Research and Development. In: Schärfe, H., Hitzler, P., Øhrstrøm, P. (eds.) ICCS 2006. LNCS (LNAI), vol. 4068, Springer, Heidelberg (2006)

Kirschner, P.A., Buckingham-Shum, S.J., Carr, C.S. (eds.): Visualizing Argumentation. Springer, Heidelberg (2003)

de Moor, A.: Improving the Testbed Development Process in Collaboratories. In: Wolff, K.E., Pfeiffer, H.D., Delugach, H.S. (eds.) ICCS 2004. LNCS (LNAI), vol. 3127, pp. 261–273. Springer, Heidelberg (2004)

Pietarinen, A.-V.: Signs of Logic: Peircean Themes on the Philosophy of Language, Games, and Communication. Springer, Heidelberg (2006)

Sowa, J.: Knowledge Representation: Logical, Philosophical, and Computational Foundations. Brooks/Cole Publishing Co. Pacific Grove, CA (2000)

Sowa, J.: Crystallizing Theories Out of Knowledge Soup. In: Ras, Z.W., Zemankova, M. (eds.) Intelligent Systems: State of the Art and Future Directions, pp. 456–487. Ellis Horwood Ltd, London (1990), http://www.jfsowa.com/pubs/challenge

Sowa, J.: Peirces Contributions to the 21st Century. In: Schärfe, H., Hitzler, P., Øhrstrøm, P. (eds.) ICCS 2006. LNCS (LNAI), vol. 4068, Springer, Heidelberg (2006)

Tursman, R.: Peirces Theory of Scientific Discovery: A System of Logic Conceived as Semiotic. Indiana University Press (1987)

Helping System Users to Be Smarter by Representing Logic in Transaction Frame Diagrams

David Cox[1] and Simon Polovina[2]

[1] www.flipp-explainers.org
djcox@fuse.net
[2] Communication & Computing Research Centre
Faculty of Arts, Computing, Engineering & Sciences
Sheffield Hallam University, UK S1 1WB
s.polovina@shu.ac.uk

Abstract. We identify a lucid way of conveying complex information to users in a highly visual, easy to follow form. As explanation, we describe several ideas about system user instructions. Several key ideas are clarified using diagrams. A direction for exploration is offered, with the view that ICCS conferees will be aware of a simple, diagrammatic way to explain use of systems dealing with very complex real world problems.

1 Introduction

The connection between diagrams and logical reasoning is well-established [2]. User instructions for new systems in health care, science, education, and government, for example, become unclear when complex choices – like complex traffic intersections -- appear in single-path text pages rather than in roadmap, diagram form.

Patterns and rules. In contrast, instructions for using complex systems often have proven clear when presented in two parts: first, as landscape or architectural views of intersection *patterns* like baseball diamonds, soccer playing fields, or chess boards -- with, second, ultra-simple *rules* on moving through the patterns on the playing field or game board [1]. People are able to deal with extraordinarily many different paths to reach football goal lines, soccer nets, and baseball home plates. But, in text form, these myriad lines of possibility are 100% invisible behind the single-line disguise text always insists on wearing.

System by game. Driving a car is a familiar example of "system by game" [1]. The game board is the streets and highways on which the car is driven by its driver. The *patterns* are formed by the painted lines which define traffic lanes. Sometimes lane patterns split like logical *ors*; sometimes they merge at intersections like logical *ands*. The driver's goal is the destination of the trip. The *rules* are the traffic laws. The driver is truly reacting to patterns on a game board while applying the rules of the particular driving game being played. Text can *describe* patterns but can't *model* them. Diagrams can do both.

U. Priss, S. Polovina, and R. Hill (Eds.): ICCS 2007, LNAI 4604, pp. 460–463, 2007.

2   A Simple Game Board Diagram Example

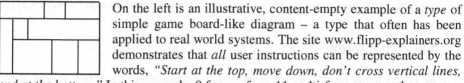

On the left is an illustrative, content-empty example of a *type* of simple game board-like diagram – a type that often has been applied to real world systems. The site www.flipp-explainers.org demonstrates that *all* user instructions can be represented by the words, *"Start at the top, move down, don't cross vertical lines, end at the bottom."* In this example, 9 frames form 11 multi-frame user pathways.

Diagrams of this type can handle vast complexity. Figure 1 highlights an example of the MS-DOS operating system's DIR command that shows 9,700,000 process variations. (At www.jfsowa.com/figs/flippdir.gif it is reproduced online; we acknowledge the support given to us by John F. Sowa in general with our work [7].)

Fig. 1. DOS DIR command; not as straightforward as you might think...

Some actual system applications have used many more scenarios. A new manufacturing plant's 26 departmental system diagrams together represented one coherent system at startup. The number of diagrammed scenarios was – probably well into the billions [1]. Even relatively simple applications have been seen by users as complex. An actual case study involving a tax calculation system with only eight scenarios had nonetheless been considered even by teachers to be quite complex [1].

3   Some Suggestions

Whilst [1] explains how to develop these panoramas of all scenarios in any given system, the following suggestions are worth noting:

Create many expectations; deliver on all of them. Build confident expectations for system users. A way to do this is by diagramming for users _all_ scenario patterns in any system on interest, not just those that answer specific situations. This philosophy has proven very helpful in quick teaming of people who did not know each other. Hundreds of temporary creative problem solving teams welcomed being given abundant clear expectations about, for example, the team processes and tools before they worked together [1].

Represent user logic without language, symbols, or formulas. They _create_ complexity. While the simple framework diagrams described above clarify logic relationships, the information _inside_ the frameworks – the _content_ -- can be in any form, any languages, any symbols, any formulas, any logic, any images, etc. Luckily, a given diagram holds content correctly even in different languages and forms. Logic is _form and connection_, not language, not symbols, not content [1].

Use diagram types that are both logical and convenient. Whilst Flow chart diagrams are perhaps the most popular means of describing complex processes, they suffer in that, among other things, they:

- don't show flow direction. Top-down is not standard, for example.
- don't have rules as to where entry and exit points are located (top, sides).
- don't reveal what, if anything, may be flowing along connector lines.
- don't always display user paths.
- don't put full information in boxes – often only one-word labels.
- don't use direct-connected frames.

Avoid throwing user instructions 'over the wall' to whom they may concern. Avoid one-way, truncated instructions. Design instructions so every available scenario is obvious to users. Instructions that work can create confidence.

See instruction frames as describing two-way transactions. The idea of frames as transactions was prompted by work by Hill [3] and Polovina [4]. Individual frames can be understood as ideally containing two-way transactions between a system and its users. Frames, as used here, seem to have no parallel in language. Since frames are not word- or sentence-limited, they are not like _phrases_ in language, music, and art. Empty frames, like intersections, have no language counterpart. Frames can hold any mix of sentences, phrases, formulas, symbols, different languages, music, images, etc. While each frame prompts user action, a text sentence may carry no such implication. This means text's capacity for transaction territory-marking is about zero – or even negative when arousing user ire.

Attach 'local' definitions of concepts to system user instructions. Such definitions could include symbols and terms like: *logic, system, content, scenario, system user* (who might be system designers; system architects; the client who pays to have the system designed and maintained; the system inheritors; the system overseers; the system accountants and bankers; the system managers the trainers of system administrators – sometimes even the public. The opportunity for wrong assumptions and confusion is real. Definitions demystify.

Emphasize user logic; soft-pedal system logic. Whilst describing to users what a system is doing during its operation is common, it is often irrelevant. Users want just whatever logic controls their success. Note that what is sequencing through user logic structures is not information but rather the attention of the user. Meaning can arise from seeing closely related alternative scenarios that may work better. Users prefer the panorama of all scenario paths experts follow. Unfortunately, this contrasts with -- for just a few examples -- Microsoft's Word 2003, Norton's 2007 Internet Security, Adobe's Reader 7, and Google's Desktop – none of which display any user scenarios at all, let alone any panoramas.

4 Concluding Remarks

Given these experiences we suggest this direction for exploration:

What seems unavailable and urgently valuable is a computer capability with which almost anyone can create and conveniently revise diagrams where contiguous logic scenario panorama structure is retained automatically. This might be a program for self-adjusting diagrams as simple as children's hopscotch game diagrams with automatic logic-rediagramming. Basing it on producing transaction frame FLIPP Explainer diagrams is one obvious approach [1].

We accept that in this short paper we cannot properly convey the potential benefits, other than highlighting some of the pertinent issues. We are nonetheless of the view that our approach will provide users with a simple framework to tackle hitherto complex real world problems. Its further exploration by a wider community would therefore be well rewarded.

References

1. Cox, D.: Explanation by Pattern Means Massive Simplification (an E-book), http://www.flipp-explainers.org
2. Dau, F.: The Logic System of Concept Graphs with Negation: And Its Relationship to Predicate Logic. In: Dau, F. (ed.) The Logic System of Concept Graphs with Negation. LNCS (LNAI), vol. 2892, Springer, Heidelberg (2003)
3. Hill, R.: A Requirements Elicitation Framework for Agent-Oriented Software Engineering – Doctoral dissertation. Sheffield Hallam University (2007)
4. Polovina, S., Hill, R.: Transactions Framework for Effective Enterprise Management. In: ICCS 2007 Workshop Proceedings. Springer, Heidelberg (2007)
5. Text vs. patterns demonstration: http://www.flipp-explainers.org/demonstration.htm
6. Case study application: http://flipp-explainers.org/casestudy1.htm
7. Sowa, J.F.: Knowledge Representation: Logical, Philosophical, and Computational Foundations. Brooks Cole Publishing (2000)

Quo Vadis, CS? – On the (non)-Impact of Conceptual Structures on the Semantic Web
(Position Paper)

Sebastian Rudolph, Markus Krötzsch, and Pascal Hitzler

Institute AIFB, Universität Karlsruhe (TH), Germany
{rudolph,kroetzsch,hitzler}@aifb.uni-karlsruhe.de

Abstract. *Conceptual Structures* is a field of research which shares abstract concepts and interests with recent work on knowledge representation for the Semantic Web. However, while the latter is an area of research and development which is rapidly expanding in recent years, the former fails to participate in these developments on a large scale. In this paper, we attempt to stimulate the Conceptual Structures community to catch the Semantic Web train.

1 Status Quo

It is a fact that mainstream Semantic Web (SW) developments currently happen with only little impact from the Conceptual Structures (CS) community. This is a curious development as Semantic Web knowledge representation is closely related to CS research. While CS certainly profits from the SW hype – as do many areas in computer science – we believe that the full potential for the transfer and use of CS methods and technologies is not given enough credit by the community.

With this somewhat provocative position paper, we intend to stimulate a controversial discussion about the possible future of Conceptual Structures as the changed situation might require to reconsider former pessimistic attitudes towards the Semantic Web as depicted in [1]. We first give a very brief history of Conceptual Structures and Semantic Web. We then analyse the two main paradigms for the CS community, namely Conceptual Graphs (CGs) and Formal Concept Analysis (FCA) in terms of their relationship to SW research. For each of the paradigms, we will give general research directions which we think would help to leverage CS for the SW.

We will substantiate some of our statements and claims by literature references, but there is too much work which is important for this position paper to explicitly refer to it all. The resulting selection is obviously very subjective.

Conceptual Structures is a term introduced by John Sowa in his 1984 book on the topic [2]. His work stimulated an interdisciplinary research community with interests in the relations between artificial intelligent knowledge representation, mathematical logic, philosophy and linguistics, manifesting itself in the annual International Conference on Conceptual Structures[1] which runs since 1993 after some workshops in the years before.

---

[1] See http://www.conceptualstructures.org.

U. Priss, S. Polovina, and R. Hill (Eds.): ICCS 2007, LNAI 4604, pp. 464–467, 2007.

The field basically comprises two interacting research communities, one of which focuses on conceptual graphs as introduced in [2], while the other comes from a tradition spawned by Rudolf Wille's work in 1982 on restructuring lattice theory [3], which led to the establishment of Formal Concept Analysis [4] as a mathematical theory and which recently finds applications in Computer Science, especially in data mining [5].

While the CS community mainly focusses on the above-mentioned areas, it also has strong mathematical and philosophical undercurrents, and also ever since has been involved in Computer Science applications, witnessed by a considerable number of implemented systems and application studies.

Semantic Web, in contrast, is a relatively new research area spawned by Tim Berners-Lee, inventor of the World Wide Web and director of the World Wide Web Consortium (W3C) [6]. Its main idea is to bring meaning to web data for intelligent processing. This is achieved by utilising knowledge representation languages for describing so-called *ontologies*, which model domains of interest in a logic-based, declarative and machine-processable way. Ontology representation languages which have been standardised by the W3C are the Resource Description Framework RDF(S)[2] and the Web Ontology Language OWL[3], both of which have concept hierarchies as their basic internal structure. While RDFS is a straightforward language which features only a simple use of inheritance for inferencing, OWL is a full-blown Description Logic (DL) [7], and as such a powerful knowledge representation language.

Semantic Web research in the last few years was driven by the W3C and by influential funding agencies such as DARPA and the European Commission. Specialised annual conferences, like the International Semantic Web Conference feature several hundred participants each year, and Semantic Web publications are present in all major conferences and journals in Artificial Intelligence, Knowledge Management, and other fields. Influential software companies like IBM, Oracle and SAP are currently starting to enter the market with products based on the underlying Semantic Technologies.

2   Quo Vadis, CG?

Both being historically founded in semantic networks, CGs and DLs share a conceptually very similar view on how to represent knowledge, as reported by Tim Berners-Lee.[4] It is thus surprising that the participation of the CG community in mainstream Semantic Web research is very limited.

While CGs have been used in some Semantic Web applications (see e.g. the Corese Semantic Web Factory[5] or [8]), such work is rarely presented at mainstream Semantic Web events and has had next to no impact on standardisation efforts. So, while CGs are the more historic approach, DLs overtook and got standardised. Considering the high impact in research and development of SW methods and technologies, it is promising to utilise the close conceptual relationship between CGs and SW languages, and to utilise

---

[2] http://www.w3.org/RDF/
[3] http://www.w3.org/2004/OWL/
[4] http://www.w3.org/DesignIssues/CG.html
[5] http://www-sop.inria.fr/acacia/corese/

CGs for the Semantic Web. In detail, we identify the following issues which seem to be particularly interesting for being taken up immediately.

- Thoroughly investigate the relationships between CGs and standardised or widely used ontology languages in order to mediate a knowledge transfer from CGs to SW languages and technologies [9].
- Employ CG technology for the visualisation of inferencing in order to explain to the naive user how implicit knowledge is being derived.
- Leverage CG technology for building visual ontology user interfaces that can be used by non-experts.

3   Quo Vadis, FCA?

FCA is very limited as a knowledge representation formalism, as it is basically restricted to concept hierarchies, with some minimal logical flavour [10,11]. Taxonomies and hierarchies, however, are fundamental to Semantic Web knowledge representation, as witnessed e.g. by the RDF(S) standard and also the fact that practical ontology modelling is usually done by initially creating concept hierarchies. Because of this, FCA can be utilised as a data mining tool for creating drafts for basic ontologies by automated means, which can subsequently be extended – see e.g. [12] or [13]. As such, FCA has indeed been established in the Semantic Web to a certain extent.

But the natural question arises, how the impact of FCA can be further pushed in order to become a more prominent basic technology for the Semantic Web. We see the following promising possible lines of development.

- Investigate expressive knowledge representation formalisms such as DLs and their extensions in their relation to FCA [14,15] in order to leverage FCA for more expressive formalisms by overcoming the fixation on concept hierarchies.
- Further investigate FCA approaches towards ontology creation and refinement [16].
- Investigate the use of FCA for user interaction with ontologies [17].

4   Quod Differtur, Non Aufertur

The CS community can draw on a rich history and well-developed methods on its path into the future. We believe that it has the potential to leverage substantial impact on the current Semantic Web trend. In order to do this, however, efforts are needed along the following lines.

- CS paradigms need to be studied in depth in their comparison with SW knowledge representation paradigms.
- CS needs to reflect about its own strengths in relation to concrete needs in Semantic Web research.
- Concerted efforts have to be undertaken to disseminate CS methods and established knowledge in the SW community.

Finally, it must not be forgotten that the Semantic Web – an extension of the current Web – also is an applied area of research, where well-specified file formats, stable software tools, standardised data types, Unicode and URIs are as important as well-founded representation formalisms.

Acknowledgements. This work is supported by the German Federal Ministry of Education and Research (BMBF) under the SmartWeb project (grant 01 IMD01 B), and by the Deutsche Forschungsgemeinschaft (DFG) under the ReaSem project.

References

1. Priss, U.: Alternatives to the "semantic web": multi strategy knowledge representation. In: Proc. ISKO 2002 (2002)
2. Sowa, J.(ed.): Conceptual Structures: Information Processing in Mind and Machine. Addison-Wesley, Reading, MA (1984)
3. Wille, R.: Restructuring lattice theory: An approach based on hierarchies of concepts. In: Rival, I. (ed.) Ordered Sets, Reidel, Dordrecht-Boston, pp. 445–470 (1982)
4. Ganter, B., Wille, R.: Formal Concept Analysis – Mathematical Foundations. Springer, Heidelberg (1999)
5. Stumme, G.: Formal concept analysis on its way from mathematics to computer science. In: Priss, U., Corbett, D.R., Angelova, G. (eds.) ICCS 2002. LNCS (LNAI), vol. 2393, pp. 2–19. Springer, Heidelberg (2002)
6. Berners-Lee, T., Hendler, J., Lassila, O.: The semantic web. Scientific American (2001)
7. Baader, F., Calvanese, D., McGuinness, D., Nardi, D., Patel-Schneider, P. (eds.): The Description Logic Handbook: Theory, Implementation, and Applications. Cambridge University Press, Cambridge (2003)
8. Dieng-Kuntz, R., Corby, O.: Conceptual graphs for Semantic Web applications. In: Dau, F., Mugnier, M.-L., Stumme, G. (eds.) ICCS 2005. LNCS (LNAI), vol. 3596, pp. 19–50. Springer, Heidelberg (2005)
9. Baader, F., Molitor, R., Tobies, S.: Tractable and decidable fragments of conceptual graphs. In: Tepfenhart, W.M. (ed.) ICCS 1999. LNCS, vol. 1640, pp. 480–493. Springer, Heidelberg (1999)
10. Ganter, B., Wille, R.: Contextual attribute logic. In: Tepfenhart, W.M. (ed.) ICCS 1999. LNCS, vol. 1640, pp. 377–388. Springer, Heidelberg (1999)
11. Hitzler, P., Krötzsch, M., Zhang, G.Q.: A categorical view on algebraic lattices in formal concept analysis. Fundamenta Informaticae 74, 301–328 (2006)
12. Cimiano, P., Stumme, G., Hotho, A., Tane, J.: Conceptual knowledge processing with formal concept analysis and ontologies. In: Eklund, P.W. (ed.) ICFCA 2004. LNCS (LNAI), vol. 2961, pp. 189–207. Springer, Heidelberg (2004)
13. Stumme, G., Hotho, A., Berendt, B.: Semantic Web mining: State of the art and future directions. Journal on Web Semantics 4, 124–143 (2006)
14. Hitzler, P., Krötzsch, M.: Querying formal contexts with answer set programs. In: Schärfe, H., Hitzler, P., Øhrstrøm, P. (eds.) ICCS 2006. LNCS (LNAI), vol. 4068, pp. 413–426. Springer, Heidelberg (2006)
15. Rudolph, S.: A deduction calculus for cumulated clauses on $\mathcal{FLE}$ concept descriptions. In: Schärfe, H., Hitzler, P., Øhrstrøm, P. (eds.) ICCS 2006. LNCS (LNAI), vol. 4068, pp. 188–201. Springer, Heidelberg (2006)
16. Rudolph, S.: Exploring relational structures via $\mathcal{FLE}$. In: Wolff, K.E., Pfeiffer, H.D., Delugach, H.S. (eds.) ICCS 2004. LNCS (LNAI), vol. 3127, pp. 196–212. Springer, Heidelberg (2004)
17. Tane, J., Cimiano, P., Hitzler, P.: Query-based multicontexts for knowledge base browsing: an evaluation. In: Schärfe, H., Hitzler, P., Øhrstrøm, P. (eds.) ICCS 2006. LNCS (LNAI), vol. 4068, Springer, Heidelberg (2006)

A Framework for Analyzing and Testing Overlapping Requirements with Actors in Conceptual Graphs

Bryan J. Smith

University of Alabama in Huntsville/Georgia Tech Research Institute
Huntsville, AL 35899
smithbj@email.uah.edu

Abstract. Requirements inconsistencies, caught early in a software lifecycle, prevents unnecessary work later in that lifecycle. Testing requirements for consistency early and automatically is a key to catching errors. This paper will share an experience with a mature software project that involved translating software requirements with overlapping definitions into a conceptual graph and recommends the use of several new actors to help automatically test a requirements consistency graph.

1 Introduction

Conceptual graphs provide a visual representation of the relationships between details of requirements that a text document does not. Some conceptual graph tools offer the visual representation of concepts that a human can inter from. With data driven software, we use certain concepts' referents, along with actors, to show consistencies within a requirement and then within the entire graph. Conceptual graphs have been shown to be helpful in modeling requirements in the past [1]. This work takes a medium scale mature project, and while graphing this project, a way was found to evaluate the consistency of certain parts of the requirements.

2 General Approach

This work is a complement to another paper, [2]. This paper addresses requirements that are defined by data sets that overlap, while the previous paper addresses requirements that are defined by data sets that are disjoint. The requirement that we will detail concerns communication and is as follows:

> *There shall be one modem per system. There shall be at least one and up to twenty different configurations of that one modem. Each configuration must have one and only one Modem Mode. Each Modem Mode must have one and only one Data Rate.*

We know that there are six different Modem Modes. Valid values for the Modem Mode are Mode 1 through Mode 6, inclusive. Furthermore, there are nine different

U. Priss, S. Polovina, and R. Hill (Eds.): ICCS 2007, LNAI 4604, pp. 468–471, 2007.
© Springer-Verlag Berlin Heidelberg 2007

Data Rates which are associated with the Modem Mode. Valid values for the Data Rate are Rate 1 through Data 9. Additionally, each Mode allows only certain Rates:

- Mode 1 has valid Rates of 2, 4 and 6.
- Mode 2 has valid Rates of 2, 4 and 6.
- Mode 3 has valid Rates of 1, 2, 3, 4, 5 and 6.
- Mode 4 has valid Rates of 7.
- Mode 5 has valid Rates of 2, 4 and 6.
- Mode 6 has valid Rates of 8 and 9.

For conciseness, we modify the consistent requirement to an inconsistent requirement by changing the definition of Modem Mode 3 to:

- Modem Mode 3 has valid Data Rates of 1, 2, 3, 4 and 6.

To check for consistency within the facts asserted in this conceptual graph, a graph needs to be created. This upcoming graph uses actors and text files (called "databases" in CharGer[3]) to check for consistency within the graph. Before we discuss the test graph of this requirement, we will summarize the <counter> and <2key_lookup> actors. These actors work together to search for all instances of a key inside of a database. <Counter> takes as input an interval, which determines how many seconds to wait before updating its output, and a Boolean, which determines whether to reset once the interval times has elapsed. If the reset Boolean is true, then the next output will reset to 1, otherwise the output will increment by one. <2key_lookup> takes three inputs. The first input is the name of a tab-delimited text file. The second and third inputs are column values found within the tab-delimited text file. When two column values match on the same row within the tab-delimited text file, the output is the third column value in that same row. These actors are vital to this test because these database files are what hold the data constraints, and we need to find every relationship between the independent and dependent data.

Here, different ModemModes can have the same or different overlapping Data-Rates. Due to this, we increment the first count *as soon as* we find a consistency. In testing this current requirement, we are checking for a consistency, and if we find one, we mark it as consistent and continue to the next controlling concept, satisfied that the iteration that we were on was consistent.

The graph represented in Fig. 1 begins by looking up each of the possible Modes from the file Modem.txt. First we get the first ModemMode from the ModemMode.txt data file. Second we get the appropriate DataRate from the ModemMode.txt data file by looking for the first instance of [ModemMode: 1], which is 1. Third, we search the definition of DataRate, found in the data file DataRate.txt, and we see what kind of ModemModes match. We use a separate counter when searching this file, finding all instances where DataRate is 1. Fourth, while searching the file DataRate.txt, we find an instance where ModemMode is 1, and we can say that this requirement is consistent. We then update [RequirementInconsistent] to 1, and the remaining <plus> actor adds 1 to the referent of [ProgramInconsistencies]. Finally, the first <counter>, after all actions downstream have completed, increments the [Count] by 1, and the loop is repeated until there are no more entries left in the file.

With Fig. 1 satisfying us that ModemModes are consistent, we now must ensure that DataRates are consistent. Fig. 2 is a modified version of Fig. 1.

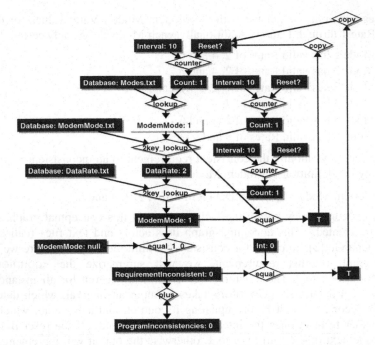

Fig. 1. The requirement checking graph where ModemMode is independent

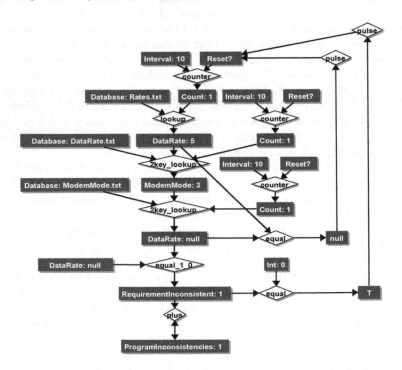

Fig. 2. The requirement checking graph where DataRate is independent

The sequence of the graph where the inconsistency is found follows. First, we get the fifth DataRate from the Rates.txt data file, which also has a value of 5. Second, we get the appropriate ModemModes from the DataRate.txt data file by looking for all instances of ModemMode that correspond to a DataRate of 5. We find a single match: a ModemMode value of 3. Finally, we now look for a corresponding DataRate to a ModemMode of 3 inside of ModemMode.txt. However, there is no such match, a null is returned, and ProgramInconsistent's reference increments.

3 Conclusion and Summary

We have shown that real requirements consistency checking can be achieved with a few modifications to current tools. This method, in conjunction with Smith and Delugach [2], ensures consistency among sets of requirements that contain disjoint data and sets of requirements that have overlapping data.

References

1. Delugach, H., Lampkin, B.: Acquiring Software Requirements as Conceptual Graphs. In: Fifth International Symposium on Requirements Engineering, pp. 296–297. IEEE Computer Society Press, Los Alamitos, California (2001)
2. Smith, B.J., Delugach, H.S.: A Framework for Analyzing and Testing Requirements with Actors in Conceptual Graphs. In: 14th International Conference on Conceptual Structures, Aalborg, Denmark, pp. 401–412. Springer, Heidelberg (2006)
3. Delugach, H.S.: CharGer Conceptual Graph Editor, version 3.5b1 (accessed) (January 2006), http://sourceforge.net/projects/charger/

Implementation of SPARQL Query Language Based on Graph Homomorphism

Olivier Corby[1] and Catherine Faron-Zucker[2]

[1] INRIA 2004 route des lucioles - BP 93
FR-06902 Sophia Antipolis cedex
`Olivier.Corby@sophia.inria.fr`
[2] I3S, Université de Nice Sophia Antipolis, CNRS
930 route des Colles - BP 145
FR-06903 Sophia Antipolis cedex
`Catherine.Faron-Zucker@unice.fr`

Abstract. The SPARQL query language is a W3C candidate recommendation for asking and answering queries against RDF data. It offers capabilities for querying by *graph patterns* and retrieval of solutions is based on *graph pattern matching*. This paper is dedicated to the implementation of the SPARQL query language and its pattern matching mechanism which is reformulated into a graph homomorphism checking constrained by filter evaluation.

1 Introduction

The SPARQL[1] query language is a W3C candidate recommendation for asking and answering queries against RDF[2] data. It offers capabilities for querying by *graph patterns* and retrieval of solutions is based on *graph pattern matching*. This paper is dedicated to the implementation of the SPARQL query language and its pattern matching mechanism in the CORESE[3] semantic search engine.

Intuitively, a SPARQL basic graph pattern P is an RDF graph whose some terms are replaced by variables; the basic graph pattern P of a query Q answered against an RDF graph G matches with pattern solution S if G entails $S(P)$, with $S(P)$ the replacement of every variable v in P by $S(v)$. This lead us to RDF graph entailment. The early development of CORESE relies on the reformulation of RDFS-entailment as graph homomorphism [2]. In CORESE last versions, SPARQL pattern matching is also reformulated as a graph homomorphism: it answers a SPARQL query by searching all the existing projections of the conceptual graph P representing the query pattern into the conceptual graph G representing the RDF(S) dataset.

We propose an efficient algorithm relying on two main principles. A first principle comes from our choice to represent the SPARQL query pattern and

---

[1] http://www.w3.org/TR/rdf-sparql-query/
[2] http://www.w3.org/TR/rdf-mt/
[3] http://www-sop.inria.fr/acacia/corese/

U. Priss, S. Polovina, and R. Hill (Eds.): ICCS 2007, LNAI 4604, pp. 472–475, 2007.

the RDF graph by conceptual graphs: we take advantage of their structure to limit the search space for node projections by dealing with relations first and ordering them so as to force node projections. Second, our algorithm integrates value constraints in the search for graph homomorphisms: SPARQL is provided with a wide range of functions and expressions to filter solutions to queries and our algorithm integrates these value constraints *during* the search process to efficiently reduce the search space. We further detail both principles in the following.

2 Solution Filtering While Pattern Matching

Value constraints and solution modifiers allow to filter solutions retrieved by pattern matching. However a sequential algorithm where filtering would succeed pattern matching would be quite unefficient. For instance, let us consider a query asking for research reports and their authors, members of the INRIA institute, after 2002. The process of retrieving *all* the reports *before* filtering them to keep the only few ones written after 2002 would be unnecessarily expensive.

Consequently, our algorithm takes into account value constraints *during* the search for a graph projection: while searching for a projection of a SPARQL query graph into an RDF graph, as soon as for instance a date in the RDF graph is rejected because it does not pass a SPARQL filter, the projection as a whole which involves this date can be rejected.

Moreover, the sooner value constraints are taken into account the smaller the search space becomes. Therefore our algorithm handles SPARQL filters as soon as they are *evaluable*, which may depend on several graph nodes. For instance, let us consider the following SPARQL query asking for the research reports written before the graduating dates of their authors.

```
SELECT ?doc ?a ?d1 ?d2 WHERE {
    ?doc rdf:type ex:ResearchReport . ?doc ex:date ?d1 .
    ?doc ex:createdBy ?a . ?a ex:graduationDate ?d2 .
    FILTER (xsd:date(?d1) >= xsd:date(?d2)) }
```

Before its filter can be evaluated, both variables $?d1$ and $?d2$ occuring in it must be projected into RDF terms.

This "as early as possible" constraint evaluation principle implicitly defines an ordering of query graph nodes and characterizes an incremental process for the construction of a projection: the set of evaluable constraints increases as fast as possible, depending on the chosen current node of the query for which a projection is searched and those for which a projection has already been found.

3 Highest Precedence for Relations in Conceptual Graphs

Our algorihm takes advantage of the hypergraph structure of our representation of RDF graphs as conceptual graphs to limit the search space for node projections.

We view conceptual graphs as hypergraphs where relation nodes have become hyperarcs, while concepts nodes remain the only nodes [1]. As a result, when searching for homomorphisms, relations no more are nodes: in our algorithm they are viewed as constraints for (concept) node projection. Nodes no more are projected in isolation but each one is projected at the same time as the other arguments of a chosen relation to which it participates; relations thus are constraints which reduce the search space of possible projections of nodes. This principle is close to the one described in [4].

Formally, we choose a first relation $r = (x_1, ..., x_i) \in U(P)$, such that $\forall t \in type(r), \exists r' = (x'_1, ..., x'_i) \in U(G)$ such that $\exists t' \in type(r')$ with $t' \leq t$. This choice determines the projections $\pi(x_1) = x'_1, ..., \pi(x_i) = x'_i$ of $x_1, ..., x_i$. While doing so the theoretical search space $V(G) \times ... \times V(G)$ has become the extension of t'. Moreover, when dealing with the next chosen relations, some of their arguments will already have projections previously chosen and the search space for the remaining arguments will even more decrease.

4   Algorithm

Ordering Relations in the Query Graph. Relations in the query graph P are heuristically ordered to constrain at best the search space. Heuristics are based on both the structure of query graph P and the RDF graph G.

Regarding the query graph structure, the ordering depends on both the connexity of relations on their arguments and the occurence of value constraints associated to relation arguments. By choosing a relation connected by the greatest number of arguments to previously chosen relations of P, these arguments already have projections which diminish the search space for the remaining arguments. Furthermore, the more value constraints on nodes are evaluated, the more the search space will diminish. At each step of the search we chose to handle the relation for which the greatest number of constraints are evaluable.

Regarding graph G against which the query is asked, the ordering depends on relation types and on how often relations of a given type (or subtype of it) occur in G. The early choice of the relations whose type occur the least in G will significantly reduce the search space.

Graph Indexing and Candidate Relations. Graph G against which the SPARQL query is asked is indexed by relation types and by each argument of the relations. Hence there is a direct access to the list of relations of a given type which involve a given node. This graph indexing is a preliminary step of our algorithm; it is preprocessed and statically stored.

Based on this static index of G, we associate to each relation $r \in U(P)$ a set $candidates(r)$ of relations of $U(G)$ candidates for arguments of r to be projected on theirs: $candidate(r) = \{s \in U(G), type(s) \leq type(r)\}$. When a candidate s is elected, each ith argument of r is projected on the ith argument of s.

The backbone of our algorithm is the stack of the ordered relations of $U(P)$ associated to their candidate lists. Candidate lists initially correspond to the static index of G; we incrementally reduce their sizes as we pile them up according to the heuristic criterions described above. Their decreasing is as follows. Let r the current relation elected to be piled up. If it is connected to some relation r' previously piled up with the ith argument of r being the jth argument of r', then relations in $candidate(r)$ can be eliminated whose i^{th} argument does not appear as j^{th} argument in $candidate(r')$. Moreover, if some value constraint is evaluable once r is piled up, $candidate(r)$ is further decreased by eliminating candidates for which the constraint evaluates to false.

As a result, let $stack(P)$ the stack where all relations of $U(P)$ are piled up; it constitutes the search space for graph projection search.

Backjump. Our algorithm incrementally search for a partial projection for nested subgraphs of P. To build these subgraphs we consider relations as they are ordered in $stack(P)$. This static ordering enables the handling of constraints during the projection search without ever and ever testing their evaluable status at each step of the algorithm, which would be too time consuming.

Based on this static ordering of relations defined by $stack(P)$, in case of failure of a partial projection search, our algorithm does not just systematically backtrack to the preceding relation in the stack but possibly goes to a deeper relation. It directly *backjumps* to the relation which solves the failure: the latest relation which binds (for the first time) one of the variables in the failing relation or the failing filter.

5   Conclusion

In this paper, we have presented the CORESE implementation of the SPARQL query language and its pattern matching mechanism. We reformulated the problem of answering SPARQL queries against RDF(S) data into a graph homomorphism checking and the CORESE algorithm takes advantage of the structure of graphs translating RDF(S) and SPARQL data and constrains graph homomorphism checking by SPARQL value constraints. Corese has proven its usability in a wide range of real world applications since 2000 [3]. Its implementation has widely evolved and it is now compliant with the core of SPARQL query language.

References

1. Baget, J.F.: Simple Conceptual Graphs Revisited: Hypergraphs and Conjunctive Types for Efficient Projection Algorithms. In: Ganter, B., de Moor, A., Lex, W. (eds.) ICCS 2003. LNCS, vol. 2746, pp. 229–242. Springer, Heidelberg (2003)
2. Corby, O., Dieng, R., Faron, C., Gandon, F.: Searching the Semantic Web: Approximate Query Processing based on Ontologies. IEEE Intelligent Systems 21(1) (2006)
3. Dieng-Kuntz, R., Corby, O.: Conceptual Graphs for Semantic Web Applications. In: Dau, F., Mugnier, M.-L., Stumme, G. (eds.) ICCS 2005. LNCS (LNAI), vol. 3596, Springer, Heidelberg (2005)
4. Croitoru, M., Compatangelo, E.: A combinatorial approach to conceptual graph projection checking. In: Webb, G.I., Yu, X. (eds.) AI 2004. LNCS (LNAI), vol. 3339, pp. 130–143. Springer, Heidelberg (2004)

Cooperative CG-Wrappers for
Web Content Extraction

Fotis Kokkoras, Nick Bassiliades, and Ioannis Vlahavas

Department of Informatics, Aristotle University,
Thessaloniki, 54124, Greece
{kokkoras,nbassili,vlahavas}@csd.auth.gr

Abstract. We use Conceptual Graphs (CGs) to model web content extraction rules (CG-Wrappers). The approach presented incorporates all major existing extraction techniques and allows the definition of synergies of cooperative wrappers for handling complex extraction task, without requiring programming.

Keywords: extraction rules, conceptual graphs, wrappers.

1 Introduction

A web content *extraction rule* (or *wrapper*) is a mapping that populates a data repository with implicit objects that exist inside a given web page [6]. The overall picture of the domain is well described in [3], [4] and [6]. Most approaches proposed are characterized by a trade-off between automation and flexibility; the more automation a method provides the less flexible it becomes [6]. For the average web user, web content extraction is better served by direct and visual wrapper construction rather than unsupervised wrapper induction. Furthermore, languages for wrapper encoding, although valuable and very flexible, can not be easily adopted by the average user. This is a serious issue because, so far, complex extraction tasks require programming.

In this paper we present CG-Wrappers, that is, extraction rules modeled with CGs. The expressiveness of the proposed encoding can help in combining the strengths of available wrapping technologies.

The rest of the paper is organized as following: Section 2 presents the CG-Wrapper paradigm. Section 3 describes how CG-Wrappers can be used in a cooperative fashion while in Section 4, a web content extraction workbench that is based on CG-Wrappers is outlined. Finally, section 5 concludes the paper and gives insight for future work.

2 CG-Wrappers

A CG-Wrapper (**Fig. 1**) is a conceptual graph consisted of four main interrelated concepts: the identifier concept *Wrapper*, the *URL* that represents the web page with the desired content, the *HTMLElement* concept (gray box in **Fig. 1**) which denotes the HTML part of the page containing an instance of the entity we want to extract and the *ExtractedData* concept which is a placeholder for the output.

U. Priss, S. Polovina, and R. Hill (Eds.): ICCS 2007, LNAI 4604, pp. 476–479, 2007.
© Springer-Verlag Berlin Heidelberg 2007

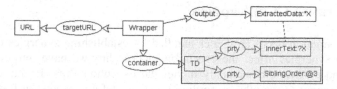

Fig. 1. A CG-Wrapper for extracting the inner text of specific table cells

The *HTMLElement* concept can be replaced (or specialized, in the CG terminology) as desired in order to perfectly describe the content we want to extract. In **Fig. 1** is has been specialized to represent an HTML table cell (*TD* concept) and has been further augmented with additional structures (concepts and relations). The desired output is encoded as the referent field X of an *InnerText* concept which, in turn, is related to the table cell with a property relation (*prty* in **Fig. 1**). At the same time, a coreference link (dotted line in **Fig. 1**) between *InnerText* and *ExtractedData*, will designate that both concepts refer to the same entity, the desired content. The coreferent link is the elegant way provided by the CG theory to ensure that, upon execution, the matched content will reach the output part of the CG-Wrapper. More that one coreference links can be used, supporting in this way multi field extraction. Additionally, all the concepts that describe some HTML or Document Object Model (DOM [2]) content can be further constrained by properly relating them to regular expressions. The set of available concepts and relations is organized in a *Cannon* as ready-to-use building blocks.

The difficulty of handling the HTML/DOM aspects of a CG-Wrapper can be alleviated by providing to the user the ability to point out with the mouse the desired chunks of data in the browser window and have almost all of the description automatically created. This is possible by properly utilizing the web browser's API. As demonstrated in [5], visually handling the HTML/DOM burden is a great time saver. Of course, pointing out a single instance of the desired data is, most of the time, not enough for achieving extraction of high recall and precision. In such cases, besides manually fine tuning the extraction rule, induction techniques can be used.

We have developed a simple generalization algorithm that given two HTML sub trees containing the entities for extraction, it produces through backtracking all their common generalizations in a breadth-first fashion. It is also capable of dropping out nodes from either the first or the second instances under consideration, or both. Thus, it can handle cases of missing HTML nodes. For example, HTML sub-trees like *<p align=left>hello</p>* and *<p align=right>world</p>* are generalized to *<p align=? >*</p>* which describes both. The common description is used to specialize the *HTMLElement* concept of the abstract CG-Wrapper description (gray box in **Fig. 1**).

When a single CG-Wrapper is executed, its container part is matched against the HTML elements of the target web page. During this pseudo-projection phase, generic concepts of the wrapper are bound to page content. If a match is achieved, then a record is extracted, the wrapped is reset and another HTML element is tested until no more elements are available.

3 Cooperative CG-Wrappers

Complex extraction needs are better handled by establishing synergies of extraction rules. Such synergies affect the execution model, thus we have introducing proper relations and concepts to manually model the execution flow. Execution is passed from one wrapper to the other according to a user defined execution flow CG, using the *follows_on_success* and *follows_on_failure* relations. Furthermore, the execution entry point and the wrapper responsible to store the extracted content are explicitly defined by using the *ExecutionEntry* and *RecordCreator* concepts (**Fig. 2**).

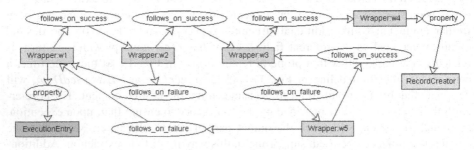

Fig. 2. An example of an execution flow CG

During execution, a CG-Wrapper can be in idle (*wsIdle*), active (*wsActive*), suspended (*wsSuspended*) or stopped (*wsStopped*) states. Initially, all wrappers are idle. An active wrapper is executed while a suspended wrapper is waiting for its turn to continue execution because execution was transferred to another wrapper. A stopped wrapper has consumed all its candidate HTML elements (search space). When all wrappers are stopped, the synergy ceases execution.

4 A Framework for CG-Wrappers

CG-Wrapper Studio is a workbench evolved out of Aggregator [5], as a more systematic approach for a general framework unveiling the potential of CG-Wrappers. The overall architecture of CG-Wrapper Studio is presented in **Fig. 3**. It is an MDI Windows application which, among others, includes a web browser component (wrapped ActiveX core of Internet Explorer) for handling the HTTP transfers, providing access to the DOM and support the visual identification of the training instances. A flexible CG-Wrapper visual editor is also included. The HTML knowledge makes the CG-Wrapper approach HTML and DOM aware and it is common across all wrapping tasks. The extraction rules, as well as the Canon, can be stored in CGXML format [1] for future use. The extracted content can be stored in a database.

In the most remarkable experimental extraction task we conducted, the synergy of **Fig. 2** processed more than 270 000 pages from Debian newsgroup archives with almost perfect results (100% recall, 99.96% precision), something usual because our extraction rules are well-engineered (we don't aim at automatic record detection).

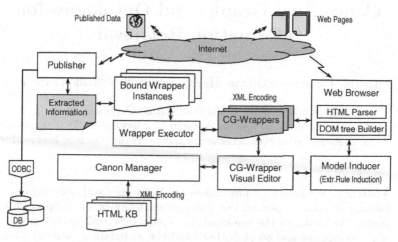

Fig. 3. CG-Wrapper Studio – System Architecture

5 Conclusions and Future Work

We presented CG-Wrappers, extraction rules encoded as CGs. They are HTML and DOM aware, can be generated either by direct modeling or by induction, can utilize regular expressions, they fit perfectly in visual information extraction environments, and finally, can be easily combined to handle complex extraction tasks without requiring programming.

Our main future plans include the separation of the execution mechanism for CG-Wrappers towards web extraction services and the utilization of domain knowledge towards more automation in the CG-Wrapper generation process.

References

[1] CoGXML: http://cogitant.sourceforge.net/cogitant_html/cogxml.html
[2] Document Object Model (DOM) Level 3 Core Specification. (2002), http://www.w3.org/ TR/ 2004/REC-DOM-Level-3-Core-20040407/
[3] Flesca, S., Manco, G., Masciari, E., Rende, E., Tagarelli, A.: Web wrapper induction: a brief survey. In: AI Communications, vol. 17, pp. 57–61. IOS Press, Amsterdam (2004)
[4] Kauchak, D., Smarr, J., Elkan, C.: Sources of Success for Boosted Wrapper Induction. Journal of Machine Learning 5, 499–527 (2004)
[5] Kokkoras, F., Bassiliades, N., Vlahavas, I.: Aggregator: A Knowledge based Comparison Chart Builder for e-Shopping. Intelligent Knowledge-Based Systems: Business and Technology in the New Millennium. In: Leondes, C.T. (ed.) Knowledge-Based Systems, ch. 6, vol.1, pp. 140–163. Kluwer Academic Publishers (2005)
[6] Laender, A., Ribeiro-Neto, B., da Silva, A.S., Teixeira, J.: A Brief Survey of Web Data Extrac-tion Tools. ACM SIGMOD Record 31(2) (2002)

Conceptual Graphs and Ontologies for Information Retrieval

Catherine Comparot, Ollivier Haemmerlé, and Nathalie Hernandez

IRIT, Université de Toulouse le Mirail, Département de
Mathématiques-Informatique, 5 allées Antonio Machado, F-31058 Toulouse Cedex
{catherine.comparot,ollivier.haemmerle,nathalie.hernandez}@univ-tlse2.fr

Abstract. We propose a mechanism for annotating and querying document collections based on the semantic modeling of the context of a search. We model on the one hand the topics concerned by the content of the document and on the other hand the metadata associated with the documents, by means of two ontologies expressed in the conceptual graph model. The semantic annotating mechanism is done by automatically building conceptual graphs.

1 Introduction

Many works in Information Retrieval (IR) aim at enhancing the classic method of document indexation with keywords by an indexation based on semantic annotation. In accordance with those which add an annotation layer based on ontologies [1] to describe the meaning of the documents, we model documents with two ontologies, a topic one and a documentary one. We propose to represent this knowledge in the conceptual graph formalism [2]. This knowledge representation model is well suited for our application for several reasons: (i) the query algorithms have been widely studied and are well suited for IR; (ii) this graphical model allows non-computer scientists to express more sophisticated queries than a simple conjunction of keywords, without the complexity of usual query languages.

Our work takes place in the French WebContent project [3], which aims at creating a software platform to accommodate the tools necessary to exploit the Semantic Web. The input of our system are news releases which are provided by news agencies. They are composed of a set of metadata and a body expressed in free text. Our goal consists in annotating each of these releases in order to enable an Information Retrieval process on that corpus.

In this article, we present successively the ontologies and their representation and the annotation mechanism.

2 An Ontology Based Model

Topic and documentary ontologies. Our system is based on two ontologies, the topic ontology and the documentary ontology. In our case study, we consider a

U. Priss, S. Polovina, and R. Hill (Eds.): ICCS 2007, LNAI 4604, pp. 480–483, 2007.

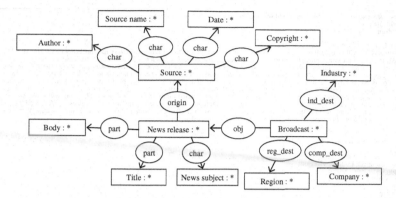

Fig. 1. An example of documentary model

topic ontology linked to the aeronautic domain, which structures the knowledge linked to an aircraft : its parts (winglet, turbojet,...), the types of persons involved in its functioning (pilot, crew,...), the tests that can be made on an aircraft (test landing, test take-off, ...). The documentary ontology represents the knowledge associated with the metadata embedded in the documents and their relations to one another. A news release has different parts (title, body), its source has several characteristics (author, news agency), its subjects refer to specific topics.

In the article, we propose to represent both ontologies by means of a global on-tology of the system expressed in the conceptual graph formalism. The concepts of both ontologies are gathered in the concept type set. Two specific concept types, *Documentary resource* and *Topic resource* are immediate subtypes of *Universal*. The concept types belonging to the documentary (resp. topic) ontology are specializations of the *Documentary resource* (resp. *Topic resource*) concept type. We also manage sets of synonyms for each concept type and individual marker. These sets are used during the annotation step.

The documentary model. In our model, the *documentary model* organizes the metadata embedded in the corpus documents. It is represented by means of the conceptual graph M_{Doc}. It is represented in Fig. 1.

A set of conceptual graphs M_{Doc}^* is associated with the documentary model. These conceptual graphs span the graph M_{Doc}. They represent the elementary pieces of information which can be returned during the semantic annotation step of the documents.

3   Semantic Annotation of the Documents

Each document is annotated by a conceptual graph which describes its content and its metadata. This conceptual graph is built by aggregating elementary conceptual graphs called *motifs* which are returned by the text analysis tools we

Airbus A380 takes off for round-the-world test flights
PublishDate : November 13, 2006
SourceName : *Agence France Presse*

TOULOUSE, France, Nov 13, 2006 (AFP) -
The Airbus A380 will take off from France on Monday for a round-the-world test mission, in the final hurdle before the superjumbo becomes the largest passenger plane in service. (...) On-board engineers and certified test pilots will put the plane through its paces under simulated commercial conditions, including test landings at key airports, refueling practices and maintenance work. The 150 hours of flying, which are expected to be the last major tests before approval from regulators next month, come at a difficult time for Airbus amid a hailstorm of bad publicity for its star project. Airbus has been forced to push back its timetable for deliveries of the A380 three times because of problems encountered when wiring the cabins, with delays now estimated at about two years. (...) Its competitor Boeing, however, has gone from strength-to-strength on the back of buoyant demand for its 787 Dreamliner jet. The A380 will leave Airbus headquarters near Toulouse on Monday, heading for Singapore then the South Korean capital Seoul on Wednesday. A second test flight will take it to Hong Kong on Saturday, then Narita in Japan on November 19, while a third test flight is to encompass airports in China, namely Guangzhou on November 22, then Beijing and Shanghai on November 23.

ar/cos/gk
© Copyright Agence France-Presse, 2006 ...
Regions:
Mediterranean Countries/Regions
Western European Countries/Regions
Companies:
Airbus S.A.S.
Industries:
Air Transport
Civil Aircraft

Fig. 2. An example of a news release used as input for our annotation system

use. We present here our annotation task on an real AFP news release partially represented in Fig. 2.

Annotation according to the documentary model. The motifs used for annotation according to the document model are specializations of the graphs belonging to the set M_{Doc}^* presented in section 2. In order to extract these motifs from the metadata explicitly present in the documents, we use an extraction mechanism based on the analysis of the tags of the documents.

Annotation according to the topic model. For each part of the document (in our case the title and the body) our topic analysis tool returns motifs in the form of:[Document part : *]–(subj)–[Topic resource]–(weight)–[Numerical_Value:x]

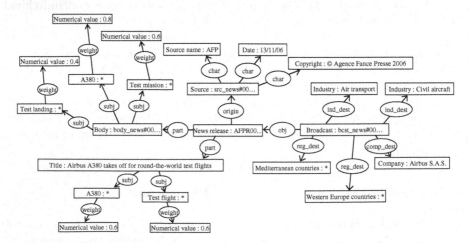

Fig. 3. The conceptual graph F which annotates our news release example

where [Topic resource] is a concept type which is a subtype of the concept type *Topic resource*, and x is a weight associated with the corresponding concept type in the document by our tool. We use the Syntex parser which extracts the set of syntagms of each document. The syntagms are then confronted to the ontology by identifying the concept type or instance they refer to. When the syntagm refers to several concept types, a disambiguisation mechanism is used [4]. When that step is completed, the identified concept types or instances are weighted according to their representativity of each document [4].

Building of the annotation conceptual graph. A non-necessarily connected conceptual graph is then built by a disjunctive sum then a normalization. The conceptual graph F presented in Fig. 3 is the result of the semantic annotation process on the example presented in this section.

4 Conclusion

We believe our work presents three originalities: (i) the annotation process combines two ontologies which can be reused separately in other contexts; (ii) we propose an automatic building of the conceptual graphs used to annotate the documents. Of course our goal is not to build conceptual graphs which represent the content of the documents entirely but which allow us to extract the main pieces of information; (iii) we propose task patterns which allow the users to express queries easily.

Our approach is currently experimented in the framework of the WebContent project on a corpus of news releases provided by EADS-Airbus[1] which is a partner of the project. The releases deal with various events that have occurred the last two years in the aeronautic field. We have focused on two tasks that can be done on this corpus: the analysis of the media coverage of a given event and the analysis of releases dealing with specific topics.

This work will be integrated in the WebContent platform and evaluated on the four application domains: economic watch in aeronautics, strategic intelligence, microbiological and chemical food risk, watch on seismic events.

References

1. Guha, R.V., McCool, R., Miller, E.: Semantic search. In: Proceedings of the 12th International World Wide Web Conference, pp. 700–709 (2003)
2. Sowa, J.F.: Conceptual structures - Information processing in Mind and Machine. Addison-Welsey, London, UK (1984)
3. WebContent. The webcontent project. Web site (2000), http://www.webcontent.fr
4. Hernandez, N., Mothe, J., Chrisment, C., Egret, D.: Modeling context through domain ontologies. Journal of Information Retrieval, Special Topic Issue on Contextual Information Retrieval, 2006 (to appear)

---

[1] We want to thank ESIS, EADS Shared Information System

Representation Levels
Within Knowledge Representation

Heather D. Pfeiffer and Joseph J. Pfeiffer Jr.

New Mexico State University
hdp@cs.nmsu.edu, pfeiffer@cs.nmsu.edu

Abstract. Representation of knowledge is used to store and retrieve informational data in a machine. Since *meaning* cannot be directly stored in the computer; this work proposes a series of levels of representation. The meaning of the data is transformed to a format that the machine can use to store and retrieve knowledge. These levels are designed to transform the knowledge from an abstract definition to a machine representation without loosing any meaning.

1 Introduction

Knowledge gives a definition or understanding of events and acts within the world; knowledge describes the world and gives it *meaning*. For the computer the description of the problem that it is to solve has become known as *knowledge representation (KR)*. The representation consists of a set of syntactic and semantic rules to describe a problem domain [1]. KR, when abstractly described as conceptual ideas or in natural language, appears very informal and without concrete machine structure.

Some of the confusion in the field of knowledge representation is what rules, syntactic or semantic, are defined when looking at an idea with an informal representation and then with a machine processable representation. In many readings, it is not made clear what knowledge can be processed directly by the computer as machine code, and what must be transformed (mapped) from another more abstract representation. It should be noted that, in general, abstract language representations are too informal for machine processing; therefore most knowledge representations must be translated to a more concrete representation using concrete structural languages in order to be coded. Then execution and analysis can be performed.

2 Background

Back in 1971, Shapiro [2] attempted to divide all representations defined by semantic networks into the following two levels: 1) item - conceptual level and 2) system - structural level. Levelization only looked at the actual semantic network represented on the page, and did not consider how to code for machine processing. The item level was concerned with the nodes that appeared in the

U. Priss, S. Polovina, and R. Hill (Eds.): ICCS 2007, LNAI 4604, pp. 484–487, 2007.

network, and the system level attempted to define the links that were present between the nodes in the network.

In 1979, Brachman [3] tried to address the confusion about representations of knowledge by defining levels for different types of representations. Brachman described one representation in terms of another; however, when levels are defined by other levels and representations are defined by other representations there is still confusion [4]. In his paper, Brachman defines a "level" as a distinctive type of network of nodes or links, and gives the following levels: 1) implementation level, 2) logical level, 3) epistemological level, 4) conceptual level, and 5) linguistic level.

For the five levels given above, Brachman saw the implementation as the lowest level; that is, the most basic type of network. The epistemological level is seen by Brachman as a missing level, located between the logical level and the conceptual level, which in a network links formal structure to conceptual units and creates a set of their interrelationships. Guarino, like Brachman, also saw missing information in the levels, and added an ontological level to Brachman's classification levels. The ontological level would give a foundation for the knowledge engineering process and depict a set of features for the computational properties of each level[5]. For Brachman and Guarino, all the levels are processed as part of the knowledge representation.

Brachman did not try to actually look at processing representation from a computer processing point of view. Then in 1982, Newell [6] began the redefinition of a "level" as needed for computer processing. He defined a level in the following way: "a level consists of a medium that is to be processed, components that provide primitive processing, laws of composition that permit components to be assembled into systems, and laws of behavior that determine how system behavior depends on the component behavior and the structure of the system" [6]. Newell referred to computer systems levels as the following bottom (highest) to top (lowest) sequence: device level, circuit level, logic level, register-transfer and program level, and configuration level. Therefore within his levels, Newell renamed the program level to symbol and added a new level just after that known as the knowledge level.

3   Representation Levels

This work expands on Newell's computer systems level idea, in particular investigating what could be the possible computational mechanisms or physical structures of the symbol level (representations), while seeing level relationships more from Brachman's definition [3]. That is: "there is a *level* of processing of representations that sees the lowest level to be a very abstract representation and then, as levels increase, the representation becomes more concrete or machine like" [3]. The highest level of representation would be processed directly by a computer (see Figure 1 [7]) because it is the actual implementation that is compiled or interpreted as machine code.

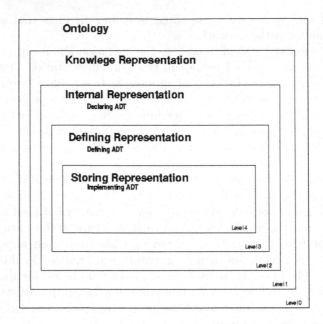

Fig. 1. Levels of Representations

Consider representation in a system to be a series of processing levels. Encapsulating the KR is the level of *ontological information* [8], level 0, with general hierarchy information[9]. This level would be considered the knowledge level under Newell's levels, part of the linguistic level for Brachman, and would be a relocation of Guarino's ontological level. The information represented is not actually part of the structure of the domain knowledge and is the most abstract of all the levels. In fact, it is more of a hierarchy of conceptual information than knowledge, so it will be called "ontology" [10].

KRs will start processing at level 1. For Newell, the *knowledge representation* level would be part of the symbol level, very close to the knowledge level. In Brachman's levels this would encompass part of the conceptual level and all of the epistemological level. Level 1 is the first real translation from a conceptual idea to actually being able to represent the *concept* to a computer.

The second level of representation, level 2, is an *internal representation* that could be viewed as a virtual machine. This is where the declaration of an abstract data type (ADT) is performed. This syntactic representation is more formal and can be used in the definition and implementation of the declared ADT. The syntactic rules are concrete and define a mapping of symbols to operators. However, in order to implement this level, there must be a third level of definitions giving more structure.

Level 3 consists of the actual *semantic definition* of the ADT declared in level 2. The semantic rules are concrete, and define a mapping of operations to functions. It defines the algorithms to be performed, and theoretical time/space analysis can be performed on these algorithms. There is a strong connection between level 2 and level 3 because the concrete rules of the representation in level 2 will work over the algorithms of level 3 during the implementation of the data structures at the next level.

The innermost level of representation, level 4, is the *actual implementation* of the ADT definition. This level is where all the data structures come together. It is at this level that a computer language, or a newly defined language is chosen [11]. The coding of data structures and algorithms will be performed, and the representation is the most concrete.

4   Conclusion

Each of the levels of representation defined here move through three categories of data: 1) meta-data, 2) abstract data, and 3) concrete data. The computer is able to process information within the concrete area even though people actually deal with most information under the abstract or meta-data area. By a knowledge engineer breaking down conceptual ideas at different representation levels, they can see how to transform knowledge from data seen by humans to data that can be processed by machines without loosing information and structure. Level 0 of representation holds meta-data to help in preventing any loss of knowledge.

References

1. Aidinejad, H.: Semantic networks as a unified model of knowledge representation. MCCS-88-117 (1988)
2. Shapiro, S.C.: A net structure for semantic information storage, deduction, and retrieval. In: Proceedings of the 2nd International Conference on Artifical Intelligence, pp. 512–523 (1971)
3. Brachman, R.: On the epistemological status of semantic networks. In: Findler, N.(ed.) Associative Networks: Representation and Use of Knowledge by Computers, pp. 3–50. Academic Press, New York (1979)
4. Charniak, E., McDermott, D.: Introduction To Artifical Intelligence. Addison-Wesley, Reading, MA (1985)
5. Guarino, N.: The Ontological Level. In: Philosophy and the Cognitive Science. Holder-Pivhler-Tempsky, Vienna, pp. 443–456 (1994)
6. Newell, A.: The knowledge level. Artifical Intelligence 18(1), 87–127 (1982)
7. Pfeiffer, H.D.: The Effect of Data Structures Modifications On Algorithms for Reasoning Operations. PhD thesis, New Mexico State University, Las Cruces, NM (2007)
8. Oehrstoem, P., Andersen, J., Scharfe, H.: What has happened to ontology. In: Dau, F., Mugnier, M.-L., Stumme, G. (eds.) ICCS 2005. LNCS (LNAI), vol. 3596, pp. 425–438. Springer, Heidelberg (2005)
9. Lehmann, F. (ed.): Semantics Networks. Pergamon Press, Oxford, England (1992)
10. Sowa, J.: Conceptual Structures: Information Processing in Mind and Machine. Addison-Wesley, Reading, MA (1984)
11. Welty, C.A.: In Integrated Representation for Software Development and Discovery. PhD thesis, Vassar College (1995)

Supporting Lexical Ontology Learning by Relational Exploration*
(Position Paper)

Sebastian Rudolph, Johanna Völker, and Pascal Hitzler

Institute AIFB, Universität Karlsruhe, Germany
{sru,jvo,phi}@aifb.uni-karlsruhe.de

Abstract. Designing and refining ontologies becomes a tedious task, once the boundary to real-world-size knowledge bases has been crossed. Hence semi-automatic methods supporting those tasks will determine the future success of ontologies in practice. In this paper we describe a way for ontology creation and refinement by combining techniques from natural language processing (NLP) and formal concept analysis (FCA). We point out how synergy between those two fields can be established thereby overcoming each other's shortcomings.

1 Introduction

Along with the evolving Semantic Web, the need for elaborated techniques for generating and employing both large and complex ontologies emerges.

Beyond the "toy-examples" mostly used in research for demonstrating and investigating the basic principles for knowledge representation, the size of knowledge bases needed in real world applications will easily exceed the capabilities of human ontology designers to completely model a domain in an undirected, manual, ad-hoc manner.

Apart from the development of suitable methodologies for ontology creation and maintenance (which we consider an important topic but will not focus on in this paper), another way of assisting the modelling process is to provide semi-automatic methods which both

- intelligently suggest the extraction of potential knowledge elements (domain axioms / facts) from certain resources such as domain relevant text corpora and
- provide guidance during the knowledge specification process by asking decisive questions in order to clarify still undefined parts of the knowledge base.

* This work is supported by the German Federal Ministry of Education and Research (BMBF) under the SmartWeb project (grant 01 IMD01 B), by the Deutsche Forschungsgemeinschaft (DFG) under the ReaSem project, and by the European Union under the SEKT project (IST-2003-506826) and the NeOn project (IST-2005-027595).

U. Priss, S. Polovina, and R. Hill (Eds.): ICCS 2007, LNAI 4604, pp. 488–491, 2007.
© Springer-Verlag Berlin Heidelberg 2007

Obviously, those two requirements complement each other. The first one clearly falls into the area of NLP. By using existing methods for knowledge extraction from texts, passages can be identified which indicate the validity of certain pieces of knowledge. For the second requirement, strictly logic-based exploration techniques are needed which yield logically crisp propositions. We argue that integrating these two directions of knowledge acquisition in one scenario will help overcoming disadvantages of either approach and complement each other.

In order to make this concrete, we will introduce Text2Onto (a tool for lexical ontology learning) and Relational Exploration (a dialogue-based method for knowledge acquisition) and sketch a way to combine those two approaches in a synergetic way.

2   Text2Onto – NLP Techniques for Ontology Learning

Text2Onto [1] is a framework for ontology learning, i.e. the automatic acquisition of ontologies from textual data by natural language processing and machine learning techniques. It relies upon an ontology metamodel [2] which allows for attaching uncertainty values and provenance information to all ontology modeling primitives. Text2Onto features algorithms for generating concepts, taxonomic and non-taxonomic relationships as well as disjointness axioms [3] based on lexical evidence.

Clearly, it is advantageous to gain suggestions for constructing or refining the ontology based on textual data: though not fully automatized (since the last decision whether a detected piece of knowledge is really valid has to be left to the ontology engineer), a lot of textual search and estimation work can be shifted to algorithms leaving merely decisions to the expert.

Yet, there is no guarantee, that the automatically generated knowledge model is correct, and precise enough for characterizing the domain in question. This is because on one hand, there might be valid and also relevant pieces of knowledge present in the text which are not properly extracted by Text2Onto. And on the other hand, the corpus itself might not contain all valid domain knowledge to the wanted extent of precision.

Exploration techniques are good means to overcome the lack of completeness and precision in learned ontologies. It is just natural to apply them in order to further specify the knowledge beyond the information extractable from the corpus, which makes relational exploration a perfect complement to automatic approaches for ontology generation based on lexical resources.

3   Relational Exploration

The technique of Relational Exploration (short: RE, introduced in [4] and thoroughly treated in [5]) is based on the well-known attribute exploration algorithm (see [6,7]) from formal concept analysis [8]. This algorithm is extended to a setting with unary and binary relations and uses description logic (DL) concept descriptions of the description logic $\mathcal{FLE}$ (see [9] for a comprehensive treatise on description logics) instead of "logically opaque" attributes. Hence it is possible to explore DL axioms (more precisely: general concept inclusion axioms, short: GCIs) with this techniques. I.e., in an interview-like process, a domain expert has to judge, whether a proposed GCI is valid in the domain he is describing. By employing a DL reasoner it can be furthermore guaranteed that during the exploration process, no redundant questions will be asked. The confirmed GCIs will then be added to the knowledge base thereby refining it. Since OWL DL [10] – the standard language for representing ontologies – is based on description logics, the RE method easily carries over to any kind of ontologies specified in that language.

The advantage of RE is that the obtained results are logically crisp and naturally consistent. Moreover, the refined knowledge base is even complete in the sense that any GCI formed out of $\mathcal{FLE}$ concept descriptions (of a certain role depth) that is valid in the described domain will be derivable from the refined knowledge base.

Yet, one major shortcoming of RE is the following: since the set of semantically different and possibly valid GCIs is growing rapidly with increasing role depth and number of atomic concept and roles, the number of asked questions will soon exceed the ontology designers resources.

4   Synergy and Conclusion

We will now sketch how an ontology refinement task can be accomplished by an intertwined application of Text2Onto and RE.

Suppose there is a knowledge base that has to be refined with respect to some specific term. Firstly, Text2Onto can be used to extract hypothetical axioms (subclass correspondences, concept instantiations, non-taxonomic relationships, and disjointness axioms)   related to this term out of a document describing the domain of interest. These potential axioms will – after having been checked for validity by the expert – be added to the knowledge base.

In the next step, RE will be applied to further specify the interdependencies of the investigated concept and the (as indicated by Text2Onto's relevance measure) important related concepts. To speed up this process, some of the confirmed axioms can be directly provided to the exploration process as a-priori knowledge: simple subclass-of relations will be trivially encoded as attribute implications, concept instantiation will be stored as object-attribute-incidences in the underlying formal context, and the disjointness of the classes, say, $A_1, \ldots, A_n$ will be encoded as implications of the form $A_1, \ldots, A_n \rightarrow \bot$. During the subsequent actual exploration process, Text2Onto will be further applied in the following way: If the RE algorithm comes up with a hypothetical axiom, Text2Onto will look for possibly relevant passages in the corpus, pinpoint to them indicating their relevance, and possibly suggest answers (and assigned possibilities) to the expert.

To put the synergy in a nutshell: Exploration helps elaborating underspecified parts of textually acquired ontologies whereas text mining can contribute to restrict the exploration space and suggest answers to single design questions brought up as the exploration goes on.

The framework sketched in this position paper still has to be evaluated for its feasibility. Therefore, we are currently implementing a prototype to integrate relational exploration with our ontology learning tools, including Text2Onto and LExO [11], which will allow us to carry out first evaluation experiments, and to obtain empirical data in the near future.

References

1. Cimiano, P., Völker, J.: A framework for ontology learning and data-driven change discovery. In: Montoyo, A., Muñoz, R., Métais, E. (eds.) NLDB 2005. LNCS, vol. 3513, Springer, Heidelberg (2005)
2. Vrandecic, D., Völker, J., Haase, P., Tran, D.T., Cimiano, P.: A metamodel for annotations of ontology elements in owl dl. In: Sure, Y., Brockmans, S., Jung, J. (eds.) Proceedings of the 2nd Workshop on Ontologies and Meta-Modeling, Karlsruhe, Germany, GI Gesellschaft für, Informatik (2006)
3. Haase, P., Völker, J.: Ontology learning and reasoning - dealing with uncertainty and inconsistency. In: Costa, P.C.G.d., Laskey, K.B., Laskey, K.J., Pool, M. (eds.) Proceedings of the Workshop on Uncertainty Reasoning for the Semantic Web (URSW), pp. 45–55 (2005)
4. Rudolph, S.: Exploring relational structures via FLE. In: Wolff, K.E., Pfeiffer, H.D., Delugach, H.S. (eds.) ICCS 2004. LNCS (LNAI), vol. 3127, p. 196–212. Springer, Heidelberg (2004)
5. Rudolph, S.: Relational Exploration - Combining Description Logics and Formal Concept Analysis for Knowledge Specification. Universitätsverlag Karlsruhe, Dissertation (2006)
6. Ganter, B.: Algorithmen zur formalen Begriffsanalyse. In: Ganter, B., Wille, R., Wolf, K.E. (eds.) Beiträge zur Begriffsanalyse, pp. 241–254. B.I. Wissenschaftsverlag, Mannheim (1987)
7. Ganter, B.: Attribute exploration with background knowledge. Theoretical Computer Science 217, 215–233 (1999)
8. Ganter, B., Wille, R.: Formal Concept Analysis: Mathematical Foundations. Springer-Verlag New York, Inc. Secaucus, NJ, USA (1997)
9. Baader, F., Calvanese, D., McGuinness, D.L., Nardi, D., Patel-Schneider, P.F.: The Description Logic Handbook: Theory, Implementation, and Applications. Cambridge University Press, Cambridge (2003)
10. McGuinness, D.L., van Harmelen, F.: OWL Web Ontology Language Overview (2004), http://www.w3.org/TR/2004/REC-owl-features-20040210/
11. Völker, J., Hitzler, P., Cimiano, P.: Acquisition of OWL DL axioms from lexical resources. In: Proceedings of the 4th European Semantic Web Conference (ESWC'07), Springer, Heidelberg (2007)

Characterizing Implications of Injective Partial Orders[*]

José L. Balcázar[1] and Gemma C. Garriga[2]

[1] Laboratori d'Algorísmica Relacional, Complexitat i Aprenentatge
Universitat Politècnica de Catalunya, Barcelona
balqui@lsi.upc.edu
[2] HIIT Basic Research Unit
Helsinki University of Technology
gemma.garriga@hut.fi

Abstract. It is known that implications in powerset-based closure systems correspond to Horn approximations in propositional logic frameworks. Here we focus on the problem of implications between injective partial orders. We set up the definitions that allow one to apply standard constructions of implications, and formally characterize the propositional theory obtained. We describe also some experimental applications of our development.

1 Introduction

One popular data representation formalism is given by binary relations, through which "objects", or "models" or "transactions", are described in terms of "attributes", or "variables" or "items". The concrete terms vary in function of the research community (concept lattices, propositional logic, relational databases...), and the standard notation changes accordingly [1,3]. A well-understood process of data analysis on these structures consists in finding pairs of sets of attributes for which the given data suggest some form of causality: each pair, frequently denoted $A \to B$, obtained along the data analysis process indicates that, due to the phenomenon where the data comes from, whenever a data object has all the attributes of A, it tends to have as well those of B However, datasets available for data mining have, in general, no guarantee at all of being a correct sample of any phenomenon. Thus, the validity of such association rules is not to be taken for granted.

When the strength of the correlation is full, that is, all objects having A also have B, the rules obtained are Horn expressions, or implications, or deterministic association rules (depending again on the community), and the validity of the process is characterized by the question of whether the phenomenon at hand does allow a Horn axiomatization ([5,7]). In case it does not, it is known that the rules obtained correspond to a "best" Horn theory in a precise sense (the empirical Horn approximation). Additional parameters have been introduced for measuring the strength of the implication for other rules ([6]) or to focus on the rules holding for a certain support, that is, large enough ratio of the objects ([1]): we will do this last pruning in our empirical validations too.

---

[*] This work is supported in part by MCYT TIC (MOISES-TA TIN2005-08832-C03,Trangram TIC2004-07925-C03-02) and by the IST Programme of the European Community, under the PASCAL Network of Excellence, IST-2002-506778.

U. Priss, S. Polovina, and R. Hill (Eds.): ICCS 2007, LNAI 4604, pp. 492–495, 2007.

Many data mining tasks proceed on the basis of structured data, instead of mere relational tuples. Here we study the extension of deterministic association rules to the same sort of data as in [4], and beyond it into partially ordered data.

2   Partial Orders and Implications

Our partial orders are formalized by labeled directed acyclic graphs. We assume that the labeling is injective; that is, the graph representing the partial order has no repeated labels. Formally, we fix our infinite set of labels L, and define our partial orders simply as pairs (V, E) where $V \subset L$ is the finite set of labeled vertices and $E \subseteq V \times V$ is antisymmetric, thus representing the reflexive and transitive closure of E. The set of all these partial orders is $\mathcal{H}$. Morphisms in $\mathcal{H}$ are defined in the standard way.

Definition 1. *We say that G is **more general** than H (denoted by $G \preceq H$), if and only if there is a morphism from G to H. Then H is also said to be **more specific** than G. (These are slight language abuses in that "or equal to" is left implicit.)*

Definition 2. *$G \cap H$ is the partial order having as vertices the intersection of the vertex sets of G and H, and where (e, e') is an edge of the intersection if and only if it is so in both G and H.*

This operation is associative and commutative, so that we can express intersections of several partial orders. This notion corresponds to a meet operation with respect to the $\preceq$ ordering, so that in fact we obtain a lower semilattice. We will also add an artificial element corresponding to the intersection of an emtpy set of partial orders (which would not be a partial order under our definition); we denote it by the unsatisfiable boolean constant $\square$, and it is "maximally specific" by convention. Therefore we obtain a lattice.

2.1   A Closure Operator

The analysis we want to attempt is made on the basis of a dataset consisting of N partial orders, identified by consecutive natural numbers from the interval $[1..N]$: $\mathcal{D} = \{G_i \mid 1 \leq i \leq N\}$. They are not necessarily different. We define the following two *derivation operators*: $\phi : 2^{[1..N]} \mapsto \mathcal{H}$ and $\psi : \mathcal{H} \mapsto 2^{[1..N]}$, as follows: $\phi(I) = \bigcap\{G_i \mid i \in I\}$ whereas for any partial order H, $\psi(H) = \{i \mid H \preceq G_i\}$. It is easy to check that they fulfill the property of the Galois connections: $H \preceq \phi(I) \iff I \subseteq \psi(H)$.

Therefore (see for instance [3]), we obtain a closure operator $\Delta = \phi \cdot \psi$ on partial orders, depending on the actual dataset $\{G_i\}$. The closure operator yields, in a fully standard manner, a notion of implication:

Definition 3. *An implication on partial orders is a pair (G, H), denoted $G \vdash H$, such that $\Delta(G) = H$.*

Now we wish to characterize precisely the rules in a purely propositional way, in terms of Horn clauses. We assign *one propositional variable $\hat{e}$ to each label $e \in L$, and one propositional variable $\widehat{ee'}$ to each pair of labels $e, e' \in L$, or edge of our graphs; and we express the elementary information that we want them to represent, in the form of the following five *background Horn axioms* (more precisely, axiom schemes):

1. $\widehat{ee'} \to \widehat{e}$
2. $\widehat{ee'} \to \widehat{e'}$
3. $\widehat{e} \to \widehat{ee}$
4. $\widehat{ee'} \wedge \widehat{e'e} \to \square$ (for different e, e')
5. $\widehat{ee'} \wedge \widehat{e'e''} \to \widehat{ee''}$

Note that indeed these axioms are Horn clauses, where the antisymmetry property is nondefinite, and written in clausal form as $\neg\widehat{ee'} \vee \neg\widehat{e'e}$. This is important in that, if such "background knowledge" cannot be expressed in that form, the correspondence between implications and Horn expressions would not hold.

Now, each input partial order $G_i \in \mathcal{D}$ in the data corresponds to a model m_i in a natural way, and we consider the set $M = \{m_i \mid 1 \le i \le N\}$, where each m_i corresponds to $G_i \in \mathcal{D}$. Similarly, each rule $G \vdash H$ obtained through the closure operator can be seen as a propositional implication: $\widehat{G} \to \widehat{H}$.

Theorem 1. *Given a set of input partial orders $\mathcal{D}$, the conjunction of all the implications constructed by the closure system, seen as propositional formulas, and together with the background Horn axioms, axiomatizes exactly the empirical Horn approximation of the theory containing the set of models M.*

The proof of this result is similar in structure to the main result proved in [2].

In the case of relational data, there are some standard methods to construct a set of axioms (or basis): one is based on pseudo-intents [3], and yields a minimal basis. We prefer here the instantaneous basis from [8] and [7], which has other advantages: we find it particularly intuitive in explaining the data analysis processes.

Lemma 1. *Let H be a closed partial order and $G \preceq H$; then $\Delta(G) = H$ if and only if $\widehat{G}$ intersects all the faces of H.*

We omit the proof. This fact allows us to reduce the problem of constructing a basis to a hypergraph transversal problem. Faces are defined as:

$$H - H' = \{\widehat{e} \mid e \in V - V'\} \cup \{\widehat{ee'} \mid e, e' \in V', (e, e') \in E - E'\}$$

Then, by intersecting a face $H - H'$ we understand the set-theoretic intersection, that is, there must be a common edge or vertex in both.

2.2   Empirical Validation

We have developed a prototype implementation, and applied it to several datasets coming from real life. We found that the running times were negligible in comparison with the computation of the lattice, with real-time responses even with quite large lattices.

In general, real-life datasets do not offer the injectiveness condition; however, this is no big inconvenient since the very proposal of searching for implications, and the proposal of using our particular approach as well, are only based on the heuristic perception that this sort of analysis can provide useful explanations.

One interesting application is the analysis of the curricula of specific students of the Computer Engineering School of our university. There, a large number of elective courses, with precedence recommendations of varying strength, give many different trajectories; we used a dataset corresponding to the courses registered along ten years, all corresponding to the same three variants of the curriculum (now superseded by a new one). Each time instant corresponds to a semester, and each of the 8793 transactions corresponds to a student and includes, for a number of consecutive semesters, the courses enrolled in each one. We restrict our analysis to the electives since compulsory courses follow a pre-established track We tried supports of 5% and 7.5% on this dataset; respectively, this gave 1689 and 585 closures, and still the number of rules, which could be huge from such a lattice, remained very manageable: 502 and 94, respectively. About one sixth of them were redundancies, such as repeated rules or transitivity. Many of the others were consequence of the precedence recommendations imposed by the School. Examples of nontrivial rules found are: if Database Design is followed by Organizative Structures, then the same student has taken Economy 2; or: each student who took Economy 1 and also took Files and Databases followed by Informatics Projects Management, also did Database Design before Informatics Projects Management.

An even more interesting dataset was obtained from the abstracts of all the 706 research reports filed into the Pascal Network of Excellence (pascal-network.org) up to a specific moment in time. Closed partial orders were computed at frequency thresholds of 10%, 5%, and 2.1%; in this last case, 954 closures were obtained, but still the number of rules was a very manageable total of 70, including still some redundancy Rules appearing include facts such as "kernel" and "support" implies "support vector"; also, if "selection" appears and "feature" appears at least twice then "feature selection" appears, and about 20 similar other rules involving "model", "error", and others.

References

1. Agrawal, R., Imielinski, T., Swami, A.N.: Mining association rules between sets of items in large databases. In: Buneman, P., Jajodia, S. (eds.) Proceedings of the 1993 ACM SIGMOD Int. Conference on Management of Data, pp. 207–216. ACM Press, New York (1993)
2. Balcázar, J.L., Garriga, G.C.: On Horn axiomatizations for sequential data. Theoretical Computer Science 371(3), 247–264 (2007)
3. Ganter, B., Wille, R.: Formal Concept Analysis. Mathematical Foundations. Springer, Heidelberg (1998)
4. Garriga, G.C., Balcázar, J.L.: Coproduct transformations on lattices of closed partial orders. In: Proceedings of 2nd. Int. Conference on Graph Transformation, pp. 336–351 (2004)
5. Kautz, H., Kearns, M., Selman, B.: Horn approximations of empirical data. Artificial Intelligence 74(1), 129–145 (1995)
6. Luxenburger, M.: Implications partielles dans un contexte. Math. Inf. Sci. hum. 29(113), 35–55 (1991)
7. Pfaltz, J.L., Taylor, C.M.: Scientific knowledge discovery through iterative transformations of concept lattices. In: SIAM Int. Workshop on Discrete Mathematics and Data Mining, pp. 65–74 (2002)
8. Wild, M.: A theory of finite closure spaces based on implications. Advances in Mathematics 108, 118–139 (1994)

DVDSleuth: A Case Study in Applied Formal Concept Analysis for Navigating Web Catalogs

Jon Ducrou

School of Computer Science and Software Engineering
University of Wollongong
Northfields Ave
Wollongong, 2522, Australia
jrd990@uow.edu.au

Abstract. Browsing images using Formal Concept Analysis (FCA) for conceptual representation, navigation and clustering was shown in the ImageSleuth projects[1,2,3]. To showcase the ideas and knowledge gained through ImageSleuth, the same techniques were applied to an information space built from a dynamic collection sourced from the Amazon.com on-line store. Using the Amazon.com catalog, conceptually similar DVDs are able to be discovered and viewed, and then used to explore the information space of their conceptual neighbourhood. A case study of the project – called DVDSleuth – is presented in this paper focusing on the history, results and difficulties encountered by the project. The shortcomings of our approach are analysed and reported as a guide for future projects using FCA techniques for information exploration using on-line Web catalogs.

1 Introduction

This paper follows a case study of a Formal Concept Analysis application, namely the DVDSleuth project. The research described focuses on combining recent successes using Formal Concept Analysis for navigating image collections by extending the paradigm to a live connection to 'real' data from a typical Web catalog. The hypothesis tested is that the paradigm for information navigation using formal concept analysis that was usefully applied to a closed collection of images could just as usefully be applied to conceptually navigate an open-ended collection of Web objects. The collection chosen to navigate was the DVD section of the Amazon.com E-Commerce Service system, which delivers information on DVDs featured for sale by Amazon.com.

The DVDSleuth project's aim was to engineer navigation, browsing and search for a conceptual landscape of DVDs. The landscape created would have a definable and mutable information space via combinable perspectives (or scales), and would allow query-by-example operations within the context of the current information space. Unlike early experiments with navigation using FCA, DVDSleuth would only process a limited set of DVD objects representing a neighbourhood of interest. The program would be designed to take advantage of the structure of

U. Priss, S. Polovina, and R. Hill (Eds.): ICCS 2007, LNAI 4604, pp. 496–500, 2007.

the lattice, presenting clusters of images sharing the same properties (extents). By using conceptual similarity ranking techniques, the program should be able to present the user with a ranked list of DVDs that closely match a given concept, search or focus of interest. Importantly, the program would be dynamic, sourcing new objects to expand the user's view over the information space, and extracting attributes and attribute hierarchies automatically, depending on the conceptual focus of user interest.

After the DVDSleuth program was completed, it was found that small faults in the incoming data from the on-line Web catalog were becoming large problems when using the software. The FCA techniques and structures used emphasised the faults in the data and compromised the conceptual integrity of the concept lattice. Previously, in the ImageSleuth projects, the data collection was well controlled with all values present, correct and consistent. Missing values, or inconsistent values in the data sourced from Amazon.com caused DVDSleuth to produce unexpected results.

2   ImageSleuth

The ImageSleuth projects experimented with navigation using concept lattices to provide pathways through image collections. As part of that research, interaction testing was performed to evaluate the usability of the software and provide feedback for future development. A well defined image collection was used which had rich semantic tagging, based on semantics of the image content, and the physical properties of the image itself.

ImageSleuth displays a single concept and its immediate neighbourhood. The neighbourhood provides paths through the collection, the context of which is mutable via selection of one or more scales. Concept navigation is provided by a list of *include* and *remove* attributes which allow movement to super- and subconcepts. Another powerful feature of ImageSleuth is the ability to rank concepts against a given concept, or semi-concept, based on conceptual similarity and distance[2,4,5].

3   Results

Given the success of ImageSleuth for conceptual clustering based on similarity, it was expected that DVDSleuth would add a new dimension to the similarity system, allowing conceptually similar DVDs to be grouped, and ranked by similarity in comparison to other DVDs.

Context creation/expansion operates successfully and the system handles many objects without significant performance decrease. Perspective creation also performs without problems allowing the information space to be altered to suit the user's needs.

However, it became apparent when navigating that there were issues with the data being returned by the ECS server. While returned information is not erroneous, there are incomplete or inconsistent entries meaning results are not found where they might be expected to be.

Some examples of inconsistent entries include:

- Multiple formats or writing conventions within a category.
- Information omitted.
- Information listed under incorrect category.
- Numbers without measurement units.

These small omissions and irregularities in the data collected from the ECS server cause much larger repercussions in the lattice structure, which in turn affect meaningful navigation and similarity metrics.

4 Analysis

To achieve the most accurate and useful lattice structure, actual values must be consistent across objects when they intend to indicate the same semantic attribute. The Amazon.com data does not follow this fundamental rule.

For example, for the attribute value 'Parental Guidance Recommended', actual values that may be associated with a DVD include; 'PG', 'P.G.', 'PGR', 'Parental Guidance', as well as others. This would not occur with a well defined control vocabulary of normalised attribute values. Another example is capitalization, which can cause a value not to match with existing attributes based solely on a small difference in case. These types of faults segregate all objects with at least one of the erroneous values. Even worse, if multiple facets of the data display this error, the structural complexity of the lattice is increased exponentially. This bloated concept lattice will in turn be less valuable and have less structural significance. Users will be presented more options for generalisation/specialisation, some of which make little sense as they appear to be semantically identical, and similarity metrics calculate longer distances between DVD objects than they should if attribute values had been normalised.

Another irregularity in the Amazon.com data is differing levels of granularity between objects. If two different objects (different movies on DVD) have differing attribute specificity, (e.g. one DVD has a complete cast listing, while another has only a single actor listed), this characteristic can cause the more specific movie's object concept to be a sub-concept of the less specific movie's object concept. This hierarchic positioning, by way of attribute implication, means that as a user drills down into the lattice structure, DVDs with low specificity, e.g. only a single actor listed, will be overlooked as they do not meet with the expected level of specificity.

Example: In a large complex lattice structure, there are two objects, g_a and g_b. g_a is high granularity and has attributes m_a, m_b, m_c, m_d, m_e. g_b is low granularity and only has the attribute m_a. A user is looking for g_b, but has a higher granularity expectation of g_b, namely g_u, which has attributes m_a, m_b, m_c, m_d, m_f.

If the user drills down via Include attribute m_a to the concept (m_a', m_a''), both g_a and g_b are in the extent, which has too many objects for a user to peruse. From this concept, the attributes provided by g_a are presented to the user as

Include attributes. This leads the user to continue their downward navigation as g_a provides many of the expected attributes from g_u. However, any downward movement in the lattice from this concept will not cause g_b to appear in the user's pathway. The user will eventually be lead to believe that the object g_b does not exist because many expected attributes were found and followed but they never lead to g_b.

Finally, Some DVDs are listed multiple times for various reasons. For example, a limited edition DVD with a collectable packaging and its standard regularly packaged edition. Fundamentally, the movies are identical. In this situation it would be expected that the two objects should have the same, or very similar, attribute set. When this is not true, they behave as independent objects found via different navigation pathways. This also decreases the objects' conceptual similarity within the lattice, in turn affecting ranking of the concept to other concepts.

5   Conclusion

Browsing of images using FCA for conceptual representation, navigation and clustering was previously engineered in the ImageSleuth projects and was found to be very successful in usability studies. However, these early projects used a purposefully created and well defined image collection with precise, correct and complete attribute sets for each object/image. When the same principles were applied to real world data in the DVDSleuth research by accessing the Amazon.com on-line catalog, the data imperfections encountered caused much larger problems to amplify into the lattice structure undermining the usefulness of the conceptual information system. Our experience demonstrates that FCA has a low tolerance to missing or incomplete attribute data, ambiguous attribute values or duplicate objects. Minor faults in the object and attribute data undermine the usefulness of the navigation paradigm. Our experience indicates that people involved with the engineering of FCA tools that use dynamic third party Web data should be wary of the pitfalls of erroneous object and attribute data. Data normalisation and scaling should be employed in these situations where possible.

References

1. Ducrou, J., Eklund, P.: Browsing and searching images using formal concept analysis. In: Proceedings of the 24th IASTED Int'l Multiconference: Artificial Intelligence and Applications, pp. 317–322 (2006)
2. Ducrou, J., Vormbrock, B., Eklund, P.: Fca-based browsing and searching of a collection of images. In: Schärfe, H., Hitzler, P., Øhrstrøm, P. (eds.) ICCS 2006. LNCS (LNAI), vol. 4068, pp. 203–214. Springer, Heidelberg (2006)
3. Ducrou, J., Eklund, P., Wilson, T.: An intelligent user interface for browsing and searching MPEG-7 images using concept lattices. In: Reviewers Note: Publish details will be known prior to camera ready date (2007)

4. Lengnink, K.: Ähnlichkeit als Distanz in Begriffsverbänden. In: Stumme, G.R.W. (ed.) Begriffliche Wissensverarbeitung: Methoden und Anwendungen, pp. 57–71. Springer, Heidelberg (2001)
5. Saquer, J., Deogun, J.S.: Concept aproximations based on rough sets and similarity measures. Int. J. Appl. Math. Comput. Sci. 11, 655–674 (2001)
6. Kim, M., Compton, P.: Formal concept analysis for domain-specific document retrieval systems. In: Stumptner, M., Corbett, D.R., Brooks, M. (eds.) AI 2001: Advances in Artificial Intelligence. LNCS (LNAI), vol. 2256, Springer, Heidelberg (2001)

Navigation in Knowledge-Based System for Helpdesk Based on FCA

Vladimír Sklenář, Martin Radvanský, and Michal Dobeš

Dept. Computer Science, Palacky University,
Tomkova 40, CZ-779 00 Olomouc, Czech Republic
{sklenar,radvansm,dobes}@inf.upol.cz
http://www.inf.upol.cz

Abstract. An important part of knowledge-based systems for helpdesk is a database of incidents that were solved in the past. A major part of the reported incidents is related to relatively small sets of repeated problems. It is important that these repeated problems are correctly recognized and their solutions are stored. In order to solve the new incident we can use the experience of solving a similar previously reported problem. Such a process is known as Case Based Reasoning (CBR). In practice however, it is not always immediately clear what the problem is about. It is due to the fact that different customer may describe the same problem in a different way. In this paper we describe an application of the helpdesk system which is based on Formal Concept Analysis (FCA), and enables to find analogous incidents by using navigation within the concept lattice.

Keywords: Formal Concept Analysis, Knowledge-Based Systems, Case Based Reasoning, Helpdesk.

1 Introduction

When solving a new problem, people usually try to reuse knowledge acquired by exploring former problem situations. If the same or analogous problem occurred in the past, they simply reuse their solution in the new situation. Such a process is known as Case Based Reasoning (CBR). When designing a software application based on this principle it is necessary to suggest and implement techniques that enable to record problem situations and their solutions, as well as mechanisms that enable to find out analogous problems recorded in the past.

In this article we describe the application that uses Formal Concept Analysis (FCA) for problem recording and retrieving. The application is a system designed for a support of helpdesk operators whose task is to find out and correctly generalize the reported incident, and suggest a proper solution to the customer.

2 Helpline Operators' Requirements

Helpdesk operator's crucial task is to identify the problem quickly and correctly on the basis of information obtained by the user during on-line communication.

U. Priss, S. Polovina, and R. Hill (Eds.): ICCS 2007, LNAI 4604, pp. 501–504, 2007.

The problem is that each customer has his/her individual form of the problem formulation. Often he/she can't recognize and distinguish the substantial facts from the marginal facts properly. The helpdesk operators are therefore required not only to master the software that they support, but also to master the art of obtaining all information relevant to the problem, identifying the problem and finding its solution. Their job could be improved by the use of an information system that collects and categorizes information about individual incidents and their solutions. Such a system should assist helpline operators to specify the reported problem clearly and correctly, to identify it, and to find out the relevant solution quickly. From this point of view, the requirements on such a system are as follows:

- Fast, efficient, and uniform recording of the reported incident that guaranties that analogous incidents are recorded in the same way.
- Fast search and retrieval of analogous incidents and their solutions.
- Assistance during incident analysis and suggestions how to obtain more precise characteristics of the recorded incident. The system should recommend queries to obtain further characteristics of the reported incident. These queries should be created on the basis of information details gleaned from the user so far, and the similarity with previously recorded incidents.
- The database ability of self correction to improve accuracy. When a new specific characteristic is found to be connected with the incident, and this characteristic is not yet entered into the database, it is necessary to add it into the system, and project it to all analogous previously stored incidents. Such a specific characteristic is usually found during reporting an incident.
- Finding of frequently repeated analogous incidents related to the same general problem, and recording a standardized solution to this problem.

The basic method used for similar applications is Case-Based Reasoning (CBR) [4]. CBR utilizes knowledge gleaned from previous specific problem situations. A new problem is solved by finding a similar problem registered in the past, and reusing its solution in the new problem situation. To fulfill our requirements for a functioning system, we use Formal Concept Analysis (FCA) as a method for recording and retrieving incidents.

3   Applying Formal Concept Analysis

Formal Concept Analysis [2] [3] enables to solve most of the requirements stated in chapter 2. There are two contexts in our application – the context of reported incidents, and the context of general problems along with their solutions. Each individual incident (general problem) is described by several attributes (characteristics of the incident). Both contexts have the same set of attributes. Objects that are in the context of general problems are created in case that the customers report the incidents with the same solution repeatedly. These objects have only the minimal set of attributes that are necessary to identify the general problem uniquely. Concrete incidents that are solved by the solution may have some additional (less important) attributes in addition to the attributes identifying the problem.

The choice of attributes is very important for the success of the FCA application. It is necessary, that the initial choice of attributes is made by experts. Also, any subsequent addition of new attributes must follow given rules in order to guarantee that each operator assigns the same attributes to the same incident. In other words it should prevent his/her from adding two attributes with different name but the same meaning. In our case, it is no problem for experienced operators. Operators try to obtain attributes following the order from general to specific. Therefore, it is suitable to display the attributes in a form of a tree hierarchy. Then it is possible to assign the attributes to the recorded incident simply by choosing the appropriate items from the menu.

In the following text, we will call all the recorded incidents that have all up to now reported attributes by the term "analogous incidents". Problem of finding analogous incidents is, therefore, a problem of finding a minimal concept with the intent comprising all given attributes. It means, if B is the set of up to now reported attributes we have to compute the concept $(B^\uparrow, B^{\uparrow\downarrow})$. The set $B^\uparrow$ is then the set of incidents that are to be found. Initially, this set can be rather large. Its cardinality is reduced by assigning another attributes to B. When all of the attributes of the reported incident are known $B^\uparrow$ contains only incidents analogous with the currently reported one. In case that $B^\uparrow$ is empty we deal with a new problem which was not recorded yet. It is possible that B is a proper subset of $B^{\uparrow\downarrow}$. It means that all previously recorded incidents have additional attributes together with all up to now reported ones. The operator should ask a customer whether it holds also in case of the reported incident. From the FCA's point of view the above described process means navigation in the concept lattice. In a given moment the application presents concept complying with the set of characteristics gained from a dialogue with the customer up to now. Assigning the additional attribute causes a move to a concept which is a direct neighbor of the current one in the concept lattice. The order in which the attributes are gained is unimportant. This navigation can be further supported by offering attributes that customer does not tell but that occurred in the past along with attributes assigned to the current incident. We can compute these "possible" attributes as the union of the attribute sets of all incidents contained in the current concept extent. The operator can utilize this piece of information during the dialogue with the customer.

Note that all described operations do not require the computation of the whole concept lattice. This is very important because of the interactive character of navigation. The database of incidents permanently increases as well as the number of concepts.

A typical scenario of the interaction with the system is as follows.

– After the call was established, the operator selects recognized characteristics of the given incident from the tree of available attributes. The system will offer three collections. The first collection contains all previously recorded (and resolved) incidents that comply with the submitted characteristics, the second one contains all described general problems having the given attributes. The operator can then choose some item from these two collections. In this way he/she specifies that the solution of the currently reported incident is the same as the solution of the chosen incident (problem).

– In case that the operator needs more information about the incident, he/she may utilize the third offered collection. It contains attributes that occurred at least in one stored incident along with already assigned attributes. The operator can use these offered attributes to ask the user aimed questions in order to obtain further characteristics of the current incident.

– When the solution is recognized the system reacts by calling the characteristics synchronization, i.e. it suggests assigning further attributes to the incident according to the chosen solution. Operator can invoke a synchronization also in case that the solved incident contains a characteristic that is not yet connected with the chosen suggested solution. The accuracy of information about the stored incidents is improved in this way.

– After synchronization, the selected solution is assigned to the current incident. At this moment, the operator can pass the solution of the incident to the customer and finish the job.

4   Experience with the System Performance and Future Development

The system is in full operation since July 2005. It fulfills all requirements applied on this system properly. Introducing this system enabled to work with the helpline system more effectively. The FCA proved to be suitable for this specific information system.

The application is developed continuously. In this time we make some experiments in order to distinguish importance of individual attributes [1]. We will assign weight to each attribute. This weight will be used to order suggested characteristics, and it will be used for finding general problems that occur repeatedly.

Acknowledgments. Supported by grant No. 1ET101370417 of GA AV ČR, grant No. 201/02/P076 of the Czech Science Foundation,

References

1. Belohlavek, R., Sklenar, V.: Formal Concept Analysis Constrained by ADF. In: Ganter, B., Godin, R. (eds.) ICFCA 2005. LNCS (LNAI), vol. 3403, Springer, Heidelberg (2005)
2. Carpineto, C., Romano, G.: Concept Data Analysis. Theory and Application. Wiley, Chichester (2004)
3. Ganter, B., Wille, R.: Formal Concept Analysis. Mathematical Foundations. Springer, Heidelberg (1999)
4. Watson, I.: Applying Case-Based Reasoning: Techniques for Enterprise Systems. Morgan Kaufmann, San Francisco (1997)

Functorial Properties of Formal Concept Analysis

Hideo Mori

Hokkaido University, School of Mathematics, Sapporo 060-0810, Japan
morih@math.sci.hokudai.ac.jp

Abstract. The concept of Chu correspondences between formal contexts is introduced. The construction of formal concepts induces a functor $\mathfrak{B}$ from the category of Chu correspondences $ChuCors$ to the category $Slat$ of sup-preserving maps between complete lattices. It turns out that the category $ChuCors$ has a $*$-autonomous category structure and the functor $\mathfrak{B}$ is shown to preserve the $*$-autonomous category structure. Details are given in [4].

Introduction

In this paper we introduce the notion of Chu Correspondence between two formal contexts and a new category $ChuCors$ of formal contexts as its objects and Chu correspondences as arrows.

In [6], $*$-autonomous category structure of the category of Chu maps are studied, but there can generally be few Chu maps between two formal contexts[3], which seems to make the category theory rather uninteresting in some field of research. In contrast, there are abundant Chu correspondences between two formal contexts. Chu correspondences give significance of the usage of the category theoretical machinery in studying formal concepts by investigating the category $Slat$ of suplattices and $\bigvee$-preserving maps.

After obtaining the result of [4], the author noticed that a weak form of Theorem 1, namely the fullness and faithfulness of Galois functor is essentially already obtained in the works of W. Xia [7], Ganter and Wille[2], Ganter [1], Krötzsch, Hitzler and Zhang [5].

The concept of Chu correspondence is new and natural and seems to give significance to the concept of dual bonds as a useful technical concept to calculate Chu correspondences.

Full discussion and proofs are shown in [4].

1 Chu Correspondence

Let $\mathbb{K}_i = (G_i, M_i, I_i)$ $(i = 1, 2)$ be formal contexts. A pair $\varphi = (L_\varphi, R_\varphi)$ is called *a correspondence from $\mathbb{K}_1$ to $\mathbb{K}_2$* if L_φ and R_φ are correspondences respectively from G_1 to G_2 and from M_2 to M_1. Here a *correspondence from a set X_1 to X_2* is a map $L : X_1 \to \mathbf{pow}(\mathbf{X_2})$. Hence a map L induces join preserving map $L_* : \mathbf{pow}(\mathbf{X_1}) \to \mathbf{pow}(\mathbf{X_2})$ by defining $L_* K_1 = \bigcup_{x \in K_1} Lx$ for $K_1 \subset X_1$.

U. Priss, S. Polovina, and R. Hill (Eds.): ICCS 2007, LNAI 4604, pp. 505–508, 2007.

Definition 1. *A correspondence φ from $\mathbb{K}_1$ to $\mathbb{K}_2$ is called a Chu correspondence if for every $g_1 \in G_1$ and $m_2 \in M_2$*

$$(L_\varphi g_1)I_2 m_2 \Leftrightarrow g_1 I_1 (R_\varphi m_2),$$

and both $L_\varphi g_1 \subset G_2$ and $R_\varphi m_2 \subset M_1$ are closed for every $g_1 \in G_1$ and $m_2 \in M_2$.

The poler set A' of $A \subset G$ is defined by $A' := \{m \in M | gIm, \forall g \in A\}$, and that of $B \subset M$ is defined similarly. For $A \subset G$, the bipolar set A'' is called the *closure* of A and will be written as $\overline{A}$. We define the relation AIB of $A \subset G$ and $B \subset M$ by aIb for all $a \in A$ and $b \in B$. Let $ChuCors(\mathbb{K}_1, \mathbb{K}_2)$ denotes the set of Chu correspondences from $\mathbb{K}_1$ to $\mathbb{K}_2$ with the order defined by $\varphi_1 \leq \varphi_2$ if and only if $L_{\varphi_1} \subset L_{\varphi_2}$. Note that this is equivalent to $R_{\varphi_1} \subset R_{\varphi_2}$.

Let $ChuCors$ be the category whose objects are formal contexts and whose arrows are Chu correspondences.

The identity Chu correspondence of a formal context $\mathbb{K}$ is defined by $g \mapsto g''$, $m \mapsto m''$ for all $g \in G$ and $m \in M$.

The composition is defined as follows. If φ and ϕ are Chu correspondences respectively from $\mathbb{K}_1$ to $\mathbb{K}_2$ and $\mathbb{K}_2$ to $\mathbb{K}_3$, their composition $\phi \circ \varphi$ is defined by

$$L_{\phi \circ \varphi} g_1 = \overline{L_{\phi *}(L_\varphi g_1)}$$

for $g_1 \in G_1$ and

$$R_{\phi \circ \varphi} m_3 = \overline{R_{\varphi *}(R_\phi m_3)}$$

for $m_3 \in M_3$.

It is easily proved that these data define a category. The homset $ChuCors(\mathbb{K}_1, \mathbb{K}_2)$ is a complete lattice.

2   Properties of *ChuCors*

2.1   Galois Functor and Its Representing Object

By $\mathfrak{B}(\mathbb{K})$ we mean the concept lattice of a formal context $\mathbb{K}$, that is the collection of formal concepts, which are closed subsets pairs such as (A, A') for closed $A \subset G$. A formal concept is also written as (B', B) for closed $B \subset M$. This induces the *Galois functor*

$$\mathfrak{B} : ChuCors \rightarrow Slat$$

in the following way.

Define $\varphi_* : \mathfrak{B}(\mathbb{K}_1) \rightarrow \mathfrak{B}(\mathbb{K}_2)$ by

$$\varphi_*(A_1, A_1') = (\overline{L_{\varphi *} A_1}, (L_{\varphi *} A_1)')$$

and $\varphi^* : \mathfrak{B}(\mathbb{K}_2) \rightarrow \mathfrak{B}(\mathbb{K}_1)$ by

$$\varphi^*(B_2', B_2) = ((R_{\varphi *} B_2)', \overline{R_{\varphi *} B_2}).$$

The pair (φ_*, φ^*) turns out to be a Galois pair and hence φ_* preserves the join and φ^* the meet.

If we define

$$\mathfrak{B}(\varphi) := \varphi_* : \mathfrak{B}(\mathbb{K}_1) \to \mathfrak{B}(\mathbb{K}_2),$$

then $\mathfrak{B}$ is a functor from $ChuCors$ to $Slat$.

Theorem 1. *The Galois functor is an equivalence between the category of the Chu correspondences and the category of join preserving maps .*

Remark. Let

$$\top := \mathbb{K}(\{ * \})$$
$$= (\{ * \}, \mathbf{pow}(\{ * \}), \in).$$

$\top$ represents functor $\mathfrak{B}$

$$ChuCors(\top, \mathbb{K}) \simeq \mathfrak{B}(\mathbb{K}). \qquad \qquad \square$$

The category $ChuCors$ is complete and cocomplete, since it is equivalent to the complete and cocomplete category $Slat$.

2.2   Operations on $ChuCors$

Internal hom functor. Define a new formal context $\mathbb{K}_1 \multimap \mathbb{K}_2$ by

$$\mathbb{K}_1 \multimap \mathbb{K}_2$$
$$:= (ChuCors(\mathbb{K}_1, \mathbb{K}_2), G_1 \times M_2, \models)$$

where

$$\varphi \models (g_1, m_2) \iff (L_\varphi \, g_1) I_2 m_2.$$

Note that since φ is a Chu correspondence, the condition of the right hand side is equivalent to $g_1 I_1 (R_\varphi \, m_2)$.

Self-duality and tensor. The category $ChuCors$ is self-dual with the dualizing functor defined by $\mathbb{K} \mapsto \mathbb{K}^d := (M, G, I^{-1})$ and for a Chu correspondence φ from $\mathbb{K}_1$ to $\mathbb{K}_2$, φ^* from $\mathbb{K}_2^d$ to $\mathbb{K}_1^d$.

The concept lattice of $\top$ is $\mathfrak{B}(\top) \simeq \mathbf{pow}(\{ * \}) \simeq \mathbf{2}$. Similarly $\bot = \top^d$ has the concept lattice $\mathbf{pow}(\{ * \})^{\mathbf{op}} \simeq \mathbf{2}^* \simeq \mathbf{2}$.

The object $\mathbf{2}$ is the dualizing object $Slat(L, \mathbf{2}) \simeq L^*$ in $Slat$. In fact, our $\bot$ is also forming *dualizing object* in $ChuCors$

$$\mathbb{K} \multimap \bot \simeq \mathbb{K}^d.$$

Since we already have the internal hom-functor and the self duality, we must define

$$\mathbb{K}_1 \boxtimes \mathbb{K}_2 := (\mathbb{K}_1 \multimap \mathbb{K}_2^d)^d$$
$$= (A_1 \times A_2, ChuCors(\mathbb{K}_1, \mathbb{K}_2^d))$$

by the monoidal closed condition in *-autonomous category.

This defines the tensor on $ChuCors$, and its tensor unit is $\top$.
There are following natural isomorphisms:

$$\mathfrak{B}(\mathbb{K}_1 \boxtimes \mathbb{K}_2) \simeq \mathfrak{B}(\mathbb{K}_1) \otimes \mathfrak{B}(\mathbb{K}_2),$$

$$\mathfrak{B}(\mathbb{K}_1 \rightarrow\!\bullet\ \mathbb{K}_2) \simeq \mathfrak{B}(\mathbb{K}_1) \multimap \mathfrak{B}(\mathbb{K}_2),$$

$$\mathfrak{B}(\top) \simeq \mathbf{2},$$

$$\mathfrak{B}(\mathbb{K}^d) \simeq \mathfrak{B}(\mathbb{K})^*.$$

Here, $\otimes$, $\multimap$ and $(-)^*$ are *tensor*, $\bigvee$*-preserving maps* and *dualizing operation* in $\mathcal{S}lat$ respectively.

Theorem 2. *The category $ChuCors$ has a structure of $*$-autonomous category and the Galois functor $\mathfrak{B}$ preserves $*$-autonomous structure.*

Acknowledgments. I very much appreciate some comments by anonymous referees and discussions by Prof. B. Ganter and Prof. T. Tsujishita.

References

1. Ganter, B.: Relational Galois connections. In: Kuznetsov, S.O., Schmidt, S. (eds.) ICFCA 2007. LNCS (LNAI), vol. 4390, Springer, Heidelberg (2007)
2. Ganter, B., Wille, R.: Formal Concept Analysis. Springer, Heidelberg (1999)
3. Mori, H.: Functorial properties of the concept lattices. In: Priss, U., Polovina, S., Hill, R., (eds.) ICCS 2007. LNCS (LNAI), vol. 4604, Springer, Heidelberg (2007) http://www.math.sci.hokudai.ac.jp/~morih/chuf.pdf
4. Mori, H.: Chu Correspondences. Hokkaido Mathematical Journal, (to appear) http://www.math.sci.hokudai.ac.jp/~morih/ccr.pdf
5. Krötzsch, M., Hitzler, P., Zhang, G-Q.: Morphisms in Context. In: Dau, F., Mugnier, M.-L., Stumme, G. (eds.) ICCS 2005. LNCS (LNAI), vol. 3596, Springer, Heidelberg (2005)
6. Pratt, V.: Chu spaces. In: School on Category Theory and its Applications, Textos de Mathematica, SerieB, No 21, Universidade de Coimbra (1999)
7. Xia, W.: Morphismen als formale Begriffe -Darstellung und Erzeugung. Verlag Shaker (1993)

Towards an Ontology to Conceptualize Solution Analysis Tasks in CSCL Environments

Rafael Duque, Crescencio Bravo, and Manuel Ortega

Department of Information Technologies and Systems
School of Computer Engineering, University of Castilla – La Mancha
Paseo de la Universidad 4 - 13071 Ciudad Real (Spain)
{Rafael.Duque,Crescencio.Bravo,Manuel.Ortega}@uclm.es

Abstract. New technologies based on meta-models and ontology engineering allow the formalization and conceptualization of the components that take part in the process of collaboration and interaction analysis in CSCL (Computer-Supported Collaborative Learning) environments. In this article, a proposal to characterize the process of analysis of solutions in CSCL environments is made by means of an ontology. These solutions are built by the learners through a process of collaboration following a problem solving approach.

1 Introduction

The CSCL (Computer-Supported Collaborative Learning) discipline tackles the problems of how to use Information Technologies to support the learners in fulfilling a common objective through collective work. Nowadays, this research area is reaching a situation of maturity, in which new software technologies based on meta-modelling, ontologies and XML-based languages are being used to automate and facilitate the development of CSCL systems. One of the present challenges is the creation of computational support to allow the automatic analysis of the work and collaboration processes, and/or of the product or artefact built by a group. From our point of view, the analysis has as its objective the study of the users' collaborative work in order to understand, evaluate and improve the processes and to confirm work and learning effectiveness. The results of the analysis are expressed commonly in the form of analysis indicators [2]. An analysis indicator gives information about either the quality of the individual activity, the mode of collaboration or the quality of the collaborative product.

In CSCL environments, the collaborative process of product building is reinterpreted in a collaborative problem solving approach. Thus, the product to be built is a solution to a problem. Some of these environments offer an analysis of the users' work according to parameters that qualify the collaboration form (e.g., C-CHENE [4]) but not according to the properties of the solution that is being constructed. Therefore, it is necessary to look in more depth at methods and models to analyze solutions in collaborative learning environments. In this article we deal with this topic by proposing a solution analysis ontology, which will be used for analysis conceptualization in a specific CSCL environment.

U. Priss, S. Polovina, and R. Hill (Eds.): ICCS 2007, LNAI 4604, pp. 509–512, 2007.
© Springer-Verlag Berlin Heidelberg 2007

2 Solution Analysis Ontology

In order to fulfil the objective of creating an ontology to conceptualize a process of solution analysis in CSCL systems, the diverse elements of the system relating to the analysis must be modelled. According to Duque et al. [3], different representation levels are identified: the application domain, the task to carry out, and the collaboration model.

In most interactive CSCL systems, the solution to a problem consists of the design of a model. This model is usually made up of a set of objects which interrelate between themselves by means of a set of relationships. This domain representation provides the necessary information to formalize meta-models to be processed by a CSCL environment, especially by its analysis support. They allow, for instance, the syntactical correction of the designs elaborated.

The task is a specification that describes the characteristics of the model that the students must build in a problem solving activity. Therefore, this specification contains the requirements that the final model must fulfil. A task representation model (see Fig. 1) is a basic information source for a process of characterization of solutions, since it depicts the elements that define and restrict the solution that the students must carry out.

A solution is an instance of the domain model (see Fig. 1). However the solution model does not only specify all the instances (objects or relationships) that make up the solution, but also the collaborative actions that produce that solution. These actions consist mainly of insertions, deletions or modifications of instances on the solution.

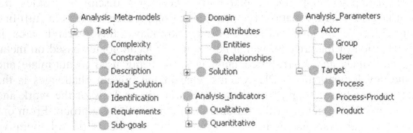

Fig. 1. Ontology to conceptualize the analysis of collaborative solutions

To carry out the analysis process, it is necessary to define in the first place two parameters defining *what* and *who* must be analyzed. To specify the target of the analysis (see Fig. 1), that is to say, the *what*, there are three aspects to analyze: (i) the work process, (ii) the final solution built, and (iii) a comprehensive approach that considers both the work and the solution. The second parameter to define is the actor to be analyzed (see Fig. 1), that is to say, the *who*. The actor whose work is under analysis can be the whole work group or, on the contrary, a specific user from the group.

Having once defined these parameters, the analysis method computes a set of analysis indicators. These indicators can be quantitative or qualitative (see Fig. 1). The quantitative indicators take their values from the domain of real numbers. They are obtained directly by processing the representation models of the solution, domain and task. The analysis indicators of a second level are qualitative. Qualitative indicators are discrete variables that are calculated starting from other indicators (either quantitative or qualitative) and/or by revising the different models and their instances.

3   A Case Study

The COLLECE (COLLaborative Edition, Compilation and Execution of programs) system [1] is a synchronous distributed groupware system to support Programming learning. The teacher organizes the work groups and sessions, and defines the task to be carried out in each work session. The task to be approached (called *problem*) consists of the creation of a program in the Java language.

The work that must be developed by the students is structured in cycles made up of three phases: (i) an edition phase, where the students create the source code with a shared editor; (ii) a compilation phase; and (iii) a execution phase, where the program is executed. These cycles are repeated until the students reach the final solution, solving the problem. The COLLECE's log files store data about the performance of each one of the users as well as of the group. Therefore, both the individual and the group work can be examined. This information consists of all the actions carried out in each phase (edition, compilation or execution) together with the coordination and communication actions. The solution analysis ontology proposed was applied to COLLECE in order to validate it. We interpret the analysis indicators of the ontology as follows:

- Quantitative analysis indicators:
  - *Time*: Total time used for creating the program with COLLECE.
  - *Size*: Number of lines of code.
  - *#Instancies*: Number of programming instructions.
  - *#Simulations*: Number of executions of the program during its creation.
  - *#SatisfiedRequirements:* A program is correct when it solves the problem by satisfying the requirements.
  - *#VerifiedConstraints*: Examples of constraints are the prohibition of the use of recursivity, the obligation to use the structured paradigm, not exceeding a specific complexity order, etc. These limitations can be checked easily.
- Qualitative analysis indicators:
  - *Well_Formed*: A program is well formed when it does not contain compilation errors.
  - *Cost*: It is calculated using the quantitative indicators and other information such as the resources used in the code (files, data structures, etc.).
  - *Accuracy*: A program is correct when *#SatisfiedRequirements = the total number of task's requirements*. The degree to which it is correct depends on the value of *#SatisfiedRequirements*.
  - *Validity*: A program is valid when *#VerifiedConstraints = the total number of task's constraints*. The degree to which it is valid depends on the value of *#VerifiedConstraints*.

- *Quality*: This is a general indicator that gives an indication of the solution quality, agglutinating the rest of the qualitative indicators.

4 Conclusions

In this article we have proposed an ontology to conceptualize the analysis of a collaborative problem solving process and, in particular, the analysis of the solution built by a group of students in such a process. The solution analysis manipulates some models that represent the application domain, the task to accomplish, and the solution built. As a result of this analysis, some indicators are generated to qualify and quantify the work process and the final solution. This analysis can focus on the group or on a particular group member.

In order to evaluate the validity of the solution analysis ontology proposed, the COLLECE groupware system was used to conceptualize the analysis of the collaborative problem solving in the domain of Programming. All the components of the ontology's models could be applied to the COLLECE system, in order to conceptualize the elements of the analysis.

Our current efforts are aimed at extending the ontology and improving the underlying meta-models. They will then be applied to new case studies and used to design a comprehensive process-product analysis framework.

Acknowledgements

This research is partially supported by the Comunidad Autónoma de Castilla-La Mancha (Spain) in the PBI-05-006 and PAC07-0020-5702 projects.

References

1. Bravo, C., Redondo, M.A., Mendes, A.J., Ortega, M.: Group Learning of Programming by means of Real-Time Distributed Collaboration Techniques. In: Lorés, J., Navarro, R. (eds.) HCI related papers of Interacción 2004, pp. 289–302. Springer, Heidelberg (2004)
2. Dimitrakopoulou, A., et al.: State of the Art on Interaction Analysis for Metacognitive Support & Diagnosis. Kaleidoscope Network of Excelence. Interaction Analysis: Supporting Participants in Technology-based Learning Activities. Jointly Executed Integrated Research Project. Deliverable D.31.1 (2005)
3. Duque, R., Bravo, C., Gallardo, J., Ortega, M.: Definición de dominios, tareas y actos conversacionales en sistemas CSCW: el caso de SPACE-DESIGN. In: Bravo, C., Redondo, M.A. (eds.) VII Congreso Internacional Interacción Persona-Ordenado, Puertollano, Spain, Puertollano, pp. 279–289 (2006)
4. Lund, K., Baker, M., Baron, M.: Modelling Dialogue and Beliefs as a Basis for Generating Guidance in a CSCL Environment. In: Proceedings of the Intelligence Tutoring Systems-96 Conference, Montreal, pp. 206–214 (1996)

Author Index

Angelova, Galia   192

Bachmeyer, Randall C.   179
Baget, Jean-François   83, 112
Balcázar, José L.   347, 492
Bassiliades, Nick   476
Bifet, Albert   347
Bravo, Crescencio   509
Buckingham Shum, Simon   29

Carpenter, Martin   360
Comparot, Catherine   480
Corby, Olivier   472
Cox, David   460
Croitoru, Madalina   126, 140, 154

Dashmapatra, Srinandan   140, 154
Dau, Frithjof   429
Delugach, Harry S.   179
Dobeš, Michal   501
Dobrev, Pavlin   192
Ducrou, Jon   496
Dupplaw, David   140, 154
Duque, Rafael   509

Faatz, Andreas   296
Faron-Zucker, Catherine   472

Galitsky, Boris A.   387
Garriga, Gemma C.   492
Ghosh, Sujata   401
Godehardt, Eicke   296
Goertz, Manuel   296

Haddad, Hedi   69
Haemmerlé, Ollivier   112, 480
Hartley, Roger T.   165
Hernandez, Nathalie   480
Hitzler, Pascal   464, 488
Hookway, Christopher   59
Hotho, Andreas   283
Hu, Bo   140, 154

Jäschke, Robert   283

Kalaydjiev, Ognian   192
Keeler, Mary   443

Kokkoras, Fotis   476
Kooi, B.P.   45
Kovalerchuk, Boris   387
Krötzsch, Markus   464
Kuznetsov, Sergei O.   241, 387

Lewis, Paul   140, 154
Liquiere, Michel   333
Lokaiczyk, Robert   296
Löwe, Benedikt   401
Lozano, Antoni   347

May, Michael   220
Mehandjiev, Nikolay   360
Mori, Hideo   505
Moulin, Bernard   69

Obiedkov, Sergei   241
Øhrstrøm, Peter   374
Old, L. John   310
Ortega, Manuel   509

Petersen, Johannes   220
Petersen, Wiebke   415
Pfeiffer, Heather D.   165, 484
Pfeiffer Jr., Joseph J.   484
Polovina, Simon   1, 460
Priss, Uta   310

Radvanský, Martin   501
Reppe, Heiko   255
Richmond, Gary   15
Roth, Camille   241
Rudolph, Sebastian   321, 464, 488

Schärfe, Henrik   374
Schiffel, Jeffrey A.   97
Schmitz, Christoph   283
Scorelle, Erik   401
Sklenář, Vladimír   501
Smith, Bryan J.   468
Stalker, Iain Duncan   360
Stumme, Gerd   283

Thomopoulos, Rallou   112
Tian, Ye   206
Troy, Adam D.   206

Uckelman, Sara L.   374

van Deemter, Kees   126
van der Hoek, W.   45
van Ditmarsch, H.P.   45
Vlahavas, Ioannis   476
Völker, Johanna   488

Werning, Markus   415
Wille, Rudolf   269
Wille-Henning, Renate   269

Xiao, Liang   140, 154

Zhang, Guo-Qiang   206

Lecture Notes in Artificial Intelligence (LNAI)

Vol. 4612: I. Miguel, W. Ruml (Eds.), Abstraction, Reformulation, and Approximation. XI, 418 pages. 2007.

Vol. 4604: U. Priss, S. Polovina, R. Hill (Eds.), Conceptual Structures: Knowledge Architectures for Smart Applications. XII, 514 pages. 2007.

Vol. 4597: P. Perner (Ed.), Advances in Data Mining. XI, 353 pages. 2007.

Vol. 4594: R. Bellazzi, A. Abu-Hanna, J. Hunter (Eds.), Artificial Intelligence in Medicine. XVI, 509 pages. 2007.

Vol. 4585: M. Kryszkiewicz, J.F. Peters, H. Rybinski, A. Skowron (Eds.), Rough Sets and Intelligent Systems Paradigms. XIX, 836 pages. 2007.

Vol. 4578: F. Masulli, S. Mitra, G. Pasi (Eds.), Applications of Fuzzy Sets Theory. XVIII, 693 pages. 2007.

Vol. 4573: M. Kauers, M. Kerber, R. Miner, W. Windsteiger (Eds.), Towards Mechanized Mathematical Assistants. XIII, 407 pages. 2007.

Vol. 4571: P. Perner (Ed.), Machine Learning and Data Mining in Pattern Recognition. XIV, 913 pages. 2007.

Vol. 4570: H.G. Okuno, M. Ali (Eds.), New Trends in Applied Artificial Intelligence. XXI, 1194 pages. 2007.

Vol. 4565: D.D. Schmorrow, L.M. Reeves (Eds.), Foundations of Augmented Cognition. XIX, 450 pages. 2007.

Vol. 4562: D. Harris (Ed.), Engineering Psychology and Cognitive Ergonomics. XXIII, 879 pages. 2007.

Vol. 4548: N. Olivetti (Ed.), Automated Reasoning with Analytic Tableaux and Related Methods. X, 245 pages. 2007.

Vol. 4539: N.H. Bshouty, C. Gentile (Eds.), Learning Theory. XII, 634 pages. 2007.

Vol. 4529: P. Melin, O. Castillo, L.T. Aguilar, J. Kacprzyk, W. Pedrycz (Eds.), Foundations of Fuzzy Logic and Soft Computing. XIX, 830 pages. 2007.

Vol. 4511: C. Conati, K. McCoy, G. Paliouras (Eds.), User Modeling 2007. XVI, 487 pages. 2007.

Vol. 4509: Z. Kobti, D. Wu (Eds.), Advances in Artificial Intelligence. XII, 552 pages. 2007.

Vol. 4496: N.T. Nguyen, A. Grzech, R.J. Howlett, L.C. Jain (Eds.), Agent and Multi-Agent Systems: Technologies and Applications. XXI, 1046 pages. 2007.

Vol. 4483: C. Baral, G. Brewka, J. Schlipf (Eds.), Logic Programming and Nonmonotonic Reasoning. IX, 327 pages. 2007.

Vol. 4482: A. An, J. Stefanowski, S. Ramanna, C.J. Butz, W. Pedrycz, G. Wang (Eds.), Rough Sets, Fuzzy Sets, Data Mining and Granular Computing. XIV, 585 pages. 2007.

Vol. 4481: J. Yao, P. Lingras, W.-Z. Wu, M. Szczuka, N.J. Cercone, D. Ślęzak (Eds.), Rough Sets and Knowledge Technology. XIV, 576 pages. 2007.

Vol. 4476: V. Gorodetsky, C. Zhang, V.A. Skormin, L. Cao (Eds.), Autonomous Intelligent Systems: Multi-Agents and Data Mining. XIII, 323 pages. 2007.

Vol. 4452: M. Fasli, O. Shehory (Eds.), Agent-Mediated Electronic Commerce. VIII, 249 pages. 2007.

Vol. 4451: T.S. Huang, A. Nijholt, M. Pantic, A. Pentland (Eds.), Artifical Intelligence for Human Computing. XVI, 359 pages. 2007.

Vol. 4438: L. Maicher, A. Sigel, L.M. Garshol (Eds.), Leveraging the Semantics of Topic Maps. X, 257 pages. 2007.

Vol. 4429: R. Lu, J.H. Siekmann, C. Ullrich (Eds.), Cognitive Systems. X, 161 pages. 2007.

Vol. 4426: Z.-H. Zhou, H. Li, Q. Yang (Eds.), Advances in Knowledge Discovery and Data Mining. XXV, 1161 pages. 2007.

Vol. 4411: R.H. Bordini, M. Dastani, J. Dix, A.E.F. Seghrouchni (Eds.), Programming Multi-Agent Systems. XIV, 249 pages. 2007.

Vol. 4410: A. Branco (Ed.), Anaphora: Analysis, Algorithms and Applications. X, 191 pages. 2007.

Vol. 4399: T. Kovacs, X. Llorà, K. Takadama, P.L. Lanzi, W. Stolzmann, S.W. Wilson (Eds.), Learning Classifier Systems. XII, 345 pages. 2007.

Vol. 4390: S.O. Kuznetsov, S. Schmidt (Eds.), Formal Concept Analysis. X, 329 pages. 2007.

Vol. 4389: D. Weyns, H.V.D. Parunak, F. Michel (Eds.), Environments for Multi-Agent Systems III. X, 273 pages. 2007.

Vol. 4384: T. Washio, K. Satoh, H. Takeda, A. Inokuchi (Eds.), New Frontiers in Artificial Intelligence. IX, 401 pages. 2007.

Vol. 4371: K. Inoue, K. Satoh, F. Toni (Eds.), Computational Logic in Multi-Agent Systems. X, 315 pages. 2007.

Vol. 4369: M. Umeda, A. Wolf, O. Bartenstein, U. Geske, D. Seipel, O. Takata (Eds.), Declarative Programming for Knowledge Management. X, 229 pages. 2006.

Vol. 4342: H. de Swart, E. Orłowska, G. Schmidt, M. Roubens (Eds.), Theory and Applications of Relational Structures as Knowledge Instruments II. X, 373 pages. 2006.

Vol. 4335: S.A. Brueckner, S. Hassas, M. Jelasity, D. Yamins (Eds.), Engineering Self-Organising Systems. XII, 212 pages. 2007.

Vol. 4334: B. Beckert, R. Hähnle, P.H. Schmitt (Eds.), Verification of Object-Oriented Software. XXIX, 658 pages. 2007.

Vol. 4333: U. Reimer, D. Karagiannis (Eds.), Practical Aspects of Knowledge Management. XII, 338 pages. 2006.

Vol. 4327: M. Baldoni, U. Endriss (Eds.), Declarative Agent Languages and Technologies IV. VIII, 257 pages. 2006.

Vol. 4314: C. Freksa, M. Kohlhase, K. Schill (Eds.), KI 2006: Advances in Artificial Intelligence. XII, 458 pages. 2007.

Vol. 4304: A. Sattar, B.-H. Kang (Eds.), AI 2006: Advances in Artificial Intelligence. XXVII, 1303 pages. 2006.

Vol. 4303: A. Hoffmann, B.-H. Kang, D. Richards, S. Tsumoto (Eds.), Advances in Knowledge Acquisition and Management. XI, 259 pages. 2006.

Vol. 4293: A. Gelbukh, C.A. Reyes-Garcia (Eds.), MICAI 2006: Advances in Artificial Intelligence. XXVIII, 1232 pages. 2006.

Vol. 4289: M. Ackermann, B. Berendt, M. Grobelnik, A. Hotho, D. Mladenič, G. Semeraro, M. Spiliopoulou, G. Stumme, V. Svátek, M. van Someren (Eds.), Semantics, Web and Mining. X, 197 pages. 2006.

Vol. 4285: Y. Matsumoto, R.W. Sproat, K.-F. Wong, M. Zhang (Eds.), Computer Processing of Oriental Languages. XVII, 544 pages. 2006.

Vol. 4274: Q. Huo, B. Ma, E.-S. Chng, H. Li (Eds.), Chinese Spoken Language Processing. XXIV, 805 pages. 2006.

Vol. 4265: L. Todorovski, N. Lavrač, K.P. Jantke (Eds.), Discovery Science. XIV, 384 pages. 2006.

Vol. 4264: J.L. Balcázar, P.M. Long, F. Stephan (Eds.), Algorithmic Learning Theory. XIII, 393 pages. 2006.

Vol. 4259: S. Greco, Y. Hata, S. Hirano, M. Inuiguchi, S. Miyamoto, H.S. Nguyen, R. Słowiński (Eds.), Rough Sets and Current Trends in Computing. XXII, 951 pages. 2006.

Vol. 4253: B. Gabrys, R.J. Howlett, L.C. Jain (Eds.), Knowledge-Based Intelligent Information and Engineering Systems, Part III. XXXII, 1301 pages. 2006.

Vol. 4252: B. Gabrys, R.J. Howlett, L.C. Jain (Eds.), Knowledge-Based Intelligent Information and Engineering Systems, Part II. XXXIII, 1335 pages. 2006.

Vol. 4251: B. Gabrys, R.J. Howlett, L.C. Jain (Eds.), Knowledge-Based Intelligent Information and Engineering Systems, Part I. LXVI, 1297 pages. 2006.

Vol. 4248: S. Staab, V. Svátek (Eds.), Managing Knowledge in a World of Networks. XIV, 400 pages. 2006.

Vol. 4246: M. Hermann, A. Voronkov (Eds.), Logic for Programming, Artificial Intelligence, and Reasoning. XIII, 588 pages. 2006.

Vol. 4223: L. Wang, L. Jiao, G. Shi, X. Li, J. Liu (Eds.), Fuzzy Systems and Knowledge Discovery. XXVIII, 1335 pages. 2006.

Vol. 4213: J. Fürnkranz, T. Scheffer, M. Spiliopoulou (Eds.), Knowledge Discovery in Databases: PKDD 2006. XXII, 660 pages. 2006.

Vol. 4212: J. Fürnkranz, T. Scheffer, M. Spiliopoulou (Eds.), Machine Learning: ECML 2006. XXIII, 851 pages. 2006.

Vol. 4211: P. Vogt, Y. Sugita, E. Tuci, C.L. Nehaniv (Eds.), Symbol Grounding and Beyond. VIII, 237 pages. 2006.

Vol. 4203: F. Esposito, Z.W. Raś, D. Malerba, G. Semeraro (Eds.), Foundations of Intelligent Systems. XVIII, 767 pages. 2006.

Vol. 4201: Y. Sakakibara, S. Kobayashi, K. Sato, T. Nishino, E. Tomita (Eds.), Grammatical Inference: Algorithms and Applications. XII, 359 pages. 2006.

Vol. 4200: I.F.C. Smith (Ed.), Intelligent Computing in Engineering and Architecture. XIII, 692 pages. 2006.

Vol. 4198: O. Nasraoui, O. Zaïane, M. Spiliopoulou, B. Mobasher, B. Masand, P.S. Yu (Eds.), Advances in Web Mining and Web Usage Analysis. IX, 177 pages. 2006.

Vol. 4196: K. Fischer, I.J. Timm, E. André, N. Zhong (Eds.), Multiagent System Technologies. X, 185 pages. 2006.

Vol. 4188: P. Sojka, I. Kopeček, K. Pala (Eds.), Text, Speech and Dialogue. XV, 721 pages. 2006.

Vol. 4183: J. Euzenat, J. Domingue (Eds.), Artificial Intelligence: Methodology, Systems, and Applications. XIII, 291 pages. 2006.

Vol. 4180: M. Kohlhase, OMDoc – An Open Markup Format for Mathematical Documents [version 1.2]. XIX, 428 pages. 2006.

Vol. 4177: R. Marín, E. Onaindía, A. Bugarín, J. Santos (Eds.), Current Topics in Artificial Intelligence. XV, 482 pages. 2006.

Vol. 4160: M. Fisher, W. van der Hoek, B. Konev, A. Lisitsa (Eds.), Logics in Artificial Intelligence. XII, 516 pages. 2006.

Vol. 4155: O. Stock, M. Schaerf (Eds.), Reasoning, Action and Interaction in AI Theories and Systems. XVIII, 343 pages. 2006.

Vol. 4149: M. Klusch, M. Rovatsos, T.R. Payne (Eds.), Cooperative Information Agents X. XII, 477 pages. 2006.

Vol. 4140: J.S. Sichman, H. Coelho, S.O. Rezende (Eds.), Advances in Artificial Intelligence - IBERAMIA-SBIA 2006. XXIII, 635 pages. 2006.

Vol. 4139: T. Salakoski, F. Ginter, S. Pyysalo, T. Pahikkala (Eds.), Advances in Natural Language Processing. XVI, 771 pages. 2006.

Vol. 4133: J. Gratch, M. Young, R. Aylett, D. Ballin, P. Olivier (Eds.), Intelligent Virtual Agents. XIV, 472 pages. 2006.

Vol. 4130: U. Furbach, N. Shankar (Eds.), Automated Reasoning. XV, 680 pages. 2006.

Vol. 4120: J. Calmet, T. Ida, D. Wang (Eds.), Artificial Intelligence and Symbolic Computation. XIII, 269 pages. 2006.

Vol. 4118: Z. Despotovic, S. Joseph, C. Sartori (Eds.), Agents and Peer-to-Peer Computing. XIV, 173 pages. 2006.